Remote Sensing in Precision Agriculture

Transforming Scientific Advancement Into Innovation

Earth Observation

Remote Sensing in Precision Agriculture

Transforming Scientific Advancement Into Innovation

Edited by

Salim Lamine

Higher School of Saharan Agriculture Adrar, Adrar, Algeria

Prashant K. Srivastava

Institute of Environment and Sustainable Development (IESD), Banaras Hindu University, Varanasi, Uttar Pradesh, India

Ahmed Kayad

Intermountain Research and Extension Center, University of California, CA, United States

Francisco Muñoz-Arriola

School of Natural Resources and Department of Biological Systems Engineering, University of Nebraska-Lincoln, Lincoln, NE, United States

Prem Chandra Pandey

Department of Life Sciences, School of Natural Sciences, Shiv Nadar Institution of Eminence (Deemed to be University), Greater Noida, Uttar Pradesh, India

Series Editor

George P. Petropoulos

ELSEVIER

ACADEMIC PRESS

An imprint of Elsevier

Academic Press is an imprint of Elsevier
125 London Wall, London EC2Y 5AS, United Kingdom
525 B Street, Suite 1650, San Diego, CA 92101, United States
50 Hampshire Street, 5th Floor, Cambridge, MA 02139, United States
The Boulevard, Langford Lane, Kidlington, Oxford OX5 1GB, United Kingdom

ISBN: 978-0-323-91068-2

For Information on all Academic Press publications
visit our website at https://www.elsevier.com/books-and-journals

Publisher: Nikki Levy
Acquisitions Editor: Nancy Maragioglio
Editorial Project Manager: Catherine Costello
Production Project Manager: Sruthi Satheesh
Cover Designer: Mark Rogers

Concept and Idea by Salim Lamine

Typeset by MPS Limited, Chennai, India

Contents

Section 3 Precision agriculture practices for irrigation water management

Section 4 Precision agriculture practices for crop yield management

Section 5 Precision agriculture practices for crop protection

List of contributors

Hasan Muhammad Abdullah
Department of Remote Sensing and GIS, Bangabandhu Sheikh Mujibur Rahman Agricultural University (BSMRAU), Gazipur, Bangladesh

Badre Alam
ICAR-Central Agroforestry Research Institute (CAFRI), Jhansi, Uttar Pradesh, India

Sandip Maruti Anpat
Department of Computer Science, Marathwada Mitramandal College of Commerce, Pune, Maharashtra, India

Argyros Argyridis
NEUROPUBLIC SA, Piraeus, Greece

Ayyandar Arunachalam
ICAR-Central Agroforestry Research Institute (CAFRI), Jhansi, Uttar Pradesh, India

Shabir Ahmed Bangroo
Division of Soil Science, Sher-e-Kashmir University of Agricultural Sciences & Technology, Shalimar, Jammu & Kashmir, India; School of Environmental Sciences, University of Guelph, Guelph, ON, Canada

Vishwa Jyoti Baruah
Centre for Biotechnology and Bioinformatics, Faculty of Biological Sciences, Dibrugarh University, Dibrugarh, Assam, India

Owais Bashir
Division of Soil Science, Sher-e-Kashmir University of Agricultural Sciences & Technology, Shalimar, Jammu & Kashmir, India

Mahima Begum
AICRP on Potato, Assam Agricultural University, Jorhat, Assam, India; Department of Agronomy, Assam Agricultural University, Jorhat, Assam, India

Raktim Bhagawati
Department of Agronomy, Assam Agricultural University, Jorhat, Assam, India

Mohammad Iqbal Bhat
Division of Soil Science, Sher-e-Kashmir University of Agricultural Sciences & Technology, Shalimar, Jammu & Kashmir, India

Md. A.H.B. Bhuiyan
Department of Remote Sensing and GIS, Bangabandhu Sheikh Mujibur Rahman Agricultural University (BSMRAU), Gazipur, Bangladesh

Gherardo Biancofiore
Institute of BioEconomy, National Research Council (IBE CNR), Florence, Italy

Asim Biswas
School of Environmental Sciences, University of Guelph, Guelph, ON, Canada

Kostas Bithas
Institute of Urban Environment and Human Resources, Department of Economic and Regional Development, Panteion University, Kallithea, Athens, Greece

Carlos Bojaca
Colombian Oil Palm Research Center—Cenipalma, Oil Palm Agronomy Research Program, Geomatics Section, Bogotá, Colombia

Dhoni Bushi
Department of Geography, Rajiv Gandhi University, Itanagar, Arunachal Pradesh, India

Ishita Chakraborty
Department of Biophysics, Manipal School of Life Sciences, Manipal Academy of Higher Education, Manipal, Karnataka, India

Christos Chalkias
Department of Geography, Harokopio University of Athens, Kallithea, Athens, Greece

Anastasia Dagla
NEUROPUBLIC SA, Piraeus, Greece

Bipul Deka
AICRP on Irrigation Water Management, Assam Agricultural University, Jorhat, Assam, India

Spyridon E. Detsikas
Department of Soil Science of Athens, Institute of Soil and Water Resources, ELGO—DIMITRA, Lycovrisi, Attiki, Greece

Marami Dutta
Department of Soil Science, Assam Agricultural University, Jorhat, Assam, India

R.P. Dwivedi
ICAR-Central Agroforestry Research Institute (CAFRI), Jhansi, Uttar Pradesh, India

Sanjai Kumar Dwivedi
Department of Horticulture, Govind Ballabh Pant University of Agriculture and Technology, Pantnagar, Uttarakhand, India

Yorgos Efstathiou
NEUROPUBLIC SA, Piraeus, Greece

Khushboo Farooq
Division of Soil Science, Sher-e-Kashmir University of Agricultural Sciences & Technology, Shalimar, Jammu & Kashmir, India

Dessislava Ganeva
Space Research and Technology Institute of the Bulgarian Academy of Sciences, Sofia, Bulgaria

Shishir Gaur
Department of Civil Engineering, Indian Institute of Technology (BHU), Varanasi, Uttar Pradesh, India

Uttam Kumar Ghosh
Department of Agronomy, Bangabandhu Sheikh Mujibur Rahman Agricultural University (BSMRAU), Gazipur, Bangladesh

Indira Govindaraju
Department of Biophysics, Manipal School of Life Sciences, Manipal Academy of Higher Education, Manipal, Karnataka, India

Carlos Guerrero
Faculty of Sciences and Technology, University of Algarve, Faro, Portugal

S.B. Hareesh
Division of Agricultural Chemistry & Soil Science, Central Coffee Research Institute, Coffee Research Station, Chikkamagaluru, Karnataka, India

Aatira Hilal
Division of Soil Science, Sher-e-Kashmir University of Agricultural Sciences & Technology, Shalimar, Jammu & Kashmir, India

Md. Saddam Hossain
Department of Agronomy, Bangabandhu Sheikh Mujibur Rahman Agricultural University (BSMRAU), Gazipur, Bangladesh

Md. Nahidul Islam
Department of Agro-Processing, Bangabandhu Sheikh Mujibur Rahman Agricultural University (BSMRAU), Gazipur, Bangladesh

Nikos Kalatzis
NEUROPUBLIC SA, Piraeus, Greece

R. Karthika
PG & Research Department of Biotechnology, Mahendra Arts & Science College (Autonomous), Kalippatti, Namakkal, Tamil Nadu, India

Md. Arifur Rahman Khan
Department of Agronomy, Bangabandhu Sheikh Mujibur Rahman Agricultural University (BSMRAU), Gazipur, Bangladesh

Nayar Afaq Kirmani
Division of Soil Science, Sher-e-Kashmir University of Agricultural Sciences & Technology, Shalimar, Jammu & Kashmir, India

Matej Knapič
Plant Protection Department, Agricultural Institute of Slovenia, Ljubljana, Slovenia

Mangesh Madhukar Kolapkar
Vidya Pratishthan's Arts Science and Commerce College, Baramati, Maharashtra, India

Sushil Kumar
ICAR-Central Agroforestry Research Institute (CAFRI), Jhansi, Uttar Pradesh, India

Samantha Lavender
Pixalytics Ltd., Plymouth Science Park, Plymouth, Devon, United Kingdom

Christina Lekka
Department of Geography, Harokopio University of Athens, Kallithea, Athens, Greece

Uğur Murat Leloğlu
Graduate School of Natural and Applied Sciences, Geodetic and Geographic Information Technologies Department, Middle East Technical University, Ankara, Turkey

Li Li
College of Land Science and Technology, China Agricultural University, Beijing, P.R. China

Romina Lorenzetti
Institute of BioEconomy, National Research Council (IBE CNR), Florence, Italy

Panagiota Louka
NEUROPUBLIC SA, Piraeus, Greece; Department of Natural Resources & Agricultural Engineering, Agricultural University of Athens, Athens, Greece

Ranjit Mahato
Department of Geography, Rajiv Gandhi University, Itanagar, Arunachal Pradesh, India

Nirmal Mazumder
Department of Biophysics, Manipal School of Life Sciences, Manipal Academy of Higher Education, Manipal, Karnataka, India

Gibji Nimasow
Department of Geography, Rajiv Gandhi University, Itanagar, Arunachal Pradesh, India

Oyi Dai Nimasow
Department of Botany, Rajiv Gandhi University, Itanagar, Arunachal Pradesh, India

Anurag Ohri
Department of Civil Engineering, Indian Institute of Technology (BHU), Varanasi, Uttar Pradesh, India

Ayşenur Özcan
Graduate School of Natural and Applied Sciences, Geodetic and Geographic Information Technologies Department, Middle East Technical University, Ankara, Turkey

Thomas Panagopoulos
Faculty of Sciences and Technology, University of Algarve, Faro, Portugal

Prem Chandra Pandey
Department of Life Sciences, School of Natural Sciences, Shiv Nadar Institution of Eminence (Deemed to be University), Greater Noida, Uttar Pradesh, India

José A. Pascual
Group of Soil Enzymology and Bioremediation and Organic Wastes, Department of Soil and Water Conservation, CEBAS-CSIC, University Campus of Espinardo, Murcia, Spain

Shantonu Paul
Department of Soil Science, Assam Agricultural University, Jorhat, Assam, India

George P. Petropoulos
Department of Geography, Harokopio University of Athens, Kallithea, Athens, Greece

Vassilis Protonotarios
NEUROPUBLIC SA, Piraeus, Greece

Rajasree Rajkhowa
Department of Plant Breeding and Genetics, Assam Agricultural University, Jorhat, Assam, India

Margarita Ros
Group of Soil Enzymology and Bioremediation and Organic Wastes, Department of Soil and Water Conservation, CEBAS-CSIC, University Campus of Espinardo, Murcia, Spain

Mehfuz Hasan Saikat
Department of Genetics and Plant Breeding, Bangabandhu Sheikh Mujibur Rahman Agricultural University (BSMRAU), Gazipur, Bangladesh

Bhaswati Sarmah
Department of Plant Breeding and Genetics, Assam Agricultural University, Jorhat, Assam, India; AICRP on Potato, Assam Agricultural University, Jorhat, Assam, India

Tanaya Sarmah
Resilience Innovation Knowledge Academy (RIKA), New Delhi, India

Shafiyoddin Badruddin Sayyad
Microwave and Imaging Spectroscopy Research Laboratory, Department of Computer and Management, Milliya College, Beed, Maharashtra, India

Michael Selvaraj
The Alliance of Bioversity International and International Center for Tropical Agriculture (CIAT), Cali, Colombia

Tajamul Islam Shah
Division of Soil Science, Sher-e-Kashmir University of Agricultural Sciences & Technology, Shalimar, Jammu & Kashmir, India

Mudassar Akhtar Shaikh
Department of Electronics, New Arts, Commerce and Science College, Ahmednagar (Autonomous), Ahmednagar, Maharashtra, India

Nikhilesh Singh
Department of Civil Engineering, Indian Institute of Technology (BHU), Varanasi, Uttar Pradesh, India

Priyanka Singh
ICAR-Central Agroforestry Research Institute (CAFRI), Jhansi, Uttar Pradesh, India

Mohit Kumar Srivastava
Department of Civil Engineering, Indian Institute of Technology (BHU), Varanasi, Uttar Pradesh, India

Prashant K. Srivastava
Institute of Environment and Sustainable Development (IESD), Banaras Hindu University, Varanasi, Uttar Pradesh, India

Mehmet Lütfi Süzen
Graduate School of Natural and Applied Sciences, Geodetic and Geographic Information Technologies Department, Middle East Technical University, Ankara, Turkey; Geological Engineering Department, Middle East Technical University, Ankara, Turkey

Sukumar Taria
ICAR-Central Agroforestry Research Institute (CAFRI), Jhansi, Uttar Pradesh, India

Dimitris Triantakonstantis
Department of Soil Science of Athens, Institute of Soil and Water Resources, ELGO—DIMITRA, Lycovrisi, Attiki, Greece

Ritvik Tyagi
School of Management, School of Management and Entrepreneurship (SME), Shiv Nadar Institution of Eminence (Deemed to be University), Greater Noida, Uttar Pradesh, India

Javaid Ahmed Wani
Division of Soil Science, Sher-e-Kashmir University of Agricultural Sciences & Technology, Shalimar, Jammu & Kashmir, India

Ashok Yadav
ICAR-Central Agroforestry Research Institute (CAFRI), Jhansi, Uttar Pradesh, India

Qianqian Zhang
Department of Earth System Science, Institute for Global Change Studies, Ministry of Education Ecological Field Station for East Asian Migratory Birds, Tsinghua University, Beijing, P.R. China

Uroš Žibrat
Plant Protection Department, Agricultural Institute of Slovenia, Ljubljana, Slovenia

About the editors

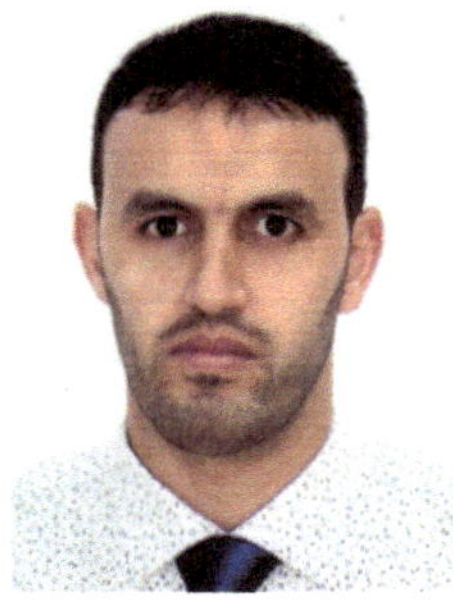

Salim Lamine is a distinguished scientist, serving as a professor and international consultant specializing in remote sensing and precision agriculture. He earned his PhD through a collaborative program between Aberystwyth University in the United Kingdom and the University of Sciences and Technology Houari Boumediene in Algeria. His academic pursuits further extended to the University of Nottingham in the UK, where he delved into GNSS applications, and he gained valuable research experience as an assistant researcher at the University of Aberystwyth. In 2017 Dr. Lamine was honored with the prestigious International Prize for the Environment, "ECOWORLD," by the Russian Academy of Natural Sciences. He holds master's degrees from both the Mediterranean Agronomic Institute of Bari in Italy and the Mediterranean Agronomic Institute of Chania in Greece. Currently, he is serving as a member of the editorial board for the prominent Taylor & Francis journal, *Geocarto International*. Additionally, he extends his expertise as a valued member of the Advisory Panel for several MDPI Journals. Dr. Lamine's research portfolio encompasses a wide range of multidisciplinary areas, including remote sensing, precision agriculture, hyperspectral imaging, drone-based data analysis, crop yield prediction, field spectroradiometry, water and soil management, agri-biosystems engineering, GIS and space mapping, machine learning and SVAT models. Dr. Lamine has made substantial contributions to the academic realm, evidenced by over 30 peer-reviewed publications in esteemed journals and participation in more than 150 international conferences. His scholarly influence also extends to the authorship of several book chapters. He authored a book entitled *REMOTE SENSING: Multispectral & Hyperspectral Applications* published in 2020.

Prashant K. Srivastava is working at IESD, Banaras Hindu University, as a faculty and was affiliated with Hydrological Sciences, NASA Goddard Space Flight Center, as research scientist on SMAP satellite soil moisture retrieval algorithm development, instrumentation, and simulation for various applications. He received his PhD degree from the Department of Civil Engineering, University of Bristol, Bristol, United Kingdom. Prashant was the recipient of several awards such as NASA Fellowship, USA; University of Maryland Fellowship, USA; Commonwealth Fellowship, UK; Early Career

Research Award (ECRA, DST, India), CSIR, as well as UGC–JRF-NET (2005, 2006). He is leading a number of projects funded from reputed agencies in India as well as world. He was also a collaborator with NASA JPL on SMAP soil moisture calibration and validation as well as Scatsat-1, NISAR, AVIRIS-NG missions of India. Prashant made more than 200+ publications in peer-reviewed journals and published 14 books with reputed publishing house such as Springer, Taylor and Francis, AGU-Wiley, and Elsevier, and several book chapters with good citation index. He presented his work in several conferences and workshops and is acting as a convener for the last few years in EGU, Hydroinformatics (HIC), and other conferences. He is also acting as Regional Editor Asia-Geocarto International (T & F), Associate Editor-Journal of Hydrology (Elsevier), GIScience and Remote Sensing (T & F), Remote Sensing Applications: Society and Environment (Elsevier), Sustainable Environment (T & F), Water Resources Management (Springer), Frontiers Remote Sensing, Associate Editor- Remote Sensing-MDPI, Associate Editor- Environment, Development and Sustainability (Springer), Environmental Processes (Springer), Bull of Env and Sci Res.

Ahmed Kayad is an agricultural engineer advisor for applied research and extension at the University of California, Agriculture & Natural Resources (UC ANR), United States. He holds a PhD in environment resources-digital agriculture from the University of Padova in Italy, awarded in 2021. His academic journey began with a BSc in agricultural engineering from Alexandria University, Egypt, followed by an MSc from King Saud University, Saudi Arabia. Ahmed's career has encompassed various roles, including serving as a farm machinery engineer at CLAAS agent in Egypt from 2010 to 2011, conducting research at the Precision Agriculture Research Chair (PARC) at King Saud University, Saudi Arabia from 2011 to 2018, and pursuing doctoral and postdoctoral research at the Digital Agriculture Lab, University of Padua, Italy, from 2018 to 2022. Additionally, he was a visiting doctoral researcher at the International Maize and Wheat Improvement Center (CIMMYT), Mexico, in 2020, and pursued postdoctoral studies at the University of California, Riverside, United States, from 2022 to 2023. Ahmed published several articles in scientific journals and international conferences, primarily focusing on remote sensing in agriculture, precision agriculture, and digital agriculture. His expertise extends to serving as an expert reviewer for prestigious funding agencies in Europe and the United States, particularly in projects related to digital and precision agriculture. Ahmed's research is centered on crop yield monitoring, weed detection, farm machinery, and the application of drones and sensors in agriculture.

Francisco Muñoz-Arriola is an associate professor in hydroinformatics and integrated hydroclimate at the University of Nebraska-Lincoln and adjunct faculty in the Department of Biosystems Engineering at Universidade do Sao Paulo. Francisco has been a volunteer for science at the US Geological Survey and national researcher at Mexico's National Council of Science and Technology. He is also a fellow of the Robert B. Daugherty Water for Food Global Institute and the Public Policy Center. He has been a member of the American Meteorological Society's Water Resources Committee for 5 years. Francisco has worked on the diagnostics and predictability of natural and altered states of biogeochemical cycles and the water continuum from the atmosphere to the aquifer. His research focuses on the predictability of hydrometeorological and climate extremes and their impact on water and agricultural resources, the development of information technologies and models. His collaborative network encompasses more than 20 countries in Asia, Europe, and the United States.

Dr. Prem Chandra Pandey received PhD from the University of Leicester, United Kingdom, under Commonwealth Scholarship and Fellowship Plan. Currently, he is working as an assistant professor at the Department of Life Sciences, School of Natural Sciences, Shiv Nadar Institution of Eminence (Deemed to be University), UP, India. Previously, he has been associated with Banaras Hindu University, India, as an SERB-NPDF fellow. He received BSc and MSc degrees from Banaras Hindu University and MTech degree (Remote Sensing) from Birla Institute of Technology, India. He has worked as a professional research fellow on remote sensing applications in the National Urban Information System funded by NRSC Government of India. He has been a recipient of several awards including Commonwealth Fellow, United Kingdom, INSPIRE fellow GoI, MHRD-UGC fellow GoI, Malviya Gold Medal from Banaras Hindu University, SERB-NPDF from Government of India, and Young Investigator Award. Dr. Pandey is working in three projects related to monitoring of wetlands/Chilika lake mainly focusing on Ramsar sites along with other natural resources–based research work funded by NGP and SERB Government of India. He has published more than 50 peer-reviewed journal papers, seven edited books, several book chapters, and presented his work in national and international conferences. He is serving as a member of editorial board for *Geocarto International* journal, Taylor & Francis, Sustainable Development (Wiley- IF 12.5) and acted as a guest editor for *Remote Sensing,*

MDPI. Additionally, he is also a member of ISG (Indian Society of Geomatics), ISRS (Indian Society of Remote Sensing), IUCN-CEM (2017–25), Society of Wetland Scientists (2021–22), SPIE, and AAG. Dr. Pandey focuses his research on remote sensing for natural resources including forestry, agriculture, urban studies, environmental pollutant modeling, and climate change.

Preface

The world of agriculture is undergoing a technological revolution, and remote sensing is at the forefront of this transformation. With the ability to collect vast amounts of data from the air and space, remote sensing is providing farmers and researchers with unprecedented insights into crop growth, soil moisture, and weather patterns. These insights are enabling farmers to make more informed decisions, optimize their yields, and reduce their environmental impact.

The book "*Remote Sensing in Precision Agriculture: Transforming Scientific Advancement into Innovation*" is a comprehensive guide to the state of the art in remote sensing for precision agriculture. It provides a thorough examination of the latest scientific advancements and how they are being translated into innovative solutions for farmers. It covers a wide range of topics, including the use of drones and satellites, machine learning and big data, and the integration of remote sensing with other precision agriculture technologies.

One of the most exciting aspects of remote sensing is its ability to support sustainable and efficient agricultural practices. The book examines in six sections on how remote sensing is being used to improve crop yields, reduce water and fertilizer usage, and detect and manage pests and diseases. It also highlights the potential of remote sensing to support precision conservation and precision forestry.

The book concludes by looking to the future of remote sensing in precision agriculture. It explores the potential of emerging technologies such as artificial intelligence to further enhance the capabilities of remote sensing. The authors also discuss the potential of remote sensing to support precision conservation and water sustainable management.

The book is edited by leading experts in the field of remote sensing and precision agriculture and is an essential resource for researchers, practitioners, and policymakers alike. Whether you are a farmer looking to improve your crop management or a researcher working on the next breakthrough in precision agriculture, this book is an essential guide to the state of the art in remote sensing for precision agriculture.

We hope that this book will provide inspiration and guidance for anyone interested in this exciting and rapidly evolving field. As the world continues to face global challenges such as climate change and population growth, the need for sustainable and efficient agricultural practices has never been greater. Remote sensing, with its ability to provide real-time data and insights, is a powerful tool for addressing these challenges and ensuring food security for future generations.

We believe that this book will be an invaluable resource for anyone looking to understand the role of remote sensing in precision agriculture and how it can be used to transform scientific advancement into innovation.

Salim Lamine
Prashant K. Srivastava
Ahmed Kayad
Francisco Muñoz-Arriola
Prem Chandra Pandey

SECTION

1

Introduction and principles of precision agriculture

CHAPTER

Precision agriculture practices from planting to postharvest: scopes, opportunities, and challenges of innovation in developing countries

1

Hasan Muhammad Abdullah[1], Md. Nahidul Islam[2], Mehfuz Hasan Saikat[3] and Md. A.H.B. Bhuiyan[1]

[1]*Department of Remote Sensing and GIS, Bangabandhu Sheikh Mujibur Rahman Agricultural University (BSMRAU), Gazipur, Bangladesh*

[2]*Department of Agro-Processing, Bangabandhu Sheikh Mujibur Rahman Agricultural University (BSMRAU), Gazipur, Bangladesh*

[3]*Department of Genetics and Plant Breeding, Bangabandhu Sheikh Mujibur Rahman Agricultural University (BSMRAU), Gazipur, Bangladesh*

1.1 Precision agriculture for smallholder farming

Precision farming is used to accurately manage field variations, reduce production costs, and improve crop yield using limited resources (Nyaga et al., 2021). Precision agriculture (PA) allows growers and farmers to work with better soil in larger fields and manage them as a group of small fields. Therefore, precision farming involves strategic ways of guiding farmers in crop rotation, optimal planting or harvesting times, and soil management to improve crop productivity and efficiency while reducing environmental impact (Kumar et al., 2020). Therefore, according to the International Society of Precision Agriculture, "It is a management strategy that gathers, processes, and analyzes temporal, spatial, individual data and combines it with other information to support management decisions according to estimated variability for improved resource use efficiency, productivity, quality, profitability and sustainability of agricultural production." PA is meant for large crop fields, and it could be equally implemented in smallholder farms with appropriate modification and technological innovation. Smallholder farms are the powerhouse for elevating poverty in developing countries, where many global populations live.

Remote Sensing in Precision Agriculture. DOI: https://doi.org/10.1016/B978-0-323-91068-2.00014-X

PA uses information technology to ensure optimum health and productivity of soil to meet crop requirements. The farmers require technical support and assistance for technology application of precision farming that combines the use of sensors, GNSS (global navigation satellite systems), robots, mapping tools, and data analytics in agricultural production such as site-specific pesticide application, pest scouting, variable-rate irrigation, or yield mapping to improve plant growth without increasing the labor (Finco et al., 2021; Modica et al., 2020). Precision farming aims to minimize environmental impact but maximize economic return by optimizing crop yield, using optimum field-level management practices and advanced crop science technology to match the needs of farmers (Onyango et al., 2021).

The success of PA depends on how well it can measure, manage, and evaluate the space-time continuum. Precision farming is tech-enabled. Therefore, integration of various technologies creates challenges and opportunities to monitor and manage variability at previously unattainable levels of depth and quality. For instance, early, rapid, and accurate detection of plant stress (biotic or abiotic) is critical in attaining higher crop yield. Detection of plant stress using a drone-based sensor, high spatiotemporal imagery, and machine learning algorithms is more effective for detecting the high sensitivity and specificity of plant stress compared with the traditional techniques (Inoue, 2020; Panday et al., 2020). Management of plant diseases is to be linked with the precise identification of diseases. Thus, PA is paving the way for obtaining increased crop production. This chapter explains PA's existing capabilities and innovation scope, focusing on developing countries.

1.2 Precision equipment, tools, and technical requirements for smallholder farming

Land preparation and plowing are the first steps in the crop production cycle. It includes opening/loosening the top layer of soil, mixing the upper soil layer with the lower soil layer, bringing fresh nutrients to the surface while burying weeds and the remains of previous crops, and allowing them to break down and decay (Peigné et al., 2018; Pierce & Fortin, 2019). Therefore, increases in the mass of agricultural machines and traffic cause soil compaction in a wet environment, especially in many developed countries. At the same time, smallholder farming has definite advantages due to smaller equipment and lighter farm machinery. Incorporation of small machinery is the first step of PA in smaller farms. GNSS enables smaller equipment such as power tillers to well serve in precision land preparation and management. One of the most significant challenges of implementing PA is fragmented cropland (mixed crop in a small piece of land), variable planting and harvest date, conflicts of interest, and profit sharing in introducing cooperative farming. Removal of the smaller farm isles by virtual fencing or geofencing could be promising in resolving the challenges of cooperative farming. Crop zoning and planting in a region at the same pace could be helpful in high-resolution drone imaging. Drone sensors capable of multi/hyperspectral imaging (HSI) could be capable of biotic and abiotic stress detection and finer-resolution prescription generation. This prescription can be used to apply nutrients and/or agrochemicals precisely. Dense or sparse network-based Internet of things (IoTs), sensors, and datalogger are widely used for crop data collection, record, and interpretation for crop monitoring and decision-making (Zhao et al., 2010). However, stable internet connectivity is a critical issue for implementing IoT in rural areas in developing countries.

1.3 Precision crop breeding

Plant breeding relies heavily on quick methods for assessing a crop's response. Fast and accurate detection of crop response could be made possible using high-throughput phenotyping (HTP). It may be possible to link genotype and phenotype using a HTP system (Sozzani & Benfey, 2011; York, 2019). Crop phenotypes typically include structural, morphological, physiological, and other performance characteristics resulting from genetic interactions with environmental conditions. Morphological, physiological, and biochemical processes can be used to characterize the crop. On the other hand, genome selection is ultimately the goal of phenotypic testing. Traditional phenotyping procedures are damaging, time-consuming, inefficient, expensive, and have many other negative aspects (Atkinson et al., 2018; Ingvarsson & Street, 2011). Plant HTP is nondestructive, and its features have made use of new approaches such as machine learning (Koh et al., 2021), imaging spectroscopy (Xie & Yang, 2020), and thermal infrared imaging (Kim et al., 2020).

Handheld and ground vehicle-mounted sensors are increasingly being used in plant phenotyping studies (Thoday-Kennedy et al., 2022). Crop types and locations sometimes limit ground vehicle phenotyping, resulting in crop damage. These approaches are time-consuming and low in efficiency, restricting their usage in large-scale applications (Costa et al., 2019; Jin et al., 2020). The weather affects satellite-based imaging. In addition, satellite images have a limited resolution, making them unsuitable for high-resolution applications (Xie & Yang, 2020). As a result, it may not be ideal for agricultural use. Remote sensing using an unmanned aerial vehicle (UAV) is less expensive, yet it can also acquire high-quality images of fields. Use of UAVs to measure and analyze crops is a novel idea. The drawbacks of the previous approaches could be overcome by using UAV-based phenotyping instead. UAV devices can offer observations with excellent resolution for field-based HTP. In addition, a UAV can acquire information from a huge area in a brief period, thus reducing the impact of meteorological conditions (Hu et al., 2020; Jiang et al., 2021).

1.4 High-throughput phenotyping to accelerate plant stress breeding

For example, HTP can be used to screen for photosynthesis, transpiration, biotic, and abiotic stress tolerance (Meacham-Hensold et al., 2019; Shakoor et al., 2017). Early and accurate diagnosis of plant stressors is essential to minimize crop output losses. The conventional approach of detecting plant stress relies on visually assessing and appraising appearances. Plant stress can be determined and quantified using approaches such as RGB imaging (Klodt et al., 2015), 3D scanning (Holman et al., 2016; Virlet et al., 2016), thermal and near-infrared (NIR) sensing (Gómez-Candón et al., 2016), multi- and hyper-spectral sensing (Devadas et al., 2015), and fluorescence imaging (Wetterich et al., 2017). An advantage of UAV-thermal imagery over conventional and manual approaches for assessing crop stress is that it enables us to collect thermal characteristics of farmlands at a plot level and provides comprehensive, quantifiable data about vegetation in a period (Sankey et al., 2019). In calm and cloud-free weather conditions, the TIR sensing can be employed at any time of the day. Thermal imagery sourced measures can be used to assess plant reactions to different stresses, which could provide essential insights into farm affliction, moisture status, foliage wellness, and biomass. They can also serve as a critical component of harvest (Ge et al., 2021; Lishawa et al., 2017).

UAVs can aid in identifying the most prospective cultivars in the face of stress. In response to stress situations, UAVs can be configured with optical and thermal infrared cameras to track and measure variations in plant temperature (Tp). Individual plants may be delineated using an object-based image processing tool, and RGB optical data makes it possible to distinguish growth from the soil context. The thermal imaging camera can acquire Tp continuously, with the difference between Tp and the surrounding temperature and the variation over time being used as a possible sign of stress (Barbedo, 2019; Jiang et al., 2021). Field phenotyping in this manner identifies stress-resistant genotypes and enables growers to mitigate crop loss by detecting plant stress upfront and intervening with improvement initiatives.

1.5 Precision agriculture technologies for plant disease management

PA is a holistic approach that includes crop information, advanced technology, and management practices. The major objective of PA is to optimize crop production. It must begin with the planting stage and continue postharvest processing. Information of the prevalence and spread of disease at the production stage is the most important resource. At the postharvest stage, information on the crop products' moisture, temperature, and humidity must be gleaned together in real time for immediate action. Production technology and operational equipment must be compatible with PA components. Equipment must be designed to apply the chemicals in the field accurately. Management is the third most crucial key to success in PA. Management practices can achieve the desired yield by using authentic information and adopting appropriate technology for modern farmers (Roberson, 2000).

1.6 Major techniques involve in plant disease management

1.6.1 Nanofungicides against field crop pathogens

Nanotechnology includes nanoparticles with single or multidimensional particles of size ≤ 100 nm. These particles control plant disease due to their small size, high surface-to-volume ratio, and exceptional optical properties (Ghormade et al., 2011). The materials used to obtain nanoparticles are metal oxide, ceramics, magnetic materials, emulsions, synthetic and natural polymers, etc. (Puoci et al., 2008).

Fungal diseases are one of the most limiting factors to crop production. A huge number of fungicides are commercially available across the globe, but their application causes detrimental effects to the environment. Global use of pesticides is about 2 million tons per year. Indiscriminate use of chemicals reduces the biodiversity in soil and water, kills the beneficial microbes, increases disease and pest resistance, declines pollinators, and destroys birds' natural habitats. The use of nanoparticles-encapsulated fungicides ensures the slow and sustained release of chemicals (Tilman et al., 2002). Therefore, the loss of chemicals in the environment is reduced.

The antifungal activity of nanoparticles is currently proved to be a suitable alternative to chemicals. Silver, zinc, and titanium-based nanoparticles are the significant elements used for their antifungal potentialities. Silver nanoparticles at low concentrations are more effective against fungi compared with zinc and titanium nanoparticles (Shyla et al., 2014).

Nanoparticles	Particle size	Pathogens	References
Silver nanoparticle	20–80 nm	*Bipolaris sorokiniana*	Mishra et al. (2014)
		Fusarium sp., *Alternaria solani*, *Pythium spinosum*, *Pythium aphanidermatum*, *Cylindrocarpon destructans*, *Cladosporium cucumerinum*, *Glomerella cingulata*, *Didymella bryoniae*, *Stemphylium lycopersici*, and *Monosporascus cannonballus*	Kim et al. (2012)
		Alternaria alternata, *Penicillium digitatum*, and *Alternaria citri*	Abdelmalek and Salaheldin (2016)

(*Continued*)

Continued

Nanoparticles	Particle size	Pathogens	References
ZnO	35–45 nm	*Aspergillus flavus* and *Aspergillus niger*	Elumalai et al. (2015)
	20–70 nm	*Botrytis cinerea* and *Penicillium expansum*	He et al. (2011)
Titanium dioxide (TiO_2)	85–100 nm	*Macrophomina phaseolina*	Shyla et al. (2014)
Cu	3–10 nm	*A. flavus*, *A. niger*	Ramyadevi et al. (2012)
		Phoma destructiva, *Curvularia lunata*, *A. alternata*, and *Fusarium oxysporum*	Kanhed et al. (2014)
		Fusarium equiseti and *F. oxysporum*	Bramhanwade et al. (2016)
Cu-Chitosan	196 nm	*A. alternata, M. phaseolina*, and *Rhizoctonia solani*	Saharan et al. (2013)
Sulfur	20–70 nm	*A. niger*, *F. oxysporum*	Choudhury et al. (2011)

1.7 Nanomaterials for plant disease detection

About 33% of crop products are lost after harvest due to postharvest pathogens (Gastavsson et al., 2011). One of the major reasons for this crop loss is the failure of timely detection of pathogens. Early and reliably detection of plant pathogens needs to monitor the plant health to curb the damage caused by microbes by using effective management practices (Martinelli et al., 2015).

The commonly used methods of plant disease detection are visual observation of symptoms, serological tests, and DNA-based assays (Bhuiyan et al., 2019; Yesmin et al., 2020). However, these techniques are not applicable at the asymptomatic stages of plant disease. Moreover, the conventional approaches need an expert opinion, labor-intensive, time-consuming, possibility of getting false-negative results from cross-contamination, costly, less sensitivity and accuracy. Another major problem is that these are unreachable to the farmers' field. Some nano-biosensors have been developed to detect plant pathogens using antibodies, DNA, and volatile compounds as biosensing receptors (Kashyap et al., 2019). The tools used to detect plant pathogens are biosensors, quantum dots (QDs), nanostructured platforms, nanoimaging, and nanopore DNA sequencing technology. These tools are used to detect pathogens onsite and large-scale monitoring (Kashyap et al., 2017). They are rapid, nondestructive, and affordable techniques to ensure the quality of food products postharvest (Mathivanan, 2021).

Table 1.1 The nano-biosensors used to detect the major foodborne pathogens.

Target organism	Nanomaterial	Detection technique	References
Salmonella enterica and *Listeria monocytogenes*	Thiol–gold NPs–PCR product	Colorimetric	Devi et al. (2013), Kashyap et al. (2019)
Brucella spp.	Fluorescent–silica nanoparticle	Florescence	Vyas et al. (2015)
Escherichia coli O157: H7	Fluorescent–silica nanoparticles	Florescence	Zhao et al. (2004)
Campylobacter jejuni and *C. coli*	Gold-nanorods	Surface plasmon resonance (SPR)	Shams et al. (2019)
Enterobacter sakazakii	Multiwalled carbon nanotubes	Electrochemical immuno sensor	Dou et al. (2013)

Major foodborne pathogens are *Campylobacter* spp., *Salmonella* spp., *Listeria monocytogenes*, and *Escherichia coli* O157:H7. These pathogens are responsible for causing major foodborne disease outbreaks in humans (Velusamy et al., 2010). These foodborne pathogens are detected using nano-biosensors listed in Table 1.1.

Nano-biosensors are becoming more popular in labeling fresh products. These biosensors include multiple chips, which are more sensitive in detecting plant disease over the conventional sensors because of the properties of good conductivity, physicochemical and electrochemical, optical, magnetic, and mechanical properties (Ruffo Roberto et al., 2019).

1.8 Deep learning models for plant disease detection

Timely detection of a particular disease is very critical for protecting plants from plant pathogens. The long-standing tradition of disease identification is performed by agricultural extension organizations, research institutes, and universities (Mohanty et al., 2016). Microscopic identification and molecular techniques have been used to diagnose pathogens. Microscopic identification is very subjective and requires time. Sometimes plant pathologists identify the pathogen incorrectly.

In contrast, although accurate, molecular detection techniques are labor-intensive and need expensive tools and specialized knowledge. Lately, artificial intelligence (AI) and PA have been widely adopted in agriculture. Image processing technologies are used extensively to identify plant diseases (Iqbal et al., 2018; Kartikeyan & Shrivastava, 2021; Pujari et al., 2014). Currently, smartphone-based approaches have been used in plant pathology (Andrianto et al., 2020). The large-scale penetration of smartphones across the globe and advances in computer vision through deep learning (DL) has been paving the way for rapid disease diagnosis by farmers. According to the recent statistics, total smartphone users are now 6.378 billion, which is expected to rise to 7.33 billion by 2025 (bankmycell.com, 2021).

Convolutional neural network (CNN) models are used to detect plant disease by using images from healthy and diseased leaves through DL methods (Ferentinos, 2018). DL models were derived from the machine learning approach, which was introduced in 1943 (McCulloch & Pitts, 1943). The evolution of DL models could be divided into two-time periods—from 1943 to 2006 and from 2012 until now. The first stages were used to resolve the training problem. The first-stage architectures were are listed as backpropagation, chain rule, neocognitron, and handwritten text recognition (LeNET architecture). In the second phase, the state-of-the-art architecture was used for many purposes: operating self-driving cars, healthcare sector, text recognition, prediction of earthquake predictions, marketing, finance, and image recognition (Saleem et al., 2019).

The steps of classifying the images using DL algorithms are given in Fig. 1.1.

The machine learning models such as CNNs are used to train the image to detect the object accurately. However, CNNs must need large datasets. The datasets of many pre- and postharvest diseases are not available. Although the extensive repository PlantVillage contains more than 50,000 images of various crop diseases, it does not contain many pre- and postharvest diseases.

Globally, farmers are aware of the cause of the plant disease. In particular, farmers in developing countries such as Bangladesh are mostly illiterate. They are often confused with the symptoms caused by either pests or pathogens. They treat

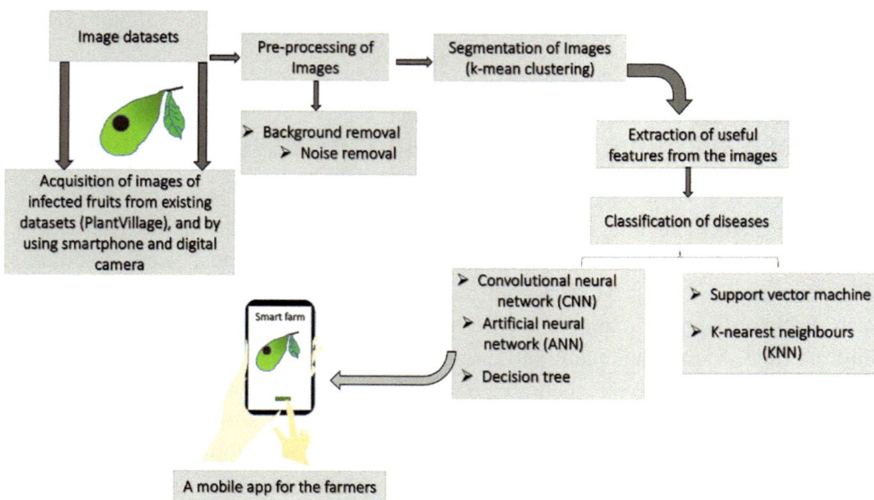

FIGURE 1.1

Flowchart of an implementation of deep learning model.

their crops with pesticides when the causal agent is a fungus and spray fungicide when the causal agents are pests. In this context, they need a quick solution using modern tools such as smartphone-assisted technology and AI to diagnose their problem properly.

Good agricultural practices are developed by the Food and Drug Administration to provide the fruit and vegetable industry guidelines to reduce the risks of microbial attacks. PA is considered the third wave of the agricultural revolution. Another word of PA is digital agriculture, also known as a technology-enabled data-driven sustainable farm management system. PA improves the decision-making ability of the farmers. The inclusion of DL approaches in plant pathology enabled farmers to rapidly and precisely identify their problems and undertake control measures.

1.9 Postharvest management

Agricultural products may get a mechanical shock during harvest and transportation. Sometimes products are invaded by pathogenic organisms from the field. After harvest, this injury accelerates metabolic activities and helps the pathogens grow and spread, and thus, the quality of the harvested products deteriorates over time and leads to losses. Therefore, assessment of the postharvest quality of the bulk produced noninvasively with higher accuracy can help the industry make quick decisions and reduce loss. Hence, NIR spectroscopy, HSI, associated data acquisition methods, and chemometrics can play a significant role.

The major issue of postharvest management of agricultural products is maintaining quality. Food providers and customers are increasingly concerned about the quality and safety of agricultural products. With the increased demand for high-throughput production, the food sector faces several challenges, the most important of which is quality management (Islam, 2018).

In general, the physicochemical characteristics of food are used to define its quality. Manual quality inspections are cost-effective for small-scale evaluations, which are very common in the urban area of Southeast Asia. Still, overlooking important quality and safety features is always the risk of overlooking (Islam et al., 2018b). Moreover, manual quality inspection becomes impractical in the case of bulk quantity. This involves the development of innovative methods for quantifying each postharvest quality indicator while also ensuring food safety. This section has discussed the exciting use of remote sensing in determining the extrinsic and intrinsic qualities of harvested fruits and vegetables.

1.10 Varietal classification

Because different cultivars have the same taste, flavor, texture, and quality, it is challenging to classify them. Accurate varietal identification ensures that the product is appropriately stored, handled, labeled, and packed to ensure product quality. Advanced remote sensing devices have allowed for quick and accurate varietal discrimination.

With a 100% recognition rate, NIR spectroscopy has been effectively applied in varietal differentiation of Thai orange cultivars using logistic regression (Suphamitmongkol et al., 2013), and with 100% classification accuracy, Chinese bayberries were discriminated using the principal component analysis–artificial neural network (PCA-ANN) approach (Li et al., 2007). For varietal classification in peaches, NIR diffuse reflectance spectra were analyzed using three different variable selection approaches uninformative variable elimination-partial least square (UVE-PLS), PCA, and successive projections algorithm (SPA), and two different modeling methods least-squares effective learning algorithm (ELM), and support vector machine (SVM) (Guo et al., 2016). The ELM has an excellent generalization performance and has successfully classified apple types (Shang et al., 2014). Using the 83–2500 nm wavelength area, PCA-LS-SVM, PCA-ELM, and UVE-PLS-LS-SVM achieved 100% accuracy in peach variety identification.

Cao et al. (2010) evaluated the viability of using visible NIR reflectance spectroscopy in the wavelength range of 325–1075 nm to discriminate the grape types using a genetic algorithm (GA) and LS-SVM. Based on the effective wavelengths compared with the entire spectrum range, the GA-LV-SVM model achieved the maximum prediction accuracy of 96.58% for variety discrimination, suggesting the reliability and practical application of chemometrics in variety discrimination. Guo et al. (2013) exhibited the successful discrimination of apples based on their geographical origin using the HSI approach in combination with the SVM classifier in discriminating Fuji apples with a high classification accuracy of 92.75% using the wavelength range of 400–1000 nm. Texture analysis was used to extract the characteristics based on a gray-level cooccurrence matrix. With a discriminating efficiency of 95.56%, a similar wavelength area was used for variety classification in pears (Sun et al., 2016).

1.11 Contaminant detection

Fruits are susceptible to pathogenic invasion, pests, and microbial contaminations starting right after harvesting, and fruit safety is critical throughout the supply chain. Both NIR spectroscopy and HSI approaches have been examined to detect pollutants, pests, and disease infections efficiently. In fruits such as fig, cherry, blueberry, jujube, mango, and chestnut, NIR spectroscopy and the HSI approach correctly predicted the existence of insects and insect infestation inside the fruits (Huang et al., 2013; Rady et al., 2017; Singh et al., 2009).

The NIR approach is effective for both direct and indirect detection of insect infestations due to the presence of lipid, hemolymph, and chitin and indirect detection due to physical changes such as dehydration and internal browning darkening or fungal contamination (Wang et al., 2011). *Bactrocera oleae* infection in olive fruit flies was diagnosed with 94% accuracy using absorbance light in the NIR band of 1100–2300 nm (Moscetti et al., 2015).

For identifying the fecal spots on intentionally infected apples, an HSI reflectance (400–700 nm) and fluorescence imaging system were utilized, with a prediction accuracy of 99.5% and 100%, respectively (Kim et al., 2002). In a full spectral range between 960 and 1700 nm, the HSI approach accurately detected the infected date fruits caused by *Aspergillus flavus* with 100% accuracy. A comparable investigation with dried figs was conducted in transmittance mode to categorize the black mold-infested fruits caused by *Aspergillus niger*. The efficiency was 93% and 99.3% in the 390–1005 nm region.

For identifying other microbial pollutants, other fungal species infection, mico-toxins, and their effect on the change of chemical composition in different fruits, more research is needed in both NIR and HSI approaches. A combination of HSI and mass spectroscopic methods could help achieve the goal. For a complete quality control system in fruits, a combination of reflectance, transmittance mood, and other detection methods must be further investigated.

1.12 Firmness

Fruit firmness is an important quality criterion for determining the stage of maturity. Firmness varies widely depending on the type of fruit; in most cases, maturity time is inversely related to firmness. Due to a decrease in cell-to-cell adhesion, the breakdown of the middle lamella, changes in the turgor pressure, and the weakening of parenchyma cell walls throughout the ripening process, the firmness of the fruit declines, eventually leading to the loss of membrane integrity via cell-wall-modifying enzymes.

The ability of NIR spectroscopy to determine firmness in three different apple cultivars, namely red and golden delicious and empire, was investigated. The diffuse reflectance acquisition mode was chosen, and spectra between 800 and 1700 nm were investigated (Lu et al., 2000). For peeled and unpeeled apples, the NIR measurements were made for firmness, with a standard error of predictions (SEP) of 10.1 and 6.6 N, respectively. Smoothing the spectra using the Savitzky–Golay spectral pre-processing method improved firmness prediction in unpeeled apples. The hardness of two sweet cherry cultivars, Hedelfinger cherry and Sam cherry, was predicted with r values of 0.8 and 0.65 and SEP values of 0.55 and 0.44 N, respectively; relatively good predictions were achieved (Lu, 2001).

Because of direct interaction with physical qualities such as particle size, tissues, and cellular structure of the fruits, the scattering acquisition technique has gained much attention for determining firmness (Lu, 2004). The HSI scattering approach for measuring fruit firmness has been the subject of numerous research publications (Lu, 2007; Peng & Lu, 2008).

Mendoza et al. (2011) used two HSI setups, an online laboratory system and a stationary system, to measure firmness in three apple cultivars in the spectral range of 500–1000 nm using a combination of two-dimensional image analysis and one-dimensional scattering outline approaches. In determining the firmness using the PLS model, for Jonagold, Golden Delicious, and Delicious apple cultivars, the SEP for hardness was 16.1%, 6.6%, and 13.7%, respectively. The addition of image and scattering outline data greatly increased the hardness prediction and proved to be a valuable tool for predicting the interior quality in fruits. Islam et al. (2018a) investigated the potential of HSI at shortwave infrared (SWIR) range to determine the firmness and dry matter content of stored onions. A proper chemometric model efficiently predicted the changes in firmness and dry matter content of onions.

Cen et al. (2011) graded blueberries based on hardness using a push-broom HSI device. To better understand the quality, reflectance photos of the fruit's calyx and stem orientation were taken. Based on partial least-squares regression (PLSR) and PCA prediction models with cross-validation, good correlation coefficients of 0.83–0.87 and root mean standard error of prediction (RMSEP) of 0.23–0.26 N/mm were found between 515 and 1000 nm. Leiva-Valenzuela et al. (2013) used a similar technique to acquire hyper-spectral pictures to assess the hardness of blueberries. Due to the accumulation of dark anthocyanin pigments, more robust absorption spectra in the range of 675–1000 nm were reported in the early stages. The blueberries were classified into two classes, firm and soft, based on this quality trait, and this method proved to be viable for online sorting and grading.

1.13 Total soluble solids

Fruit harvest ripeness is determined by chemical composition and internal and external fruit characteristics. Total soluble solid (TSS) is one of the essential compositional parameters. TSS is a naturally occurring component that includes soluble sugars, acids, and other water-soluble compounds. The ratio of TSS to acid is a good indication of the fruit's flavor.

Kawano et al. (1992) used interactants acquisition mode in the spectral region of 680–1235 nm to quantify the sugar in intact peaches. Four wavelengths (870, 878, 889, and 906 nm) were selected, which predicted sugar content with a correlation coefficient (r) of 0.97. Kawano et al. (1993) estimated the sugar content in intact Satsuma mandarins using NIR transmittance mode with high accuracy ($r = 0.989$; SEP = 0.32°Brix). Slaughter (1995) predicted the sugar content of intact nectarines and peaches using a multiple linear regression (MLR)-based model in the region between 400 and 1100 nm (SEP = 1.0°Brix; $r = 0.90$) using interacting mode. Lammertyn et al. (1998) used visible NIR spectroscopy (reflectance mode) in the 380–1650 nm wavelength region to predict the sugar content in the Jonagold apple variety. They observed that spectral information in the VIS/NIR region up to 900 nm was not informative in predicting the sugar content and conveyed that apple skin had a significant effect upon the spectral reflectance pattern.

The ability of NIR spectroscopy to determine the sugar content of three different apple kinds, namely red and golden delicious and empire apples, was investigated. In the range of 800–1700 nm, diffuse reflectance acquisition mode was chosen and studied (Lu et al., 2000). The raw reflectance spectra were smoothed (pre-processed) using the Savitzky–Golay method. Sugar content was accurately predicted with *r* of 0.93–0.97 and SEP of 0.37–0.42°Brix.

Wang et al. (2011) used NIR spectroscopy to predict the TSS content in jujube fruit by comparing interactance and transmittance measurements based on stepwise discriminant analysis (SDA) and PLSR models in the 500–1000 nm spectral region. The RMSEP of the interactance spectra was 2.018–3.2°Brix, with a correlation coefficient (*r*) of 0.74–0.91. In the reflectance mode, the SDA model was more effective at distinguishing infested jujubes from healthy intact fruits. Interaction mode offered a considerably better prediction compared with transmission measurements (RMSEP = 3.517–3.863 and $r = 0.63–0.73$). Wang et al. (2014) used visible shortwave NIR spectroscopy to examine the reflectance (460–1000 nm), interacting (520–1000 nm), and transmittance (550–930 nm) modes of image collection to assess the TSS of oranges.

TSS in fruits is an important quality parameter and can be assessed accurately by HSI technique in a shorter time (Fan et al., 2016; Tian et al., 2018; Zhu et al., 2012). Elmasry et al. (2007) used an integrated visible (VIS)/NIR HSI technique in 400–1000 nm wavelength region for the prediction of TSS content using PLSR in strawberry fruits. TSS was predicted with an SEP of 0.184°Brix and *r*-value of 0.80. Further, MLR models were developed using some selected optimal wavelengths of 500–960 nm considering the reflectance spectra. The MLR model gave better results compared with PLS models with good prediction values for TSS in strawberries. Mendoza et al. (2011) tried to improve TSS prediction in apples by integrating spectral scattering and image acquisition and achieved a 3% reduction of SEP in the spectral region of 500–1000 nm. Spatially resolved reflectance images were collected to predict TSS in peach fruits within 515–1000 nm wavelength (Cen et al., 2011). Prediction models were developed using PLS, and integration of PCA with the LS-SVM yielded prediction accuracy with an *r*-value of 0.504.

Li et al. (2018) examined two hyper-spectral cameras to assess TSS in plum fruits in the VIS/NIR region (wavelengths between 600 and 975 nm) and the SWIR range (wavelengths between 865 and 1610 nm). The research looked at how the light landing on the fruit's spherical surface affects prediction accuracy. Gómez-Sanchis's automatic shape correction method was explored, which resulted in a marginal improvement in prediction accuracy (Gómez-Sanchis et al., 2008). The information from the chemical vibrations of N–H, C–H, and O–H organic molecules was gathered by the SWIR hyper-spectral influencing zones, which accurately predicted TSS with an *r*-value of 0.83. Similar results were reported for apple (Park et al., 2003), mango (Marques et al., 2016), lychee (Pu et al., 2016), and kiwifruit (McGlone & Kawano, 1998).

1.14 Moisture and dry matter

The moisture and dry matter content of fruits play a big role in consumer acceptance. It also plays a significant role in deciding different processing factors. Fruit with a higher moisture content has a higher freshness rating. Moisture loss during long-term storage causes weight loss, shriveling of the fruit peel, and a greasy or glossy texture, all of which cost the manufacturer money. On the other hand, ripening accelerates the increase in dry matter content, which correlates with the fruit's starch content and makes it more palatable (Cozzolino et al., 2008; McGlone & Kawano, 1998).

Blakey et al. (2009) used the PLSR model to measure the moisture content of avocados using a handheld NIR spectrometer with reflectance spectra between 1000 and 1800 nm. The Savitzky–Golay derivative and the standard normal variate transformation were carried out to get the best spectral pre-processing treatments for the prediction model (50%–87%, with an r-value of 0.96 and an SEP of 2.78% moisture content). A comparable approach to estimating the moisture content of golden delicious apples using NIR spectroscopy found that the density of apples in the linear regression model was strongly influenced by the reduction in moisture content over time (Vesali et al., 2011).

Bag et al. (2011) confirmed that Fourier transform NIR spectroscopy can be used to quickly estimate the moisture content in bael pulp using the PLSR model and a wavelength range of 800–2500 nm in diffused reflectance mode. Based on the obtained maximum coefficient of determination value ($r^2 = 0.993$), NIR spectra pre-processed with the min–max normalization method were found to be more acceptable. This approach was found to be the quickest for determining the moisture content in bael pulp, taking less than 10 seconds for each sample

The dry matter content of intact banana and mango was assessed using SWIR spectrophotometry with interacting mode data acquisition in the 500 and 1050 nm spectral area (Subedi & Walsh, 2011). Marques et al. (2016) experimented with a handheld ultra-compact NIR spectrometer based on linear-variable filter technology for predicting dry matter content and depth of penetration of NIR spectra into mango tissues in the 950–1650 nm spectral region. With a radiation penetration depth of 7.4 mm into the mango pericarp tissue, the RMSEP and coefficient of determination were 0.17 and 0.5, respectively.

For the determination of the moisture content in bananas at different maturation stages with three storage temperatures (20°C, 25°C, and 30°C), Rajkumar et al. (2012) used HSI using reflectance acquisition mode. The link between spectral reflectance and moisture content was established using the PLS calibration model at a wavelength between 400 and 1000 nm. As the fruit's moisture content grew with maturity, the moisture content changed followed a linear pattern. Elmasry et al. (2007) found similar peaks in the spectral absorption bands in strawberries with an r-value of 0.87. It is challenging to determine the moisture and dry matter content when the scales of fruits and vegetables are thick.

In addition, if the product is covered with a dry scale, then the NIR-based quality evaluation may become more challenging (Islam et al., 2018a).

1.15 Defect identification

The chilling injury affects the majority of tropical and subtropical fruits. Fruit tissue is vulnerable to injury at low temperatures, resulting in altered metabolic activity. Surface pitting, browning, failure to ripen, wilting, flavor loss, and rotting have all been recorded in sensitive species.

Bruising is a mechanical injury that occurs on the surface of the fruit during harvest and transportation. It has an impact on the appearance, texture, and color of the fruit and promotes water loss. Moreover, bruising accelerates secondary infections. The bruised patches raise the possibility of bacterial and fungal infestation, making the product less appealing to customers. Studies have been reported on the potentials of NIR spectroscopy and HSI techniques for the detection of bruises in mango (Raghavendra et al., 2021), guava (Nila et al., 2017), coconut (Noypitak et al., 2019), and jujube (Wu et al., 2016).

In the wavelength range of 350–2500 nm, Zhang et al. (2013) discovered mild bruising in jujube fruits. The LS-SVM-SPA-PCA model correctly diagnosed bruising in fresh samples with a 100% accuracy rate, proving that VIS/NIR spectroscopy with reflectance mode is feasible. The degree of bruising in fruits can be inspected and classified to provide helpful information for proper postharvest handling and storage.

Based on HSI data in the wavelength range of 450–930 nm, a multispectral classification approach termed spectral information divergence was developed for the inspection of citrus cankers (fruit disease) in grapefruits (Qin et al., 2009). The model projected a 96% overall classification accuracy. Wu et al. (2016) investigated the potential of using a visible NIR HSI system to detect bruising, insect infestation, and cuticle cracks in jujubes simultaneously. A region of interest was carefully selected from the spectral band 400–1600 nm using PCA. The soft independent modeling of class analogy models outperformed SVM models with a classification accuracy of 96% for bruises, 94% for insect infestation, and 96% for cuticle cracks when compared with SVM models.

Using the 450–1000 nm wavelength area, a transmittance mode-based VIS/NIR HSI system was employed to develop an improved approach for differentiating cherry with pits, pit pieces, and no pits. Supervised classification models were built using a correlation-based feature selection approach and the Savitsky–Golay second derivative spectral pre-processing technique from the recorded images. The backpropagation neural network model was determined to be the most acceptable among the several classification models tested, with a prediction accuracy of 95%. According to this study, future research should focus on the effects of fruit orientation, surface and internal mechanical flaws, and insect infestation (Siedliska et al., 2017).

Based on the spectral information for each pixel inside the NIR images at a given wavelength, the aforementioned studies show that NIR and HSI systems are promising in differentiating the bruising of fruits.

References

Abdelmalek, G. A. M., & Salaheldin, T. A. (2016). Silver nanoparticles as a potent fungicide for citrus phytopathogenic fungi. *Journal of Nanomedicine Research*, *3*(5), 00065.

Andrianto, H., Faizal, A., & Armandika, F. (2020). *Smartphone application for deep learning-based rice plant disease detection. 2020 international conference on information technology systems and innovation (ICITSI)* (pp. 387–392). IEEE.

Atkinson, J. A., Jackson, R. J., Bentley, A. R., Ober, E., & Wells, D. M. (2018). Field phenotyping for the future. In J. A. Roberts (Ed.), *Annual plant reviews online* (pp. 719–736). Wiley. Available from https://doi.org/10.1002/9781119312994.apr0651.

Bag, S. K., Srivastav, P. P., & Mishra, H. N. (2011). FT-NIR spectroscopy: a rapid method for estimation of moisture content in bael pulp. *British Food Journal*, *113*, 494–504. Available from https://doi.org/10.1108/00070701111123970.

Barbedo, J. G. A. (2019). A review on the use of unmanned aerial vehicles and imaging sensors for monitoring and assessing plant stresses. *Drones*, *3*(2), 40.

Bhuiyan, M. A. H. B., Vaghefi, N., & Taylor, P. W. J. (2019). Ray blight of pyrethrum in Australia: a review of the current status and future opportunities. *Plant Pathology*, *68*(4), 620–627.

Blakey, R. J., Bower, J. P., & Bertling, I. (2009). Influence of water and ABA supply on the ripening pattern of avocado (*Persea americana* Mill.) fruit and the prediction of water content using near infrared spectroscopy. *Postharvest Biology and Technology*, *53*, 72–76. Available from https://doi.org/10.1016/j.postharvbio.2009.03.004.

Bramhanwade, K., Shende, S., Bonde, S., Gade, A., & Rai, M. (2016). Fungicidal activity of Cu nanoparticles against Fusarium causing crop diseases. *Environmental Chemistry Letters*, *14*(2), 229–235.

Cao, F., Wu, D., & He, Y. (2010). Soluble solids content and pH prediction and varieties discrimination of grapes based on visible–near infrared spectroscopy. *Computers and Electronics in Agriculture*, *71*, S15–S18. Available from https://doi.org/10.1016/j.compag.2009.05.011.

Cen, H., Lu, R., Mendoza, F. A., & Ariana, D. P. (2011). Peach maturity/quality assessment using hyperspectral imaging-based spatially resolved technique. *SPIE Proceedings*. Available from https://doi.org/10.1117/12.883573.

Choudhury, S. R., Ghosh, M., Mandal, A., Chakravorty, D., Pal, M., Pradhan, S., & Goswami, A. (2011). Surface-modified sulfur nanoparticles: an effective antifungal agent against *Aspergillus niger* and *Fusarium oxysporum*. *Applied Microbiology and Biotechnology*, *90*(2), 733–743.

Costa, J. M., Marques da Silva, J., Pinheiro, C., Barón, M., Mylona, P., Centritto, M., & Oliveira, M. M. (2019). Opportunities and limitations of crop phenotyping in southern European countries. *Frontiers in Plant Science*, *10*, 1125.

Cozzolino, D., Cynkar, W. U., Dambergs, R. G., Mercurio, M. D., & Smith, P. A. (2008). Measurement of condensed tannins and dry matter in red grape homogenates using near infrared spectroscopy and partial least squares. *Journal of Agricultural and Food Chemistry*, *56*, 7631–7636. Available from https://doi.org/10.1021/jf801563z.

Devadas, R., Lamb, D. W., Backhouse, D., & Simpfendorfer, S. (2015). Sequential application of hyperspectral indices for delineation of stripe rust infection and nitrogen deficiency in wheat. *Precision Agriculture*, *16*(5), 477–491.

Devi, R., Yadav, S., Nehra, R., Yadav, S., & Pundir, C. S. (2013). Electrochemical biosensor based on gold coated iron nanoparticles/chitosan composite bound xanthine oxidase for detection of xanthine in fish meat. *Journal of Food Engineering*, *115*(2), 207–214.

Dou, W., Tang, W., & Zhao, G. (2013). A disposable electrochemical immunosensor arrays using 4-channel screen-printed carbon electrode for simultaneous detection of *Escherichia coli* O157:H7 and *Enterobacter sakazakii*. *Electrochimica Acta*, *97*, 79–85.

Elmasry, G., Wang, N., Elsayed, A., & Ngadi, M. (2007). Hyperspectral imaging for nondestructive determination of some quality attributes for strawberry. *Journal of Food Engineering*, *81*, 98–107. Available from https://doi.org/10.1016/j.jfoodeng.2006.10.016.

Elumalai, K., Velmurugan, S., Ravi, S., Kathiravan, V., & Ashokkumar, S. (2015). RETRACTED: green synthesis of zinc oxide nanoparticles using Moringa oleifera leaf extract and evaluation of its antimicrobial activity. *Spectrochimica Acta Part A: Molecular and Biomolecular Spectroscopy*, *143*, 158–164.

Fan, S., Zhang, B., Li, J., Liu, C., Huang, W., & Tian, X. (2016). Prediction of soluble solids content of apple using the combination of spectra and textural features of hyperspectral reflectance imaging data. *Postharvest Biology and Technology*, *121*, 51–61. Available from https://doi.org/10.1016/j.postharvbio.2016.07.007.

Ferentinos, K. P. (2018). Deep learning models for plant disease detection and diagnosis. *Computers and Electronics in Agriculture*, *145*, 311–318.

Finco, A., Bucci, G., Belletti, M., & Bentivoglio, D. (2021). The economic results of investing in precision agriculture in durum wheat production: A case study in central Italy. *Agronomy*, *11*(8), 1520.

Gastavsson, J., Cederberg, C., & Sonesson, U. (2011). *Global food losses and food waste*. Rome, Italy: Food and Agriculture Organization (FAO) of the United Nations.

Ge, H., Ma, F., Li, Z., & Du, C. (2021). Grain yield estimation in rice breeding using phenological data and vegetation indices derived from UAV images. *Agronomy*, *11*(12), 2439.

Ghormade, V., Deshpande, M. V., & Paknikar, K. M. (2011). Perspectives for nanobiotechnology enabled protection and nutrition of plants. *Biotechnology Advances*, *29* (6), 792–803.

Gómez-Candón, D., Virlet, N., Labbé, S., Jolivot, A., & Regnard, J. L. (2016). Field phenotyping of water stress at tree scale by UAV-sensed imagery: new insights for thermal acquisition and calibration. *Precision Agriculture*, *17*(6), 786–800.

Gómez-Sanchis, J., Moltó, E., Camps-Valls, G., Gómez-Chova, L., Aleixos, N., & Blasco, J. (2008). Automatic correction of the effects of the light source on spherical objects. An application to the analysis of hyperspectral images of citrus fruits. *Journal of Food Engineering*, *85*, 191–200. Available from https://doi.org/10.1016/j.jfoodeng.2007.06.036.

Guo, W., Gu, J., Liu, D., & Shang, L. (2016). Peach variety identification using near-infrared diffuse reflectance spectroscopy. *Computers and Electronics in Agriculture*, *123*, 297–303. Available from https://doi.org/10.1016/j.compag.2016.03.005.

Guo, Z., Huang, W., Chen, L., Zhao, C., & Peng, Y. (2013). Geographical classification of apple based on hyperspectral imaging. *SPIE Proceedings*. Available from https://doi.org/10.1117/12.2015559.

He, L., Liu, Y., Mustapha, A., & Lin, M. (2011). Antifungal activity of zinc oxide nanoparticles against *Botrytis cinerea* and *Penicillium expansum*. *Microbiological Research*, *166*(3), 207–215.

Holman, F. H., Riche, A. B., Michalski, A., Castle, M., Wooster, M. J., & Hawkesford, M. J. (2016). High throughput field phenotyping of wheat plant height and growth rate in field plot trials using UAV based remote sensing. *Remote Sensing*, *8*(12), 1031.

Hu, Y., Knapp, S., & Schmidhalter, U. (2020). Advancing high-throughput phenotyping of wheat in early selection cycles. *Remote Sensing*, *12*(3), 574.

Huang, M., Wan, X., Zhang, M., & Zhu, Q. (2013). Detection of insect-damaged vegetable soybeans using hyperspectral transmittance image. *Journal of Food Engineering*, *116*, 45–49.

Ingvarsson, P. K., & Street, N. R. (2011). Association genetics of complex traits in plants. *New Phytologist*, *189*(4), 909–922.

Inoue, Y. (2020). Satellite-and drone-based remote sensing of crops and soils for smart farming: A review. *Soil Science and Plant Nutrition*, *66*(6), 798–810.

Iqbal, Z., Khan, M. A., Sharif, M., Shah, J. H., Ur Rehman, M. H., & Javed, K. (2018). An automated detection and classification of citrus plant diseases using image processing techniques: A review. *Computers and Electronics in Agriculture*, *153*, 12–32.

Islam, M.N. (2018). *Postharvest quality changes of onions during long-term storage. Non-destructive quality assessment and modeling of bulb quality* [PhD thesis]. Aarhus University.

Islam, M. N., Nielsen, G., Stærke, S., Kjær, A., Jørgensen, B., & Edelenbos, M. (2018a). Noninvasive determination of firmness and dry matter content of stored onion bulbs using shortwave infrared imaging with whole spectra and selected wavelengths. *Applied Spectroscopy*, *72*, 1467–1478. Available from https://doi.org/10.1177/0003702818792282.

Islam, M. N., Nielsen, G., Stærke, S., Kjær, A., Jørgensen, B., & Edelenbos, M. (2018b). Novel non-destructive quality assessment techniques of onion bulbs: a comparative study. *Journal of Food Science and Technology*, *55*, 3314–3324. Available from https://doi.org/10.1007/s13197-018-3268-x.

Jiang, Z., Tu, H., Bai, B., Yang, C., Zhao, B., Guo, Z., ... Zhang, J. (2021). Combining UAV-RGB high-throughput field phenotyping and genome-wide association study to reveal genetic variation of rice germplasms in dynamic response to drought stress. *New Phytologist*, *232*(1), 440–455.

Jin, X., Zarco-Tejada, P. J., Schmidhalter, U., Reynolds, M. P., Hawkesford, M. J., Varshney, R. K., ... Li, S. (2020). High-throughput estimation of crop traits: A review of ground and aerial phenotyping platforms. *IEEE Geoscience and Remote Sensing Magazine*, *9*(1), 200–231.

Kanhed, P., Birla, S., Gaikwad, S., Gade, A., Seabra, A. B., Rubilar, O., Duran, N., & Rai, M. (2014). In vitro antifungal efficacy of copper nanoparticles against selected crop pathogenic fungi. *Materials Letters*, *115*, 13–17.

Kartikeyan, P., & Shrivastava, G. (2021). Review on emerging trends in detection of plant diseases using image processing with machine learning. *International Journal of Computer Applications*, *975*, 8887.

Kashyap, P. L., Kumar, S., Jasrotia, P., Singh, D. P., & Singh, G. P. (2019). Nanosensors for plant disease diagnosis: Current understanding and future perspectives. *Nanoscience for Sustainable Agriculture*, 189–205.

Kashyap, P. L., Kumar, S., & Srivastava, A. K. (2017). Nanodiagnostics for plant pathogens. *Environmental Chemistry Letters*, *15*(1), 7–13.

Kawano, S., Fujiwara, T., & Iwamoto, M. (1993). Non-destructive determination of sugar content in satsuma mandarin using near infrared (NIR) transmittance. *Engei Gakkai zasshi*, *62*, 465–470. Available from https://doi.org/10.2503/jjshs.62.465.

Kawano, S., Watanabe, H., & Iwamoto, M. (1992). Determination of sugar content in intact peaches by near infrared spectroscopy with fiber optics in interactance mode. *Engei Gakkai zasshi*, *61*, 445–451. Available from https://doi.org/10.2503/jjshs.61.445.

Kim, M. S., Lefcourt, A. M., Chen, Y. R., Kim, I., Chan, D. E., & Chao, K. (2002). Multispectral detection of fecal contamination on apples based on hyperspectral imagery: Part II. Application of hyperspectral fluorescence imaging. *Transactions of the ASAE*, *45*. Available from https://doi.org/10.13031/2013.11416.

Kim, S. L., Kim, N., Lee, H., Lee, E., Cheon, K. S., Kim, M., ... Kim, K. H. (2020). High-throughput phenotyping platform for analyzing drought tolerance in rice. *Planta*, *252*(3), 1–18.

Kim, S. W., Jung, J. H., Lamsal, K., Kim, Y. S., Min, J. S., & Lee, Y. S. (2012). Antifungal effects of silver nanoparticles (AgNPs) against various plant pathogenic fungi. *Mycobiology*, *40*(1), 53–58.

Klodt, M., Herzog, K., Töpfer, R., & Cremers, D. (2015). Field phenotyping of grapevine growth using dense stereo reconstruction. *BMC Bioinformatics*, *16*(1), 1–11.

Koh, J. C., Spangenberg, G., & Kant, S. (2021). Automated machine learning for high-throughput image-based plant phenotyping. *Remote Sensing*, *13*(5), 858.

Kumar, R., Mishra, J. S., Rao, K. K., Mondal, S., Hazra, K. K., Choudhary, J. S., & Bhatt, B. P. (2020). Crop rotation and tillage management options for sustainable intensification of rice-fallow agro-ecosystem in eastern India. *Scientific Reports*, *10*(1), 1–15.

Lammertyn, J., Nicolaï, B., Ooms, K., Smedt, V. D., & Baerdemaeker, J. D. (1998). Non-destructive measurement of acidity, soluble solids, and firmness of jonagold apples using nir-spectroscopy. *Transactions of the ASAE*, *41*, 1089–1094. Available from https://doi.org/10.13031/2013.17238.

Leiva-Valenzuela, G. A., Lu, R., & Aguilera, J. M. (2013). Prediction of firmness and soluble solids content of blueberries using hyperspectral reflectance imaging. *Journal of Food Engineering*, *115*, 91–98. Available from https://doi.org/10.1016/j.jfoodeng.2012.10.001.

Li, B., Cobo-Medina, M., Lecourt, J., Harrison, N., Harrison, R. J., & Cross, J. V. (2018). Application of hyperspectral imaging for non-destructive measurement of plum quality attributes. *Postharvest Biology and Technology*, *141*, 8–15. Available from https://doi.org/10.1016/j.postharvbio.2018.03.008.

Li, X., He, Y., & Fang, H. (2007). Non-destructive discrimination of Chinese bayberry varieties using Vis/NIR spectroscopy. *Journal of Food Engineering*, *81*, 357–363. Available from https://doi.org/10.1016/j.jfoodeng.2006.10.033.

Lishawa, S. C., Carson, B. D., Brandt, J. S., Tallant, J. M., Reo, N. J., Albert, D. A., ... Clark, E. (2017). Mechanical harvesting effectively controls young *Typha* spp. invasion and unmanned aerial vehicle data enhances post-treatment monitoring. *Frontiers in Plant Science*, *8*, 619.

Lu, R. (2001). Predicting firmness and sugar content of sweet cherries using near–infrared diffuse reflectance spectroscopy. *Transactions of the ASAE*, *44*. Available from https://doi.org/10.13031/2013.6421.

Lu, R. (2004). Multispectral imaging for predicting firmness and soluble solids content of apple fruit. *Postharvest Biology and Technology*, *31*, 147–157. Available from https://doi.org/10.1016/j.postharvbio.2003.08.006.

Lu, R. (2007). Non-destructive measurement of firmness and soluble solids content for apple fruit using hyperspectral scattering images. *Sensing and Instrumentation for Food Quality and Safety*, *1*, 19–27. Available from https://doi.org/10.1007/s11694-006-9002-9.

Lu, R., Guyer, D. E., & Beaudry, R. M. (2000). Determination of firmness and sugar content of apples using near-infrared diffuse reflectance. *Journal of Texture Studies*, *31*, 615–630. Available from https://doi.org/10.1111/j.1745-4603.2000.tb01024.x.

Marques, E. J. N., De Freitas, S. T., Pimentel, M. F., & Pasquini, C. (2016). Rapid and non-destructive determination of quality parameters in the 'Tommy Atkins' mango using a novel handheld near infrared spectrometer. *Food Chemistry*, *197*, 1207–1214. Available from https://doi.org/10.1016/j.foodchem.2015.11.080.

Martinelli, F., Scalenghe, R., Davino, S., Panno, S., Scuderi, G., Ruisi, P., Villa, P., Stroppiana, D., Boschetti, M., Goulart, L. R., & Davis, C. E. (2015). Advanced methods of plant disease detection. A review. *Agronomy for Sustainable Development*, *35*(1), 1–25.

Mathivanan, S. (2021). Perspectives of nano-materials and nanobiosensors in food safety and agriculture. *Novel Nanomaterials*, 197.

McCulloch, W. S., & Pitts, W. (1943). A logical calculus of the ideas immanent in nervous activity. *The Bulletin of Mathematical Biophysics*, *5*(4), 115–133. Available from https://www.bankmycell.com/blog/how-many-phones-are-in-the-world.

McGlone, V. A., & Kawano, S. (1998). Firmness, dry-matter and soluble-solids assessment of postharvest kiwifruit by NIR spectroscopy. *Postharvest Biology and Technology*, *13*, 131–141. Available from https://doi.org/10.1016/s0925-5214(98)00007-6.

Meacham-Hensold, K., Montes, C. M., Wu, J., Guan, K., Fu, P., Ainsworth, E. A., ... Bernacchi, C. J. (2019). High-throughput field phenotyping using hyperspectral reflectance and partial least squares regression (PLSR) reveals genetic modifications to photosynthetic capacity. *Remote Sensing of Environment*, *231*, 111176.

Mendoza, F., Lu, R., Ariana, D., Cen, H., & Bailey, B. (2011). Integrated spectral and image analysis of hyperspectral scattering data for prediction of apple fruit firmness and soluble solids content. *Postharvest Biology and Technology*. Available from https://doi.org/10.1016/j.postharvbio.2011.05.009.

Mishra, S., Singh, B. R., Singh, A., Keswani, C., Naqvi, A. H., & Singh, H. B. (2014). Biofabricated silver nanoparticles act as a strong fungicide against *Bipolaris sorokiniana* causing spot blotch disease in wheat. *PLoS One*, *9*(5), e97881.

Modica, G., Messina, G., De Luca, G., Fiozzo, V., & Praticò, S. (2020). Monitoring the vegetation vigor in heterogeneous citrus and olive orchards. A multiscale object-based approach to extract trees' crowns from UAV multispectral Imagery. *Computers and Electronics in Agriculture*, *175*, 105500.

Mohanty, S. P., Hughes, D. P., & Salathé, M. (2016). Using deep learning for image-based plant disease detection. *Frontiers in Plant Science*, *7*, 1419.

Moscetti, R., Haff, R. P., Stella, E., Contini, M., Monarca, D., Cecchini, M., & Massantini, R. (2015). Feasibility of NIR spectroscopy to detect olive fruit infested by *Bactrocera oleae*. *Postharvest Biology and Technology*, *99*, 58–62. Available from https://doi.org/10.1016/j.postharvbio.2014.07.015.

Nila, I. R., Saputro, A. H., & Imawan, C. (2017). *The prediction system of bruising depth of guava (*Psidium guajava *L.) based on Vis-NIR imaging. 2017 international conference on sustainable information engineering and technology (SIET)* (pp. 420–424). IEEE.

Noypitak, S., Imsabai, W., Noknoi, W., Karoojee, S., Terdwongworakul, A., & Kobori, H. (2019). Detection of cracked shell in intact aromatic young coconut using near infrared spectroscopy and acoustic response methods. *Journal of Food Measurement and Characterization*, *13*, 1991–1999.

Nyaga, J. M., Onyango, C. M., Wetterlind, J., & Söderström, M. (2021). Precision agriculture research in sub-Saharan Africa countries: A systematic map. *Precision Agriculture*, 1–20.

Onyango, C. M., Nyaga, J. M., Wetterlind, J., Söderström, M., & Piikki, K. (2021). Precision agriculture for resource use efficiency in smallholder farming systems in Sub-Saharan Africa: A systematic review. *Sustainability*, *13*(3), 1158.

Panday, U. S., Pratihast, A. K., Aryal, J., & Kayastha, R. B. (2020). A review on drone-based data solutions for cereal crops. *Drones*, *4*(3), 41.

Park, B., Abbott, J. A., Lee, K.-J., Choi, C., & Choi, K. (2003). Near-infrared diffuse reflectance for quantitative and qualitative measurement of soluble solids and firmness of Delicious and Gala apples. *Transactions of the ASAE*, *46*, 1721.

Peigné, J., Vian, J. F., Payet, V., & Saby, N. P. (2018). Soil fertility after 10 years of conservation tillage in organic farming. *Soil and Tillage Research*, *175*, 194–204.

Peng, Y., & Lu, R. (2008). Analysis of spatially resolved hyperspectral scattering images for assessing apple fruit firmness and soluble solids content. *Postharvest Biology and Technology*, *48*, 52–62. Available from https://doi.org/10.1016/j.postharvbio.2007.09.019.

Pierce, F. J., & Fortin, M. C. (2019). Long-term tillage and periodic plowing of a no-tilled soil in Michigan: Impacts, yield, and soil organic matter. In A. Paul, Keith H. Paustian, E. T. Elliott, & C. Vernon Cole (Eds.), *Soil organic matter in temperate agroecosystems* (pp. 141–149). CRC Press.

Pu, H., Liu, D., Wang, L., & Sun, D.-W. (2016). Soluble solids content and pH prediction and maturity discrimination of lychee fruits using visible and near infrared hyperspectral imaging. *Food Analytical Methods*, *9*, 235–244.

Pujari, J. D., Yakkundimath, R., & Byadgi, A. S. (2014). *Identification and classification of fungal disease affected on agriculture/horticulture crops using image processing techniques. 2014 IEEE international conference on computational intelligence and computing research* (pp. 1–4). IEEE.

Puoci, F., Iemma, F., Spizzirri, U. G., Cirillo, G., Curcio, M., & Picci, N. (2008). Polymer in agriculture: A review. *American Journal of Agricultural and Biological Sciences*, *3*(1), 299–314.

Qin, J., Burks, T. F., Ritenour, M. A., & Bonn, W. G. (2009). Detection of citrus canker using hyperspectral reflectance imaging with spectral information divergence. *Journal of Food Engineering*, *93*, 183–191. Available from https://doi.org/10.1016/j.jfoodeng.2009.01.014.

Rady, A., Ekramirad, N., Adedeji, A., Li, M., & Alimardani, R. (2017). Hyperspectral imaging for detection of codling moth infestation in GoldRush apples. *Postharvest Biology and Technology*, *129*, 37–44.

Raghavendra, A., Guru, D., & Rao, M. K. (2021). Mango internal defect detection based on optimal wavelength selection method using NIR spectroscopy. *Artificial Intelligence in Agriculture*, *5*, 43–51.

Rajkumar, P., Wang, N., Eimasry, G., Raghavan, G. S. V., & Gariepy, Y. (2012). Studies on banana fruit quality and maturity stages using hyperspectral imaging. *Journal of Food Engineering*, *108*, 194–200. Available from https://doi.org/10.1016/j.jfoodeng.2011.05.002.

Ramyadevi, J., Jeyasubramanian, K., Marikani, A., Rajakumar, G., & Rahuman, A. A. (2012). Synthesis and antimicrobial activity of copper nanoparticles. *Materials Letters*, *71*, 114–116.

Roberson, G. T. (2000). Precision agriculture technology for horticultural crop production. *Horttechnology*, *10*(3), 448–451.

Ruffo Roberto, S., Youssef, K., Hashim, A. F., & Ippolito, A. (2019). Nanomaterials as alternative control means against postharvest diseases in fruit crops. *Nanomaterials*, *9*(12), 1752.

Saharan, V., Mehrotra, A., Khatik, R., Rawal, P., Sharma, S. S., & Pal, A. (2013). Synthesis of chitosan based nanoparticles and their in vitro evaluation against phytopathogenic fungi. *International Journal of Biological Macromolecules*, *62*, 677–683.

Saleem, M. H., Potgieter, J., & Arif, K. M. (2019). Plant disease detection and classification by deep learning. *Plants*, *8*(11), 468.

Sankey, T. T., Leonard, J. M., & Moore, M. M. (2019). Unmanned aerial vehicle – based rangeland monitoring: Examining a century of vegetation changes. *Rangeland Ecology & Management*, *72*(5), 858–863.

Shakoor, N., Lee, S., & Mockler, T. C. (2017). High throughput phenotyping to accelerate crop breeding and monitoring of diseases in the field. *Current Opinion in Plant Biology*, *38*, 184–192.

Shams, S., Bakhshi, B., Moghadam, T. T., & Behmanesh, M. (2019). A sensitive gold-nanorods-based nanobiosensor for specific detection of *Campylobacter jejuni* and *Campylobacter coli*. *Journal of Nanobiotechnology*, *17*(1), 1–13.

Shang, L., Guo, W., & Nelson, S. O. (2014). Apple variety identification based on dielectric spectra and chemometric methods. *Food Analytical Methods*, *8*, 1042–1052. Available from https://doi.org/10.1007/s12161-014-9985-5.

Shyla, K. K., Natarajan, N., & Nakkeeran, S. (2014). Antifungal activity of zinc oxide, silver and titanium dioxide nanoparticles against *Macrophomina phaseolina*. *Journal of Mycology and Plant Pathology*, *44*(3), 268–273.

Siedliska, A., Baranowski, P., Zubik, M., & Mazurek, W. (2017). Detection of pits in fresh and frozen cherries using a hyperspectral system in transmittance mode. *Journal of Food Engineering*, *215*, 61–71. Available from https://doi.org/10.1016/j.jfoodeng.2017.07.028.

Singh, C., Jayas, D., Paliwal, J., & White, N. (2009). Detection of insect-damaged wheat kernels using near-infrared hyperspectral imaging. *Journal of Stored Products Research*, *45*, 151–158.

Slaughter, D. C. (1995). Non-destructive determination of internal quality in peaches and nectarines. *Transactions of the ASAE*, *38*, 617–623. Available from https://doi.org/10.13031/2013.27874.

Sozzani, R., & Benfey, P. N. (2011). High-throughput phenotyping of multicellular organisms: finding the link between genotype and phenotype. *Genome Biology*, *12*(3), 1–7.

Subedi, P. P., & Walsh, K. B. (2011). Assessment of sugar and starch in intact banana and mango fruit by SWNIR spectroscopy. *Postharvest Biology and Technology*, *62*, 238–245. Available from https://doi.org/10.1016/j.postharvbio.2011.06.014.

Sun, X., Liu, Y., Li, Y., Wu, M., & Zhu, D. (2016). Simultaneous measurement of brown core and soluble solids content in pear by online visible and near infrared spectroscopy. *Postharvest Biology and Technology*, *116*, 80–87. Available from https://doi.org/10.1016/j.postharvbio.2016.01.009.

Suphamitmongkol, W., Nie, G., Liu, R., Kasemsumran, S., & Shi, Y. (2013). An alternative approach for the classification of orange varieties based on near infrared spectroscopy. *Computers and Electronics in Agriculture*, *91*, 87–93. Available from https://doi.org/10.1016/j.compag.2012.11.014.

Thoday-Kennedy, E., Good, N., & Kant, S. (2022). Basics of sensor-based phenotyping in wheat. In A. Bilichak, & J. D. Laurie (Eds.), *Accelerated breeding of cereal crops. Springer Protocols Handbooks* (pp. 305–331). New York, NY: Humana. Available from https://doi.org/10.1007/978-1-0716-1526-3_16.

Tian, X., Li, J., Wang, Q., Fan, S., & Huang, W. (2018). A bi-layer model for nondestructive prediction of soluble solids content in apple based on reflectance spectra and peel pigments. *Food Chemistry*, *239*, 1055–1063. Available from https://doi.org/10.1016/j.foodchem.2017.07.045.

Tilman, D., Cassman, K. G., Matson, P. A., Naylor, R., & Polasky, S. (2002). Agricultural sustainability and intensive production practices. *Nature*, *418*(6898), 671–677.

Velusamy, V., Arshak, K., Korostynska, O., Oliwa, K., & Adley, C. (2010). An overview of foodborne pathogen detection: In the perspective of biosensors. *Biotechnology Advances*, *28*(2), 232–254.

Vesali, F., Gharibkhani, M., & Komarizadeh, M. H. (2011). An approach to estimate moisture content of apple with image processing method. *Australian Journal of Crop Science*, *5*, 111–115.

Virlet, N., Sabermanesh, K., Sadeghi-Tehran, P., & Hawkesford, M. J. (2016). Field Scanalyzer: An automated robotic field phenotyping platform for detailed crop monitoring. *Functional Plant Biology*, *44*(1), 143–153.

Vyas, S. S., Jadhav, S. V., Majee, S. B., Shastri, J. S., & Patravale, V. B. (2015). Development of immunochromatographic strip test using fluorescent, micellar silica nanosensors for rapid detection of *B. abortus* antibodies in milk samples. *Biosensors and Bioelectronics*, *70*, 254–260.

Wang, A., Hu, D., & Xie, L. (2014). Comparison of detection modes in terms of the necessity of visible region (VIS) and influence of the peel on soluble solids content (SSC) determination of navel orange using VIS–SWNIR spectroscopy. *Journal of Food Engineering*, *126*, 126–132. Available from https://doi.org/10.1016/j.jfoodeng.2013.11.011.

Wang, J., Nakano, K., & Ohashi, S. (2011). Non-destructive evaluation of jujube quality by visible and near-infrared spectroscopy. *LWT - Food Science and Technology*, *44*, 1119–1125. Available from https://doi.org/10.1016/j.lwt.2010.11.012.

Wetterich, C. B., De Oliveira Neves, R. F., Belasque, J., Ehsani, R., & Marcassa, L. G. (2017). Detection of Huanglongbing in Florida using fluorescence imaging spectroscopy and machine-learning methods. *Applied Optics*, *56*(1), 15–23.

Wu, L., He, J., Liu, G., Wang, S., & He, X. (2016). Detection of common defects on jujube using Vis-NIR and NIR hyperspectral imaging. *Postharvest Biology and Technology*, *112*, 134–142.

Xie, C., & Yang, C. (2020). A review on plant high-throughput phenotyping traits using UAV-based sensors. *Computers and Electronics in Agriculture*, *178*, 105731.

Yesmin, N., Jenny, F., Abdullah, H. M., Hossain, M. M., Kader, M. A., Solomon, P. S., & Bhuiyan, M. A. (2020). A review on South Asian wheat blast: The present status and future perspective. *Plant Pathology*, *69*(9), 1618–1629.

York, L. M. (2019). Functional phenomics: An emerging field integrating high-throughput phenotyping, physiology, and bioinformatics. *Journal of Experimental Botany*, *70*(2), 379–386.

Zhang, S., Zhang, H., Zhao, Y., Guo, W., & Zhao, H. (2013). A simple identification model for subtle bruises on the fresh jujube based on NIR spectroscopy. *Mathematical and Computer Modelling*, *58*, 545–550. Available from https://doi.org/10.1016/j.mcm.2011.10.067.

Zhao, J. C., Zhang, J. F., Feng, Y., & Guo, J. X. (2010). *The study and application of the IOT technology in agriculture, 2010 3rd international conference on computer science and information technology* (Vol. 2, pp. 462–465). IEEE.

Zhao, X., Hilliard, L. R., Mechery, S. J., Wang, Y., Bagwe, R. P., Jin, S., & Tan, W. (2004). A rapid bioassay for single bacterial cell quantitation using bioconjugated nanoparticles. *Proceedings of the National Academy of Sciences*, *101*(42), 15027–15032.

Zhu, Q., Huang, M., Zhao, X., & Wang, S. (2012). Wavelength selection of hyperspectral scattering image using new semi-supervised affinity propagation for prediction of firmness and soluble solid content in apples. *Food Analytical Methods*, *6*, 334–342. Available from https://doi.org/10.1007/s12161-012-9442-2.

CHAPTER

Scope, importance, and limitations of precision-agriculture-based practices for crop management: Indian perspective

2

Sushil Kumar, Ashok Yadav, Sukumar Taria, Badre Alam, Priyanka Singh, R.P. Dwivedi and Ayyandar Arunachalam

ICAR-Central Agroforestry Research Institute (CAFRI), Jhansi, Uttar Pradesh, India

2.1 Introduction

Since ages, agriculture is the backbone of Indian economy. At present, its contribution to a country's GDP is about 18.3% besides providing employment to the 70% of the population directly and indirectly. It is believed that traditional farming will not be capable of feeding the burgeoning population. Hence, new generation technologies, namely precision agriculture (PA), need to be adopted (Mondal & Basu, 2009) to meet the forthcoming demand and challenges while enhancing crop productivity, resources use efficiency (RUE), and sustainability of the agricultural systems. With passing time, PA has become an attractive technique of farm management among the stakeholders for natural farm resources management and realizing sustainable growth (Far & Rezaei, 2018; Liaghat & Balsundaram, 2010). PA has created scope of transforming the traditional agriculture through proper resource utilization and management practices to sustainable agriculture with rapid socioeconomic changes in some developing countries, including India (Gorai et al., 2015; Mondal & Basu, 2009). Many a times it is seen that crops are not uniform within a field due to variability in the soils. The sincere efforts of the researchers over the last decade have developed technical methods, namely spatially variable crop production, global positioning system (GPS)-based agriculture, site-specific management, and PA to use modern electronics to deal with field variability. The concept of PA uses the recent developments in sensors, greenhouse, and protected agriculture structures. However, the basic goal of PA is not only to optimize yield with minimum input but also to reduce environmental pollution. In PA, the spatial variability occurring within a field is noted, mapped, and then management actions are taken by adopting site-specific management systems applying remote sensing (RS), GPS, and geographical information system (GIS). The special tools, techniques, and resources are required for recognizing the inherent spatial variability associated with soil

Remote Sensing in Precision Agriculture. DOI: https://doi.org/10.1016/B978-0-323-91068-2.00005-9

characteristics, crop growth in PA to prescribe suitable management strategy on a site-specific basis. PA is a technology-based crop management system that attempts to match the kind and amount of inputs with the actual crop needs for small areas within a farm field to increase the agriculture productivity (Hakkim et al., 2016). This information and technology-based farm management system not only identifies and analyzes but also manages variability in fields, which ultimately helps in undertaking crop production practices appropriately at the right place and time and in the right way for better profitability, sustainability, and conservation of resources. Despite considerable research efforts and myriad benefits of technology-driven PA technologies, its adoption is still limited to only a small chunk of farmers in developing countries (Mwangi & Kariuki, 2015) to reorganize the total system of agriculture toward low-input, high-efficiency, and sustainable agriculture. Nowadays, rapid socioeconomic changes in some developing countries, including India, are creating new scopes for application of PA (Mondal & Basu, 2009). PA is an approach wherein precise/need-based amount of inputs are applied with a sole objective to achieve higher average yields and RUE besides maintaining the sustainability of the agriculture systems (Shibusawa, 1998) as compared with traditional farming. PA is a farming management concept that applies advanced information technology and the corresponding principles to increase production and economic returns, often also intending to reduce the impact on the environment (Groeneveld et al., 2021; Rains & Thomas, 2009). PA in the Indian context can be defined more precisely as application of agricultural inputs based on soil, weather, and crop requirement to maximize sustainable productivity, quality, and profitability. The wider adoption of PA practices has not been registered in India yet due to small size of agricultural fields and involvement of higher initial cost. At present, about 58% of operational land holdings have less than 1 hectare size in the country. Farmers in few Indian states, namely Punjab, Rajasthan, Haryana, and Gujarat, where agricultural lands have suitable operational holding size, are happily getting engaged in PA. The adoption of PA in commercial and horticultural crops is viewed to be highly advantageous and economically viable in countries such as India wherein more than population depends on it for their livelihood. Application of PA technologies is currently at the nascent stage in India (Mondal & Basu, 2009). PA is purely information and communication technologies-based farm management innovation to increase profitability, crop production, and RUE while overcoming the environmental problems arising due to indiscriminate use of inputs in agriculture. In the present scenarios, when input costs are increasing and commodity prices are decreasing, the farmers are looking for new ways to increase efficiency and cut costs. Hence, the PA seems to be a viable approach to not only increase RUE but also to improve profitability and productivity.

2.2 Components of precision agriculture and their use

As PA is completely a technology-driven farm management approach, it includes a vast array of tools of hardware, software. and equipment, namely GPS receivers, differential global positioning system, GIS, RS, variable-rate applicator, and combine harvesters with yield monitors (Fig. 2.1). Based on certain factors, namely soil pH, nutritional status, pest infestation, and yield rates, which affect crop production, the fields are divided into management zones in PA, and management decisions are adopted as per the requirements of each management using PA tools to control the zone inputs. The precise location is identified with the help of GPS, while application on the exact location can be made with the help of available advanced equipments. Thus, the mounting of a GPS receiver on the tractor applying the chemical is needed to know its location in the field. An in-vehicle computer must contain the inputs–need map, which compares with the field position data recorded from the GPS receiver.

2.2.1 Mapping

The generation of maps for crop and soil properties is the most important and first step in PA. These maps measure the spatial variability and provide the basis for controlling spatial variability. Data collection occurs both before and during crop production and is enhanced by collecting precise location coordinates using the GPS. The data collection technologies are grid soil sampling, yield monitoring,

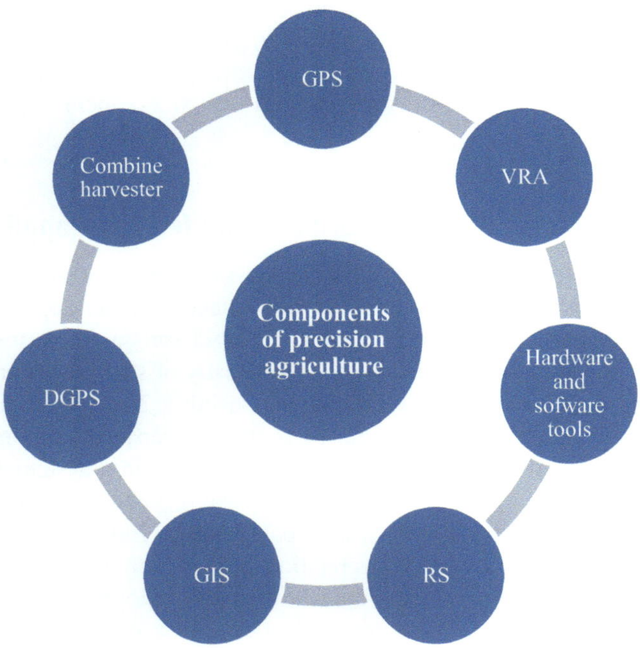

FIGURE 2.1

Major components of precision agriculture.

RS, and crop scouting. During crop production, the data are collected through sensing instruments such as soil probes, electrical conductivity (EC), and soil nutrient status. The mapping can be done using RS, GIS, and manually as well during field operations.

2.2.2 Global positioning system receivers

GPS satellites broadcast signals that allow GPS receivers to compute their location. This information is provided in real time, meaning that continuous position information is provided while in motion. Having precise location information at any time allows soil and crop measurements to be mapped. GPS receivers, either carried to the field or mounted on implements, allow users to return to specific locations to sample or treat those areas.

2.2.3 Yield monitoring and mapping

In highly mechanized systems, grain yield monitors continuously measure and record the flow of grain in the clean-grain elevator of a combine. When linked with a GPS receiver, yield monitors can provide data necessary for yield maps. Yield measurements are essential for making sound management decisions. However, soil, landscape, and other environmental factors should also be weighed when interpreting a yield map. Used properly, yield information provides important feedback in determining the effects of managed inputs such as fertilizer amendments, seed, pesticides, and cultural practices including tillage and irrigation.

2.2.4 Grid soil sampling and variable-rate fertilizer application

Soil cores taken from random locations in the sampling area are combined and sent to a laboratory to be tested. Crop advisors make fertilizer application recommendations from the soil test information. The grid soil sampling uses the same principles of soil sampling but increases the intensity of sampling. For example, a 20-acre sampling area would have 10 samples using a 2-acre grid sampling system (samples are spaced 300 feet from each other) compared with one sample in the traditional recommendations. Soil samples collected in a systematic grid also have location information that allows the data to be mapped. The goal of grid soil sampling is to generate a map of nutrient requirement. Grid soil samples are analyzed in the laboratory, and an interpretation of crop nutrient needs is made for each soil sample. Then the fertilizer application map is plotted using the entire set of soil samples. The application map is loaded into a computer mounted on a variable-rate fertilizer spreader. The computer uses the application map and a GPS receiver to direct a product-delivery controller that changes the amount and/or kind of fertilizer product, according to the application map.

2.2.5 Remote sensing

RS is collection of data from a distance. Data sensors can simply be handheld devices, mounted on aircraft, or satellite-based. Remotely sensed data provide a tool for evaluating crop health. Plant stress related to moisture, nutrients, compaction, crop diseases, and other plant health concerns are often easily detected in overhead images. Electronic cameras can also record near-infrared images that are highly correlated with healthy plant tissue. RS can reveal in-season variability that affects crop yield and can be timely enough to make management decisions that improve profitability for the current crop. Remotely sensed images can help determine the location and extent of crop stress. Analysis of such images can help determine the cause of certain components of crop stress. The images can then be used to develop and implement a spot treatment plan that optimizes the use of agricultural chemicals. Satellite RS has provided a tool for acreage estimation 1 month in advance, with more than 95% accuracy and in mono-crop area yield estimation with more than 90% accuracy 10 days in advance. The most popular procedure is to take images from satellites such as LANDSAT or SPOT. Finally, images are used for generating maps and calibration of the measurement, assuming that measurements are taken in field to ground-truth accuracy. These images allow mapping of crop, pest, and soil properties for monitoring seasonally variable crop production, stress, weed infestation, and extent within a field.

2.2.6 Geographic information systems

GISs are computer hardware- and software-based tool that is specifically used to feature attributes and location data to produce maps. An important function of an agricultural GIS is to store multidimensional information, namely yields, soil survey maps, remotely sensed data, crop scouting reports, and soil nutrient levels. Geographically referenced data are displayed in the GIS, which provide a visual perspective for interpretation. In addition to data storage and display, the GIS can also be used to evaluate present and alternative management by combining and manipulating data layers to produce an analysis of management scenarios. Currently GIS technology is being adapted for use on small Asian farms, in Japan, the Republic of Korea, and the Taiwan Province of China, where government programs are developing the use of web-based GIS systems (Mondal & Basu, 2009). GIS application encourages farmers to use the Internet and to obtain free information on the soil properties of their farms, including soil fertility and nutrient levels. GIS is being used in Indonesia to reevaluate appropriate agricultural land use (Mondal & Basu, 2009). Besides identifying the best crop for a particular region, GIS technique is also used to identify areas suitable for arable land (Anonymous, 2004).

2.2.7 Quantifying farm variability

Every farm presents a unique management puzzle. Not all the tools described earlier will help determine the causes of variability in a field, and it would be cost-prohibitive to implement all of them immediately. An incremental approach is a wiser strategy, using one or two of the tools at a time and carefully evaluating the results. Following are the components where variability can be assessed for adoption of PA.

1. *Soil*: physical properties (texture, structure, moisture holding capacity, bulk, and particle density); chemical properties (pH, EC, available plant nutrients)
2. *Crop*: planting geometry (row-to-row and plant-to-plant spacing, plant stand); nutrient composition of standing plant, plant stress due to biotic and abiotic factors, weed, insect and disease, potential economic and biological yield;
3. *Climate*: air temperature, temperature around plant canopy, relative humidity, rainfall, solar radiation, day length, wind velocity.

2.2.8 Soil variation

Soil variation is a spatial variable. Water-holding capacity or organic matter variation, along with topography, provides even a more interesting view of a field in which a producer places inputs or disturbs the soil. Other variables could be layered within this field to create a series of interacting elements. The central question is how to quantify soil variation. Collection and analysis of the samples only provide one portion of the base layer of information. Individual samples represent points; to be of value, they must be interpolated. There are as many interpolation schemes as there are sampling schemes. These schemes include central tendency, proximal, inverse distance methods, splines, and geostatistical. None of these methods or schemes are described herein, but all have been used to determine the variability across a field. Topographic variation within fields can be collected from topographic maps, but the resolution on these maps is often insufficient to provide the necessary detail about variations within fields. Topographic maps can be generated from differential or kinematic GPSs. The role of topographic variations on water use, plant growth, soil processes, yield, surface runoff, and groundwater hydrology has not been quantified for agricultural fields.

2.2.9 Variability of soil water content

It is a well-established fact that soil water content in a field varies over time and location, and this temporal and spatial variability in soil water content patterns may have profound implications for PA in general and water management in particular. Knowledge of the underlying stable soil water distribution could provide a useful basis for precision water management and lead to savings in energy, water, equipment cost, labor, and improved production efficiency.

2.2.10 Time and space scales

PA requires an understanding of time and space scales. Time scales are critical because operations occur when they will benefit the crop most. Space scales become a fundamental principle of field management because inputs and cultural practices are varied with soil type, pest population, or crop maturity. The challenge is to determine how to use time and space scales to advantage in developing an improved understanding of agricultural management. To fully achieve the goals of PA, management must be applied in a space and time context. The challenge of monitoring in space and time is important to document the changes that are naturally occurring within a field. To fully realize the potential impact of principles of precision agriculture on environmental quality, however, will require the design and implementation of experiments in space and time.

2.3 Why precision farming

Nowadays, agricultural sector faces the challenge of fulfilling several goals and societal values at the same time while also expecting to be sustainable and economically viable on a long-term basis (Groeneveld et al., 2021). Further, the rapidly growing human population has increased food demands for human survival on the Earth (Shafi et al., 2019). Thus, introduction and adoption of modern technology such as PA in Indian agriculture are inevitable. The popularity of PA in developed countries results in maximization of agricultural productivity with application of different technologies such as satellite using technologies and GIS (Meena et al., 2021). PA visualizes economical and environmental benefits through reduced use of water, fertilizers, herbicides, and pesticides besides the farm equipment. Instead of managing/applying inputs to an entire field based upon some hypothetical average condition, the PA approach recognizes site-specific differences within fields and adjusts need-based management actions. The adoption of PA techniques enhances economically and environmentally benefits by way of optimizing utilization of water, fertilizers, herbicides, and pesticides other than farm equipment. PA offers the potential not only to automate and simplify the collection and analysis of information but also allows management decisions to be made and quickly implemented on small areas within larger fields. Hence, there is a wider scope of implementing precision agriculture techniques in India: (1) to increase production efficiency, (2) to improve product quality, (3) to use chemicals more efficiently, (4) to conserve energy, (5) to increase input use efficiency, (6) to protect soil and groundwater, and (7) to improve soil structure.

2.4 Precision-agriculture-based practices for crop management

PA techniques adopted in the developed countries may not be adopted as such in the developing countries including India due to some constraints, namely economical, educational, and fragmented land holdings. Thus, as per the suitability and feasibility, variable rate technology (VRT), precision land leveling (PLL), precision planting (PL), and precision nutrient management by using leaf color chart (LCC), soil plant analysis development (SPAD), green seeker, site-specific nutrient management (SSNM), and soil test crop response (STCR) are the main PA techniques used in India for crop management. Though these practices are not common across the length and width of the country and are mostly concentrated in the fields of farmers in the states of Haryana and Punjab. These techniques have a lot of potential to enhance crop yield and input use efficiency under field conditions while reducing the cost of production and deleterious impacts on the environment (Kumar et al., 2018). The implementation of PA techniques increases the productivity of agricultural crops. The use of PA technique in sugar cane crops in Brazil showed that the output of the sugar cane yield monitored and the harvested load weight presented a correlation of 0.66 with the system performance being stable and reliable (Magalhaes & Cerri, 2007). In another study conducted in Uruguay, the site-specific application of zinc (Zn) proved to be beneficial and enhanced profit level up to US$50 per hectare (Gimenez & Lamothe, 2000). The PLL by using of laser leveler increases rice yield by 15%–20% (Rajput & Patel, 2006) and pigeon pea yield by 32% (Chaudhuri et al., 2006). Dai and Dong (2014) reported that the precision seeding technology reduces the dosage of seed, decreases the operating cost of agricultural machinery, and increases the crop yield and farmer's income. The adoption of integrated PA techniques, namely precise seeding and laser land leveling, enhances wheat yield and reduces residual soil N (Li et al., 2018). LCC, an effective and easy-to-use low-cost technology of PA, is commonly getting popularity in India for the need-based nitrogen (N) management in field crops especially rice. The use of LCC for N management enhances productivity, nitrogen use efficiency (NUE), and farmer's profit (Moharana et al., 2017; Premalatha & Angadi, 2017; Singh et al., 2007) besides preventing fertilizer contamination of surface- and groundwater leading to nitrous oxide emission (Bhatia et al., 2012; Ladha et al., 2005). Thus, use of LCC may help to rationalize the use of urea and other nitrogenous fertilizers as per the need of the crop (Powlson et al., 2011). In a study conducted on rice, LCC-based real-time N management recorded almost similar average yield to farmer's practice, but meantime reduces the application of N about 40 kg/ha as compared with the farmer's practice (Mondal & Basu, 2009). Singh et al. (2006) conducted LCC-based N management experiments on rice varieties commonly grown by the farmers in north-western India and reported less consumption of N (16–43 kg/ha) in the LCC-based management as compared with the farmer's practice. The adoption of LCC by the farmers in rice crop for N management in West Bengal, India, saved N in the range of 19.4%–21.0% (Islam et al., 2007). They also opined that the combined effects of reduced N and insecticide use may help enormously in protecting the environment. The use of customized LCC for N management in rice in Orissa, India, not only delayed the N fertilizer application day (4–5 days) but also saved N fertilizer and decreased disease incidence as compared with farmer's practice (Sarangi et al., 2020).

They also reported improvement in net benefit by about Rs. 5569 ha^{-1} in customized LCC compared with farmer's practice. The SSNM is another important commonly used PA technique for need-based management of nutrients. The SSNM encourages judicious utilization of chemical fertilizers for sustainable and economic crop productivity in intensively cultivated areas of India (Gorai et al., 2015). Laser land leveling is a process of smoothening the land surface from its average elevation using laser-equipped drag buckets to achieve precision in land leveling (Jat et al., 2006). Laser land leveling not only increases the input use efficiency but also reduces the cost of production. Precise leveling of fields with laser leveler significantly increases the irrigation efficiencies, namely average application efficiency (65%), storage efficiency (70%), and water distribution efficiency (80%) (Mondal & Basu, 2009). The adoption of Precision N management practices in rice–wheat cropping system results higher N, P, and K harvest index and agronomic efficiency, apparent recovery, and physiological efficiency (Kumar et al., 2018). The application of PA techniques, namely PLL, PL, real-time N application using LCC, SSNM in rice–wheat cropping system, enhances crop production, input use efficiency, and minimizes the cost of production and environmental impacts (Kumar et al., 2017). Jat et al. (2021) reported that precision nutrient management approaches not only increase crop yields, NUE, and profitability but also reduce the emission of greenhouse gases (GHGs) from maize–wheat systems in eastern Indo Gangetic Plains.

2.5 Scope of precision agriculture in India

The scope of PA is different for the different countries due to many factors, namely socioeconomic condition, technical manpower, and size of landholding. The economic growth is driving an unprecedented shift in population to urban centers especially in low- and middle-income nations such India (Mondal & Basu, 2009). The implications of such dramatic shifts for economic development, poverty reduction, and energy consumption are immense (Mondal & Basu, 2009). Among the developing countries, Argentina, Brazil, China, India, Malaysia, and others have begun to adopt some components of PA, especially on research farms, but their adoption is still very limited (Srinivasan, 2006). The scope of application of PA in Indian conditions ranges from plantation and cash crops to small-scale farms. Further, its adoption and success purely depend on design of adoption strategies (Mondal & Basu, 2009). Some of the common strategies suggested for the application of PA in developing countries including India are listed in Table 2.1. PA is basically categorized into two parts such as soft and hard (Mondal & Basu, 2009; Nowak, 1997). The balanced use of both the parts of PA decides its success in India. Among the constraints, the land fragmentation is considered the main obstacle for large-scale adoption of mechanized agriculture system including PA in India. Since ages, these fragmented lands have been

Table 2.1 Proposed common precision agriculture adoption strategies for developing countries.

Strategic precision agriculture adoption component	Technologies	Target sectors
Single precision agriculture technology	Single low-level precision agriculture technologies, LCC, small machine-based VRT, etc.	Small-scale farms
Precision agriculture technology package	SPAD, LCC, DSS, GIS, VRT, GPS, etc.	Consolidated plots, plantation crops, cash Crops, cooperative farming, etc.
Integrated precision agriculture techniques	Online sensor, image processing, remote sensing, yield monitoring system, VRT, GPS, etc.	Organized farming sector

Source: *From Mondal, P., & Basu, M. (2009). Adoption of precision agriculture technologies in India and in some developing countries: Scope, present status and strategies.* Progress in Natural Science, 19*(6), 659–666.*

cultivated by small farmers following soft PA technologies (visual observation of crop and soil and management based on experience and intuition, rather than on statistical and scientific analysis). At present, India produces about 316 million tonnes (Mt) of food grain, which makes the country self-sufficient in fulfilling its food demand. However, this quantity cannot meet the need of the globalized agricultural market. Hence, excellent quality and high productivity are needed to compete with other countries in the globalized era, and that's why the huge scope of PA lies in India (Mondal & Basu, 2009). Worldwide, India ranks first in pulses and cotton, second in wheat and rice production. However, the productivity these crops is poor as compared with the rest of the world. The poor scale of mechanization with very small average land-holding sizes including other reasons seems to be the main factor behind poor productivity of agricultural crops in India. The overall fertilizer consumption rate of India is lower in comparison with other countries (Anonymous, 2004). It has already been emphasized that systematic soil testing followed by proper application of NPK fertilizers can not only increase the crop productivity level by 2–3 times in most of the Indian states but also reduce the cost of production. Some of the Indian states are already using large-scale mechanization and high doses of fertilizers and pesticides. The over-exploitation of land and excessive use of agricultural inputs in these areas have declined the natural resources base and environmental quality, which has brought the attractions of the policymakers and planners. Thus, keeping all these points in view, it is suggested that PA techniques are having greater scope for wider adoption in India for higher-quality production, sensible resources utilization, and environment protection.

2.6 Challenges and limitations of precision agriculture

Ostensibly educational and economic challenges are the most important challenges in the application of PA in India. The educational challenges involve lack of local experts, knowledgeable research, and extension personnel. However, initial costs and funds involved in the application of PA are the main hurdles among the economic challenges compared with the other issues. Besides the aforementioned, fragmented and small-size land holdings, heterogeneity of cropping systems, specific software, poor economic condition of general Indian farmers, and market imperfections are other factors limiting the adoption of PA in the country. In addition to the aforementioned, some other challenges in the adoption of precision agriculture are as follows.

1. **Technology factors**
   Complexity of technology usage
   Limitation of technology use
   Lack of installation/training assistance
   Availability and accessibility in sale
2. **Economic factors**
   Higher initial cost
   Higher operational cost
   Lack of institutional and government assistance
3. **Behavioral factors**

   Lack of technology awareness/knowledge
   Rigidity to adopt new technology/believe in old traditional factors
   Lack of awareness of government/institutional support

2.7 The policy approach to promote precision farming at farm level

The following policy approaches are needed to promote PA at farm level generally in developing countries and particularly in India.

1. The identification of the niche areas for the wider promotion of crop-specific PA practices.
2. The constitution of multidisciplinary teams involving agricultural scientists of various fields, engineers, manufacturers, and economists to study the overall scope and feasibility of PA.
3. To provide complete technical backstopping to the farmers for devising and developing precision agriculture models, which can further be replicated on a large scale.
4. To conduct pilot study on farmers' fields to show them the results of PA implementation and adoption.

5. To create awareness among farmers and other stakeholders through different mechanisms about the consequences of applying imbalanced doses of farm inputs such as irrigation, fertilizers, insecticides, pesticides, and others.
6. To incentivize tools and implements basically used in PA.

2.8 Conclusions

PA is an information and technology-based farm management system that not only identifies and analyzes but also manages variability in fields, which ultimately helps in undertaking crop production practices appropriately at the right place and time and in the right way for better profitability, sustainability, and conservation of resources. It uses crop inputs, namely fertilizers, pesticides, tillage, and irrigation water, more effectively and efficiently. The more effective use of inputs in PA is meant as higher-quality crop yield without any detrimental effect on the environment. The adoption of PA is different in different countries due to some location-specific constraints. As per the suitability and feasibility, VRT, PLL, PL, and precision nutrient management are commonly used PA techniques in India. Hence, a proper policy planning is needed to create great scope for wider adoption of PA in India for higher-quality production, sensible resources utilization, and environment protection.

References

Anonymous. (2004). *Agricultural statistics at a glance* (pp. 1–221). Directorate of Economics & Statistics, Govt. of India.

Bhatia, A., Pathak, H., Jain, N., Singh, P. K., & Tomer, R. (2012). Greenhouse gas mitigation in rice–wheat system with leaf color chart-based urea application. *Environmental Monitoring and Assessment*, *184*(5), 3095–3107.

Chaudhuri, D., Mathankar, S.K., Singh, V.V., et al. (2006). Effect of precision land levelling on production of pigeon pea crop: A case study. In: *Proceedings of 40th annual convention and symposium of ISAE.*

Dai, J., & Dong, H. (2014). Intensive cotton farming technologies in China: Achievements, challenges and countermeasures. *Field Crops Research*, *155*, 99–110. Available from https://doi.org/10.1016/j.fcr.2013.09.017.

Far, S. T., & Rezaei, K. (2018). Impacts of the precision agricultural technologies in Iran: An analysis experts' perception & their determinants. *Information Processing in Agriculture*, *5*(1), 173–184.

Gimenez, A.E., & Lamothe, A.G. (2000). Site specific management: Zinc deficit in an irrigated corn field. In: *Proceedings of the 5th international conference on precision agriculture (CD)*, July 16–19, Bloomington, MN, USA.

Gorai, T., Ahmed, N., Patra, A. K., Sahoo, R. N., Sarangi, A., Meena, M. C., & Sharma, R. K. (2015). Site specific nutrient management of an intensively cultivated farm using geostatistical approach. *Proceedings of the National Academy of Sciences, India Section B*. Available from https://doi.org/10.1007/s40011-015-0590-1.

Groeneveld, D., Tekinerdogan, B., Garousi, V., et al. (2021). A domain-specific language framework for farm management information systems in precision agriculture. *Precision Agriculture*, *22*, 1067–1106.

Hakkim, A. V. M., Joseph, A. E., Gokul, A. A. J., & Mufeedha, M. (2016). Precision farming: The future of Indian agriculture. *Journal of Applied Biology and Biotechnology*, *4*, 68–72.

Islam, Z., Bagchi, B., & Hossain, M. (2007). Adoption of leaf color chart for nitrogen use efficiency in rice: Impact assessment of a farmer participatory experiment in West Bengal, India. *Field Crops Research*, *103*(1), 70–75.

Jat, M. L., Chandna, P., Gupta, R., Sharma, S., & Gill, M. (2006). Laser land leveling: A precursor technology for resource conservation. *Rice-Wheat Consortium Technical Bulletin Series*, *7*, 48.

Jat, R. K., Bijarniya, D., Kakraliya, S. K., et al. (2021). Precision nutrient rates and placement in conservation maize-wheat system: Effects on crop productivity, profitability, nutrient-use efficiency, and environmental footprint. *Agronomy*, *11*, 2320.

Kumar, S., Karaliya, S. K., & Chaudhary, S. (2017). Precision farming technologies towards enhancing productivity and sustainability of rice-wheat cropping system. *International Journal of Current Microbiology and Applied Sciences*, *6*(3), 142–151.

Kumar, S., Panwar, A. S., Naresh, R. K., et al. (2018). Improving rice-wheat cropping system through precision nitrogen management: A review. *Journal of Pharmacognosy and Phytochemistry*, *7*(2), 1119–1128.

Ladha, J. K., Pathak, H., Krupnik, T. J., Six, J., & van Kessel, C. (2005). Efficiency of fertilizer nitrogen in cereal production: Retrospects and prospects. *Advances in Agronomy*, *87*, 85–176.

Li, C., Wang, Y., Lu, C., & Huai, H. (2018). Effects of precision seeding and laser land leveling on winter wheat yield and residual soil nitrogen. *International Journal of Agriculture & Biology*. Available from https://doi.org/10.17957/IJAB/15.0820.

Liaghat, S., & Balsundaram, S. K. (2010). A review: Role of remote sensing in precision farming. *American Journal of Agriculture and Biological Science*, *5*(1), 5055.

Magalhaes, P. S. G., & Cerri, D. G. P. (2007). Yield monitoring of sugar cane. *Biosystems Engineering*, *96*(10), 1–6.

Meena, H., Kumar, D., & Prasad, K. (2021). Precision agriculture: Present status & scope: A review. *Just Agriculture*, *1*(5), 213–218.

Moharana, S., Gulati, J. M. L., & Jena, S. N. (2017). Effect of LCC-based nitrogen application on growth and yield of rice (*Oryza sativa* L.) varieties during dry season. *Indian Journal of Agricultural Research*, *51*(1), 49–53.

Mondal, P., & Basu, M. (2009). Adoption of precision agriculture technologies in India and in some developing countries: Scope, present status and strategies. *Progress in Natural Science*, *19*(6), 659–666.

Mwangi, M., & Kariuki, S. (2015). Factors determining adoption of new agricultural technology by smallholder farmers in developing countries. *Journal of Economics and Sustainable Development*, *6*(5), 208–216.

Nowak, P. (1997). *A sociological analysis of site specific management. In F.J. Pierce, & E.J. Sadler (Eds). The state of site-specific management for agriculture* (pp. 397–422). American society of Agronomy.

Powlson, D. S., Gregory, P. J., Whalley, W. R., Quinton, J. N., Hopkins, D. W., Whitmore, A. P., Hirsch, P. R., & Goulding, K. W. T. (2011). Soil management in relation to sustainable agriculture and ecosystem services. *Food Policy*, *36*, S72–S87.

Premalatha, B. R., & Angadi, V. V. (2017). Influence of integrated nitrogen management and leaf color chart (LCC) based nitrogen management on rice (*Oryza sativa* L.) yield and yield attributes, nitrogen uptake and soil available nitrogen status. *Environment and Ecology*, *35*(3B), 2012–2020.

Rains, G.C., & Thomas, D.L. (2009). *Precision farming: An introduction*. University of Georgia Cooperative Extension, University of Georgia.

Rajput, T.B. S., & Patel, N. (2006). Laser land leveling: A step towards precision agriculture. In: *Proceedings of 19th national convention of agricultural engineers on role of information technology in high-tech agriculture and horticulture*, Bangalore, India, pp. 120–130.

Sarangi, D. R., Chourasia, M., Sahoo, T. R., Prasad, S. M., Sethy, S., Das, S. K., & Mohanta, R. K. (2020). Use of customized leaf colour chart for input saving and disease occurrence in rice var. Swarna (MTU-7029). *e-planet*, *18*(1), 13–17.

Shafi, U., Mumtaz, R., García-Nieto, J., Hassan, S. A., Zaidi, S. A. R., & Iqbal, N. (2019). Precision agriculture techniques and practices: From considerations to applications. *Sensors*, *19*(17), 3796. Available from https://doi.org/10.3390/s19173796.

Shibusawa, S. (1998). Precision farming and terra-mechanics. In: *The 5th ISTVS Asia-Pacific regional conference in Korea*, October 20–22.

Singh, B., Singh, Y., Singh, M., et al. (2006). Plant-need based real time nitrogen management in rice grown by small farmers in Asia. In: *18th world congress of soil science*, July 9–15, Philadelphia, PA, USA.

Singh, Y., Singh, B., Ladha, J. K., Bains, J. S., Gupta, R. K., Singh, J., & Balasubramanian, V. (2007). On farm evaluation of leaf color chart for need-based nitrogen management in irrigated transplanted rice in North Western India. *Nutrient Cycling in Agroecosystems*, *78*, 167–176.

Srinivasan, A. (2006). *Handbook of precision agriculture: Principles and applications*. The Haworth Press.

SECTION

2

Precision agriculture practices for soil management

CHAPTER

IoT based soil monitoring for precision agriculture 3

Shafiyoddin Badruddin Sayyad[1], Mudassar Akhtar Shaikh[2], Sandip Maruti Anpat[3] and Mangesh Madhukar Kolapkar[4]

[1]*Microwave and Imaging Spectroscopy Research Laboratory, Department of Computer and Management, Milliya College, Beed, Maharashtra, India*

[2]*Department of Electronics, New Arts, Commerce and Science College, Ahmednagar (Autonomous), Ahmednagar, Maharashtra, India*

[3]*Department of Computer Science, Marathwada Mitramandal College of Commerce, Pune, Maharashtra, India*

[4]*Vidya Pratishthan's Arts Science and Commerce College, Baramati, Maharashtra, India*

3.1 Introduction

Agriculture is an important factor in global development. Healthy, sustainable, and inclusive food systems are vital to achieving the world's development goals. Agricultural development is a key factor in alleviating poverty, raising social and economic standards, and solving the food problem of billions of people. Considering the global population, the number of people who depend on agriculture is higher than that in other sectors (Awokuse & Xie, 2015). Therefore the development in the agricultural sector is complementary to global development. Global climate change is having an adverse effect on agriculture. It shows declining agricultural yields due to declining rainfall, unseasonal rains, hailstorms, and extreme weather conditions (Hendricks et al., 2019). Consequently, there is a need for controlled use of modern tools in agriculture to monitor the soil parameters, water parameters, fertilizer planning, identification of diseases, and proper use of pesticides.

Precision agriculture (PA) is the term for the use of smart tools for monitoring and managing crops in terms of soil, water, nutrients, plant indices, diseases, and pest control (Berry et al., 2003; Delgado et al., 2019). PA is also called satellite farming, which includes the concept of agricultural management based on observation, measurement, and response of inter- and inter-regional variability in crops. In recent years, the demand for quality food has increased in view of the growing population, and advanced technologies such as PA methods are being used globally to meet this need. Accurate farming (PA) provides a comprehensive data source with accurate and cost-effective advanced techniques for monitoring, analyzing, and controlling crop and soil-related components. It plays an important

Remote Sensing in Precision Agriculture. DOI: https://doi.org/10.1016/B978-0-323-91068-2.00026-6

role in identifying and predicting potential problems related to soil health, crop diseases and pest systems, irrigation management, nutrient use, and production forecasting. PA helps in increasing agricultural production and controlling wastage of water and nutrients (Aubert et al., 2012; Berry et al., 2003; Bongiovanni & Lowenberg-DeBoer, 2004; Delgado et al., 2019; Gebbers & Adamchuk, 2010; Pierpaolia et al., 2013; Srinivasan, 2006; Zhang et al., 2002). PA is incorporated with growing technologies such as remote sensing, Internet of things (IoT), artificial intelligence (AI) to enhance agricultural quality production.

The remote sensing technique can be applied in the monitoring of crops, management of water, nutrients, and pests and can also help to predict the crops' growth. Usually, remote sensing systems generate a huge amount of spectral data (Huang et al., 2018). To obtain useful information from RS-generated big data, some techniques are used such as big data analysis, AI, and machine learning (Kamilaris et al., 2017). The cloud computing system is also useful in PA to store, process, and distribute big data (Khattab et al., 2016; Pavón-Pulido et al., 2017; Zhou et al., 2016).

In the last decade, a variety of works were done in the application of remote sensing in PA. Some studies focused on specific application areas such as estimation of soil parameters (Ge et al., 2011; Petropoulos et al., 2018), irrigation management (Petropoulos et al., 2018), evapotranspiration estimation (Courault et al., 2005; Ge et al., 2011; Maes & Steppe, 2012; Petropoulos et al., 2018), crop diseases, and pest control (Zhang et al., 2019), and some included more than one area of application (Atzberger, 2013; Bégué et al., 2015; Mulla, 2013; Triantafyllou et al., 2019; Weiss et al., 2020). The wireless sensor network (WSN) has played a vital role in the PA. WSN helps to develop a flexible network, which connects base stations widely through the Internet called IoT. The IoT platform helps small farmers to manage their greenhouses (Aiello et al., 2018). IoT platforms with corresponding sensor nodes such as soil moisture sensor, humidity sensor, dielectric sensor, and temperature sensor are used (Ceballos et al., 2015; Cevallos Barriga, 2011; Culman et al., 2017; Erazo et al., 2015; Ferrández-Pastor et al., 2016; Flores et al., 2016; Flores-Medina et al., 2015; García et al., 2016; Grupo Banco Mundial, n.d.; Grupo Banco Mundial, 2018; Hernández Rojas et al., 2015; INEC, 2018; Ipanaqué et al., 2017; Karimi et al., 2018; Kodali et al., 2016).

3.2 Proposed system

The present system is a combination of three elements such as sensor node unit, remote sensing platform, and data analysis unit. The sensor node unit is embedded with sensor nodes, ATmega328 Arduino Nano microcontroller, and ESP8266 Wi-Fi module. The sensors used are temperature sensor, electric conductivity (EC) sensor, moisture sensor, and pH sensor. The Sentinel-1 data are used for remote sensing data collection of the study area. The data collected are further processed using image analysis technique. The developed system is quite useful in the monitoring of soil parameters, namely pH, EC, temperature, humidity, and moisture of the soil. The IoT-based sensor network has been designed to communicate data from sensor node to cloud for further process. The system is miniaturized, portable, and reliable to mount on multi field areas (Fig. 3.1).

3.2.1 Arduino

The Arduino Nano is a breadboard friendly ATmega328 microcontroller. It is an open-source hardware and software platform. It consists of 14 digital input/output pins, 8 analog inputs, 32 KB of Flash, 1 KB of EEPROM, and 2 KB of SRAM. The Arduino Nano has three parts as Arduino hardware, Arduino software, and Arduino expansion element. The software is called the sketch, and sketch code consists of two parts, void setup () and void loop (). First one is placed inn Arduino hardware initialization code within the function's scope, and its role is to

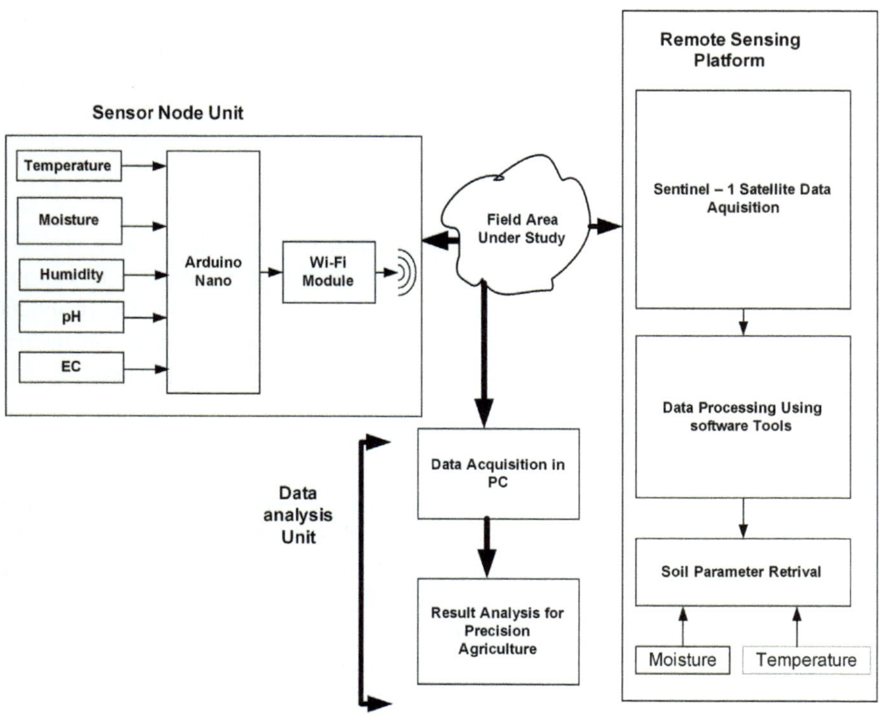

FIGURE 3.1

Block diagram of the IoT-based soil parameters monitoring system for PA.

start writing before the main program for initialization parameters and define the pin. And another is placed in Arduino sketch; this repeats the execution (Huang et al., 2015).

3.2.2 ThingSpeak-IoT platform

ThingSpeak is an open-source IoT application and API to store and retrieve data from things using the HTTP protocol over the Internet or via a Local Area Network. ThingSpeak is an analytic IoT platform service that allows you to aggregate, visualize, and analyze live data streams in the cloud. It offers free data storage and analysis of time-stamped numeric or alphanumeric data. The advantage of using Thingspeak compared with any other platform is its convenience of using MATLAB® analysis and MATLAB visualizations. In the present system to upload data, a ThingSpeak channel is created. As this is an open-source platform, to create the new channel, sign-up to ThingSpeak website is required. Further, one can able to add necessary fields to the created channel. Each channel has up to eight fields. Channels may be public, where anyone can see the data, or private, where only the owner and permitted users can read the data. Each channel has an associated Write API Key that is used to control the writing permission to a channel. In addition, private channels have one or more Read API Keys to control who can read from private channel. An API Key is not required to read from public channels. The built-in version of MATLAB is used to visualize and analyze the data received (ThingSpeak IoT platform, n.d.).

3.2.3 ESP8266 Wi-Fi module

ESP8266 Wi-Fi module is a wireless module, which supports 802.11 b/g/n protocol. It is simple to interface using serial port. This module has integrated TCP/IP protocol stack, efficient power usage, compact design, and reliable performance (ESP8266 Wi-Fi module information, n.d.). This module helps to send data over the Internet using TCP/IP protocol. These data are stored on a ThingSpeak cloud system.

3.2.4 Sensors

In the present study, different soil parameters have been observed such as pH, EC, humidity, temperature, and moisture. The pH is a measure of acidity or alkalinity of a solution, and the pH scale ranges from 0 to 14. The pH indicates the concentration of hydrogen $[H]+$ ions present in certain solutions. It can accurately be quantified by a sensor that measures the potential difference between two electrodes: a reference electrode (silver/silver chloride) and a glass electrode that is sensitive to hydrogen ion (pH sensor, n.d.). The EC shows the level of ability the soil water must carry an electrical current. The EC level of the soil water is a good indicator of the amount of nutrients available for crops to absorb. Generally, all the major and minor nutrients important for plant growth take the form of either cations (positively charged ions) or anions (negatively charged ions). These ions that are dissolved in the soil water carry electrical charge and thus determine the EC level of the soil and how many nutrients are available in the crops to take in. Knowing the soil's EC can allow to make more educated farming decisions (Srishti Rawal, 2017). The DHT11 is a basic, low-cost digital humidity sensor. It gives digital output and hence can be used to output directly to data pin instead of ADC. DHT11 sensor includes resistive-type humidity measurement component (DHT11 sensor, n.d.). This sensor can be also used to measure the temperature. Soil moisture sensors are usually used to detect water content in the soil. The relative relationship between the measured property and soil moisture is calibrated and varies depending on the environmental conditions such as soil type and EC (Vijayakumar & Ramya, 2015).

3.3 Satellite data processing

The microwave SAR dataset of the selected study area is processed using the image analysis techniques. The image analysis techniques are geometric and radiometric correction, speckle noise filtering, decomposition, and classification. Particularly the classification technique includes unsupervised classification such as H A Alpha and Wishart classifier and supervised classification such as Wishart and support vector machine. The area of interest (AOI) is selected from the simulation results of image analysis techniques. The same AOI is used for ground study, which is verified by using GPS meter. The sensor nodes are deployed at five different sites to monitor the parameters of soil, namely pH, EC, temperature, moisture, and humidity. A number of samples have been collected to quantify them for validation by soil testing kit.

The vegetation and soil parameters such as soil moisture and backscattering coefficient (σ_0) are retrieved using image analysis techniques and ground study. Surfaces are characterized in terms of their dielectric and geometrical properties with the help of texture and roughness of the AOI of the study area. The microwave C band Sentinel-1 SAR satellite dataset is used. The software tools PolSARPro, NEST, and SNAP are used for SAR dataset processing.

3.4 Study area

The study area of the proposed project work is Ahmednagar Taluka Region, Maharashtra, India, (latitude 190 10′10″ N and longitude 740 74′06″ E). Ahmednagar is one of the largest areas covered district in Maharashtra state, India. The total area covered is 1605.74 km^2, out of that around 85.73% area (1376.64 km^2) is covered by agriculture, 8.2% area (131.65 km^2) is covered by forest, and remaining 6.07% (97.45 km^2) is covered by industries and households.

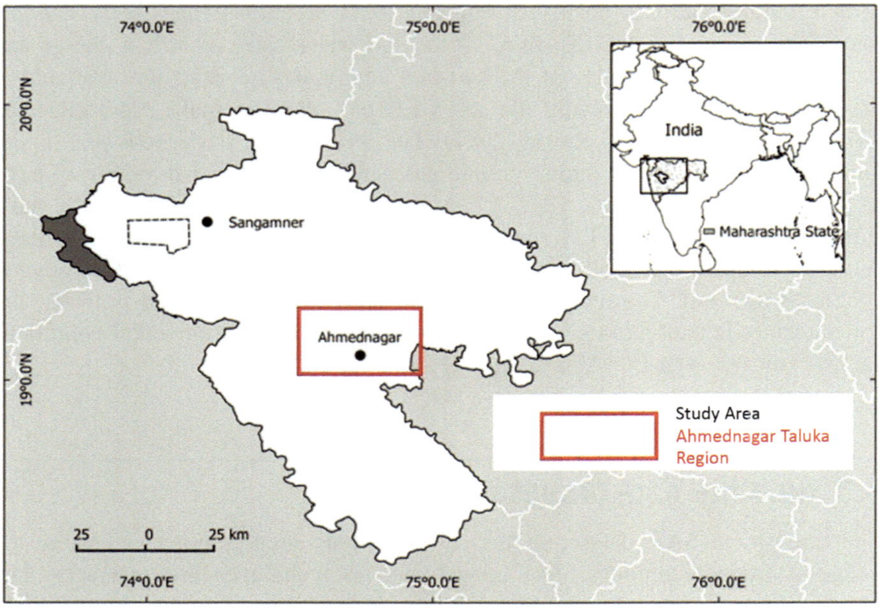

FIGURE 3.2

Outline map of study area (Ahmednagar Taluka Region, Maharashtra, India).

Sugarcane, soybean, and wheat are the major crops grown in selected agricultural areas. For the same reason, it has been decided to study these crops (Fig. 3.2).

In the present study, microwave Sentinel-1 SAR satellite dataset is selected. The sentinel-1 is the microwave C band SAR data freely available on European Space Agency (ESA) sentinel online hub (https://scihub.copernicus.eu/). Table 3.1 shows detailed specification for Sentinel-1 satellite dataset, and Fig. 3.3 shows the original Sentinel-1 dataset for selected study region.

In the present study, five sites are chosen, and accordingly samples are selected, the selected site details are shown in Table 3.2. And its actual site marking is shown in Fig. 3.4.

3.5 Result and discussion

3.5.1 IoT-based results

The IoT-based system is designed for measurement of real-time soil parameters. Five sensor nodes are implemented and installed at five different places as shown in Table 3.2. The samples are taken in the morning, and the parameters, namely pH, EC, humidity and moisture are measured at 31°C temperature. The details of the test are shown in Table 3.3, and its graphical representation is shown in

Table 3.1 Satellite dataset with detailed specification.

Specification	Particular
Satellite	Sentinel-1
Microwave band	C
Format of data	SLC
Date and time	2021-10-25T07:45:15.723Z
Polarization	VV VH
Product class description	SAR Standard L1 Product
Pass direction	DESCENDING
Product type	GRD
Resolution	High

FIGURE 3.3

Sentinel-1 SAR dataset for Ahmednagar Taluka Region Maharashtra India.

Table 3.2 Ahmednagar Taluka region study site.

Sample number	Latitude	Longitude	Site name
1	19.101106	74.733417	MNC Area, Ahmednagar, M.S., India
2	19.156031	74.703111	MIDC, Ahmednagar, M.S., India
3	19.107729	74.762419	Burhanagar Road, Ahmednagar, M.S., India
4	19.081488	74.747705	Sarasnagar, Ahmednagar, M.S., India
5	19.037621	74.652510	Chas, Ahmednagar, M.S., India

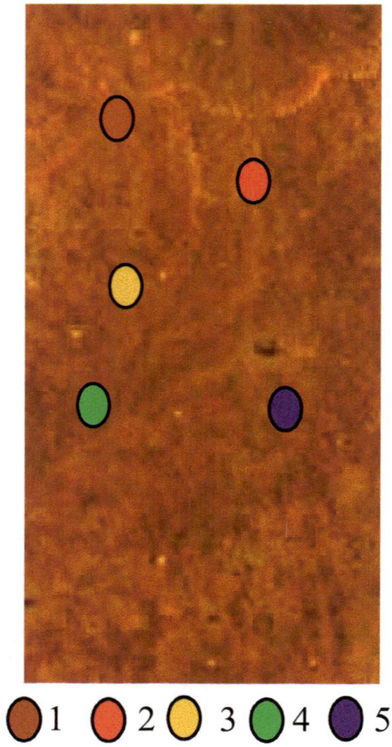

FIGURE 3.4

Study site in Ahmednagar Taluka region.

Table 3.3 At temperature of 31°C.

Soil parameter	Sample 1	Sample 2	Sample 3	Sample 4	Sample 5
pH (ppt)	08.32	08.40	08.34	08.31	08.42
EC (mS/cm)	00.54	00.86	00.68	00.78	00.83
Humidity (%)	19.40	20.22	19.89	20.21	21.56
Moisture (%)	71.00	78.00	71.22	71.58	75.42

Fig. 3.5. It is observed that the pH and humidity are slightly vary in the five samples. The conductivity increases in proportion to moisture of the corresponding sample.

Similarly, the samples are taken in the afternoon for 37°C at the same five sites. Table 3.4 shows the corresponding readings of the five samples. The pH value remained unchanged, and a the slight variation was found in the EC and humidity. However, in case of moisture, large variation is noted, as moisture is temperature-dependent (Figs. 3.6–3.8).

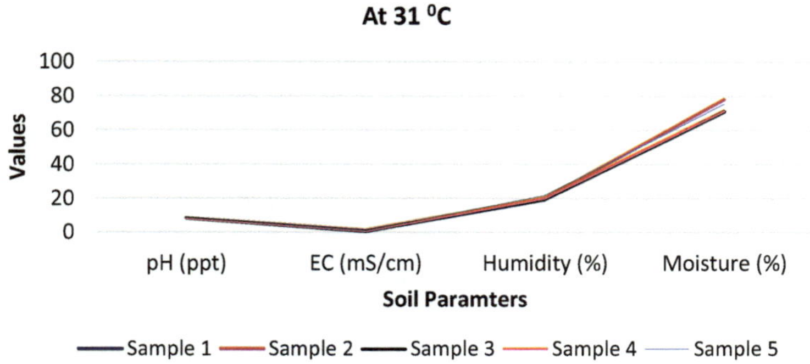

FIGURE 3.5

Representation of soil parameters for samples at 31°C temperature.

Table 3.4 At temperature of 37°C.

Soil parameter	Sample 1	Sample 2	Sample 3	Sample 4	Sample 5
pH (ppt)	08.32	08.40	08.34	08.31	08.42
EC (mS/cm)	00.71	00.74	00.69	00.72	00.64
Humidity (%)	18.33	19.20	18.66	18.53	19.62
Moisture (%)	68.66	70.40	68.12	68.78	62.21

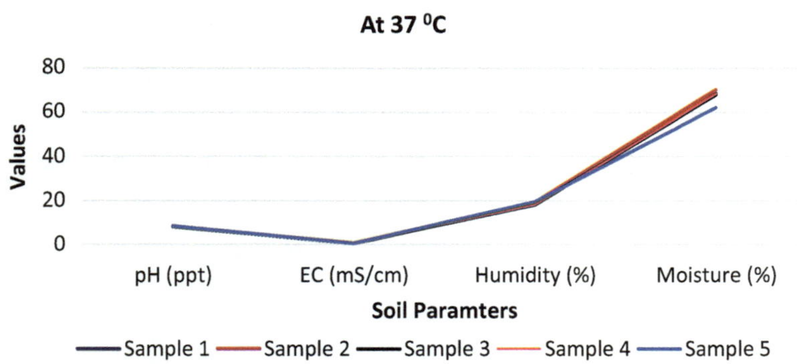

FIGURE 3.6

Representation of soil parameter for samples at 37°C temperature.

The two sets of sample have been retrieved by IoT-based sensor nodes and further validated with a laboratory-based soil testing kit; no significant difference is found in both the measured values (Tables 3.5 and 3.6).

Fig. 3.9 shows actual setup of designed and developed Arduino-based system, the soil testing kit used at laboratory, IoT system for sensor data acquisition, agricultural site, field work with team and soil sample on field.

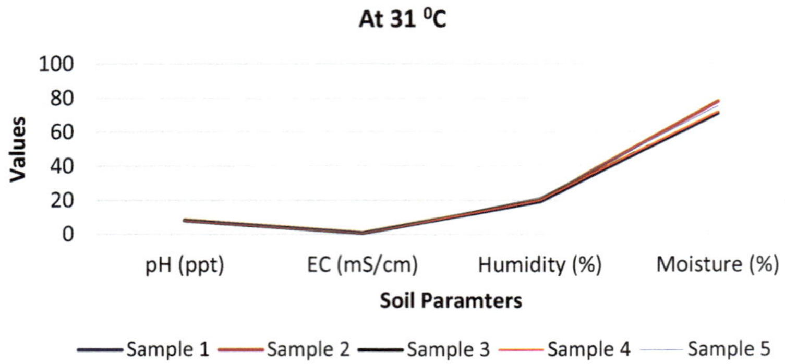

FIGURE 3.7

Sample testing results by soil testing kit at temperature of 31°C.

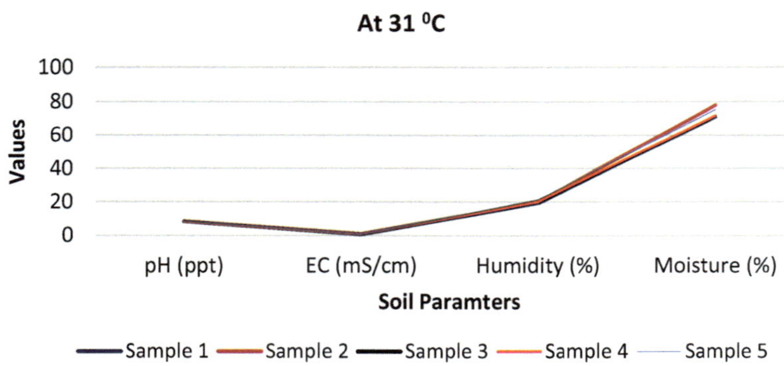

FIGURE 3.8

Sample testing results by soil testing kit at temperature of 37°C.

Table 3.5 Sample testing results by soil testing kit at temperature of 31°C.

Soil parameter	Sample 1	Sample 2	Sample 3	Sample 4	Sample 5
pH (ppt)	08.32	08.40	08.34	08.31	08.42
EC (mS/cm)	00.54	00.86	00.68	00.78	00.83
Humidity (%)	19.40	20.22	19.89	20.21	21.56
Moisture (%)	71.00	78.00	71.22	71.58	75.42

Table 3.6 Sample testing results by soil testing kit at temperature of 37°C.

Soil parameter	Sample 1	Sample 2	Sample 3	Sample 4	Sample 5
pH (ppt)	08.32	08.40	08.34	08.31	08.42
EC (mS/cm)	00.54	00.86	00.68	00.78	00.83
Humidity (%)	19.40	20.22	19.89	20.21	21.56
Moisture (%)	71.00	78.00	71.22	71.58	75.42

FIGURE 3.9

(A) Experimental setup for Arduino-based system, (B) soil testing kit setup, (C) IoT system for sensor data acquisition, (D) agricultural site, (E) field work with team, and (F) soil sample on field.

3.5.2 Satellite-based results

The microwave Sentinel-1 SAR satellite-retrieved dataset is imported in the PolSARpro software. Initially the data preprocessing such as noise removal and geometrical correction is carried out. Further, the overall dataset is classified, and study area has been finalized. Fig. 3.10 shows site mapping to deploy sensor node.

Once the data are classified by using PolSARpro and SNAP software tools, the backscattering coefficient (σ_0) and moisture are retrieved. The moisture values received form the sensor node and retrieved from satellite data are compared. Initially the data retrieval is taken in the morning time with temperature of 31°C as shown in Table 3.7.

From Fig. 3.11, it is observed that the moisture value received by IoT-based sensor node is in good agreement with the moisture values retrieved by satellite. Similarly, data retrieval is taken in the afternoon time with temperature of 37°C as shown in Table 3.8.

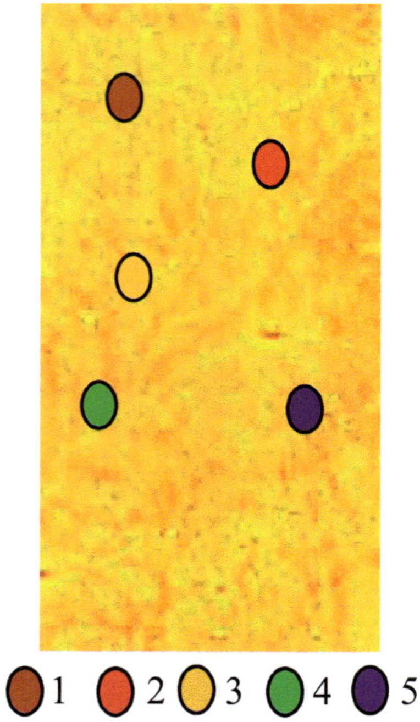

FIGURE 3.10

Soil moisture retrieval from study sites.

Table 3.7 At temperature of 31°C.

Samples	Standard moisture (%)	Observed moisture from satellite (%)	Observed moisture from IoT system (%)
Sample-1	50–70	69.56	71.00
Sample-2		75.29	78.00
Sample-3		70.01	71.22
Sample-4		68.95	71.58
Sample-5		72.55	75.42

From Fig. 3.12, it is observed that the moisture value received by IoT-based sensor node is in good agreement with the moisture values retrieved by the satellite.

From the satellite data processing results, the moisture and temperature retrieval is initiated for real-time soil parameter retrieval. All the soil parameters such as pH, EC, humidity, moisture, and temperature are equivalently important for planning sowing, fertilization, crop growth, etc.

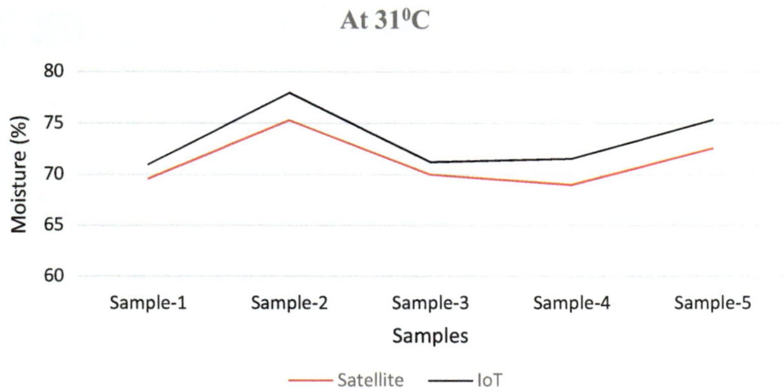

FIGURE 3.11

Soil moisture satellite versus IoT at 31°C.

Table 3.8 At temperature of 37°C.

Samples	Standard moisture (%)	Observed moisture from satellite (%)	Observed moisture from IoT system (%)
Sample-1	50–70	67.30	68.66
Sample-2		69.21	70.40
Sample-3		66.52	68.12
Sample-4		67.80	68.78
Sample-5		61.56	62.21

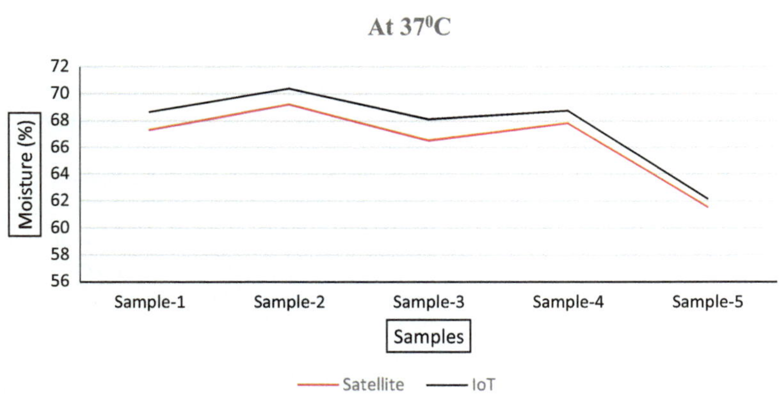

FIGURE 3.12

Soil moisture satellite versus IoT at 37°C.

3.6 Conclusion

The IoT-based soil parameter monitor system is successfully implemented. It is significantly useful for application in PA. The sensor nodes for pH, EC, temperature, humidity, and moisture are successfully interfaced to the wireless IoT. The retrieved parameter values are validated by cross-testing with a laboratory-based soil testing kit. Additionally, the moisture and temperature are retrieved from microwave sentinel-1 SAR satellite dataset and compared with the data received by the sensor nodes. The results obtained from the measurement have shown that the system performance is quite reliable and accurate.

Acknowledgments

Sincere thanks to Anjuman Ishat-E-Taleem, Beed Society, Principal Milliya Arts, Science and Management Science College, Beed, and Dr. Sayyad Shafiyoddin for providing soil testing facility at his Indian Space Research Organisation (ISRO) funded soil testing laboratory. Special thanks to Ahmednagar Jilha Maratha Vidya Prasarak Samaj, Ahmednagar Society, and Principal Dr. B. H. Zaware of New Arts, Commerce and Science College, Ahmednagar (Autonomous), for sanctioned Golden Jubilee Research Fellowship (GJRF) for financial support to author Dr. Shaikh Mudassar.

References

Aiello, G., Giovino, I., Vallone, M., Catania, P., & Argento, A. (2018). A decision support system based on multisensor data fusion for sustainable greenhouse management. *Journal of Cleaner Production.*, *172*, 4057–4065.

Atzberger, C. (2013). Advances in remote sensing of agriculture: Context description, existing operational monitoring systems and major information needs. *Remote Sensing of Environment.*, *5*, 949–981.

Aubert, B. A., Schroeder, A., & Grimaudo, J. (2012). IT as enabler of sustainable farming: An empirical analysis of farmers' adoption decision of precision agriculture technology. *Decision Support System.*, *54*, 510–520.

Awokuse, T. O., & Xie, R. (2015). Does agriculture really matter for economic growth in developing countries? *Canadian Journal of Agricultural Economics.*, *63*, 77–99.

Bégué, A., Arvor, D., Lelong, C., Vintrou, E., & Simoes, M. (2015). *Agricultural systems studies using remote sensing. Remote sensing handbook, Vol. 2: Land resources monitoring, modeling, and mapping with remote sensing*. CRC Press.

Berry, J. K., Delgado, J. A., Khosla, R., & Pierce, F. J. (2003). Precision conservation for environmental sustainability. *Journal of Soil and Water Conservation*, *58*, 332–339.

Bongiovanni, R., & Lowenberg-DeBoer, J. (2004). Precision agriculture and sustainability. *Precision Agriculture.*, *5*, 359–387.

Ceballos, M. R., Gorricho, J. L., Palma Gamboa, O., Huerta, M. K., Rivas, D., & Erazo Rodas, M. (2015). Fuzzy system of irrigation applied to the growth of habanero pepper

(*Capsicum chinense* Jacq.) under protected conditions in Yucatan, Mexico. *International Journal of Distributed Sensor Networks*, *11*(6), 123543.

Cevallos Barriga, J.M. (2011). *Producción y comercialización del cacao en el Ecuador período 2009–2010* (B.S. thesis). Universidad de Guayaquil facultad de Ciencias Económicas.

Courault, D., Seguin, B., & Olioso, A. (2005). Review on estimation of evapotranspiration from remote sensing data: From empirical to numerical modeling approaches. *Irrigation and Drainage Systems.*, *19*, 223–249.

Culman, M., Portocarrero, J.M. T., Guerrero, C.D., Bayona, C., Torres, J.L., & Farias, C. M. D. (2017). *PalmNET: An open-source wireless sensor network for oil palm plantations*. In: Proceedings of the IEEE fourteenth international conference on networking, sensing and control (ICNSC), May, pp. 783–788.

Delgado, J., Short, N. M., Roberts, D. P., & Vandenberg, B. (2019). Big data analysis for sustainable agriculture on a geospatial cloud framework. *Frontiers in Sustainable Food Systems*, *3*, 54.

DHT11 sensor. (n.d.). https://www.mouser.com/ds/2/758/DHT11-Technical-Data-Sheet-Translated-Version-1143054.pdf.

Erazo, M., Rivas, D., Pérez, M., Galarza, O., Bautista, V., Huerta, M., & Rojo, J.L. (2015). *Design and implementation of a wireless sensor network for rose greenhouse monitoring*. In: Proceedings of the sixth international conference on automation, robotics and applications (ICARA), IEEE, pp. 256–261.

ESP8266 Wi-Fi module information. (n.d.). https://www.espressif.com/sites/default/files/documentation/0a-esp8266ex_datasheet_en.pdf.

Ferrández-Pastor, F. J., García-Chamizo, J. M., Nieto-Hidalgo, M., Mora-Pascual, J., & Mora-Martínez, J. (2016). Developing ubiquitous sensor network platform using internet of things: Application in precision agriculture. *Sensors*, *16*(7), 1141.

Flores, K. O., Butaslac, I. M., Gonzales, J. E. M., Dumlao, S. M. G., & Reyes, R. S. (2016). *Precision agriculture monitoring system using wireless sensor network and Raspberry Pi local server* (pp. 3018–3021). IEEE.

Flores-Medina, M., Flores-García, F., Velasco-Martínez, V., Gonźalez-Cervantes, G., & Jurado-Zamarripa, F. (2015). Monitoreo de humedad en suelo a través de red inalámbrica de sensores. *Tecnologiá y Ciencias del Agua*, *6*(5), 75–88.

García, S., Ramírez-Gallego, S., Luengo, J., & Herrera, F. (2016). Big data: Preprocesamiento y calidad de datos. *Novática*, *237*, 17.

Ge, Y., Thomasson, J. A., & Sui, R. (2011). Remote sensing of soil properties in precision agriculture: A review. *Frontiers in Earth Science.*, *5*, 229–238.

Gebbers, R., & Adamchuk, V. (2010). Precision agriculture and food security. *Science (New York, N.Y.)*, *327*, 828–831.

Grupo Banco Mundial. (n.d.). *Agricultura, valor agregado*. https://datos.bancomundial.org/indicador/NV.AGR.TOTL.ZS.

Grupo Banco Mundial. (2018). *Empleos en agricultura*. https://datos.bancomundial.org/indicador/SL.AGR.EMPL.ZS?locations = EC.

Hendricks, G. S., Shukla, S., Roka, F. M., Sishodia, R. P., Obreza, T. A., Hochmuth, G. J., & Colee, J. (2019). Economic and environmental consequences of over fertilization under extreme weather conditions. *Journal of Soil and Water Conservation*, *74*, 160–171.

Hernández Rojas, D.L., Mazón Olivo, B.E., & Campoverde Marca, A.M. (2015). *Cloud computing para el internet de las cosas*. Caso de estudio orientado a la agricultura de precisión. I Congreso Internacional de Ciencia y Tecnología UTMACH. IoT Architecture Based on Wireless Sensor Network 57.

Huang, C.H., Shen, P.Y., & Ceng Huang, Y. (2015). *IoT-based physiological and environmental monitoring system in animal shelter*. In: Proceeding of the seventh IEEE international conference on ubiquitous and future networks, July, pp. 317–322.

Huang, Y., Chen, Z., Yu, T., Huang, X., & Gu, X. (2018). Agricultural remote sensing big data: Management and applications. *Journal of Integrative Agriculture*, *7*, 1915–1931.

INEC. (2018). *Estadísticas agropecuarias*. http://www.ecuadorencifras.gob.ec/estadisticas-agropecuarias-2/.

Ipanaqué, W., Belupú, I., Castillo, J., & Salazar, J. (2017). *Internet of things applied to monitoring fermentation process of cocoa at the piura's mountain range*. In: Proceedings of the CHILEAN conference on electrical, electronics engineering, information and communication technologies (CHILECON), IEEE, pp. 1–5.

Kamilaris, A., Kartakoullis, A., & Prenafeta-Boldú, F. X. (2017). A review on the practice of big data analysis in agriculture. *Computers and Electronics in Agriculture.*, *143*, 23–37.

Karimi, N., Arabhosseini, A., Karimi, M., & Kianmehr, M. H. (2018). Web-based monitoring system using wireless sensor networks for traditional vineyards and grape drying buildings. *Computers and Electronics in Agriculture.*, *144*, 269–283. Available from https://doi.org/10.1016/j.compag.2017.12.018.

Khattab, A., Abdelgawad, A., & Yelmarthi, K. (2016). *Design and implementation of a cloud-based IoT scheme for precision agriculture*. In: Proceedings of the twenty-eighth international conference on microelectronics (ICM), Giza, Egypt, 17 December, Vol. 4, pp. 201–204.

Kodali, R.K., Soratkal, S., & Boppana, L. (2016). *WSN in coffee cultivation*. In: Proceedings of the international conference on computing, communication and automation (ICCCA), April, pp. 661–666. https://doi.org/10.1109/CCAA.2016.7813804.

Maes, W. H., & Steppe, K. (2012). Estimating evapotranspiration and drought stress with ground-based thermal remote sensing in agriculture: A review. *Journal of Experimental Botany*, *63*, 4671–4712.

Mulla, D. J. (2013). Twenty-five years of remote sensing in precision agriculture: Key advances and remaining knowledge gaps. *Biosystems Engineering.*, *114*, 358–371.

Pavón-Pulido, N., López-Riquelme, J. A., Torres, R., Morais, R., & Pastor, J. A. (2017). New trends in precision agriculture: A novel cloud-based system for enabling data storage and agricultural task planning and automation. *Precision Agriculture.*, *18*, 1038–1068.

Petropoulos, G. P., Srivastava, P. K., Piles, M., & Pearson, S. (2018). Earth observation-based operational estimation of soil moisture and evapotranspiration for agricultural crops in support of sustainable water management. *Sustainability*, *10*, 181.

pH sensor. (n.d.). https://scidle.com/how-to-use-a-ph-sensor-with-arduino.

Pierpaolia, E., Carlia, G., Pignattia, E., & Canavaria, M. (2013). Drivers of precision agriculture technologies adoption: A literature review. *Procedia Technology.*, *8*, 61–69.

Srinivasan, A. (Ed.), (2006). *Handbook of precision agriculture: principles and applications*. Food Products Press, Haworth Press.

Rawal, S. (2017). IOT based smart irrigation system. *International Journal of Computer Applications*, *159*(8), 7–11.

ThingSpeak IoT platform. (n.d.). https://thingspeak.com.

Triantafyllou, A., Sarigiannidis, P., & Bibi, S. (2019). Precision agriculture: A remote sensing monitoring system architecture. *Information*, *10*, 348.

Vijayakumar, N., & Ramya, R. (2015). *The real time monitoring of water quality in IoT environment*. In: IEEE sponsored second international conference on innovations in information, embedded and communication systems (ICIIECS), pp. 1–5.

Weiss, M., Jacob, F., & Duveillerc, G. (2020). Remote sensing for agricultural applications: A meta-review. *Remote Sensing of Environment.*, *236*, 111402.

Zhang, J., Huang, Y., Pu, R., Gonzalez-Moreno, P., Yuan, L., Wu, K., & Huang, W. (2019). Monitoring plant diseases and pests through remote sensing technology: A review. *Computers and Electronics in Agriculture.*, *165*, 104943.

Zhang, N., Wang, M., & Wang, N. (2002). Precision agriculture—A worldwide overview. *Computers and Electronics in Agriculture.*, *36*, 113–132.

Zhou, L., Chen, N., Chen, Z., & Xing, C. (2016). ROSCC: An efficient remote sensing observation-sharing method based on cloud computing for soil moisture mapping in precision agriculture. *IEEE Journal of Selected Topics in Applied Earth Observations and Remote Sensing*, *9*, 5588–5598.

CHAPTER

LIFE GEOCARBON: carbon farming geolocation support by establishing a spatial soil database management system

4

Dimitris Triantakonstantis[1], Kostas Bithas[2], Spyridon E. Detsikas[1], Gherardo Biancofiore[3], Romina Lorenzetti[3], José A. Pascual[4], Margarita Ros[4], Carlos Guerrero[5] and Thomas Panagopoulos[5]

[1]Department of Soil Science of Athens, Institute of Soil and Water Resources, ELGO—DIMITRA, Lycovrisi, Attiki, Greece

[2]Institute of Urban Environment and Human Resources, Department of Economic and Regional Development, Panteion University, Kallithea, Athens, Greece

[3]Institute of BioEconomy, National Research Council (IBE CNR), Florence, Italy

[4]Group of Soil Enzymology and Bioremediation and Organic Wastes, Department of Soil and Water Conservation, CEBAS-CSIC, University Campus of Espinardo, Murcia, Spain

[5]Faculty of Sciences and Technology, University of Algarve, Faro, Portugal

4.1 Introduction

Climate change constitutes a major natural hazard to agricultural systems, endangering the global food security. A major source of greenhouse gas (GHG) emissions is the agricultural sector, being responsible for almost a quarter (24%) of the total GHG emissions (IPCC, 2014). Carbon farming has been proposed by the European Commission and European Parliament as one of the major pilots of the forthcoming Common Agricultural Policy (CAP) aiming to reduce the GHG emissions (Khan & Johansson, 2022; Mattila et al., 2022; McDonald et al., 2021). The development of the Carbon Farming Information platform will incorporate the relevant information, which is related to monitoring, reporting, controlling, and verifying the existing and new payment solutions, under development in the context of the CAP starting in 2023.

Subsidies will be linked to carbon farming, necessitating the development of operational methods for determining the farmer's carbon balance (European Parliament, 2021). One of the most typical knowledge gaps impeding a complete assessment of carbon balance at both the micro and macro levels is the so-called

Remote Sensing in Precision Agriculture. DOI: https://doi.org/10.1016/B978-0-323-91068-2.00010-2

Activity Data. Furthermore, the quality and completeness of existing estimations of agricultural soil carbon sink capacity remain unaddressed until today. These impediments are now reflected in the National GHG Inventory Reports, which apply default emission factors for most Mediterranean countries, resulting in Tier 1 reporting status (Baccour et al., 2021; IPCC, 2006). As a result, a more exact assessment of changes in soil carbon balance in relation to agricultural management approaches could be beneficial for climate change mitigation.

4.1.1 Introducing LIFE GEOCARBON and its framework

LIFE GEOCARBON is a preparatory project that aims to contribute to the implementation of a neutral carbon balance farm-level system. GEOCARBON intends to use farmers' knowledge to strengthen the so-called Activity data for the Land Use, Land-Use Change, and Forestry (LULUCF) sector. Farmers will use a mobile application on the field to record their actual practices (management techniques and cultivation methods). This is the most cost-effective method of filling the Activity Data gap. To allow the development of an interactive Carbon Farming Calculation tool, Activity Data will be merged with all current available datasets.

Moreover, GEOCARBON aims to use a farmer-based network to bridge the gap between farmers, the scientific community, and decision-makers by allowing the farmers to participate in and drive geospatial data. The geospatial data will be accessed via a web-GIS application for smartphones, which will be built specifically for the project's needs. After easy IT operations, the web-GIS will interact with farmers, giving visualization of the farms and acquisition of activity data. As a result, a crowdsourcing monitoring app will be created to collect farmers' responses. The GEOCARBON crowdsourcing application will be developed under the canopy of a Community of Practice (CoP) strategy, with farmers serving as the primary nodes of the CoP network. The use of a crowdsourcing program will help researchers better understand farmers' perceptions of climate change mitigation.

4.1.2 Aims and objectives

The main goal of the GEOCARBON project is to address the lack of adequate farming-level information by supplementing current earth-based data with a structured, harmonized geospatial framework system that is ready to be used as the input to the Carbon Farming Calculation Tool. A future carbon-related project will complete the Carbon Farming Calculation Tool. Nonetheless, the GEOCARBON project's geospatial framework system will be a crucial step toward exact carbon computation at the farm level, using IT-based decision support tools as part of a climate change mitigation strategy.

The GEOCARBON project's main deliverable will be a demonstration of a high-resolution geospatial information system that collects crucial data (e.g., soil

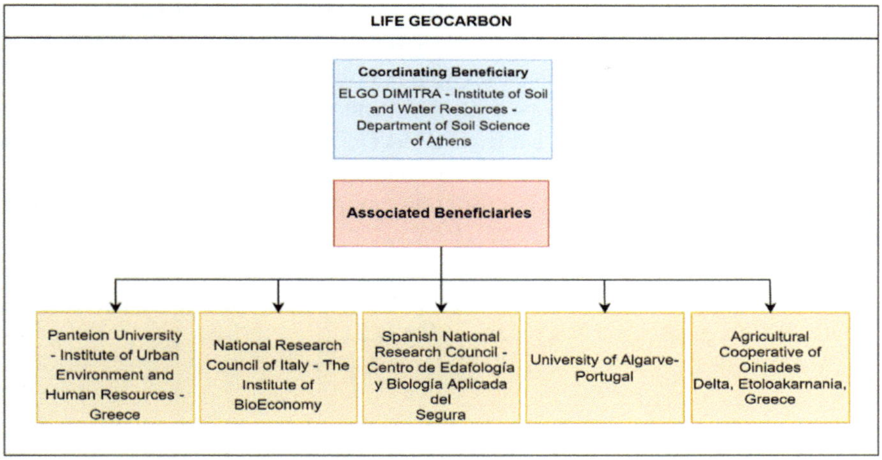

FIGURE 4.1

The LIFE GEOCARBON project structure.

and climatic conditions, landscape elements, and management practices) to assess the potential of farm-level climate change mitigation and to develop targeted Carbon Farming plans. The GEOCARBON project will geo-locate a representative sample of the Mediterranean area farms (at the agricultural parcel level) and provide simulations of alternative management practices under different environmental policy scenarios.

4.1.3 Project beneficiaries

The beneficiaries of the GEOCARBON project are research institutes and universities from countries located in southern Europe. Fig. 4.1 depicts the project's structure with the project's coordinating and associated beneficiaries.

4.2 Materials and methods

4.2.1 Data sources

As it was previously mentioned, GEOCARBON intends to enrich the spatial databases of agricultural management practices on a local or national level with contemporary IT and GIS web-based tools. In this context, GEOCARBON will exploit all the available preexisting sources of soil and agriculture management practices databases. The GEOCARBON database is going to be enriched with databases from all the countries that GEOCARBON beneficiaries originate. Such databases are the ELGO DIMITRA Greek soil database, the National Soil Map of

Greece, scale: 1.30.000 (Misopolinos, 2015), the Spanish CARBOSOL database (Llorente et al., 2018), the Andalusian SeisNET Database (De la Rosa et al., 2001), and the INFOSOLO database (Ramos et al., 2017), containing data from Portugal. In addition, data will be collected from ongoing/ended EU-funded projects (e.g., LIFE MediNet, LIFE CLIMATREE; LIFE OLIVE4CLIMATE; ClimaMED). Last but not least, it will also include data from the LUCAS soil survey. In Fig. 4.2, a summary of the different components of GEOCARBON database is presented.

The Activity Data, which are already exist in the abovementioned soil datasets, will be enriched by the contribution of the farmers. Some Activity Data that GEOCARBON project focuses on are in Fig. 4.3.

4.2.2 Data acquisition

GEOCARBON aims to collect the activity data of agricultural management practices through crowdsourcing. Crowdsourcing is a revolutionary activity that is

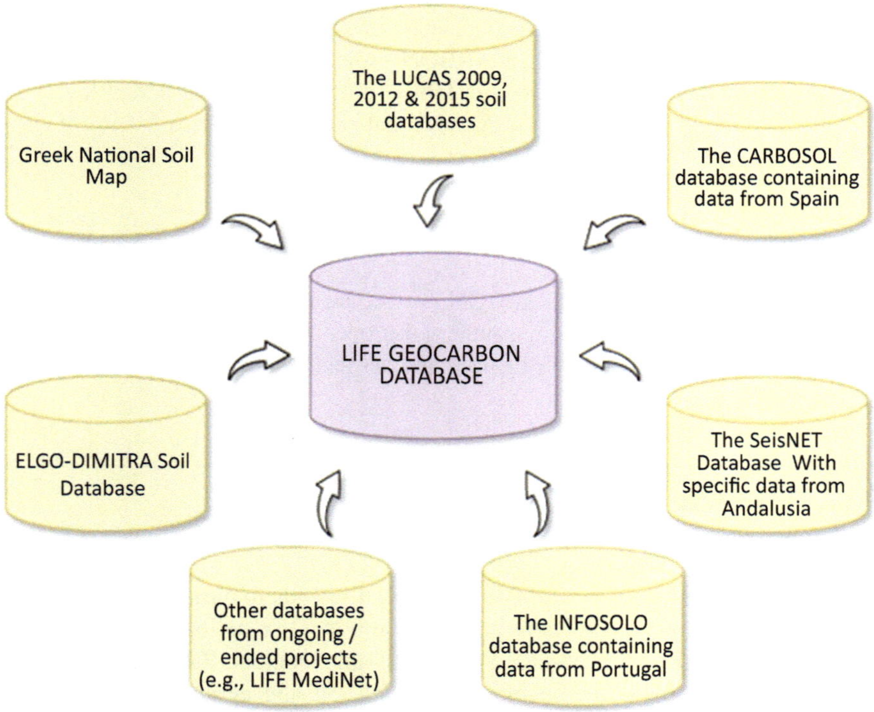

FIGURE 4.2

The contributing databases in LIFE GEOCARBON.

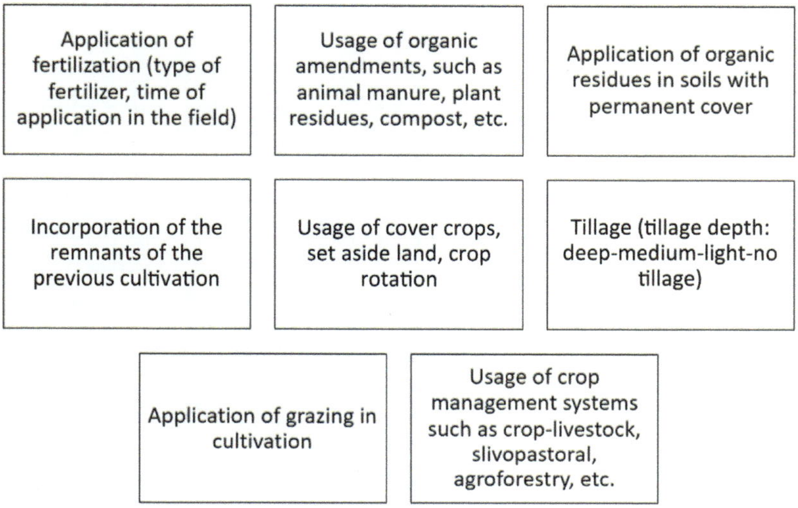

FIGURE 4.3

The main Activity Data of LIFE GEOCARBON.

transforming how people interact and make decisions (Žabota & Kobal, 2020). Crowdsourcing gives access to a large number of individuals over the Internet, making a human workforce more accessible (Krupowicz et al., 2020). The following components must be considered before beginning the crowdsourcing activity: establishing the research goal, defining the target audience, identifying appropriate engagement mechanisms that will encourage crowd participation, and finally, determining the technical platform to support the activity (Nevo & Kotlarsky, 2020).

An operational web-GIS tool will be developed as the crowdsourcing collection platform of the project. Web-GIS tools allow easier and wider access to geospatial data, data analysis tools, and information for more effective decision-making. Web-GIS tools can be developed for a wide range of applications, such as for soil environmental quality monitoring (Xu et al., 2021), air quality monitoring (Lopes et al., 2019), landslide risk calculation (Capolupo et al., 2021), and coastal monitoring (Randazzo et al., 2021).

The development of the web-GIS will be carried out by leveraging the tools offered by ESRI's ArcGIS Online, a cloud-based platform for map creation, geospatial data analysis, and web-based tool or application development. Among the tools that will be exploited are the ArcGIS WebApp Builder, the ArcGIS Survey123, and the ArcGIS Story maps. In particular, the ArcGIS Online Web App builder tool will be used to develop the web-based application. The application will include widgets with capabilities for basic data analysis (interpolation, hot spot analysis, etc.) and user interaction with maps of the parameters studied. A beta version of GEOCARBON's web app is presented in Fig. 4.4.

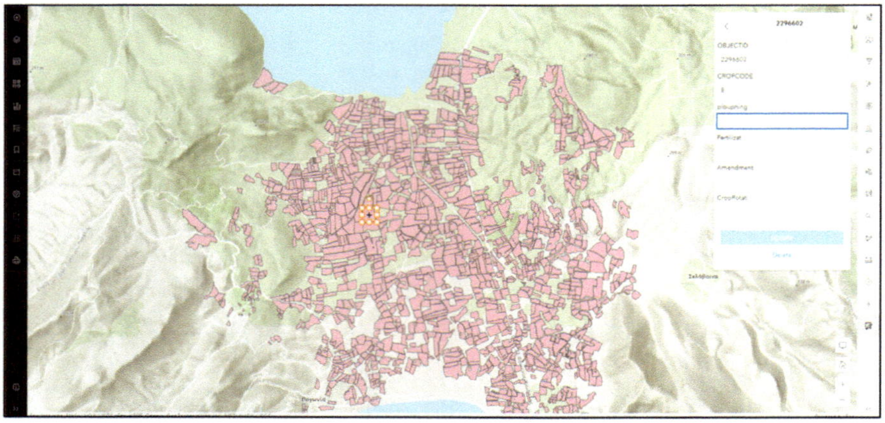

FIGURE 4.4

The beta version of the LIFE GEOCARBON WebGIS platform.

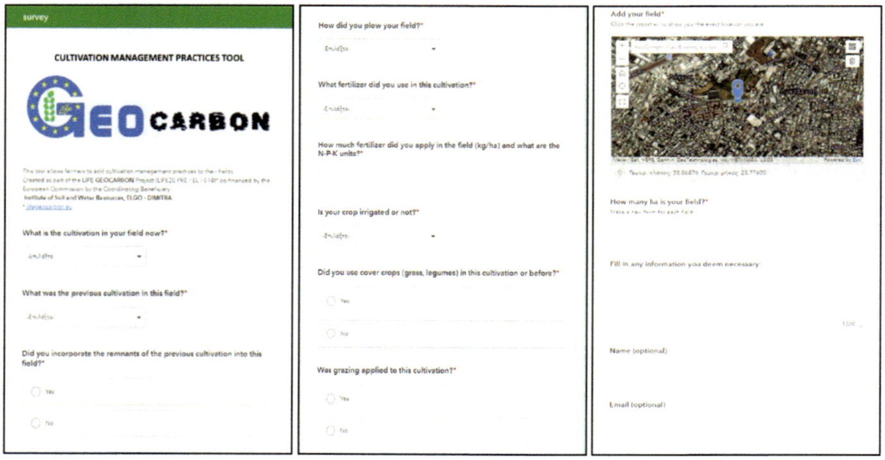

FIGURE 4.5

The GEOCARBON ArcGIS Survey 123 tool.

Additionally, to obtain the activity data that the project aims to collect, a web-based form was created using the ArcGIS Survey123. Through this application, it is possible to create web-based smart forms with multiple-language support. Exploiting the capabilities of ArcGIS Survey123, it was possible to create an activity data collecting survey. In Fig. 4.5, the GEOCARBON ArcGIS Survey 123 tool is presented.

Finally, the ArcGIS Story Maps will be used for the presentation of the results and for the demonstration of the active involvement of the program participants, especially the farmers in the collection of the data of the research project. The Story Maps tool has been successfully used as a dissemination tool in various types of studies (Antoniou et al., 2018; Castro & Rifai, 2021). The use of Story Maps will achieve a connection between the research program and the community, which is an important element for the success of any planning and management (Opdam, 2010).

4.3 Expected results

The GEOCARBON expected results are to prepare a farming-level knowledge and recording system leveraging the spatial web-based tools to assist the decision-makers and farmers in an upcoming carbon farming scheme.

A web-based tool, named Farmer Activity Data Information Tool, will be created where the users will interact to insert, organize, analyze, and share the input data by obtaining maps, geographic information, and analyses available on any device. This mapping and analytic platform include a powerful GIS web services server with dedicated web-GIS infrastructure. All these components are used to establish a web-based data-sharing system that can be accessed via the GEOCARBON official website. In the ArcGIS portal of Farm Activity Data Information Tool, we will incorporate the following functions: (1) create, save/locate, and share web maps; and (2) create and host web-mapping applications. The data that will be accessed through Farm Activity Data Information Tool are the following: (1) soil databases, which will be acquired from different sources; (2) LPIS/IACS will be used to ensure the compartmentalization of land parcels; and (3) Activity Data, which will be entered by the farmers.

The Farmers Activity Data Information Tool will be available to be used by stakeholders and decision-makers. Research and education could be invited to exploit the relevant potentials. The tool will be also available to the Ministries of Agriculture and Environment—Climate for supporting the design of climate and agricultural policies. The free access and the close collaboration with the CAP payment Authority in Greece (OPEKEPE) will be ensured as our workplan for our common activities with OPEKEPE and systematic consideration of the process underlining the payment solutions. In a nutshell, the project has been designed to support the operational application of carbon farming schemes.

Therefore they will add these data from the implementation areas of GEOCARBON countries through a website/app/portal. It will include a built-in map of each area with high-resolution satellite images as a basemap, in which the farmers could select their land parcels. For the compartmentalization of the land parcels, LPIS will be used to insure the entitlement of each land. After the selection, the farmers would report their practices. In such a way, farmers will update the database each time they carry out a soil management practice via the app.

4.4 Conclusions

This chapter introduced the EU-funded GEOCARBON project (lifegeocarbon.eu), providing a detailed overview of the project's research aims and objectives. The geospatial framework system developed in GEOCARBON will contribute to a more precise carbon farming calculation, using IT-based tools as a backbone of a climate change mitigation plan. Using a friendly mobile application, GEOCARBON is expected to enhance farmers' awareness and participation toward strengthening the acquisition of activity data for the LULUCF sector. Farmers will use a cost-effective mobile application on the field to record the farming practices that are applied in their own farms. Data are stored in a GIS cloud environment to be used for further analysis under different environmental policy scenarios. GEOCARBON, as a preparatory step before the development of a carbon farming calculation tool, will provide accurate field-based knowledge by farmers. This information along with additional soil/land datasets will play an important role in monitoring the policies on carbon farming and assist in climate change mitigation.

References

Antoniou, V., Ragia, L., Nomikou, P., Bardouli, P., Lampridou, D., Ioannou, T., Kalisperakis, I., & Stentoumis, C. (2018). Creating a story map using geographic information systems to explore geomorphology and history of Methana Peninsula. *ISPRS International Journal of Geo-Information*, *7*(12), 484.

Baccour, S., Albiac, J., Kahil, T., Esteban, E., Crespo, D., & Dinar, A. (2021). Hydroeconomic modeling for assessing water scarcity and agricultural pollution abatement policies in the Ebro River Basin, Spain. *Journal of Cleaner Production*, *327*, 129459.

Capolupo, A., Monterisi, C., Saponieri, A., Addona, F., Damiani, L., Archetti, R., & Tarantino, E. (2021). An interactive WebGIS framework for coastal erosion risk management. *Journal of Marine Science and Engineering*, *9*(6), 567.

Castro, C. V., & Rifai, H. S. (2021). Development and assessment of a web-based national spatial data infrastructure for nature-based solutions and their social, hydrological, ecological, and environmental co-benefits. *Sustainability.*, *13*(19), 11018.

De la Rosa, D., Mayol, F., Fernandez, M., Moreno, D., Ruiz, J., Moreno, J. A., Rosales, A., Castillo, V., Moreno, F., Cabreba, F., Giron, I., Cordon, R., Diaz-Pereira, E., Sánchez, J., Colomer, J. C., Ano, C., Recatala, L., Antoine, J., Masui, S., ... Prange, N. (2001). SEIS. net: Sistema Español de Información de Suelos en Internet. *Edafologia*, *8*, 45–56.

European Parliament. (2021). *Common agricultural policy* (package of legislation adopted on 23 November 2021, to be published in OJ). https://www.europarl.europa.eu/doceo/document/TA-9-2021-11-23-TOC_EN.html.

IPCC. (2006). *IPCC guidelines for national greenhouse gas inventories*. Intergovernmental Panel on Climate Change.

IPCC. (2014). Climate Change 2014: Synthesis report. In Core Writing Team, R. K. Pachauri, & L. A. Meyer (Eds.), *Contribution of Working Groups I, II and III to the 5th assessment report of the Intergovernmental Panel on Climate Change*, IPCC, Geneva, Switzerland, 151 pp.

Khan, J., & Johansson, B. (2022). Adoption, implementation and design of carbon pricing policy instruments. *Energy Strategy Reviews*, *40*, 100801.

Krupowicz, W., Czarnecka, A., & Grus, M. (2020). Implementing crowdsourcing initiatives in land consolidation procedures in Poland. *Land Use Policy*, *99*, 105015.

Llorente, M., Rovira, P., Merino, A., Rubio, A., Turrion, M., Badía, D., Romanya, J., & González, J.C. J.A. (2018). *The CARBOSOL Database: A georeferenced soil profile analytical database for Spain*. PANGAEA.

Lopes, S.I., Moreira, P.M., Cruz, A.M., Martins, P., Pereira, F., & Curado, A. (2019). RnMonitor: A WebGIS-based platform for expedite in situ deployment of IoT edge devices and effective Radon Risk Management. *IEEE International Smart Cities Conference (ISC2)*, pp. 451–457.

Mattila, T. J., Hagelberg, E., Söderlund, S., & Joona, J. (2022). How farmers approach soil carbon sequestration? Lessons learned from 105 carbon-farming plans. *Soil and Tillage Research*, *215*, 105204.

McDonald, H., Frelih-Larsen, A., Lóránt, A., Duin, L., Pyndt Andersen, S., Costa, G., and Bradley, H. (2021). Carbon farming – Making agriculture fit for 2030, Study for the committee on Environment, Public Health and Food Safety (ENVI), Policy Department for Economic, Scientific and Quality of Life Policies, European Parliament, Luxembourg.

Misopolinos, N. (2015). *Greek national soil map (scale: 1:30,000)*. Greek Payment and Control Agency for Guidance and Guarantee Community Aid (OPEKEPE).

Nevo, D., & Kotlarsky, J. (2020). Crowdsourcing as a strategic is sourcing phenomenon: Critical review and insights for future research. *The Journal of Strategic Information Systems*, *29*(4), 101593.

Opdam, P. (2010). Learning science from practice. *Landscape Ecology*, *25*, 821–823.

Ramos, T. B., Horta, A., Gonçalves, M. C., Pires, F. P., Duffy, D., & Martins, J. C. (2017). The INFOSOLO database as a first step towards the development of a soil information system in Portugal. *Catena*, *158*, 390–412.

Randazzo, G., Italiano, F., Micallef, A., Tomasello, A., Cassetti, F. P., Zammit, A., D'Amico, S., Saliba, O., Cascio, M., Cavallaro, F., Crupi, A., Fontana, M., Gregorio, F., Lanza, S., Colica, E., & Muzirafuti, A. (2021). WebGIS implementation for dynamic mapping and visualization of coastal geospatial data: A case study of BESS project. *Applied Sciences*, *11*(17), 8233.

Xu, W., Liang, H., Luo, W., & Kang, X. (2021). Design of Yunnan Province soil environmental quality monitoring and analysis platform based on WebGIS. *IOP Conference Series: Earth and Environmental Science*, *687*, 012043.

Žabota, B., & Kobal, M. (2020). A new methodology for mapping past rockfall events: From mobile crowdsourcing to rockfall simulation validation. *ISPRS International Journal of Geo-Information*, *9*, 514.

CHAPTER

Mapping and monitoring of salt-affected soils: the contribution of geoinformation

5

Christina Lekka[1], Spyridon E. Detsikas[2], George P. Petropoulos[1] and Christos Chalkias[1]

[1]Department of Geography, Harokopio University of Athens, Kallithea, Athens, Greece
[2]Department of Soil Science of Athens, Institute of Soil and Water Resources, ELGO—DIMITRA, Lycovrisi, Attiki, Greece

5.1 Introduction

Soils are not only of key importance in the food security of an increasing world population, but they function also as regulatory factors on major global biogeochemical cycles (carbon, nutrient, and water) critical to Earth's ecosystems and climate balance. Soils play a crucial role in delineating GHG emissions, and hence, soils' capability to sequester carbon and consequently reduce the atmospheric CO_2 could potentially attribute decisively for mitigating climate change effects (Paustian et al., 2016). Protecting soils' health is as relevant as ever. In particular, the soil organic carbon (SOC) content is approximately about two-thirds of the terrestrial biosphere carbon pool (Eswaran et al., 1995). The healthy, productive, and nonconstraint factors of soils, usually, have already reached their equilibrium in SOC content, making it very challenging to increase the SOC content. Therefore their potential to sequester carbon is not as much as in the case of degraded soils. Any form of soil degradation increases carbon losses into the atmosphere. SOC reserves are an outcome of photosynthesis inputs and losses due to decomposition, leaching, and erosion (Setia et al., 2013). Thus it is clear that not every soil has the same carbon sequestration potential. The presence of soil organic matter (SOM) preserves the soils' structure, reduces evaporation, and improves water penetration, while preventing condensation and protecting it from erosion.

Soil degradation, especially that of salt-affected soils (SAS), significantly affects the potential of carbon (C) sequestration in biomass and in the SOC stocks. In landscapes affected by salts, salinity and limited carbon availability determine CO_2 emissions (Setia et al., 2011). In addition, land degradation due to aggregation of salts is an important hazard for soils' quality and

Remote Sensing in Precision Agriculture. DOI: https://doi.org/10.1016/B978-0-323-91068-2.00016-3

agricultural activity around the globe. Soil salinization prompts changes in natural, chemical, and physical characteristics of soil properties (Singh, 2022). In a dynamic phenomenon such as soil salinity, changes in variation occur due to temperature or precipitation and soil moisture evaporation (Azabdaftari & Sunar, 2016). Under current climatic conditions, soil salinization is induced in arid regions, where evaporation exceeds precipitation for most of the year (Lal, 2008). Future changes in temperature and rainfall patterns resulting from climate change can pose a direct threat to saline soils extend (Rengasamy, 2008a, b), mainly in semiarid and arid regions. Additionally, their significant impacts on carbon loss and increased soil salinity feedback represent an indirect threat on global, long-term scale for soil productivity and food security (Allbed & Kumar, 2013). According to various studies, the primary soil salinity worldwide is approximately 955 Mha, while secondary salinization due to human-induced processes is about 77 Mha, with more than half (58%) being in the areas with systematic irrigation (Metternicht & Zinck, 2003). Soil salinity is acknowledged today as one of the main threats to soil degradation and desertification with damaging effects on agricultural production, soil and water quality. Indeed, SAS presents promising economic and environmental interest through the implementation of sustainable management practices to restore and reclaim these lands (Wicke et al., 2011).

Thus it is understood that information on spatiotemporal variability of SAS is essential for accurate decision-making to halt soil salinity and soil degradation. To achieve such a goal, a suitable mapping method must be selected. Traditional soil mapping methods are based on field sampling approaches to measure the soil properties of an area. Yet, the cost, time, and excessive needs of labor constitute this approach ineffective, particularly so if the objective is to fully capture the dynamic changes of salts in the soil over large areas. Geoinformation technology can effectively map SAS and their spatiotemporal changes for the exact global extent delineation of SAS (Ivushkin et al., 2019). A major component of geoinformation technologies includes the analysis of Earth Observation (EO) data. Several studies have demonstrated the promising potential of this technology in the detection and the mapping of saline and sodic soils (Kumar et al., 2021; Mashimbye, 2013; Mougenot et al., 1993). EO and remote sensing are two terms used interchangeably and provide qualitative and quantitative geographical information on SAS through cost-effective and fast techniques (Gorji et al., 2017, 2019). There are also several studies that have shown the added value of the synergy between EO and Geographic Information System (GIS) in the context of mapping of soils' properties (e.g., Singh, 2022). In this context, various geospatial data analysis and interpretation tools, data processes, and modeling techniques have been developed and continue to rapidly develop nowadays (e.g., web-GIS, web-based services, cloud computing). The rapid technological developments in all the fields of geoinformation over the last decade have revealed its promising potential as the way forward to map and monitor soils' properties worldwide, including the spatiotemporal distribution of SAS.

In this context, the main objectives of this chapter include: (1) to furnish a brief overview of the global status of SAS with an emphasis on the Mediterranean region; (2) to describe the relationship and the dynamics of SAS with climate change and SOC content, and (3) to present an overview of the use of geoinformation in mapping and monitoring SAS, summarizing the key challenges and future research directions in regard to the use of this technology in this context.

5.2 Salt-affected soils: driving factors and present status

5.2.1 A general overview

Salts have anions and cations that are positively and negatively charged ions, respectively, which can be either dissolved in water or be present as solids. Saline soils are soils with an increased accumulation of neutral soluble salts in the soil layers. These salt compounds are mainly chloride or sulfate salts of calcium, magnesium, potassium, and to lesser extent, sodium (Table 5.1).

Salts in a soil profile can originate from water used for irrigation; from fertilizers, manures, or other amendments; or from soil parent material and natural processes (Horneck et al., 2007). The US Salinity Laboratory Staff (Richards, 1954) suggests the distribution of SAS into three classes of saline, sodic, and saline-sodic soils (Table 5.2). These two main groups of soils affected by the accumulation of salts show significant differences in their physical and chemical characteristics, in their physical and biological properties as well as in their geographical and geochemical distribution. In addition, these two group soils require dissimilar approaches of land and farm management as well as reclamation techniques (Abrol et al., 1988). In saline-sodic soils, the domination of either sodium or calcium defines the soils' characteristics as saline or sodic (Horneck et al., 2007).

1. *Saline soils*: Due to exchangeable cation sites, these soils are dominated by calcium and magnesium. In general, saline soils with high calcium have stable soil structure (Szabolcs, 1974). Saline soils often have visible salt deposits on the top surface of the ground due to the evaporation of water on it, giving them the name "white alkali." Salts can increase the soil osmotic potential. In addition, high salt accumulation in soil can cause plants to wilt even when moisture in the soil is sufficient (Horneck et al., 2007).

Table 5.1 Most common salt compounds in salt-affected soils (Ullman, 2013).

Cation (+)	Anion (−)
Sodium (Na^+)	Chlorides (Cl^-)
Magnesium (Mg^{2+})	Sulfates (SO_4^{2-})
Calcium (Ca^{2+})	Carbonate (CO_3^{2-})
	Bicarbonate (HCO_3^-)

Table 5.2 Synopsis of soil properties for SAS classification.

Soil property	Units	Saline	Saline-sodic	Sodic
pH	–	<8.5	<8.5	>8.5
EC_e	ds/m	>4	>4	<4
ESP	%	<15	>15	>15
SAR	%	<13	>13	>13

Source: *From Richards, L. (1954). Diagnosis and improvement of saline and alkali soils.* Soil Science, *782, 154, http://doi.org/10.1097/00010694-195408000-00012*

Groundwater salt-affected landscapes commonly have electrolyte high concentration and a future risk of salinization (Szabolcs, 1974).

2. *Sodic soils*: In these soils, considerable amounts of neutral soluble salts are commonly absent. Substantial salts amount capable of alkaline hydrolysis, for example, Na_2CO_3, is present (Szabolcs, 1974). The soil structure is destroyed because the dispersion of soil particles, and water movement into and through the soil is prevented (Horneck et al., 2007). These soils have also low infiltration rates and poor aeration (Ullman, 2013). Sodic soils negatively affect the plant growth chiefly as a result of the additional exchangeable sodium in poor properties and the consequence of high soil pH on nutritional imbalances. Groundwater in the areas that are influenced by soil sodicity has generally low-to-medium electrolyte concentration, and in certain cases, residual sodicity also has a potential sodicity hazard (Szabolcs, 1974).

5.2.2 Drivers of salt-affected soils

According to Abrol et al. (1988), salinity could be classified into primary and secondary salinization. Primary salinity factors are closely correlated with the natural processes of the soil–water–climate system such as parent material and rock weathering, aquifer rise, and climatic conditions. Whereas, the secondary drivers are influenced by human activity, unsustainable soil management and agriculture practices, changes in land use/land cover, extensive irrigation with poor quality irrigation water, and seawater intrusion by the overexploitation and the fall of water tables. A more detailed discussion on primary and secondary salinity and on the geographical areas in which those are often located is furnished next.

1. *Primary salinity*: Primary salinity is induced during the physical or chemical weathering process, which involves processes such as hydrolysis, hydration, solution, oxidation, and carbonation. Thus primary salinity is referred to the presence of excessive amount of salts through natural common processes, including geologic deposition and transport from parent material; as a result of saline groundwater, or through wind transport or underground aquifers or seawater intrusion. In the areas with xerothermic climatic conditions, salt

aggregation in soils can occur in nonirrigated landscapes or be developed through aquifer capillary rise and insufficient water evapotranspiration (Shahid et al., 2018). Due to the submersion of the coastal area under seawater after geological events, salts that remained trapped as marine sediments even after the seawater receded could have been aggregated in soils (Wendland et al., 2008). Salts could also accumulate in coastal areas from tidal events (tsunami waves, tides, etc.), usually accompanied by simultaneous intrusion of seawater. Such phenomenon has been observed in areas such as northern France, western Netherlands, Belgium, Denmark, Belgium, and the Netherlands (Trnka et al., 2013; Van Weert et al., 2009). Primary salinization of soils can be found in Austria, Hungary, Bosnia, Slovakia, Croatia, Ukraine, Serbia, Romania, Bulgaria, Spain, and Greece (Daliakopoulos et al., 2016).

2. *Secondary salinity*: Secondary salinity is directly related to human activity. Irrigation is a regulating driver of secondary soil salinization. Irrigating with brackish water or insufficiently leaching the soil to remove excess salts can lead to the development of SAS. Extensive irrigation from drilling wells in fields near coastal areas could have consequence in water table depletion and seawater intrusion. Secondary drivers can also be related to point sources of salinity. Improper use of fertilizers and other soil amendments can cause salinity issues. Fertilizers with a high salinity index may lead to further intensification of salinity problems. Furthermore, industrial and mining activities and improper use of wastewater are potential sources of salinization. Finally, land use or land cover change impacts the amount of salt accumulation. A change in vegetation type can disturb the water and evapotranspiration balance, resulting in a gradual reduction of the water availability and salt concentration. In arid and semiarid areas, reduced vegetation is associated with an increased risk of saline soils (Perri et al., 2018).

5.2.3 Salinity impact on soil organic carbon stocks

SOM is a component used to measure SOC. Organic matter has a crucial function in the physical, chemical, and biological properties of agricultural soils. The worldwide soils contain 1550 billion tonnes of organic carbon compared with the atmospheric reservoir, which contains 760 billion tonnes of carbon, and 560 billion tones contained in microorganisms and plants (EEA, 2011). Soils, therefore, contain twice the amount of carbon that is retained in the atmosphere and over three times that in vegetation. The carbon sequestration in soils is considered as a potential method for climate change mitigation by reducing carbon dioxide from the atmosphere. Any form of soil degradation can reduce the inputs of carbon and release of C to the atmosphere. Therefore carbon dioxide (CO_2) emission from soils is regulated by the organic material input and the decomposition rates; SOC stocks are a function of plant inputs through photosynthesis and release due to decomposition, leaching, and erosion (Setia et al., 2011, 2013).

As soil salinization affects the biological, physical, and chemical properties of soils and negatively affects plant health and biomass, sequestration of C is limited, particularly in regard to soil structure and vegetation health. Alternation affects soil osmotic potential, plant growth, ionic balance, toxicity, and consequently, the nutrient cycle and organic matter decomposition process. Extensive accumulation of salts degrades the soil, reduces vegetation productivity, and herein decreases biomass inputs. Therefore the biological activity of the soil is also hindered. Lal (2001) reports that SAS have substantial potential for C sequestration in the SOC and also in ground biomass. Additionally, the losses of C are associated with increasing salinity, especially secondary salinization, in sodic soils. SOC stocks are modulated according to the changes in soil biomass, biological load, and nutrient availability. Moreover, SOC losses potentially might be higher from dispersed aggregates because of sodicity as a result of salinity (Wong et al., 2004). Changes in the soil physicochemical properties combined with changes in vegetation directly affect the microbe activity and SOC decomposition. Yet, the potential effects of SAS on C mineralization remain quite uncertain.

Several studies have investigated the negative effect of SAS in carbon mineralization (Setia et al., 2011; Wong et al., 2006). According to Setia et al. (2011), overall CO_2 release in salt-affected landscapes is affected by biological, chemical, and physical properties of soils. Also, among the soil properties examined, the prominent parameter was electrical conductivity with a negative effect on CO_2 release in soils and thus should be taken into consideration when estimating C losses in landscapes affected by salts. Additionally, Setia et al. (2013) in their study demonstrated that soil salinity has an impact on decreasing plant productivity and C inputs to the soil but also increases SOC decomposition rates due to changes in the microbial activity. Setia et al. (2013), based on modeling estimation, suggested that global soils may lose over 6.8 pg SOC due to SAS extent by the year 2100, and current models are probable to have overestimated the global SOC content and underestimated the carbon outputs from soils and plants. In another study, Wong et al. (2006) compared the SOC stocks between SAS and soils recovered by vegetation restoration, and a very close relationship between carbon sequestration and vegetation is observed. The authors reported in their study on SAS that the SOC stocks are three times lower than those in the soils with natural vegetation. Results obtained from this study indicate that in restored and bioremediation areas, SOC levels in the topsoil were comparable with those of soils with natural vegetation.

5.2.4 Geographical distribution of salt-affected soils worldwide

SAS are distributed across all the continents in various extents and are found in almost all climatic conditions (Szabolcs, 1974). The worldwide extent of primary soil salinity is approximately 955 Mha, while secondary affects 77 Mha (Metternicht & Zinck, 2003). More than 932 Mha of the world's soils are degraded due to salinity, 23% of cultivable land is affected by salinization, and another 10% by salinity-sodicity (FAO, 1988). According to Food and Agricultural Organization (FAO, 2005), globally 20% of irrigated areas are covered by SAS. Based on previous estimations, the overall extent of saline soils was 397 Mha, while the area of sodic soils was 434 Mha. SAS covers 3.1% of worldwide soils, and based on current projections concerning the global temperature, population growth, and future soil degradation, an increase is expected in the geographical spread of SAS in the future (Rengasamy, 2008a, 2008b). Every year 10 Mha of agricultural land get degraded due to salinity impacts. Among the 230 Mha of the global irrigated land, 45 Mha or 19% of them was classified as SAS, and of the 1500 Mha of agricultural land, 32 million (2.1%) was SAS. Currently over 310 Mha of worldwide irrigated area, approximately 20%–33% is estimated to be salt-affected (Shahid et al., 2018). FAO estimated that in Europe, using the FAO/UNESCO Soil Map of the World (1970–1980), saline soils cover 6.7 Mha or 0.3% and sodic soils 72.7 Mha or 3.6% of the total area (Table 5.3).

The Global Map of Salt-Affected Soils (GSAS map) is a country-driven initiative by FAO and Global Soil Partnership (GSP) to record the current state of world's SAS. This global scale map provides information on spatial distribution of SAS at two depth intervals of 0–30 and 30–100 cm. The GSAS map also includes EC, ESP, pH soil parameters, and SAS into classes. According to the GSAS map, approximately 424 Mha of topsoil (0–30 cm) and 833 Mha of subsoil (30–100 cm) are affected by salts. Based on the GSAS map, higher than 3% of worldwide soils in the top layer and up to 6% of total subsoils worldwide are affected by salinity or sodicity. In addition, over two-thirds of SAS worldwide are found under arid and semiarid climatic zones. As derived by GSAS map, 85% of SAS in the top layer (0–30 cm) are saline, 10% are sodic, and 5% are saline-sodic. Also, 62% of salt-affected subsoils are saline, 24% are sodic, and 14% are saline-sodic. In Fig. 5.1, the current GSAS map (2021) released recently by Food and Agricultural Organization of the United Nations (FAO, 2021) is illustrated.

Table 5.3 Globally distribution of salt-affected soils according to FAO/UNESCO soil map of the world (1970–80).

Region	Total area (Mha)	Saline soils	%	Sodic soils	%
Africa	1899.1	38.7	2.0	33.5	1.8
Asia-Pacific – AU	3107.2	195.1	6.3	248.6	8.0
Europe	2010.8	6.7	0.3	72.7	3.6
Latin America	2038.6	60.5	3.0	50.9	2.5
Near East	1801.9	91.5	5.1	14.1	0.8
North America	1923.7	4.6	0.2	14.5	0.8
Total	12,781.3	397.1	3.1	434.3	3.4

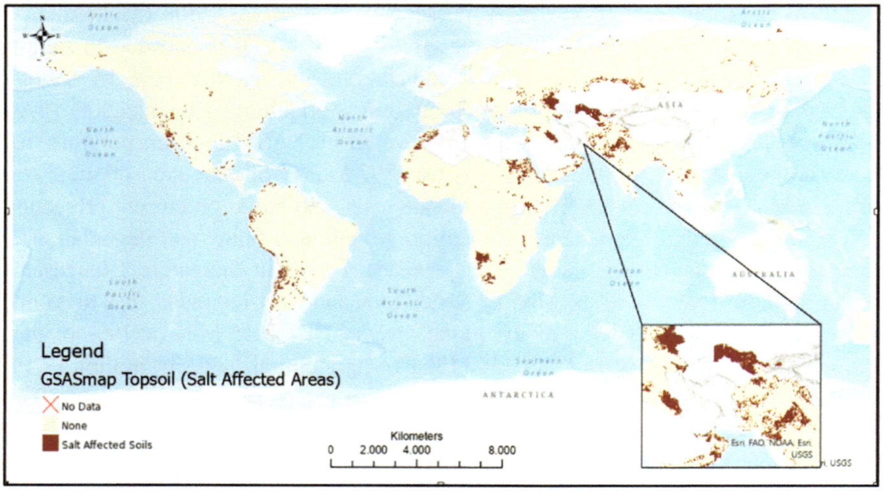

FIGURE 5.1

Global Map of Salt-Affected Soils.

Reproduced with permission FAO. (2021). Global map of salt-affected soils (GSASmap). Retrieved January 20, 2022, from https://www.fao.org/soilsportal/data-hub.

As can also be observed from Fig. 5.1, SAS are found in areas with varying climatic, soil, and topographical conditions. In Europe, many regions characterized by Mediterranean climatic conditions with high temperatures combined with irrational agricultural soil management are particularly prone to secondary salinity. In the Mediterranean, it is estimated that soil salinity affects 25% of irrigated land under agriculture exploitation (Geeson et al., 2003). Many countries in southern Europe such as Portugal, Greece, Italy, and Spain are facing the effects of saline soils due to saltwater intrusion into aquifers (Shahid et al., 2018) and evapotranspiration. In southern Mediterranean areas, which are indicated by prolonged dry-thermal conditions during hot season and wet winters during cold season, soil salinity formation is favored. Due to the occurrence of primary salinity in many landscapes, the phenomenon is exacerbated as it coincides with salt-affected areas due to secondary salinization. In the irrigated agricultural lands of Spain, about 3% of the 3.5 Mha of systematically irrigated area has significantly reduced agricultural activity due to salinization, while another 15% is vulnerable to soil salinity. In Greece, salinity is a major process of soil degradation; it is estimated that almost 35% of Greek soils are at high risk of desertification (Kosmas et al., 2000). According to surveys, SAS were estimated to cover 88,123 ha. Saline soils were estimated at 35,388 ha, sodic soils at 34,916 ha, and saline-sodic soils at 17,815 ha. However, also regions in the northern part of Europe such as Denmark, Poland, Latvia, and Estonia are experiencing similar salinization problems. In the EU coastal zones, through overexploitation of groundwater caused by the demand for increasing industrial exploitation and agriculture, seawater can seep into the soil and cause serious salinity problems.

5.3 Salt-affected soils assessment and the use of geoinformation

The urgent need for intensive and direct monitoring the quality of natural resources and atmosphere and dynamic phenomena such as soil salinity has led to the development of new sciences to record and evaluate the main object that serves as the target of study (Anand et al., 2021; Omuto et al., 2020). Geoinformation involves the combined use of remote-based technology, global positioning system (GPS), and GIS. These technologies can be used as prominent modern tools for acquiring accurate information of SAS spatiotemporal extent (Gorji et al., 2019; Pandey et al., 2020; Petropoulos et al., 2012; Singh et al., 2017). RS and GIS provide today the way forward for monitoring the distribution of SAS through fast and low-cost methods (Gorji et al., 2015). Commonly used methods based on soil field sampling require extensive soil sampling in large areas and laboratory analysis, which are labor-intensive, expensive, and time-consuming (Kumar et al., 2021; Manevski et al., 2012). RS technology has the ability to capture the spectral properties of soils in a variety of spatial and temporal extents and relate it with soils' properties such as their concentration in salts. In addition, the use of RS in the monitoring and assessment of SAS has been proven to be a useful implement for detecting the extent, the intensity, and the spatiotemporal changes of the topsoil salt aggregation on large-scale areas (Abbas et al., 2013). Thus the approaches of RS and GIS synergistically in mapping and monitoring SAS are more accurate, rapid, economically, and quantitative (Allbed & Kumar, 2013). EO products, especially when they are used in combination with field data, have a promising potential in the delineation of SAS due to their ability to provide systematic worldwide cover, time-efficient, simple-to-use, and free-of-cost data (Abbas et al., 2013). Many approaches that use the tools of EO and GIS have shown promising results on a global scale in the delineation of SAS extent (e.g., Gorji et al., 2019; Ivushkin et al., 2019; Pandey et al., 2021).

5.3.1 Remote sensing of salt-affected soils

The limitations of traditional methods for obtaining information on the spatiotemporal variability of soils' properties were already addressed in this chapter. It became evident that the use of such approaches in detecting and monitoring the physical characteristics of soils over large areas can be quite challenging, especially in otherwise inaccessible regions. It was also already noted that geoinformation, and in particular EO, is regarded today as the way forward in that respect. Indeed, RS allows for easy collection of data remotely (without physical contact) for an object or phenomenon by measuring its reflected and emitted radiation based on the physical and chemical characteristics. This approach also provides a fast and cost-effective approach of remotely acquiring information over a variety of scales within repetitive coverage and multiple spatial resolutions. Thus EO can be used for coverage over large areas where field surveys are limited or out of reach or in combination with the field sampling approach (Lamine et al., 2019; Petropoulos et al., 2015).

EO sensors can also be carried out in airborne platforms or drones. This solution offers certain advantages over satellite platforms such as user flexibility in the selection of target area and at the frequency (repeatability) of derived data. In addition to high temporal resolution, using UAVs, platform that carries EO sensors, can also provide very-high-resolution data, free-to-cloud cover. Yet, there are some limitations that should be considered when using this solution in EO data collection, namely the cost of own equipment and the legislation process.

A wide range of EO data are available today from sophisticated spaceborne sensors that offer a great potential in monitoring landscape changes and dynamic phenomena such as the extent and variability of SAS. Satellite platforms carrying RS imagers have been used in many studies combined with a variety of techniques for mapping and monitoring SAS (Abdullah et al., 2019; Asfaw et al., 2018; Gorji et al., 2017; Wang et al., 2019; Kumar et al., 2021).

5.3.1.1 Multispectral versus hyperspectral imaging in soils properties characerization

Independently of the platform used, EO imaging systems are characterized as multispectral or hyperspectral, which depends on the number of spectral bands used to record the reflected energy (spectral response) from the soils. According to the spectral band number, an imagine sensor acquires this spectral information and is divided into multispectral and hyperspectral sensor. Multispectral sensors acquire spectral information at a small number of discrete spectral bands, which allows for the discrimination of various soil properties (Pandey et al., 2019). However, often and depending on the soil property that is targeted to be extracted in a small number of broad spectral bands, features with spectrally similarities are difficult to distinguish.

On the other hand, hyperspectral sensors have the ability to record reflected energy from objects on ground surface from various ranges of the electromagnetic spectrum from visible to the shortwave infrared in numerous narrow continuous spectral bands, acquiring various amounts of spectral information (Elatawneh et al., 2012; Gorji et al., 2019; Pandey et al., 2020). As a result, an enhanced level of spectral information recorded by selective channels of these sensors is provided accordingly to the characteristics of target under analysis (Metternicht & Zinck, 2003). Many investigators have underlined the ability of hyperspectral images in potentially better discrimination between similar ground-cover features compared with traditional multispectral images (Allbed et al., 2014). In recent years, hyperspectral remote sensing is considered as a powerful tool of EO sources for acquiring imagery data (Qian, 2020). As a result of the merits aforementioned, use of hyperspectral imagery has become more and more widespread in various applications linked to the characterization of soil properties (Pandey et al., 2020). The remainder of this section focuses on presenting some studies exploiting both multispectral and hyperspectral imaging in detecting and monitoring saline and sodic soils.

5.3.1.2 Use of multispectral imaging

Multispectral satellite sensors have a critical role to play in SAS detection. Multispectral and spatiotemporal data are crucial in the rapid depiction of the dynamic changes of the salt accumulation in the surface soil. Extensive studies have been conducted by researchers attempted to monitor SAS using RS multispectral optimal data with moderate-to-high resolution such as Landsat Thematic Mapper (TM) and Enhanced TM or IRS-1D LIS-III (or LIS-II) and SPOT XS (Douaoui et al., 2006; Koshal, 2012; Yossif & Ebied, 2016). However, several studies have pointed out that multispectral data with high resolution (<5 m) such as IKONOS (Allbed et al., 2014; Morgan et al., 2018) are more accurate for studying the severity levels of salinity in a small region (Dwivedi et al., 2008). Though, limited studies have been conducted in mapping the soil salinity in a farm region due to cost of high-resolution data (Singh et al., 2017).

Variations in the color and surface layer of saline soils during different seasons are easily detected by the satellite sensor and are a useful and direct approach for detecting soil salinity, providing high accuracy. Additionally, findings suggest that salinity in the root zone can be identified based on vegetation canopy reflectance. For example, when a crop is affected by the root zone salinity, the reflectance of the crop in the blue (B), green (G), and red (R) ranges of the electromagnetic spectrum and in the near-infrared (NIR) range is characterized by an increase and then a decrease, respectively (Scudiero et al., 2017). Furthermore, many studies indicate the usage of thermal band allowing for better distribution and detection of SAS (Metternicht & Zinck, 1996; Singh et al., 2017). In general, soils affected by salts are sensitive to reflected solar energy (400–2500 nm), to emitted thermal energy (8000–14,000 nm), or radiation at longer wavelengths known as microwaves (Farifteh et al., 2006). Metternicht and Zinck (1996) obtained results according to which the visible narrow bands performed well in distinguishing saline and nonsaline soils; the infrared range can provide good results in separating saline-sodic areas from sodic areas, and the thermal band is useful for separability between sodic and saline-sodic soils (Metternicht & Zinck, 2003). As spectral reflectance of the soils depends on different physicochemical properties of soils, it provides an alternative approach for applying a supervised or unsupervised classification of SAS based on moderate- and high-resolution satellite imagery. Different spectral bands of individual or combined multispectral channels such as false composite color (Abbas et al., 2013) and vegetation and salinity indices have been developed and used successfully used for the evaluation and monitoring of SAS using remote sensing data (Abbas & Khan, 2007; Daliakopoulos et al., 2016; Khan et al., 2005).

Alternatively, many image transformations of satellite data such as principal components have been used very efficiently for identifying the SAS (Abbas & Khan, 2007; Bouaziz et al., 2018; Khan et al., 2005).

Abuelgasim & Ammad (2019) proceeded to develop an empirical model for mapping and monitoring soil salinity in agricultural soils in dry and semiarid lands using regression analysis of soil spectral reflectance. In this research, the spectral channels from 2 to 7 and NIR and SWIR regions of the electromagnetic spectrum from Landsat 8 OLI satellite images, as well as sampled soil salinity values from the field survey, were mainly used. The NSI index was developed using the NIR-SWIR-SWIR2 spectral regions and provided very high performance (60% accuracy) in detecting soil salinity compared with the ground truth data. The NSI index can accurately identify soils with high salinity and excessive salt accumulation in the soil. The reflectance of saline seems to peak in the SWIR1 spectral region (1.57−1.65 μm). Soils affected by salinity show higher values of reflectance in the regions of the visible-to-infrared part of the spectrum compared with nonsaline soils, while in the SWIR2 spectral band (2.11−2.29 μm), the reflectance of soils affected by salinity is extremely reduced compared with nonsaline soils, indicating significant absorption (Abuelgasim & Ammad, 2019).

As soil salinity affects crop yield and vegetation health, several vegetation and salinity indices have been extensively used for prediction and mapping of SAS (Gorji et al., 2017, 2019). Later, Nguyen et al. (2020) conducted a study to assess the salinity in the Mekong Delta, Vietnam, using Landsat 8 OLI satellite imagery focusing on specific spectral regions of visible (green and red) and NIR. It was concluded that the NIR range can be sufficiently used to acquire valid results in salt extent delineation. They used the Vegetation Soil salinity Index (VSSI), Normalized Difference Vegetation Index (NDVI), Normalized Difference Salinity Index (NDSI), and Soil Adjusted Vegetation Index (SAVI). These indices are approved to be closely related to saline soil reflectance and vegetation health as well as in situ measurements of EC in the study area. Statistical analysis between the EC and satellite indices showed that spectral responses in the NIR region and VSSI were best correlated (r^2 0.89 and r^2 0.77, respectively). After comparison with EC measurements indicated the aforementioned spectral regions are suitable for the study of saline soils. NDVI, NDSI, SI4, and SAVI were also correlated with the measurements. In another study, Gorji et al. (2020) compared soil salinity indices regression results with EC measurement, derived from Landsat 8 OLI and Sentinel-2A satellite imagery, for assessing the SAS in the western part of Urmia Lake in Iran. In this study, three different soil salinity indices (S1, S2, and S3) were used and coupled with EC data. Both models obtained from linear and multiple regression analysis were approved to be acceptable and accurate with almost similar high r^2 values in both regression approaches. This study concluded that RS data of Sentinel-2A and Landsat 8 OLI imageries can be used for mapping soil salinity effectively. Therefore sensors with the same characteristic of multispectral and spatiotemporal data could also be used for soil salinity identification purposes (Gorji et al., 2020).

According to the aforementioned studies, the usage of RS indices to extract targeted spectral information from selected wavelengths is very useful for processing and analyzing the spatiotemporal dimension of the information. The combination of data of different attributes often provides more reliable information about the object of study or the properties of a surface. Through the computation of radiometric indices from spectral bands, images with specific features to highlight specific properties are generated. The properties of the soil become more distinct if they are combined with channels from other spectral regions selected according to the nature of the phenomenon and the object of study. However, even though multispectral satellite images with medium-to-high spatial resolution provide great potential and can be used for the identification and mapping of the extent of SAS, by providing qualitative and quantitative spatial information, the use of these data for the development of soil salinity prediction models is constrained by the limited number of bands. The development progress of hyperspectral imagers based on satellites or airborne has overcome the restrictions of limited spectral information of multispectral satellite data (Kumar et al., 2021).

5.3.1.3 Use of hyperspectral imaging

Although multispectral RS data can be used efficiently for delineation and examining the distribution of SAS, there is a restricted analytic ability due to low spatial resolution and confounding reflectance among different saline soils' properties (Singh, 2022). Hyperspectral RS imagers have overcome the limits of multispectral data due to high-spatiotemporal-resolution data (Pandey et al., 2021). All the studies discussed further indicate the overall hyperspectral RS data as most promising, which can provide high accuracy in the identification and mapping of soil salinity severity.

For example, Kumar et al. (2014) used RS satellite data of hyperspectral remote sensor (Hyperion EO-1) to generate the severity maps of SAS. Within this study, 6 (band 9, 20, 22, 28, 29, and 46) out of 148 spectral bands were pointed out as the most sensitive bands to the soil salinity indicators of pH, EC, SAR, and ESP. These bands were used for the generation of spectral indices related to soil salinization (such as NDVI, COSRI, NDSI, CI, BI, and SI). The high correlation of vegetation and salinity indices and soil salinity indicators showed a promising potential of hyperspectral data in characterizing and mapping different levels in SAS. Later, Qian et al. (2019) using hyperspectral data derived from a spectroradiometer proceed to the development of a linear model for the ascertainment of soil salinity content indicated that wavelengths at 1358 and 2382 nm can be used sufficiently as the most suitable wavebands for quantifying the accumulation of different predominant salts (chlorine and sulfate compounds) at Minqin Oasis. In another study, Mahajan et al. (2020) using regression techniques based on hyperspectral data and fieldwork samples estimated different properties of SAS (such as pH and EC) in a region of India. Their results indicated a downward tendency of spectral reflectance when the levels of salinity in SAS are increasing. In the same study, a conspicuous absorption was noticed in the region close to wavelengths of 427, 487, and 1917 nm; the absorption in these wavelengths varies with the salinity level in soil. Furthermore, results obtained from this study suggested that the VNIR (visible, NIR) spectroscopy has the potential to predict the properties of SAS.

5.3.2 Geographic information system applications in salt-affected soil monitoring and assessment

GIS is a key technology for analysis, processing storage, and retrieval of large geographical data, which is important for natural resources management and effective decision-making (Mandal et al., 2009; Tsatsaris et al., 2021).

The integration of GIS in SAS assessment allows incorporating a wealth of options in visualization, overlaying, processing, and transformation of spatial data. For example, Mandal and Sharma (2008, 2009) developed regional and zonal digital databases using GIS, describing the characteristics and soil salinity distribution in India. Later, Gorji et al. (2017) used GIS techniques and RS data to produce multitemporal maps of SAS, in Turkey's data-scarce region. GIS sufficiently contributed to the assessment of spatial variability and in understating and explaining different soils' properties due to the ability of storage and processing of different attribute data (Abdel-Fattah, 2020). The increasing need of monitoring the variability of SAS has led to the conception of an array of techniques that are putting together modeling, statistical analysis, and RS data. Many studies have demonstrated the significant added value of using GIS techniques in combination with RS data for surveying and mapping of soil salinity extent from field to regional scale (Allbed et al., 2014; Gorji et al., 2019; Singh, 2022; Singh et al., 2017). Different approaches of digital image classification and various GIS-based models of soil salinity have been developed using RS data, soil salinity indicators' properties (such as EC, ESP, and pH), and other environmental variables (Asfaw et al., 2018; Azabdaftari & Sunar, 2016; Yu et al., 2018).

The incorporation of RS data and techniques with ancillary data (field data, environmental factors, and spatial models) within GIS environment can enhance the ability to generate high-quality and accurate soil salinity maps (Allbed et al., 2014; Douaoui et al., 2006; Farifteh et al., 2006). Additionally, GIS techniques are commonly used for exploration and visual interpretation of the RS data, but an additional usage is the analysis and interpretation of data collected physically. Furthermore, the interfacing of GPS logger with GIS permits effective data input and direct access to GIS layers and spatial database on the field (Ojo & Ilunga, 2020).

5.4 Concluding remarks

SAS are a major threat around the world, which are formed consequential to anthropogenic activity and natural factors, mainly occurring in semiarid and arid regions and land under agricultural exploitation. Soil salinity is a dynamic phenomenon with various impacts on the environment, climate, and economy, especially in agricultural sector. The implementation of sustainable soil management will enhance crop yield productivity production and ensure food security. Hence, there is an urgent need for frequent detection of SAS spatial variability in local and regional scale for management processes and reclamation planning of SAS, particularly so in the context of climate change impact mitigation globally.

Geoinformation technology is considered as a great promising potential mean to access accurate information on spatiotemporal distribution of SAS and SOC worldwide. Geoinformation is an important aid in the protection of soils' quality and physical environment from climate change impacts and mitigation of those effects. The rapid technological development in geoinformation fields has revealed its potential as the way forward to address the SAS effects and inform today's policy- and decision-makers. Many researchers indicated the contribution of GIS in utilization of RS data in soil salinity assessment as a more accurately, rapidly, and quantitatively method (Kumar et al., 2021). This chapter aimed in providing an overview on soil salinity process with a briefly reference on its impacts on SOC process, a key soil parameter related to climate change mitigation. Within this framework, information has been given about various methods that have been developed for delineating SAS' distribution using combined modern tools of RS-based and GIS technology.

RS can be used sufficiently as an alternative for mapping the variability of soils' salinity using different approaches of process and analysis of RS images. However, using optical RS data for detection, the accurate extent of salinity in soils consists of some restrictions regarding accurate identification and spatial resolution. The direct detection of SAS is constrained by low spectral resolution of RS multispectral images, which limits the identification and quantification of distinct types of SAS. The use of hyperspectral RS imagers allows assessing SAS in more detail because of higher spatial and spectral resolution. Hyperspectral data have overcome these limits by being able to detect the special features associated with specific salt minerals (Singh et al., 2017). Nevertheless, hyperspectral data are marked by the drawback of low signal-to-noise proportion according to less strength that includes small wavebands (Singh, 2022). Furthermore, RS data are able to provide spatiotemporal information on the variability of SAS based on soil surface features reflectance and/or other soil characteristics such as vegetation health, moisture, surface roughness, and color. However, limited information can be extracted about the whole soil profile (Sethi et al., 2010) and characteristics of topsoil surface may not be representative of the deeper layers of the soil profile (Allbed et al., 2014). Therefore, using field measurements for assessment and validation is an effective approach for retrieving more reliable findings on SAS extents (Metternicht & Zinck, 2003). Thus a comprehensive approach for accurate identification and quantification of different variations in salts of soils would be the use of both RS and GIS technologies combined with ancillary data for the development of the appropriate model (Allbed et al., 2014; Singh et al., 2017).

References

Abbas, A., & Khan, S. (2007). Using remote sensing techniques for appraisal of irrigated soil salinity. In L. Oxley, & D. Kulasiri (Eds.), *Advances and applications for management and decision making land, water and environmental management: Integrated systems for sustainability MODSIM07* (pp. 632–2638). Modelling and Simulation Society of Australia and New Zealand.

Abbas, A., Khan, S., Hussain, N., Hanjra, M. A., & Akbar, S. (2013). Characterizing soil salinity in irrigated agriculture using a remote sensing approach. *Physics and Chemistry of the Earth, Parts A/B/C, 55–57*(January), 43–52. Available from https://doi.org/10.1016/J.PCE.2010.12.004.

Abdel-Fattah, M. K. (2020). A GIS-based approach to identify the spatial variability of salt-affected soil properties and delineation of site-specific management zones: A case study from Egypt. *Soil Science Annual, 71*(1), 76–85. Available from https://doi.org/10.37501/soilsa/121495.

Abdullah, A. Y. M., Biswas, R. K., Chowdhury, A. I., & Billah, S. M. (2019). Modeling soil salinity using direct and indirect measurement techniques: A comparative analysis. *Environmental Development, 29*, 67–80. Available from https://doi.org/10.1016/j.envdev.2018.12.007.

Abrol, I. P., Yadav, J. S. P., & Massoud, F. I. (1988). *Salt-affected soils and their management. Soil resources, management and conservation service*. FAO Land and Water Development Division. ISBN 92-5-102686-6.

Abuelgasim, A., & Ammad, R. (2019). Mapping soil salinity in arid and semi-arid regions using Landsat 8 OLI satellite data. *Remote Sensing Applications: Society and Environment, 13*, 415–425. Available from https://doi.org/10.1016/j.rsase.2018.12.010.

Allbed, A., & Kumar, L. (2013). Soil salinity mapping and monitoring in arid and semi-arid regions using remote sensing technology: A review. *Advance Remote Sensing, 2*, 373–385. Available from https://doi.org/10.4236/ars.2013.24040.

Allbed, A., Kumar, L., & Aldakheel, Y. Y. (2014). Assessing soil salinity using soil salinity and vegetation indices derived from IKONOS high-spatial resolution imageries: Applications in a date palm dominated region. *Geoderma, 230–231*, 1–8. Available from https://doi.org/10.1016/j.geoderma.2014.03.025.

Anand, A. R. K. M. Malhi, Srivastava, P. K., Singh, P., Mudaliar, A. N., Petropoulos, G. P., & Kiramn, C. S. (2021). Optimal band characterisation in reformation of hyperspectral indices for species diversity estimation. *Physics & Chemistry of the Earth*, 1030–1040. Available from https://doi.org/10.1016/j.pce.2021.103040.

Asfaw, E., Suryabhagavan, K. V., & Argaw, M. (2018). Soil salinity modeling and mapping using remote sensing and GIS: The case of Wonji sugar cane irrigation farm, Ethiopia. *Journal of the Saudi Society of Agricultural Sciences, 17*(3), 250–258. Available from https://doi.org/10.1016/j.jssas.2016.05.003.

Azabdaftari, A., & Sunar, F. (2016). Soil Salinity Mapping using multitemporal Landsat data. In *The international archives of the photogrammetry, remote sensing and spatial information sciences, Volume XLI-B7* (pp. 3–9). XXIII ISPRS Congress, 12–19 July, Prague. Available from https://doi.org/10.5194/isprs-archives-XLI-B7-3-2016.

Bouaziz, M., Chtourou, M. Y., Triki, I., Mezner, S., & Bouaziz, S. (2018). Prediction of soil salinity using multivariate statistical techniques and remote sensing tools. *Advances in Remote Sensing, 7*, 313–326. Available from https://doi.org/10.4236/ars.2018.74021.

Daliakopoulos, I. N., Tsanis, I. K., Koutroulis, A., Kourgialas, N. N., Varouchakis, A. E., Karatzas, G. P., & Ritsema, C. J. (2016). The threat of soil salinity: A European scale review. *The Science of the Total Environment*, *573*(December), 727–739. Available from https://doi.org/10.1016/J.SCITOTENV.2016.08.177.

Douaoui, A. E. K., Nicolasb, H., & Walter, C. (2006). Detecting salinity hazards within a semiarid context by means of combining soil and remote-sensing data. *Geoderma*, *134* (1-2), 217–230. Available from https://doi.org/10.1016/j.geoderma.2005.10.009.

Dwivedi, R., Kothapalli, R., & Singh, A. (2008). Generation of farm level information on salt-affected soils using IKONOS-II multispectral data. In D. G. Metternicht, & D. A. Zinck (Eds.), *Remote Sensing of Soil Salinization: Impact on Land Management* (1st ed.). CRC Press. Available from https://doi.org/10.1201/9781420065039.

Elatawneh, A., Kalaitzidis, C., Petropoulos, G. P., & Schneider, T. (2012). Evaluation of diverse classification approaches for land use/cover mapping in a Mediterranean region utilizing hyperion data. *International Journal of Digital Earth*, 1–23. Available from https://doi.org/10.1080/17538947.2012.671378.

Eswaran, H., van den Berg, E., Reich, P., & Kimble, J. (1995). Global soil carbon resources. In R. Lal, J. Kimble, E. Levine, & B. A. Steward (Eds.), *Soils and global change* (pp. 27–43). Available from https://www.routledge.com/Soils-and-Global-Change/Kimble-Levine-Stewart/p/book/9781566701181.

European Commission, Directorate-General for Environment. (2011). Soil: the hidden part of the climate cycle, Publications Office. Available from https://data.europa.eu/doi/10.2779/30669.

FAO. (1988). Salt-affected soils and their management. In *Soils bulletin* (39). Rome: FAO.

FAO. (2005). *Global network on integrated soil management for sustainable use of salt-affected soils*. Rome, Italy: FAO Land and Plant Nutrition Management Service. Available from http://www.fao.org/ag/AGL/agII/spush/.

FAO. (2021). Global map of salt-affected soils (GSASmap). Retrieved January 20, 2022. Available from https://www.fao.org/soilsportal/data-hub.

Farifteh, J., Farshad, A., & George, R. J. (2006). Assessing salt-affected soils using remote sensing, solute modeling, and geophysics. *Geoderma*, *130*(3-4), 191–206. Available from https://doi.org/10.1016/j.geoderma.2005.02.003.

Geeson, N., Brandt, J., & Imeson, A. (2003). A Desertification Indicator System for Mediterranean Europe (DIS4ME). Available from https://doi.org/10.1002/9781118649596.ch6.

Gorji, T., Sertel, E., & Tanik, A. (2017). Monitoring soil salinity via remote sensing technology under data scarce conditions: A case study from Turkey. *Ecological Indicators*, *74*, 384–391. Available from https://doi.org/10.1016/j.ecolind.2016.11.043.

Gorji, T., Tanik, A., & Sertel, E. (2015). Soil salinity prediction, monitoring and mapping using modern technologies. *Procedia Earth and Planetary Science*, *15*, 507–512. Available from https://doi.org/10.1016/j.proeps.2015.08.062.

Gorji, T., Yildirim, A., Hamzehpour, N., Tanik, A., & Sertel, E. (2020). Soil salinity analysis of Urmia Lake basin using Landsat-8 OLI and Sentinel-2A based spectral indices and electrical conductivity measurements. *Ecological Indicators*, *112*, 106–173. Available from https://doi.org/10.1016/j.ecolind.2020.106173.

Gorji, T., Yildirim, A., Sertel, E., & Tanik, A. (2019). Remote sensing approaches and mapping methods for monitoring soil salinity under different climate regimes. *International Journal of Environment and Geoinformatics*, *6*(1), 33–49. Available from https://doi.org/10.30897/IJEGEO.500452.

Horneck, D. A., Ellsworth, J. W., Hopkins, B. G., Sullivan, D. M., & Stevens, R. G. (2007). Managing salt-affected soils for crop production. https://ir.library.oregonstate.edu/concern/administrative_report_or_publications/pr76f371k.

Ivushkin, K., Bartholomeus, H., Bregt, A. K., Pulatov, A., Kempen, B., & De Sousa, L. (2019). Global mapping of soil salinity change. *Remote Sensing of Environment*, *231*, 111260. Available from https://doi.org/10.1016/J.RSE.2019.111260.

Khan, N. M., Rastoskuev, V. V., Sato, Y., & Shiozawa, S. (2005). Assessment of hydrosaline land degradation by using a simple approach of remote sensing indicators. *Agricultural Water Management*, *77*(1–3), 96–109. Available from https://doi.org/10.1016/j.agwat.2004.09.038.

Koshal, A. K. (2012). *Satellite image analysis of salinity areas through GPS. Remote sensing and GIS*. ACS Publications.

Kosmas, C., Danalatos, N. G., & Gerontidis, S. (2000). The effect of land parameters on vegetation performance and degree of erosion under Mediterranean conditions. *Catena*, *40*, 3–17. Available from https://doi.org/10.1016/S0341-8162(99)00061-2.

Kumar, N., Singh, S. K., Reddy, G. P. O., Mishra, V. N., & Bajpai, R. K. (2021). Remote sensing applications in mapping salt affected soils. *Agricultural Reviews*. Available from https://doi.org/10.18805/ag.R-2008.

Kumar, S., Gautam, G., & Saha, S. K. (2014). Hyperspectral remote sensing data derived spectral indices in characterizing salt-affected soils: a case study of Indo-Gangetic plains of India. *Environmental Earth Sciences*, *73*(7), 3299–3308. Available from https://doi.org/10.1007/s12665-014-3613-y.

Lal, R. (2001). Potential of desertification control to sequester carbon and mitigate the greenhouse effect. *Climate Change*, *51*, 35–72. Available from https://doi.org/10.1023/A:1017529816140.

Lal, R. (2008). Carbon sequestration. *Philosophical Transactions of the Royal Society B: Biological Sciences*, *363*, 815–830. Available from https://doi.org/10.1098/rstb.2007.2185.

Lamine, S., Petropoulos, G. P., Brewer, P. A., Bachari, N. I., Srivastava, P. K., Manevski, K., Kalaitzidis, C., & Macklin, M. G. (2019). Heavy metal soil contamination detection using combined geochemistry and field spectroradiometry in the United Kingdom. *Sensors MDPI*, *19*, 762. Available from https://doi.org/10.3390/s19040762.

Mahajan, G., Das, B., Gaikwad, B., Desai, A., Morajkar, S., Patel, K., Kulkarni, R., & Murgaonkar, D. (2020). Using hyperspectral remote sensing to monitor the properties of salt-affected soils. *Authorea*. Available from https://doi.org/10.22541/au.158212912.27052084.

Mandal, A. K., & Sharma, R. C. (2008). Computerized database on salt affected soils in western and central India using GIS. *Geocarto International*, *23*(5), 373–391. Available from https://doi.org/10.1080/10106040701207589.

Mandal, A. K., Sharma, R. C., & Singh, G. (2009). Assessment of salt affected soils in India using GIS. *Journal GeocardoInternational*, *24*(6), 437–456. Available from https://doi.org/10.1080/10106040902781002.

Manevski, K., Manakos, I., Petropoulos, G. P., & Kalaitzidis, C. (2012). Spectral discrimination of Mediterranean maquis and phrygana vegetation: Results from a case study in Greece. *IEEE Journal of Selected Topics in Applied Earth Observations and Remote Sensing*, *5*(2), 604–616. Available from https://doi.org/10.1109/JSTARS.2012.2190044.

Mashimbye, Z. E. (2013). Remote sensing of salt-affected soil. In *Ph.D. Thesis* (p. 151) South Africa: Faculty of AgriSciences, Stellenbosch University.

Metternicht, G. I., & Zinck, J. A. (1996). Modelling salinity – alkalinity classes for mapping salt-affected topsoils in the semi-arid valleys of Cochabamba (Bolivia). *ITC Journal*, *1996-2*, 125–135.

Metternicht, G. I., & Zinck, J. A. (2003). Remote sensing of soil salinity: Potentials and constraints. *Remote Sensing of Environment*, *85*(1), 1–20. Available from https://doi.org/10.1016/S0034-4257(02)00188-8.

Morgan, R. S., Abd El-Hady, M., & Rahim, I. S. (2018). Soil salinity mapping utilizing sentinel-2 and neural networks. *Indian Journal of Agricultural Research*, *52*, 524–529. Available from https://doi.org/10.18805/IJARe.A-316.

Mougenot, B. M., Pouget., & Epema, G. F. (1993). Remote sensing of salt affected soils. *Remote Sensing Reviews*, *7*(3-4), 241–259. Available from https://doi.org/10.1080/02757259309532180.

Nguyen, A. K., Liou, A. Y., Tran, P. H., Hoang, P. P., & Nguyen, H. T. (2020). Soil salinity assessment by using near infrared channel and Vegetation Soil Salinity Index derived from Landsat 8 OLI data: A case study in the TraVinh Province, Mekong Delta, Vietnam. *Progress in Earth and Planetary Science*, *7*, 1. Available from https://doi.org/10.1186/s40645-019-0311-0.

Ojo, O. I., & Ilunga, M. F. (2020). *Use of geoinformatics techniques for the assessment and mapping of soil salinity: Concepts and applications. Natural resources management and biological sciences*. IntechOpen. Available from http://doi.org/10.5772/intechopen.92443.

Omuto, C. T., Vargas, R. R., El Mobarak, A. M., Mohamed, N., Viatkin, K., & Yigini, Y. (2020). Mapping of salt-affected soils: Technical manual. Rome, FAO. Available from https://doi.org/10.4060/ca9215en.

Pandey, M. K., Srivastava, P. K., & Petropoulos, G. P. (2021). Future pathway for research and emerging applications in GPS/GNSS. In G. P. Petropoulos, & P. K. Srivastava (Eds.), *GPS and GNSS technology in geosciences* (pp. 429–436). Elsevier, ISBN: 9780128186176.

Pandey, P. C., Koutsias, N., Petropoulos, G. P., Srivastava, P. K., & Dor, E. B. (2019). Land use/land cover in view of Earth Observation: Data sources, input dimensions and classifiers—A review of the state of the art. *Geocarto International*, *36*(3), 1–38, https//doi.org/10.1080/10106049.2019.162964.

Pandey, P. C., Srivastava, P. K., Bhattacgarya, B., & Petropoulos, G. P. (2020). *Hyperspectral remote sensing: Theory & applications*. Elsevier, ISBN: 978-0-08-102894-0.

Paustian, K., Lehmann, J., Ogle, S., Reay, D., Philip Robertson, G., & Smith, P. (2016). Climate-smart soils. *Nature*, *532*, 49–57. Available from https://doi.org/10.1038/nature17174.

Perri, S., Suweis, S., Entekhabi, D., & Molini, A. (2018). Vegetation controls on dryland salinity. *Geophysical Research Letters*, *45*, 11669–11682. Available from https://doi.org/10.1029/2018GL079766.

Petropoulos, G. P., Kalivas, D. P., Georgopoulou, H., & Srivastava, P. K. (2015). Urban vegetation cover extraction from hyperspectral remote sensing imagery & GIS spatial analysis techniques: The case of Athens, Greece. *Journal of Applied Remote Sensing*, *9*, 1–18. Available from https://doi.org/10.1117/1.JRS.9.096088.

Petropoulos, G. P., Arvanitis, K., & Sigrimis, N. (2012). Hyperion hyperspectral imagery analysis combined with machine learning classifiers for land use/cover mapping. *Expert Systems with Applications*, *39*, 3800–3809. Available from https://doi.org/10.1016/j.eswa.2011.09.083.

Qian, S. E. (2020). *Hyperspectral satellites and system design* (1st ed.). CRC Press. Available from https://doi.org/10.1201/9780429266201.

Qian, T., Tsunekawa, A., Peng, F., Masunaga, T., Wang, T., & Li, R. (2019). Derivation of salt content in salinized soil from hyperspectral reflectance data: A case study at Minqin Oasis, Northwest China. *Journal of Arid Land*, *11*(1), 111–122. Available from https://doi.org/10.1007/s40333-019-0091-9.

Rengasamy, P. (2008a). *Soil salinity and sodicity. Growing crops with reclaimed and undefined*. Cabdirect.Org Accessed December 30, 2021. Available from https://www.cabdirect.org/cabdirect/abstract/20063129536.

Rengasamy, P. (2008b). Salinity in the landscape: A growing problem in Australia. *Geotimes*, *53*, 34–39. Available from https://hdl.handle.net/2440/53569.

Richards, L. (1954). Diagnosis and improvement of saline and alkali soils. *Soil Science*, *78* (2), 154. Available from https://doi.org/10.1097/00010694-195408000-00012.

Scudiero, E., Corwin, D. L., Anderson, R. G., Yemoto, K., Clary, W., Wang, Z. L., & Skaggs, T. H. (2017). Remote sensing is a viable tool for mapping soil salinity in agricultural lands. *California Agriculture*, 71. Available from https://doi.org/10.3733/ca.2017a0009.

Sethi, M., Bundela, D. S., Sethi, M., Lal, K., & Kamra, S. K. (2010). Remote sensing and geographic information system for appraisal of salt-affected soils in India. *Journal of Environmental Quality*, *39*(1), 5–15. Available from https://doi.org/10.2134/jeq2009.0032.

Setia, R., Gottschalk, P., Smith, P., Marschner, P., Baldock, J., Setia, D., & Smith, J. (2013). Soil salinity decreased global soil organic carbon stocks. *The Science of the Total Environment*, *465*, 267–272. Available from https://doi.org/10.1016/j.scitotenv.2012.08.028.

Setia, R., Marschner, P., Baldock, J., Chittleborough, D., & Verma, V. (2011). Relationships between carbon dioxide emission and soil properties in salt-affected landscapes. *Soil Biology and Biochemistry*, *43*, 667–674. Available from https://doi.org/10.1016/j.soilbio.2010.12.004.

Shahid, S. A., Zaman, M., & Heng, L. (2018). *Soil salinity: Historical perspectives and a world overview of the problem. Guideline for salinity assessment, mitigation and adaptation using nuclear and related techniques*. Springer. Available from https://doi.org/10.1007/978-3-319-96190-3_2.

Singh, A. (2022). Soil salinity: A global threat to sustainable development. *Soil Use and Management*, *38*, 39–67. Available from https://doi.org/10.1111/sum.12772.

Singh, R., Setia, P., Arora, R., Verma, S., Kumar, V., & Pateriya, B., P. (2017). Satellite remote sensing of salt-affected soils: Potential and limitations. *Journal of Soil and Water Conservation*, *16*(2), 97–107. Available from https://doi.org/10.5958/2455-7145.2017.00015.7.

Szabolcs, I. (1974). *Salt affected soils in Europe*. Springer. Available from https://doi.org/10.1007/978-94-011-942

Trnka, M., Kersebaum, K. C., Eitzinger, J., Hayes, M., Hlavinka, P., Svoboda, M., Dubrovský, M., Semerádová, D., Wardlow, B., Pokorný, E., Možný, M., Wilhite, D., & Žalud, Z. (2013). Consequences of climate change for the soil climate in Central Europe and the central plains of the United States. *Climatic Change*, *120*(1–2), 405–418. Available from https://doi.org/10.1007/s10584-013-0786-4.

Tsatsaris, A., Kalogeropoulos, K., Stathopoulos, N., Louka, P., Tsanakas, K., Tsesmelis, D. E., Krassanakis, V., Petropoulos, G. P., Pappas, V., & Chalkias, C. (2021). Geoinformation technologies in support of environmental hazards monitoring under climate change: An extensive review. *ISPRS International Jouranl of Geo-Information*, *10*(94), 1–32. Available from https://doi.org/10.3390/ijgi10020094.

Ullman, J. L. (2013). March 26 *Strategies for minimizing salinity problems and optimizing crop production*. Hastings, FL: In-Service Training.

Van Weert, F., van der Gun, J., & Reckman, J. (2009). Global overview of saline groundwater occurrence and genesis, IGRAC (International Groundwater Resources Assessment Centre), Utrecht, The Netherlands. http://www.un-igrac.org/downloads.

Wang, F., Yang, S., Yang, W., Yang, X., & Jianli, D. (2019). Comparison of machine learning algorithms for soil salinity predictions in three dryland oases located in Xinjiang Uyghur Autonomous Region (XJUAR) of China. *EU Journal of Remote Sensing*, *52*(1), 256–276. Available from https://doi.org/10.1080/22797254.2019.1596756.

Wendland, F., Blum, A. A., Coetsiers, A. M., Gorova, A. R., Griffioen, A. J., Grima, A. J., Hinsby, A. K., et al. (2008). European aquifer typology: A practical framework for an overview of major groundwater composition at European scale. *Environmental Geology*, *55*, 77–85. Available from https://doi.org/10.1007/s00254-007-0966-5.

Wicke, B., Smeets, E., Dornburg, V., Vashev, B., Gaiser, T., Turkenburg, W., & Faaij, A. (2011). The global technical and economic potential of bioenergy from salt-affected soils. *Energy & Environmental Science*, *4*(8), 2669–2681. Available from https://doi.org/10.1039/C1EE01029H.

Wong, V. N., Greene, R. S. B., Murphy, B., & Dalal, R. (2004). The effects of salinity and sodicity on soil carbon turnover. In 3rd Australian/New Zealand Soils Conference, 5th-9th December.

Wong, V. N., Greene, R. S., Murphy, B. W., Dalal, R., Mann, S., & Farquhar, G. (2006). The effects of salinity and sodicity on soil organic carbon stocks and fluxes: An overview. *Regolith Consolidation and Dispersion of Ideas*, *7*, 367–371.

Yossif, T. M. H., & Ebied, G. M. (2016). Monitor of changes of land uses and soil characteristics at North West area of Sinai using satellite imagery, Egypt. *Egyptian Journal of Soil Science.*, *56*(3), 471–487. Available from https://doi.org/10.21608/EJSS.2016.3245.

Yu, H., Mingyue, L., Baojia, D., Zongming, W., Liangjun, H., & Bai, Z. (2018). Mapping soil salinity/sodicity by using Landsat OLI imagery and PLSR algorithm over semiarid West Jilin Province, China. *Journal Sensors*, *18*, 4. Available from https://www.mdpi.com/1424-8220/18/4/1048.

CHAPTER

6 Soil organic carbon sequestration potential dynamics in saline and sodic soils in Greece

Dimitris Triantakonstantis and Spyridon E. Detsikas

Department of Soil Science of Athens, Institute of Soil and Water Resources, ELGO—DIMITRA, Lycovrisi, Attiki, Greece

6.1 Introduction

Salinity and sodicity degrade about 932 million ha of land worldwide. Of this area, salinity affects 23% of arable land, while saline-sodic soils affect a further 10% (Szabolcs, 1989). In Europe, salt-affected soils are situated at altitudes up to 500 m in both arid and humid climates (Szabolcs, 1996). In Greece, salt-affected soils are formed (1) in dry and semiarid conditions, (2) in flat low-elevated areas, (3) in seaside areas, (4) in estuaries or river deltas, (5) in areas with shallow level of groundwater, (6) in low-hydraulic-conductivity soils, and (7) in soils where their irrigation contains high concentration in soluble salts. In Greece, the exact extent of salt-affected soils is still unknown. According to the Greek literature, their extent is about 100,000–150,000 ha (Dellios, 1989). These areas are located in different climate, soil, and topographic conditions. Koukoulakis et al. (2000) estimated the saline areas to 35,389 ha, the alkaline ones to 17,819 ha, and the saline-alkaline areas to 34,917 ha (total 88,123 ha). In saline-sodic soils, the aggregated salt concentration directly affects the osmotic dynamics, thus hindering plant growth. In particularly, the increased salinity decreases the activity of soil microorganisms and consequently affects the soil organic carbon (SOC) decomposition (Omuto et al., 2020; Panagos et al., 2012; Setia et al., 2013). With soil containing more than the half of the global carbon pool, salinity and sodicity regulate the SOC levels and as a result, the soil's greenhouse gas emissions (Ardon et al., 2016).

The objective of this chapter is to compare the content of the current (2020) and the projected (2040) SOC levels with different salinity levels. In this scope, the Greek Saline-Sodic soils Map (FAO, 2021a) is compared with the SOC Map of Greece for 2020 (FAO, 2021c) and the projected SOC map of 2040 (FAO, 2021b).

Remote Sensing in Precision Agriculture. DOI: https://doi.org/10.1016/B978-0-323-91068-2.00022-9

6.2 Materials and methods

6.2.1 Study area

Greece was selected as the investigated area of this chapter. Greece is in the south-eastern part of Mediterranean and Europe between 34 and 42 parallel and 19–28 East with a total area of 130,049 km^2. Greece's climate can be defined as Mediterranean with cool wet winters and hot dry summers. Eleftheriou et al. (2018) have pointed out that different climate variations were created due to complex topography of the country. Moreover, the extensive coastline has a significant impact on Greek climate.

6.2.2 Input data

6.2.2.1 Greek National Map of salt-affected soils

The Greek National Map of salt-affected soils (0–30 cm of the surface soil) was created by combining georeferenced data from the laboratory analyses of the Institute of Soil and Water Resources of ELGO-DIMITRA for the years 2015–2020 and the Greek National Soil Map following the proposed methodology of Omuto et al. (2020). The FAO methodology uses Digital Soil Mapping (DSM) techniques to predict and map the soil property under consideration. The DSM methodological framework combines ground truthing soil property data, in this case pH, electrical conductivity, with environmental variables in raster format and multispectral satellite imagery to calculate the soil property within the boundaries of the study area.

The data used were land-use data from ESA, the GDEM digital soil model, topographic indicators generated from the DEM model (slope, valley depth, and longitudinal curvature), and 14 spectral indices from long-term means of multispectral MODIS images (Table 6.1). Principal component analysis was also

Table 6.1 Spectral indices and their formulas.

Spectral indices	Formula
Normalized salinity index	NSI = (SWIR1–SWIR2)/(SWIR1–NIR)
Salinity index	$SI = \sqrt{Green \times Red}$
Soil-adjusted vegetation index	SAVI = (NIR–Red)/((NIR + Red + 0.5) × 1.5)
Normalized difference vegetation index	NDVI = (NIR–Red)/(NIR + Red)
Canopy response salinity index	$CRSI = \sqrt{\frac{NIR \times Red - Green \times Blue}{NIR \times Red + Green \times Blue}}$
Brightness index	$BI = \sqrt{Red^2 + NIR^2 + Gr^2}$
Salinity ratio	SR = (Green–Red)/(Blue + Red)
Vegetation soil salinity index	VSSI = (2 × Green)–5 × (NIR + Red)
Normalized difference salinity index	NDSI = (Red–NIR)/(Red + NIR)

applied among the spectral indices, and principal components were selected. The prediction of values is carried out using prediction models such as machine learning and data mining. The raster data set was combined with the soil points for each considered property separately, and the most accurate model was selected.

The methodology of mapping of salt-affected soils (Omuto et al., 2020) was applied to produce the final Greek national pap of salt-affected soils. The Greek national map of salt-affected soils is given in Fig. 6.1 (FAO, 2021a). Soils with moderate salinity cover the larger area of croplands. Strong salinity soils seem to exist in areas near the coastline, while slight sodicity problems appear in Greek islands. In Fig. 6.2, the areas of different classes of salt-affected soils are given.

6.2.2.2 Greek National Map of Soil Organic Carbon in 2020

The Greek National Soil Organic Carbon Map was created from georeferenced data of the laboratory analyses of the Institute of Soil and Water Resources of ELGO Demetra following the proposed methodology of the World Organization for Agriculture and Food. The creation of the National Soil Organic Carbon Map was carried out in the framework of the cooperation of the Institute of Soil Science of ELGO-DIMITRA with FAO as a member of the Global Soil Partnership, where the Institute is the National Focal Point.

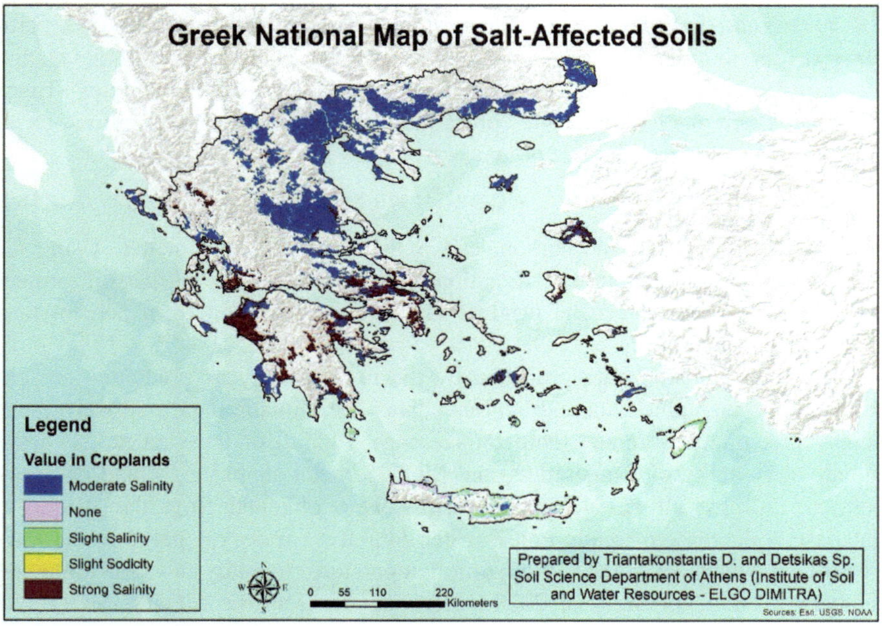

FIGURE 6.1

The Greek National Map of salt-affected soils.

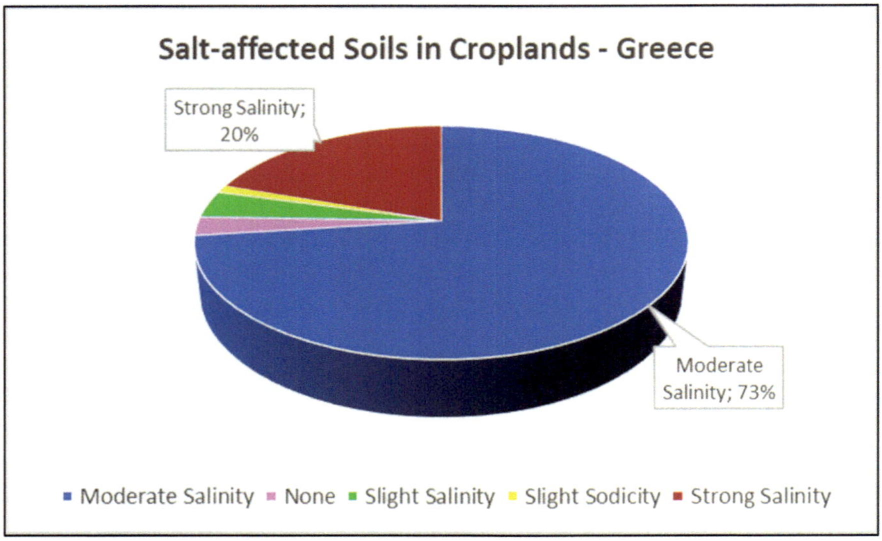

FIGURE 6.2

Proportions of different classes of salt-affected soils in croplands.

Soil data were collected during the period 2015–2020 from the topsoil (0–30 cm) and describe the main physical and chemical soil properties (pH, particle size distribution, electrical conductivity at paste saturation, organic matter, etc.). In the laboratory data, organic matter is calculated by the Walkley–Black wet oxidation method, so the proposed Van-Bemmelen conversion factor Eq. (6.1) was used to calculate organic carbon.

$$SOC = SOM \times 0,58 \tag{6.1}$$

where SOC is soil organic carbon, and SOM is soil organic matter. Using the Honey–Ratkowsky pedotransfer function, the apparent density of the soil samples was calculated to convert the original units of measurement from percent to g SOC per kg soil.

The environmental data proposed by the methodology as predictors are climatic and topographic data and land-use data. The climatic data used were mean annual precipitation, annual temperature range, rainfall of the wettest and driest month, seasonality of temperature, and MODIS mean annual night and day temperature. Almost all environmental factors were obtained from the Wordclim database, a database of ready-to-use raster data at a variety of spatial resolutions. For the annual mean night and daytime temperature, monthly day and night soil temperature data for the year 2020 were acquired from the MODIS MOD11C3 V.6 product, and then the annual mean was calculated.

In addition, the topographic data used include the MERIT digital elevation model. The MERIT DEM was chosen due to the absence of various noise sources.

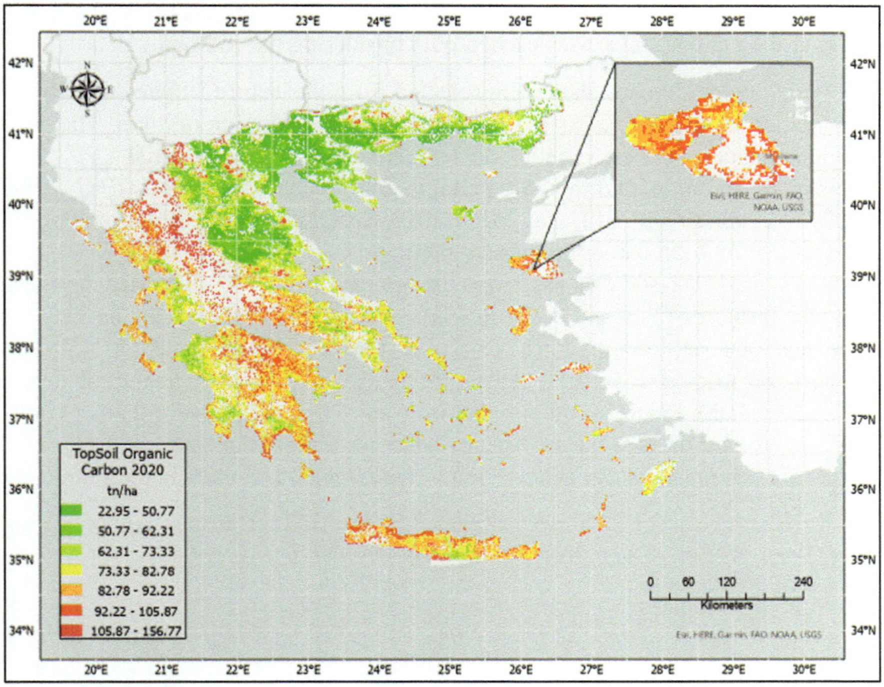

FIGURE 6.3

The Greek National Map of soil organic carbon.

Using SAGA software, the topographic data of slope, valley depth, wetness index, and multiresolution index of valley bottom flatness were produced. Finally, land-use data from the climate change initiative land cover of the European Space Agency were used.

The proposed prediction models used to predict SOC were Regression-Kriging, Random Forests (RF), and Support Vector Machines. RF yielded the highest accuracy in the evaluation indicators used. Fig. 6.3 gives the Greek National SOC Map.

6.2.2.3 Greek National Map of Soil Organic Carbon in 2040

The Soil Organic Carbon Sequestration Potential National Map for the year 2040 was developed, as mentioned earlier, by the Institute of Soil and Water Resources of ELGO-DIMITRA in the framework of its participation in the efforts of FAO and GSP for global mapping.

The methodology followed to generate the maps is the proposed FAO methodology as described in the Technical Manual for Carbon Sequestration (FAO, 2020). The RothC model and its R programming language version were applied to calculate the different organic carbon fractions. The application of the RothC model at the national level was carried out considering the following data:

- *Climate data*: Climatic data were selected from the Terra Climate database, which provides climate raster files with a spatial resolution of 4 km. The climate data used were monthly air temperature, monthly potential evapotranspiration, and monthly precipitation for the time periods 1981–1990, 1991–2000, 2000–2011, and 2011–2018. Still, the spatial resolution was redefined to 1 km using the bilinear method.
- *Terrain data*: The soil data used were the National Organic Carbon Map developed by the Institute of Soil and Water Resources of ELGO-DIMITRA and the clay content, obtained from the ISRIC databases at a spatial resolution of 250 × 250 m (a bilinear method was used to downscale the data to a 30-arc-second resolution).
- *Land-use data*: Finally, land-use data from the Climate Change Initiative Land Cover of the European Space Agency were used. From the land-use data, specific categories of agricultural soils were selected and used as the study area.

The SOC stocks for 2040 were obtained by the BAU (Business As Usual) scenario (FAO, 2021b). The corresponding map is given in Fig. 6.4.

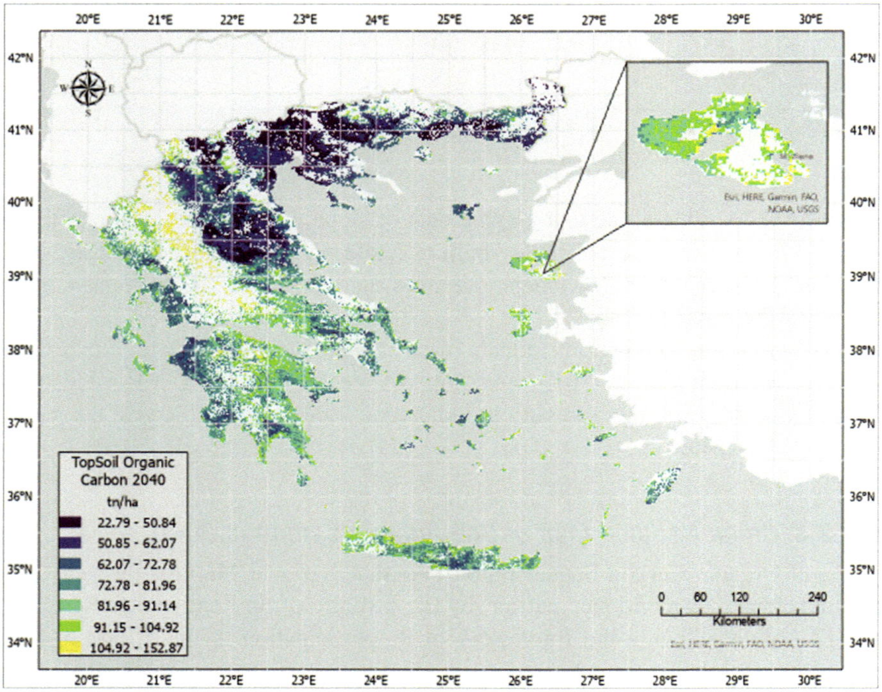

FIGURE 6.4

The projected soil organic carbon in 2040.

6.2.3 Methodology

To investigate and compare the relationship, statistical analysis was conducted between SOC, 2040 predicted SOC, and the salt-affected soil classes. The statistical parameters that were estimated were the following:

1. Descriptive statistics and boxplots
2. Spearman's ρ correlation coefficient

   Spearman's ρ is a non-parametric correlation test, and the formula of Spearman's ρ correlation coefficient is the following:

$$r_s = 1 - \frac{6\Sigma d^2}{n(n^2 - 1)} \quad (6.2)$$

3. Kendall's τ B correlation coefficient

   Kendall's τ B correlation coefficient is another non-parametric test for evaluating the strength of association between two variables measured on at least an ordinal scale. The equation of Kendall's τ B correlation coefficient is the following:

$$t_b = \frac{P - Q}{\sqrt{(P + Q + X_o) \times (P + Q + Y_o)}} \quad (6.3)$$

6.3 Results and discussion

The results showed a significant correlation between SOC and the saline-sodic soils of Greece. In Table 6.2, the results of correlation analysis are presented.

The results of the correlation coefficients indicate a negative correlation between the salinity and the predicted SOC of 2020 and 2040. The correlation coefficients of the predicted SOC in 2040 under a BAU scenario appear to be lower than the ones of the predicted SOC in 2020. At a significance level of 0.001, both the correlation coefficients Spearman's ρ and Kendall's τ B indicate the importance of the relationship between SOC and the different classes of salt-affected soils, which is obviously shown in the boxplots of Figs. 6.5 and 6.6.

Table 6.2 Correlation analysis between the different classes of salt-affected soils for SOC in 2020 and 2040.

Correlation coefficient	SOC in 2020	SOC in 2040
Spearman's ρ	−0.754	−0.335
P-value	<.001	<.001
Kendall's τ B	−0.563	−0.263
P-value	<.001	<.001

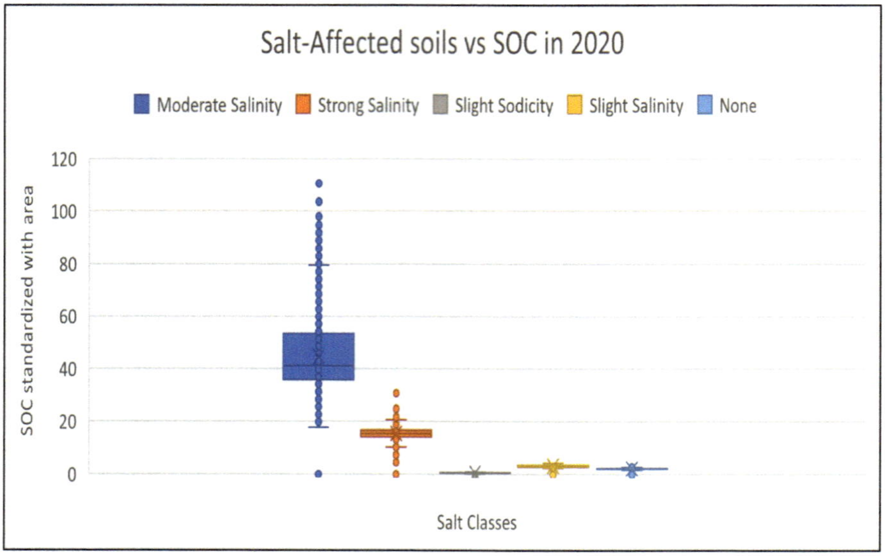

FIGURE 6.5

Predicted SOC of 2020 in different classes of salt-affected soils.

More specifically, the areas with Moderate Salinity appear to have more SOC compared with those of strong salinity. It can also be concluded from these boxplots that the salts decrease the available organic carbon in soils. The other classes have no interesting results due to small areas covered. Moreover, a slight decrease in all classes was presented, comparing the two SOC content predictions. The higher decrease was observed in the strong salinity class, with mean SOC reduced to 9.84 from 15.19 tn/ha. Table 6.3 presents the descriptive statistics of each class.

6.4 Conclusions

In this study, the relationship between different salt-affected soils and current and future SOC content is examined with the use of geographic information system. Our results depicted a strong negative correlation between saline soils and their SOC content.

The current study can be used as an important tool for the decision-makers and policy stakeholders, assisting them in sustainable management. The sequestrated carbon in the soil not only increases the SOC but also improves the soil quality and productivity. Moreover, it is important for agricultural policies to know that salt-affected soils in Greece, as the Greek

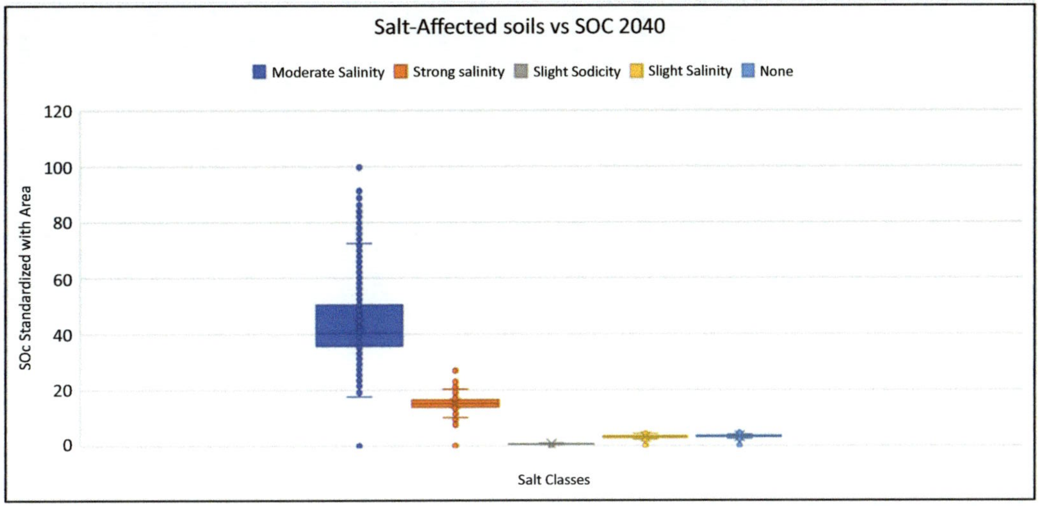

FIGURE 6.6

Predicted SOC of 2040 in different classes of saline soils.

Table 6.3 Descriptive statistics of predicted SOC in 2020 and 2040.

	Moderate salinity	Strong salinity	Slight salinity	Slight sodicity	None
SOC in 2040					
Mean	41.46	9.84	1.67	0.30	1.27
SD	22.27	7.42	1.59	0.30	1.13
Maximum	94.15	25.54	4.36	1.11	2.94
SOC in 2020					
Mean	43.46	15.19	0.57	3.12	3.32
SD	10.33	2.03	0.13	0.38	0.31
Maximum	101.48	4.6	4.28	1.12	26.97

National Map created by FAO (2021a) showed, need to appropriately be treated to sequestrate more organic carbon and mitigate climate change. According to FAO (2021b), the wide adoption of sustainable soil practices could mitigate about 14%–56% of current national agriculture emissions. Farmers should apply sustainable agricultural practices, which are considered necessary and helpful to this direction, and therefore, strong motivations should be given.

The agricultural practices for managing salt-affected soils can be focused on the draining systems, salt leaching, suitable fertilizers (with low salt index), and salt resilient cultivations (e.g., barley and cotton). Usual check for capillary ascension of salt from lower horizons and stern scrutiny of pumping underground water from aquifers that contact with seawater are some of the actions the Greek National Plan for Combating Desertification 19 (Greek National Committee for Combating Desertification, 2001) suggests.

Greek agriculture areas do not appear to be under major threat due to saltiness. Nevertheless, according to the scenarios of global warming (raise between 1.5°C and 2°C, RCP 2.6 model, Paris Climate Agreement), the current situation is expected to change with unpredictable consequences. Moreover, the need for healthy soils for intense cultivation is a necessity because of the increase of global population. Therefore, applying sustainable agricultural management practices in salt-affected soils is more urgent than ever.

The study can be seen under the policy context set by the forthcoming Common Agricultural Policy whose eco-schemes and especially carbon farming will be among its major priorities. The coupling of agricultural and climate objectives seems beyond a necessity as a feasible. The exploitation of mitigation potentials of agriculture can contribute to the achievement of European Union climatic targets while resulting in substantial co-benefits of soil formation and health, sustainable agriculture, food security, and rural development (Bithas & Latinopoulos, 2021).

References

Ardon, M., Helton, A. M., & Bernhardt, E. S. (2016). Drought and saltwater incursion synergistically reduce dissolved organic carbon export from coastal freshwater wetlands. *Biogeochemistry*, *127*(2–3), 411–426.

Bithas, K., & Latinopoulos, D. (2021). Managing tree-crops for climate mitigation. An economic evaluation trading-off carbon sequestration with market goods. *Sustainable Production and Consumption*, *27*, 667–678.

Dellios, M. (1989). *Long term effects of intensive cultivation and irrigation on physical and chemical properties of saline and alkali soils*. Meeting held in Sofia, Bulgaria, October.

Eleftheriou, D., Kiachidis, K., Kalmintzis, G., Kalea, A., Bantasis, C., Koumadoraki, P., Spathara, M. E., & Tsolaki, A. (2018). Determination of annual and seasonal daytime and nighttime trends of MODIS LST over Greece – Climate change implications. *Science of the Total Environment*, *616*, 937–947.

FAO. (2020). *Technical specifications and country guidelines for Global Soil Organic Carbon Sequestration Potential Map (GSOCseq)*.

FAO. (2021a). *Salt-affected soils in Greece*. http://54.229.242.119/GloSIS/.

FAO. (2021b). *Soil organic carbon sequestration potential national map*. http://54.229.242.119/GloSIS/.

FAO. (2021c). *Green National Map of soil organic carbon*. http://54.229.242.119/GloSIS/.

Greek National Committee for Combating Desertification. (2001). *Greek National Plan for Combating Desertification*.

Koukoulakis, P.H., Simonis, A.D., Gertsis, A.C., Paschalidis, Chr, & Rigas, G. (2000). Maps of the salt-affected soils of Greece and basic principles of their fertilization. In *Proceedings of the 8th Hellenic Soil Congress*, Kavala, pp. 178–189.

Omuto, C. T., Vargas, R. R., El Mobarak, A. M., Mohamed, N., Viatkin, K., & Yigini, Y. (2020). *Mapping of salt-affected soils: Technical manual*. FAO.

Panagos, P., Van Liedekerke, M., Jones, A., & Montanarella, L. (2012). European Soil Data Centre: Response to European policy support and public data requirements. *Land Use Policy*, *29*(2), 329–338.

Setia, R., Gottschalk, P., Smith, P., Marschner, P., Baldock, J., Setia, D., & Smith, J. (2013). Soil salinity decreases global soil organic carbon stocks. *Science of the Total Environment*, *465*, 267–272.

Szabolcs, I. (1989). *Salt-affected soils*. CRC Press, 274 pp.

Szabolcs, I. (1996). An overview on soil salinity and alkalinity in Europe. In N. Misopolinos, & I. Szabolcs (Eds.), *Soil salinization and alkalization in Europe* (pp. 1–12). ESSC Editions.

CHAPTER

7

The latest applications of remote sensing technologies for soil management in precision agriculture practices

S.B. Hareesh

Division of Agricultural Chemistry & Soil Science, Central Coffee Research Institute, Coffee Research Station, Chikkamagaluru, Karnataka, India

7.1 Introduction

Successful agriculture production system is a consequence of a composite interaction of seed, soil, water and various agrochemical inputs. Soil is a thin layer over the earth surface and accomplishes many processes essential to life. It also serves as a medium for supporting plant growth, as a nutrient reservoir, and as the abode for many biological processes involved in decomposition and recycling of plant and animal products (Wienhold et al., 2006). Soil integrates, transforms, stores, and filters materials significant to its environmental and management environments in the spatial context (Dalia, 2008). It is also a medium that is challenged by changing environmental and management conditions (Toth et al., 2007). From the time immemorial, it has been well accepted that crops and soils within a field or region are both spatially and temporally variable. Accordingly, farmers/growers around the world are attempted traditionally to manage such variability to a limited extent primarily by perception. The basis of precision agriculture, the spatial and temporal variability in soil and crop factors within a field has been cherished for centuries. The capability of a soil to function within ecosystem boundaries and interact with the environment, external to that system, forms the responsible source for potential impact of soil management systems on the environment. Consequently, the measurement of soil quality should be a quantitative tool to help guide management decisions. Among the several evaluating parameters, nutrient balancing of soil is one of the imperative approaches to presume a good input return. Hence, soil sampling undertakes considerably superior significance when precision or site-specific farming is implemented to achieve good input return. Subsequently, the precision and representation required the variable rates of nutrient calculation and application and the economics of the technology as a

Remote Sensing in Precision Agriculture. DOI: https://doi.org/10.1016/B978-0-323-91068-2.00008-4

whole. In this line, precision management is the accurate use of high-technology equipment and software to quantify field conditions and micromanage the elements or resources at small or large land areas to achieve farmer's optimum income (Mishra et al., 2019).

Precision farming (PF) has evolved through many stages; it began with farming by soil and progressed to site-specific soil management based on grid sampling and management zones. Further recently, there has been increasing emphasis on real-time on-the-go monitoring with fully automated aerial or ground-based sensors (Mulla, 2013). Of late, satellite-guided procedures and computer-aided decision support systems are forecasted to revolutionize the way agricultural farms are managed. As estimation of soil nutrients and residue cover is achievable by using aerial remote sensing, in real time, high-resolution nutrient and residue covermaps can be generated. Further, these maps might be used to determine the rate of nutrient deficiencies, which differ within diverse areas of the agricultural field. These nutrient deficiencies could be ameliorated by varying the rate of application among the deficient spots in the map (Sarkar & Jha, 2020). This variable rate application is the basis of precision agriculture (PA), and new studies have delineated the competence of aerial remote sensing for faster andautomated data collection and analysis (Sarkar et al., 2021). Site-specific nutrient management or PF is conceivably the most recent technique among the series of nutrient management methods specifically recognized by the worldwide farmers. Throughout the past 15 years, geospatial technology as applied to agricultural crop production has improved enormously. Further, this technology has expanded rapidly from a mere practice to operational reality in several million hectares all over the world (Berry et al., 2003). Blanket application of fertilizer-based nutrient supply creates a mismatch and results in unequal nutrient removal from patches of soil. Thus, persistence of unevensoil fertility endures in the entire farm. Consequently, crops tend to be unevenwith regard to productivity. Therefore, precision approaches such as closely spaced grid sampling of soils and their analysis are a necessity. In this line, discrimination of a large field into a "management zone" is a good substitute. Similarly, maps depicting soil fertility variations can be also used to apply fertilizers accurately resulting in better economics, in addition to providing uniformity to soil fertility (Krishna, 2013).

Over the past decade, the development of several new technologies has been emerged with the potential to significantly alter soil fertility evaluation and management. PA is the application of modern technological tools to apply the right amount of inputs at the right time and the right location systematically. It is also discussed as site-specific management, which has become a progressively accepted and widespread approach to farming that allows us to (1) specifically map soil fertility in a field or on a farm at a great detail (characteristically in 1 ha grids) by using handheld or equipment-mounted GPS that uses signals from a network of satellites to instantaneously locate a position on the earth's surface; (2) usage of variable rate application equipment, armed with GPS and "on-the-go" sensors of soil nutrient status, to supply nutrients in accordance with these maps;

and (3) generate GPS-based maps of crop yields. Aspects associated to yield that cannot easily be measured by soil and plant testing can be acquired from other computerized databases such as soil series from soil surveys; otherwise, they can be evaluated using remote sensing such as plant health, soil moisture, irrigation, drainage, and pest pressure parameters. The most commonly used remote sensing devices today include cameras and other imaging systems mounted on aircraft or satellites. The electromagnetic (EM) energy that is emitted or reflected from plants (or bare soils) is acquired by remote sensing devices, and they convert this energy into data that can be used in soil fertility evaluation. Combination of soils, plants, environmental conditions, and management practices distinctly has a characteristic spectral "signature" a specific spectrum of radiation that can be used to diagnose soil fertility problems (Thomas Sims & McGrath, 2012).

Further, it has been predicted that, based on soil, crop, and climate, future machineries used in agricultural production will be automatically controlled to recommend cultural practices. The key persistence of the remote sensing technologies in soil management is to augment returns on inputs, sustainable crop production capabilities accompanied by ensuring environmental stewardship though optimum application rates of agricultural inputs conferring to local needs. Solicitations of highly advanced technologies in PA involve constant access to detailed information for characterizing the environmental conditions under which the crop production takes place. Such information may be acquired from airborne and satellite images at the farm/field scale. The capability of using remote sensing data to govern fertilization needs of plants is based on the nutrient content of crops and soils, which aids to increase crop yields and increase the quality of harvested seeds and fruits, which is important for improving the crop profitability and in turn farmer's income. Accurate fortitude of the nutritional requirements of plants at critical stages during a particular season under field conditions helps to optimize fertilization application rates in addition to reducing the potential adverse impacts associated with excess use of agrochemicals. Besides, remote sensing has been also used to assess the water needs of plants and determine the date of commencement of irrigation, making it easier to manage crop production under conditions of water stress.

The most promising approaches of PA technology are management of soil pH and macronutrients at variable rates. Generally, different layers of spatial field data are integrated to develop application algorithms, recognized as prescription maps. However, results of systematic soil sampling followed by standardized laboratory analyses usually serve as the best assessment of soil nutrients and pH distribution throughout the field. Contrastingly, the two foremost limitations of contemporary methods are: (1) inadequate recommendation algorithms, which were developed for macroscale agricultural practices; and (2) high cost of soil sampling and analysis resulting in maps with inadequate spatial resolution. To overcome these limitations, high-density soil data layers such as bare soil aerial photography, electrical conductivity (EC) maps, and digital elevation models have been used to adapt soil sampling patterns and develop management zones requiring similar soil treatments. Though, a methodology for reliably obtaining higher = density direct measurements of soil properties possibly will provide capabilities for resolving both the limitations of variable rate technology. The on-the-go system for mapping soil pH and macronutrient levels could provide additional data layers for prescribing variable-rate application maps or conduct research studying local yield response to certain soil treatments. Nevertheless, more research is required to evaluate the application of direct soil measurements (DSMs) by using the stability of ion-selective electrodes for mapping soil pH, plant-available K and NO_3, both in terms of improving sensor performance and interpretation of the results for site-specific management decisions. The major drawback of this methodology is that the values measured on-the-go represent snapshots of pH or nutrient availability with little information regarding soil buffering and nutrient release over time (Adamchuk et al., 2003).

7.2 Active and passive approaches of remote sensing

Recently, techniques such as active soil sensing, satellite photography, and multispectral imaging that are generally associated with futuristic space and satellite-based communication are being increasingly introduced into farm management. Nowadays, they are actually considered valuable and futuristic aspects of agriculture (Sonka et al., 1998). Remote sensing is an advanced approach that allows us to assess soil, crop growth and evaluate precise requirements of fertilizer, irrigation, and other agronomic procedures. Both passive and active remote sensors are being used during PF. Passive remote sensors are usually mounted on satellites, airplanes, or balloons. However, the passive remote sensing satellites cannot acquire data on soils or crops during night. Meantime, weather parameters such as clouds may also affect its accuracy. Studying the imagery and interpretation is yet another problem involved with passive remote sensing. Consequently, passive remote sensing usually uses light wavelength in the visible band for assessing the crops. The NIR light is also used to detect soil water status, crop disease, or nutrient status (Krishna, 2013).

Therefore, nowadays "Active Remote Sensing" or sensor-based crop and soil assessment is widespread among the farmers adopting PF. It is a latest technique that overcomes certain limitations involved with passive systems (mounted on satellites). Primarily, an active sensor has its own light source, hence it is entitled "active." It allows data accumulation both during day and night and measures crop/soil reflectance mutually in visible and non-visible ranges of the EM spectrum. Further, active sensors are handheld, inexpensive gadgets, which can be easily procurable even by poor-resource farmers. Further most notably, the spatial resolution of handheld sensors (active) is relatively high, easy to decode, interpret, and quantify. Subsequently these sensors are held closer to crop canopy or soil surface, and details on imagery are outstanding and most accurate (Krishna, 2013).

There is indeed a wide range of crop canopy sensors available in the market to suit the purpose and monetary resources of the farmers (Khosla, 2008). Recently, several of the models are available in the market, which can be easily mountable on tractors and other farm vehicles (Fig. 7.1).

7.3 Sensor technologies for soil management

Enhanced management of essential soil nutrients is a vital goal in achieving sustainable agriculture and preserving essential increases in food production while minimizing economic losses and environmental impacts (Goulding et al., 2008). Technological advancements play a catalytic role and inconspicuous mutual ground between environmental and economic goals. Modern advances designate that efficient nutrient management in crop fields can be accomplished through the application of PA-based geospatial technologies such as global positioning system (GPS), geographical information system (GIS), remote sensing, geostatistics, and variable-rate application (Gebbers & Adamchuk, 2010). Variable-rate fertilizer

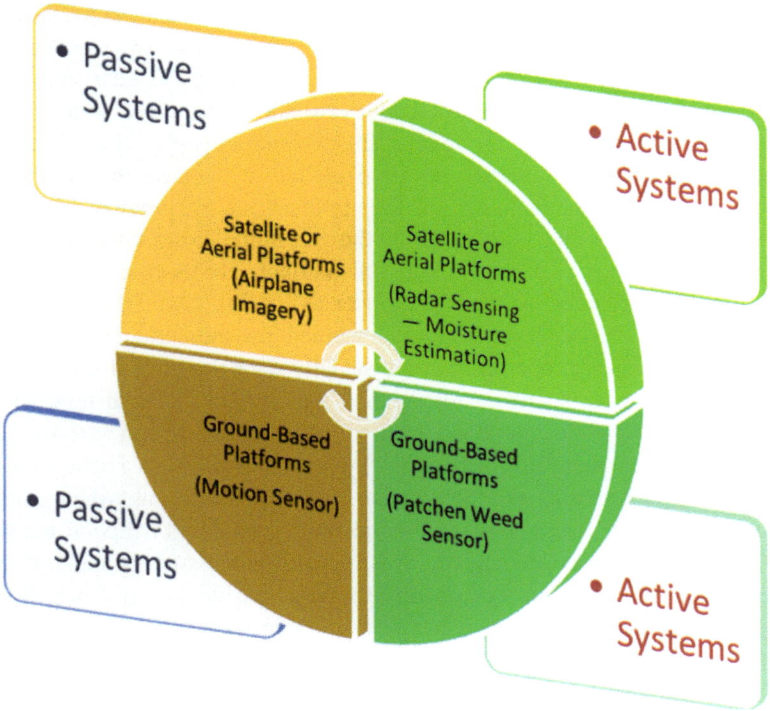

FIGURE 7.1

Assortment of remote sensing systems.

application is one of the basic standards of PA and has been revealed to optimize fertilizer use efficiency by overcoming the problem of over- and under-fertilization (Schirrmann & Domsch, 2011). Both spatial and temporal variabilities in crop and/or soil productivity are influenced mutually by intrinsic (soil forming factors such as parent material, climate, topography, fauna/flora, and time) and extrinsic factors (such as farm management practices and maintenance operations) (Sun et al., 2003). Quantification of the spatial and temporal variabilities of soil properties and reclamation of this variability through prudently designed site- and time-specific input application are believed to enhance nutrient assimilation in crops (Bah et al., 2012).

Generally, in PA, intensive grid sampling is considered as one of the best accurate methods of mapping the variability of crop and soil attributes. Nevertheless, intensive grid sampling is laborious, time-consuming, expensive, and consequently, impractical for implementation on a large scale (Srinivasan, 2006). Thus, development a more rapid means of obtaining spatial and temporal data for detailed variability mapping is indispensable. By using low-cost sensors to estimate soil properties, the efficiency of site- and time-specific crop-soil management and monitoring strategies can be improved. In this scenario, to quantify the soil properties, various sensors have been developed recently, prerequisite for accurate mapping within-field variability as per necessity. Ideally, sensor devices are fitted with a GPS to allow for soil data to be captured on-the-go and instantaneously transformed into distribution maps. This would enable real-time monitoring and intervention of soil nutrient status, which can potentially counterbalance restrictions imposed by the inherent spatial and temporal variabilities in soil nutrient supply (Bah et al., 2012). Real-time soil sensor information can be exploited in manners that increase farm production and profit, maintain and increase product quality, promote food security, and ensure environmental protection. In the meantime, researchers have already endeavored to develop real-time in situ soil nutrient sensors based on optical and electrochemical techniques (Lamar Burton et al., 2020). Similarly, on-the-go soil sensor technologies can serve as a rapid method for measuring soil mechanical, physical, and chemical properties that are steadily developing. Soil sensors may be used to vary application rates in response to sensor output in real time without a GPS receiver (Morgan & Ess, 1997). Consequently, on-the-go soil sensors can be a part of either "map-based" or "real-time" systems. Further, this application would enable real-time monitoring and intervention of soil nutrient status, which can potentially offset the limitations imposed by the inherent spatial and temporal variabilities in soil nutrient supply. Fundamentally, these sensors harness the potential to revolutionize decision management systems in agriculture as Internet of things (IoT) soil nutrient sensors (Lamar Burton et al., 2020). While there are vast assortments of design concepts, most on-the-go soil sensors described in the literature comprise one of the following measurement methods:

- **Electrical and EM sensors**: quantity electrical resistivity (ER)/conductivity, capacitance, or inductance affected by the composition of tested soil.

- **Optical and radiometric sensors:** use EM waves to distinguish the level of energy absorbed/reflected by soil particles.
- **Mechanical sensors:** measure forces resulting from a tool engaged with the soil.
- **Acoustic sensors**: quantify the sound produced by a tool interacting with the soil.
- **Pneumatic sensors**: assess the ability to inject air into the soil.
- **Electrochemical sensors**: use ion-selective membranes that produce a voltage output in response to the activity of selected ions (H^+, K^+, NO_3^-, Na^+, etc.).

(Adamchuk et al., 2004)

Moreover, soil sensors can be used to generate real-time soil data, such as pH (soil reaction), EC, salinity, dissolved oxygen (DO), and nutrient concentration, which are successively turned into geo-referenced maps to facilitate site-specific nutrient application (Mishra et al., 2019). Recently, several on-the-go sensors have been manufactured by many scientific firms to quantity soil mechanical, physical, and chemical properties, and majority of these devices are based on electrical and EM, optical and radiometric, mechanical, acoustic, pneumatic, and electrochemical measurement concepts (Adamchuk et al., 2004), as presented in Table 7.1.

In general, the main concerns in sensor performance efficiency are the issues of precision and accuracy (PPI, 1999). Precision refers to the ability of the sensor to repeat its own measurement in the same location and time, whereas accuracy refers to how well the sensor measurements correlate to an actual soil property that is determined using the conventional (reference) measurement technique. A given soil property can be reliably predicted based on optimum precision and accuracy of the sensor output. As per the most sensor-based studies conducted earlier, the goodness of fit between sensor outputs and conventional measurements are expressed either as Pearson coefficient of correlation (r) or as coefficient of determination (R^2). Further, on-the-go sensors have the advantage of providing nondestructive and rapid quantification of soil variability to empower precision soil nutrient management and monitoring. Further, the prospects of electrochemical, electrical, EM, optical, and radiometric sensors for real-time mapping of vital soil chemical and physical properties to facilitate precision soil nutrient management and monitoring are promising (Bah et al., 2012).

7.3.1 Soil moisture sensing technology methods

Recent trends of demand for smart irrigation and water-saving practices in agriculture have triggered the development of diverse soil moisture sensing techniques that can operate under harsh field conditions. Uninterrupted monitoring of soil moisture is a common practice in digital agriculture to reduce the effects of water logging, drought, increase yield and profitability (Yu et al., 2021). Soil moisture can be determined by remote sensing technology, where soil surface is

Table 7.1 The different categories of on-the-go soil sensors and their applications.

Sensor types	Applications	References
(1) Electrochemical—capable of assessing spatial variability of different soil chemical properties directly or indirectly.	Soil pH, nitrate, potassium (K)	Adamchuk et al. (2007), Adamchuk et al. (2004)
(2) Electrical and electromagnetic—uses various measurement systems based on electrical circuits to determine the ability of soil media to conduct or accumulate electrical charge.	Soil texture, soil moisture content, soil depth variability (depth of topsoil, depth to hardpan), cation exchange capacity	Kim et al. (2009), King et al. (2005), and Sudduth et al. (2003)
(3) Optical and radiometric—optical sensing technology uses visible and near-infrared wavelength ranges to rapidly quantify soil properties.	Soil organic matter, soil moisture	Rossel et al. (2006), Chang et al. (2001)
(4) Acoustic—equipped with a sound-recording device (microphone) that records sound produced through interaction of the soil and the shank having a rough surface and hollow cavity.	Soil texture, soil bulk density, soil depth variability	Grift et al. (2002)
(5) Mechanical—designed to measure soil strength, which is conventionally done by measuring mechanical resistance.	Soil compaction, compacted soil layers	Stafford and Werner (2003), Manor and Clark (2001)

monitored using remote sensor measurements of reflected or emitted EM energy, which is analyzed to establish the relationship between the soil moisture and soil water content. This technology has several advantages such as suitable for all-day, all-weather, and multi-polar applications. In this line, previously several sensors were tested in different soils to evaluate the quantity accuracy of each sensor in different soils. Yin et al. (2003) used remote sensing technology such as near-infrared reflectance sensor for soil moisture measurement. Similarly, Asghari et al. (2019) conducted a comparative evaluation of the IoT technology applications considering the technical characteristics (quality of service, case studies, and evaluation environment) to acquire indicators reflecting the application level of IoT in macro-agriculture from different perspectives. Recently, Yu et al. (2021) introduced a comprehensive Delphi method and analytic hierarchy process to evaluate the performance of a soil moisture sensing technology applied to a field environment. They concluded that comprehensive methodology provides agricultural operators with guidance and suggestions in terms of the selection of an appropriate soil moisture sensing technology. Different remote sensing technologies used for soil moisture evaluation are detailed in Table 7.2.

7.3.2 Effective soil use and management

To manage the soil resource effectively, basic information of soil spatial distribution is indispensable. For effective soil use and management, the role of clay content is one of the most important mandatory attributes monitored by farmers/agricultural community. Additionally, clay minerals affect soil physical properties primarily through the formation of micro aggregates, which in turn influence nutrient availability and plant growth. As clay minerals constitute an important component of the soil system, acquaintance of their role in soil fertility is imperative for sustainable soil management and productivity. This is particularly the case in the topsoil because, (1) clay content greatly affects the water-holding capacity and the hydraulic properties of a soil (Jabro, 1992), (2) clay content is associated to the cation exchange capacity of the soil (Russell, 1973), and (3) clay content influences the fertility and hence productivity of the soil (Davey, 1990). Most imperative roles of clay minerals vis-à-vis soil fertility have been well documented in literature. Soil fertility is detailed as "the quality of a soil that enables it to provide nutrients in adequate amounts and in proper balance for the growth of specific plants or crops" (Buresh et al., 1997). Further, the soil fertility is dependent on several soil properties comprising chemical, physical, biological, and mineralogical, all of which directly or indirectly affect the nutrient dynamics and availability (Roy et al., 2006). As per the hydrological perception, knowledge

Table 7.2 Soil moisture sensing remote sensing technology methods.

Methods/ technical name	Soil moistureevaluating principle	Technical features
Hyperspectral remote sensing method (M6)	This technology can directly establish relationship between the soil moisture and soil reflectance to monitor the soil moisture.	**Advantages:** High spectral resolution within a specificband (range)
Microwave remote sensing method (M7)	This technology uses the soil moisture contrast in the dielectric constant to invert the soil moisture.	**Advantages:** Suitable for all-day, all-weather, multi-polarapplications, high resolution, penetrability, sensitivity to the soil moisture content. **Disadvantages:** The vegetation and soil moisture monitoring precision are low.

of the subsoil and vadose zone clay content is also important because large amounts can reduce permeability, inhibit deep drainage, and potentially lead to waterlogged soil conditions (Triantafilis et al., 2001). Therefore, future research must explore the relationships between clay minerals and soil nitrogen vis-à-vis sequestration and stabilization. Further, clay minerals should be deliberated in studies dealing with soil quality assessment, especially in the choice of soil quality indicators (Kome et al., 2019).

Soil sampling for geostatistical mapping can be laborious and expensive, predominantly to identifying and mapping subsurface clay content. By considering this information, many studies have incorporated ancillary variables to enhance prediction; among which, the most commonly used method is electromagnetic induction (EMI), where instruments measure the apparent soil electrical conductivity (ECa), which is a function of various soil properties including salinity, clay content, moisture content, and mineralogy (Corwin & Lesch, 2003; Triantafilis et al., 2000). Similarly, airborne EM systems could be deployed to increase the EM survey resolution, or supplementary types of ancillary data (gamma radiometric, LANDSAT, RADARSAT, etc.) may be incorporated into the modeling process. Further, a collective FKM (fuzzy *k*-means) analysis of the EM34, EM38, and remotely sensed data might provide better distinction of surface sediments and delineation of robust soil management units using a quantitative approach (Triantafilis & Lesch, 2005).

7.3.3 Site-specific management zones

As a more economical approach to define regions within a farm field for variable application of crop inputs, contemporary research areas in PA have focused on the use of management zones. In this line, many studies have concluded that management zones could be used as an alternative method to grid soil sampling to develop nutrient maps for variable-rate fertilizer applications (Khosla, 2008; Khosla et al., 2002). In the modern eras for developing prescription maps and managing variability, many scientists have well-documented and verified several management zone techniques for their effectiveness. A unique technique that has been proposed by Franzen et al. (2000) includes use of soil survey maps to define N management zones. One more technique was adopted by Kravchenko et al. (2000) for varying crop inputs within farm fields by using distinctions in topography or landscape positions. Similarly, an in situ method was evaluated for use of application of EM induction to measure ECa to delineate management zones (Sudduth et al., 2001). Further, Fleming et al. (1999) evaluated the management zone techniques based on soil color reflectance, topography, and the farmer's past production experience. The technique was evaluated by Fleming et al. (1999) and is commercially available as Centrak. Similarly, Hornung et al. (2003) developed and evaluated a new technique "site-specific management zones" intended for delineating production-level management zones of maize (*Zea mays* L.) grain yield and NUE in Colorado field.

7.4 Digital soil mapping

7.4.1 Proximal sensing and digital terrain models

The topography is one of the most common soil-forming factors used in the predictions of soil classes and properties by analyses of a digital elevation model and its derivatives. Predictive soil mapping has been widely used in recent years, as soils occur in response to water movement all the way through the landscape, which is controlled by local relief (Scull et al., 2003). Worldwide, digital terrain models (DTMs) have been extensively used for soil mapping and as advanced tools for adequate predictors of soil properties. Moreover, considering the continuous nature of DTM variation (raster-based distribution), they also have been used in soil predictive models for providing spatially exhaustive auxiliary variables (De Menezes et al., 2016). Subsequently, soils are the consequence of complex interaction of soil-forming factors, thus DTM is deliberated as very useful in environments where topography is strongly related to the processes driving soil formation (Scull et al., 2003). By using DTM model, improved predictions are frequently achieved with the contribution of extra variables provided by the use of proximal sensors (magnetometers and portable X-ray fluorescence scanners—pXRF). In this line, Silva et al. (2016) used an ordinary least square regression model to predict sand and clay contents using DTM, pXRF, and the magnetometer as predicting variables. The results of this study revealed that variables obtained with pXRF showed a greater ability for predicting soil classes (overall accuracy of 78% with 0.67 kappa index), additionally for estimating sand and clay contents, compared with those acquired only with DTM and the magnetometer. Further, this study also established that pXRF offers additional variables that are key for mapping soils and predicting soil properties at a detailed scale.

Meanwhile, recent studies are examining for new tools associated with soil attributes, especially those concerning chemical features. Relatively soil chemical elements or properties could function as tracers or indicators of different parent materials, these properties sequentially could be related to soil classes and properties. Similarly, these data could potentially advance soil mapping and modeling processes. Therefore, equipment that accomplish fast analyses in the field and deliver a large spectrum of data, for instance, proximal sensors, have been recently adopted to support soil mapping. The proximal sensing device comprises proximal or remote in situ and ex situ (for both field and laboratory), noninvasive or intrusive, and mobile or stationary devices (McBratney et al., 2011). Some examples are magnetometers, which quantify the magnetic susceptibility of different materials, and pXRF scanners, which are used to identify and quantify chemical elements and compounds present in soil samples (Horta et al., 2015). The XRF spectrometry is a nondestructive, rapid, simultaneous multielement analytical methodology used for both solid and liquid samples. Its applications are comprehensive, and XRF spectra cover most elements, with a dynamic range from 100% down to the μg/g level. Further, XRF is a well-established laboratory-based method, but it is also one of the few atomic spectrometric techniques that can be used for field portable instrumentation (Hou et al., 2004). The recent innovations are presented in Table 7.3.

7.4.2 Automated methods for soil mapping—electromagnetic induction and direct current or galvanic methods

Site-specific management in agriculture entails spatial information on soil properties within a field. However, identifying this variation reliably comprises extensive soil sampling effort. Determining the spatial variation of soil properties using ancillary data would be more practical approach with less expensive to obtain

Table 7.3 Applications of the proximal sensing device in remote sensing.

S. no.	References	Method/instrument	Applications
1	Maher (1986)	Mineral magnetic measurements	Characterization of soils
2	De Jong et al. (2000)	Magnetic susceptibility of soils	Different slope positions in Saskatchewan, Canada
3	Hanesch and Scholger (2005)	The influence of soil type on the magnetic susceptibility.	Measurement throughout soil profiles
4	Hanesch et al. (2007)	Lithological and pedological influences on the magnetic susceptibility of soil	Their consideration in magnetic pollution mapping.
5	Guzmán et al. (2010)	Evaluation of magnetic iron oxides as sediment tracers	Water erosion experiments
6	Karimi et al. (2011)	Relationships between magnetic susceptibility and heavy metals	Urban top soils in the arid region of Isfahan, central Iran
7	Wang (2013)	Magnetic susceptibility measurements	Assessment of heavy metal pollution in Xuzhou urban top soils
8	Jordanova et al. (2014)	Pattern of cumulative soil erosion and redistribution pinpointed through magnetic signature.	Chernozem soils
9	Siqueira et al. (2015)	Soil-landscape relation and spatial variability of magnetic susceptibility and soil color.	Detailed mapping unit design
10	Horta et al. (2015)	Integrated field spectroscopy and spatial analysis	Enhanced assessment of soil contamination
11	Liu et al. (2016)	Spatial distribution of soil magnetic susceptibility and correlation	As a proxy method for Heavy metal pollution in Kaifeng, China

information. The electrical methods used for noninvasive measurements of soils are EMI method to measure EC and direct current (DC) or Galvanic method to measure the ER. These powerful tools play the most important role for the characterization of soils in agriculture, several studies concluded a strong relationship between agronomical significant parameters such as clay or available water content, porosity and texture, and ER/EC (Sudduth et al., 2001). From a physical point of view, ER and EC in soils display a wide dynamic range (1–20, minimum). They serve as a basis for a precise location of soil sampling and demarcation of homogeneous zones needed in the definition of management zones. Further, they also fetch a spatial accuracy, which is compatible with the prerequisite of PA machines. Both electrical methods perform on the principle that an electrical current is injected in the soil by means of a pair of electrodes. This current is either a DC current or a slow alternating current (several Hertz) to avoid polarization effects and eddy currents. The current flow in the whole volume of soil and subsoil and its spatial distribution are functions of the spatial distribution of the ER. As the soil is rarely uniform, geophysicists use the term apparent ER, the resistivity of a uniform ground giving the same measure. Further, the voltage resulting from this spatial distribution is measured at the ground surface by two or more electrodes. The details of automated methods for soil mapping are detailed in Table 7.4.

In EMI method, eddy currents are induced in the soil by means of a low-frequency magnetic field (Hp) originating from an oscillating current in a coil above the soil as per Faraday's law. The spatial distribution of these currents is a function of the EC (σ) of the soil, and conductivity is the reciprocal of the ER. These currents could be measured by electrodes on the ground surface; nevertheless, in most of the EMI methods, a second coil (Rx) measures the secondary magnetic field (Hs) generated by the eddy currents. However, it has been observed that ER methods have drawbacks such as labor-intensive measurement and problems when high contact resistances are encountered (dry soils or frozen zone). Therefore, DC method is advantageous compared with EC method for soil mapping in-depth sensitivity and the investigation depth (Dabas & Tabbagh, 2003). In recent times, ECa measurement by EMI sensors is becoming more prevalent, since it provides large amounts of data relatively economically. Preceding investigations has shown and concluded the value of ECa data as an investigative tool for soil salinity assessments (Triantafilis et al., 2000), soil water content (Sheets & Hendrickx, 1995) and aid to conduct soil survey (Inman et al., 2002; King et al., 2001). Contrastingly, the strength of the relation between ECa and some soil properties, such as clay and water content, appears complex. There is a negative relation for ECa with some soil properties (sand and bulk density) and a positive one for others (silt and clay) (Frogbrook et al., 2003). However, this emphasizes that a map of ECa is not a substitute for sampling the soil. Though, ECa and other related ancillary variables could be used to delineate management zones within which soil sampling could be targeted. For the adoption of PF in agriculture, the amount of soil variability across a farm and within a field serves

Table 7.4 Recent innovations in automated methods for soil mapping.

S. no.	References	Method/instrument	Application
1	Zhu et al. (2001).	Fuzzy logic—soil-land inference model (SoLIM)	Soil mapping using GIS, expert knowledge
2	Triantafilis and Lesch (2005)	Electromagnetic induction techniques	Mapping clay content variation
3	Shi et al. (2009)	Integrating different types of knowledge for digital soil mapping—soil inference engine (SIE), was developed to facilitate an eight-step integrated RBR-CBR DSM process	DSM will impact the traditional soil survey and mapping paradigm, standards, processes, and applications
4	Cavazzi et al. (2013)	Fine resolution digital elevation models	Digital soil mapping
5	Jasiewicz and Stepinski (2013)	Geomorphons—pattern recognition approach	Classification and mapping of landforms
6	Cambule et al. (2013)	Spatial model calibrated on the basis of limited soil sampling and explanatory covariables related to soil-forming factors, developed from readily available secondary information from accessible areas.	Digital soil mapping in poorly accessible areas
7	Mosleh et al. (2016)	The effectiveness of digital soil mapping	Prediction of soil properties over low-relief areas.

as a strategic prominence for defining potential benefits. Similarly, many previous soil science researchers have already confirmed that the variability of soil physical properties within fields has a crucial role in evaluating the potential for variable-rate management of agrochemical inputs for some crops and for demarcating within-field management zones. Similarly, various authors have concluded that an EMI method sensor can provide useful information on soil patterns; however, this method substitutes for field inspection of soil properties but improves the targeting and reduces the time and cost of soil examination.

7.4.3 Multispectral satellite sensors and their advantages

Advantages of using remote sensing technology include saving time, wide coverage (i.e., satellite remote sensing delivers the only source when data are mandatory over vast areas or regions), rapid data acquisition compared with ground/traditional methods, and enables long-term monitoring of specified location. In the current scenario, multispectral satellite sensors are the considered as an ideal method for mapping and monitoring soil salinity. This is predominantly due to the low cost of the imagery (Landsat, SPOT, etc.) and their capability to map extreme surface expressions of salinity. Further, these techniques are also afford multispectral image with resolutions ranging from medium to high, in addition to hyperspectral image. Therefore, these remotely sensed data have been successfully used for monitoring and mapping of soil salinity for decades with mixed results (Allbed & Kumar, 2013). Over the last three decades, extensive research has been accompanied by using satellite imagery for mapping and monitoring soil salinity frequently with multispectral sensors. These comprise Landsat Thematic Mapper (TM), Landsat Multispectral Scanner System (MSS), Landsat Enhanced Thematic Mapper Plus (ETM +), SPOT, Advanced Space borne Thermal Emission & Reflection Radiometer (Terra-ASTER), Linear imaging self-scanning sensor (LISS-III), and IKONOS (Dwivedi, 2001; Dwivedi et al., 2008; Verma, et al., 1994). These research findings on multispectral satellite sensors for mapping and monitoring soil salinity with conclusions are presented in Table 7.5.

7.4.3.1 Limitations of multispectral satellite sensors

Multispectral data have limited diagnostic capability because of its coarse spatial and spectral resolutions (Spies & Woodgate, 2005). The direct mapping of soil salinity with multispectral imagery had major limitations that arise especially where there are no salt features on the soil surface and where saline soils are dominated by halophyte plants (Furby et al., 1995; Howari, 2003). Similarly, multispectral satellite sensors have also triggered confusing reflectance, as they found that non-saline soils were confused with extremely saline soils (Furby et al., 1995). While the application of Landsat imagery and the spectral resolution of Landsat was insufficient due to the difference between the spectra of saline land, and waterlogged land was not being sufficient to allow spectral separation, as well as the variable spectral response of saline soil (Fraser & Joseph, 1998). Furthermore, Hick and Russell (1987) opined that the ability of discriminating plant species and plant health conditions is challenging through Landsat imagery owing to the lack of narrow bands in the range of 700–730, 730–760, and 900–1100 nm.

7.4.4 Hyperspectral remote sensing

The energy incident on the Earth's surface is absorbed, transmitted, or reflected reliant on the wavelength and characteristics of the surface features such as soil, vegetation, or water body. However, these three progressions are not mutually exclusive; either energy incident on a surface may be partially reflected, absorbed, or transmitted. Which process takes place on a surface depends on the following factors: (1) wavelength of the radiation, (2) the characteristics of the incident radiation, (3) angle at which the radiation intersects the surface, and (4) composition and physical properties of the surface. After interaction with the surface features, energy that is reflected or re-emitted from the features is recorded at the sensors and

Table 7.5 The different categories of multispectral satellite sensors for mapping and monitoring of soil salinity.

S. no.	Authors	Method/instrument	Applications
1	Elnaggar and Noller (2009)	Landsat TM imagery integration with decision tree analysis (DTA)—by maximum likelihood supervised classification, classified the image into non-saline soils (EC < 4 dS/m) and saline soils (accuracies of 97% and 60%), whereas DTA predicted five classes of soil salinity (overall accuracy of approximately 99%)	Use of Landsat TM imagery effectively identified in bare soils characterized by high spectral reflectance due to a high salt content on the surface, and integrating DTA with remote sensing data was more accurate and effective compared to remote sensing analysis alone
2	Katawatin and Kotrapat (2005)	Landsat-7 ETM + with three different combinations of sources of ancillary data (topography, geology, and underground water quality)—maximum likelihood classification method was employed for soil salinity mapping	The use of Landsat ETM + data bands 4, 5 and 7 in combination with all three types of ancillary data yielded the most accurate (83.6% overall accuracy) soil salinity map.
3	Eldeiry and Garcia (2008)	Detecting soil salinity in alfalfa fields using spatial modeling and remote sensing—integrated approach using remote sensing techniques and spatial models geophysical surveys	Improve the development of high-quality soil salinity maps
4	Goossens et al. (1993)	Multispectral sensors for soil salinity research—accuracy of Land sat TM, MSS, and SPOT XS imagery for soil salinity mapping	Landsat TM was found to be optimal for soil salinity mapping
5	Elhaddad and Garcia (2006)	IKONOS satellite imagery and crop reflectance—identified the severity level of soil salinity and its effect on crop yield	The superior effectiveness due to the use of high spatial resolution imagery.
6	Dwivedi et al. (2008)	Performance of IKONOS imagery, mapping salt-affected soils—overall accuracy of 92.4% was gained by using IKONOS data compared to 78.4% and 84.3% obtained by using the IRS-ID LISS-III multispectral sensor	Indicated the great potential of high spatial resolution IKONOS images for soil salinity mapping and detection

analyzed to identify the target features and interpret their characteristics (Zibayiwa, 2020). Further, the incident electromagnetic energy may interact with the earth surface features in three possible ways: Reflection, Absorption, and Transmission. The energy that is reflected by topographies on the earth's surface over a variety of different wavelengths gives their spectral responses in the remote sensing systems. Each type of feature/object has a unique spectral response/reflectance characteristic, also recognized as spectral signature, which can be used to identify the respective surface features and to study their properties. The majority of radiation incident on a soil surface is either reflected or absorbed, and little is transmitted. The characteristics of the soil that determine its reflectance properties are its moisture content, organic matter content, texture (proportion of sand, silt, and clay), structure, and presence of iron oxide content. These factors are complex, variable, and interrelated. The presence of moisture in soil decreases its reflectance. This effect is greatest in the water absorption bands at 1.4, 1.9, and 2.1 μm. On the other hand, similar absorption characteristics are displayed by the clay soils. Clay soils have hydroxyl ion absorption bands at 1.4 and 2.2 μm. Two other factors that reduce soil reflectance are surface roughness and the content of organic matter. Presence of iron oxide in a soil also significantly decreases reflectance, at least in the visible region of wavelengths (Aggarwal, 2003).

A spectroradiometer is a type of instrument that has the capability to include bumble bee (UV, 10–380 nm), human eye (visible, 380–700 nm), gold fish (infrared, 700–1 mm), and so many others in a single instrument. Hyperspectral: hyper (excessive): thousands of narrow bands as compared with multispectral: 3–10 broad bands only. Reflectance of the surface of a material is its effectiveness in reflecting radiant energy; it is the fraction of incident electromagnetic radiation that is reflected at an interface. The reflectance spectrum or spectral reflectance curve is the plot of the reflectance as a function of wavelength. Reflectance spectroscopy is a nondestructive technique that examines the interactions between incident light and surface of the material. This is a rapid nondestructive and less-expensive technique for determination of various soil properties such as (1) mineral identification; (2) organic matter characterization; (3) prediction of secondary properties by calibrating spectral models; and (4) reflectance values at different wavelength (λ) used as predictor variables. Rapid, reliable, and cost-effective assessment of soil properties has become a focus area for monitoring and management of agricultural and natural resources. Over the years, spectroscopic techniques have been contributing a ray of hope since these techniques are very rapid, nondestructive, inexpensive, user-friendly, and also eco-friendly. Various spectroscopic techniques such as visible–near infrared spectroscopy (VIS–NIR spectroscopy) and mid-infrared spectroscopy (MIR spectroscopy) are mainly used for prediction and analysis of various soil properties. In the prevailing conditions, there is a huge potential for applications of VIS–NIR and MIR for the rapid, reliable, and cost-effective prediction of soil physical, chemical, and biological properties both in laboratory and field conditions. Further, the diffuse reflectance spectroscopy has the prospective for prediction of various soil properties by using VIS, NIR, MIR and combined spectra or from a single spectrum, therefore leading to increase in the efficiency over conventional/traditional soil analytical techniques (Mondal, 2008). The recent advancements of spectroscopic techniques for prediction of soil properties are detailed in Table 7.6.

Table 7.6 Spectroscopic techniques for prediction of soil properties.

S. no.	References	Method/instrument	Application
Visible-near infrared spectroscopy (VNIR) spectroscopy			
1	Reeves et al. (2002)	The potential of diffuse reflectance spectroscopy for the determination of carbon inventories in soils	Predicted different soil properties such as organic carbon and nitrogen
2	Moron and Cozzolino (2002)	Analysis of organic C, total N, and pH in soils of Uruguay	Analysis of potentially mineralizable nitrogen
3	Ludwig et al. (2002)	C:N ratio and soil biological activities	To soil sustainability
4	Cozzolino and Moron (2003)	VNIR spectroscopy	Analysis of heavy metals, macro and micronutrients
5	Brown et al. (2006)	Analyzed more than 4000 soil samples to evaluate the accuracy of VNIR empirical models for global soil characterization	Strong prediction for soil minerals such as kaolinite, montmorillonite, clay content, CEC, soil organic and inorganic carbon and extractable iron.
Mid-infrared (MIR) diffuse reflectance spectroscopy			
1	Shepherd and Walsh (2002)	Development of reflectance spectral libraries for characterization of soil properties	The strong absorbance in the 3600–3800 cm^{-1} region is due to hydroxyl stretching vibrations associated with clay minerals.
2	Urselmans et al. (2010)	Prediction of soil fertility properties from a globally distributed soil mid-infrared spectral library	MIR is reliable and fast analytical tool and can be used to predict a number of soil properties for a globally distributed soil spectral library

7.4.5 Soil salinity mapping and monitoring

The process of land degradation through accumulation of excess salts in the root zone resulting in partial or complete loss of soil productivity and eventual disappearance of the vegetation is defined as soil salinization. The soil salinity problems occur in all continents and under all climate conditions. Soil salinity is a serious environmental problem and most widespread in the arid and semiarid regions; nevertheless; salt-affected soils also exist extensively in sub-humid climates, predominantly in coastal regions where intrusion of seawater through estuaries and rivers, and through groundwater, causes large-scale salinization. The soil salinity condition is either primary naturally occurring or secondary it may be human-induced. Soil salinity is a prevalent environmental hazard in arid and semiarid regions around the world (Hillel, 2000). Similarly, the United Nations

Food and Agriculture Organization (FAO) has estimated that saline soil covered 397 million hectares of the aggregate land area of the world (Koohafkan, 2012). Soil salinity adversely affects plant growth, crop production, soil and water quality, and it eventually results in soil erosion and land degradation (Corwin & Lesch, 2003). Thus, it is important to monitor and map soil salinity at an early stage to enact an effective soil reclamation program that helps lessen or prevent future increase in soil salinity. The detection of soil salinity by the traditional way is time-consuming; conversely, remote sensing data and techniques offer more efficient and economical rapid tools and techniques for monitoring and mapping of soil salinity. Therefore, in recent years, remote sensing has surpassed the traditional method for monitoring and mapping of soil salinity. Generally, remote sensing uses the electromagnetic energy reflected from targets to obtain information about the Earth's surface with diverse levels of detail. Successively based on this concept, the spectral reflectance of the salt features at the soil surface has been widely studied using remote sensing and used as a direct indicator for soil salinity detection and mapping. However, when the soil moisture is more or the crust salt is invisible on the soil surface or mixed with other soil constituents, this direct approach becomes complicated and may yield unreliable results as these factors influence the soil spectral reflectance (Allbed & Kumar, 2013). Since from 1960s when black-and-white and color aerial photographs were used to delineate salt-affected soils, the remote sensing information and techniques have been progressively pragmatic to monitor and map soil salinity (Dale, et al., 1986). Soil salinity can be detected directly from remotely sensed data through salt features that are visible at the soil surface, for instance, bare soil with white salt crusts on the surface (Matinfar et al., 2013) or indirectly from indicators such as the presence of halophytic plant and the performance level of salt-tolerant crops (Alhammadi & Glenn. 2008). Notwithstanding the effects of salt features on the soil surface on the spectral reflectance, several scientists have been considered remote-sensed data through salt features that are visible at the soil surface and are good direct indicators of soil salinity. Similarly, Fernandez-Buces et al. (2006) used surface features to predict soil salinity. Further, they also established that the correlation coefficient between surface colors, EC, and the sodium adsorption ratio was statistically significant, which suggested that efflorescence color is a promising surface indicator with which to estimate soil salinity. The recent innovations of digital soil salinity mapping by using applications of remote sensing technologies and their endeavors are listed in Table 7.7.

Mainstream studies deliberated earlier have endeavored to map and monitor soil salinity by using different multispectral satellite sensors. Though, the use of such sensors is found to be limiting, as spectral resolution delivered by these applications impacts both the quality and quantity of the data. Therefore, development of airborne and satellite-based hyperspectral sensors has currently overcome some of the spatial and spectral limitations of multispectral satellite imagery for monitoring and mapping soil salinity, mutually regionally and locally (Allbed & Kumar, 2013). Subsequently, hyperspectral sensors offer a large number of

Table 7.7 Application of remote sensing technologies for soil salinity mapping and monitoring.

S. no.	Authors	Method/instrument	Application
Salt features at the soil surface			
1	Metternicht and Zinck (2008)	Remote sensing of soil salinization—impact on land management	The dynamic processes at the surface of saline soil limit the monitoring and assessment of the salinization process because they influence the spectral, spatial and temporal behavior of the salt features
2	Brown et al. (2006)	VNIR diffuse reflectance spectroscopy—global soil characterization	Via the physicochemical properties of soil such as soil moisture content, organic matter, soil texture, types of clay color, and surface roughness, soil spectral reflectance is determined
3	Shrestha et al. (2005)	Analysis and classification of hyperspectral data	Mapping of land degradation in Southern Spain
4	Schmid et al. (2008), and Thomas (2011)	Application of hyperspectral imagery and arid zone geomorphology	Soil salinity mapping and process, form and change in drylands—due to salinity, soil properties change, which affect the spectral reflectance of features that occur at the soil surface, including salt crusts and efflorescence besides variations in surface texture and structure
5	Schmid et al. (2008)	Application of hyperspectral imagery—soil salinity mapping	Crusted saline soil reflects strongly in the visible and near-infrared (NIR) bands
6	Singh and Sirohi (1994) and Rao et al. (1995)	Spectral reflectance and spectral behavior—properties of different types of soil surfaces and salt-affected soils	Crusted saline soil surface is generally smoother than a non-saline surface and exhibits high reflectance in the visible and NIR bands
7	Metternicht and Zinck (1997)	Spatial discrimination—salt and sodium affected soil surfaces	The reflectance in the visible and NIR bands is highly affected by both the crust color and surface roughness factors

spectral bands through high spatial resolution that allow the discrimination of halophyte plants from non-halophyte plants besides the identification of surface salt features in more detail compared with multispectral sensors (Dutkiewicz, 2006;

Gupta, 2003). Similarly, Hamzeh et al. (2012) explored the ability of Hyperion space-borne hyperspectral data for mapping salinity stress in sugarcane fields. They adopted different classifications such as support vector machine (SVM), spectral angle mapper, minimum distance, and maximum likelihood with different band combinations to classify soil salinity into low, moderate, and high salinity classes. Further, they concluded that SVM classification using all bands as input data yielded a salinity map with good accuracy, with an overall accuracy and kappa coefficient of 78.7% and 0.68, respectively.

7.4.5.1 Impediments in remote sensing technologies

- Low-spectral-resolution satellite images limit the direct detection and mapping of soil salinity, principally due to their incompetence to detect particular absorption bands of some salt types and frequent incidences of complications with mixed spectral signatures due to a variety of surface components (Dehaan & Taylor, 2003).
- The spatial, temporal, and vertical variabilities in the soil profile remain the limiting factors in the assessment and mapping of soil salinity by means of remote sensing data, as the spectral reflectance only perceives the soil surface and consequently unable to provide information on the whole soil profile (Sethi et al., 2010).
- The direct deduction of soil salinity in arid and semiarid regions characterized by dense vegetation cover would be difficult, as the vegetation leads to spectral mixing (Ding et al., 2011).
- The soil surface characteristics in innumerablecircumstances may not be representative of the deeper soil profile (Dewitte et al., 2012).
- The possibility of on-the-go sensor fusion that would allow simultaneous spatial variability quantification of important crop-soil properties under diverse growing conditions is still unclear (Bah et al., 2012).
- When the soil moisture is high or the crust salt is invisible on the soil surface or mixed with other soil constituents, the direct approach of remote sensing to monitor and map soil salinity becomes complicated and may produce unreliable results, as these factors influence the soil spectral reflectance (Allbed & Kumar, 2013).
- Therefore, application of remote sensing for detecting and mapping of soil salinity is challenging in arid and semiarid regions. This is predominantly attributed to the weakness of the spectral signals from saline soils compared with the noise caused by other factors (Allbed & Kumar, 2013).

7.4.6 Hybrid strategy on the development of a pedological map

Digital map can be generated by a combination of three variables: remote sensing (Landsat data), proximal sensing (laboratory spectroscopy), and relief. The Pedological map is allied to the specialization of soil class units and reveals the variation of several characteristics, such as texture, fertility, organic matter, depth, pedregosity, and water dynamics. These are the most important types of information for crop growing, environment monitoring, and land use planning.

Remote sensing is the process of accomplishing information about objects without coming into direct contact with the object. The carrier of data in remote sensing is electromagnetic radiation, which travels in vacuum at the speed of light in the form of waves of different lengths. The most advantageous wavelengths in remote sensing cover VIS light and extend through the NIR and shortwave infrared, to thermal infrared and microwave bands. The active sensors used in remote sensing emit their own radiation, which further interacts with the designated target identified for investigation and returns to the measuring instrument. While passive sensors record incident radiation reflected or emitted from the objects (Marek et al., 2016).

7.4.6.1 LEADS

- Handheld or in situ sensors and remote sensing techniques that aid in soil fertility and grain yield mapping have again improved precision. Rapid "on-the-go" sensors and computer-guided decision support systems have revolutionized fertilizer and irrigation prescription (Krishna, 2013).
- Hyperspectral remote sensing is nondestructive technique evading the use of environmentally detrimental chemicals or reagents for analysis and consequently conserving the sustainability of the agricultural ecosystem and environment.
- Diffuse reflectance spectroscopy technique can be used effectively in soil fertility evaluation, land management, soil amelioration, soil health risk assessment, sustainable soil health management, carbon sequestration, geological survey, environmental monitoring, modeling, sustainable agriculture as well as for precision agriculture (Mondal, 2008).
- The detection soil salinity by the traditional way is time-consuming; conversely, remote sensing data and techniques offer more efficient and economical rapid tools and techniques for monitoring and mapping soil salinity.
- Despite the effects of salt features on the soil surface on the spectral reflectance, many scientist have considered remote-sensed data through salt features that are visible at the soil surface and are good direct indicators of soil salinity; further, the amalgamation of remote sensing data with geophysical surveys and simulation models can be an alternative viable option (Farifteh et al., 2006).
- Surface reflectance is greatly affected by soil's moisture content, salt content, color, and surface roughness. High salt concentrations can be recognized through the existence of characteristic vegetation types and growth patterns or by the salt efflorescence and crust that are present on bare soils. Therefore, consideration of the spectral reflectance of halophytic plants is necessary for soil salinity detection and mapping, particularly in highly saline areas (Allbed & Kumar, 2013).

7.5 Conclusion and way forward

Increasing world population growth coupled with the increasing risks associated with climate change inevitably necessitates a proportionate increase in agricultural productivity. To ensure sustainable soil productivity while conserving high crop yields with reducing environmental pollution is a key to this challenging task. Interim, widespread multi-nutrient deficiencies coupled with gradual deterioration of soil health are prominent causes of low nutrient use efficiency, productivity, and profitability. Apart from this, the prevailing issues related to climate change have created the enhanced depletion of soil quality, availability of irrigation water and use efficiency of resources and inputs, and crop productivity worldwide. In this juncture, the implementation of sensor technologies for soil nutrient management and monitoring is a step in the right direction. Therefore, remote sensing has and enormous prospective for assessing and mapping the soil salinity for soil management. Precision technique involves diverse aspects such as soil analysis, mapping, decision support system, variable-rate applicators, and yield monitoring. Multispectral satellite sensors are the preferred method for mapping and monitoring soil salinity, largely due to the low cost of such imagery and the ability to map extreme surface expressions of salinity. However, multispectral data have limited capabilities due to their spatial and spectral resolution. Hyperspectral imagery, with its fine spatial and spectral resolutions, allows soil salinity mapping in greater detail and represents another alternative. Major advantages from precision techniques are established to reduction in fertilizer consumption, reduction in loss of precious soil nutrients, and marginal improvement in grain yield. Extensive research and development of spectral indices are required to estimate multiple soil characteristics. Future technological advancement in radiation measuring sensors mounted on ground-based vehicles, unmanned aerial vehicles (UAVs), or satellites UAVs, data processing and statistical technics (machine learning, computer vision, and artificial neural networks) in conjunction with soil management practices may lead to successful agricultural farming systems. Further, the adoption of climate resilient practices along with the application of remote sensing, GIS, and advanced restoration technologies is imperative for restoring the fertility of the soil health for sustainable development. This would not only help in combating climate change issues but also help in formulating strong policies for managing soil resources.

References

Adamchuk, V. I., Hummel, J. W., Morgan, M. T., & Upadhyaya, S. K. (2004). On-the-go soil sensors for precision agriculture. *Computers and Electronics in Agriculture.*, *44*, 71–91, 10113/16856/1/IND43638848.

Adamchuk, V. I., Lund, E., Dobermann, A., & Morgan, M. T. (2003). On-the-go mapping of soil properties using ion-selective electrodes. In J. Stafford, & A. Werner (Eds.),

Precision agriculture (Precision agriculture: Papers from the 4th European Conference on Precision Agriculture, Berlin, Germany, 15–19 June 2003). Wageningen Academic Publishers.

Adamchuk, V. I., Lund, E. D., Reed, T. M., & Ferguson, R. B. (2007). Evaluation of an on-the-go technology for soil pH mapping. *Precision Agriculture*, *8*, 139–149. Available from https://doi.org/10.1007/s11119-007-9034-0.

Aggarwal, S. (2003). Principles of remote sensing. Satellite remote sensing and GIS applications in agricultural meteorology. In *Satellite remote sensing and GIS applications in agricultural meteorology proceedings of the training workshop*, July 7–11, 2003, Dehra Dun, India, pp. 23–38.

Alhammadi, M. S., & Glenn, E. P. (2008). Detecting date palm trees health and vegetation greenness change on the Eastern Coast of the United Arab Emirates using SAVI. *International Journal of Remote Sensing*, *29*(6), 1745–1765. Available from https://doi.org/10.1080/01431160701395195.

Allbed, A., & Kumar, L. (2013). Soil salinity mapping and monitoring in arid and semi-arid regions using remote sensing technology: A review. *Advances in Remote Sensing*, *2*, 373–385.

Asghari, P., Rahmani, A. M., & Javadi, H. (2019). Internet of things applications: A systematic review. *Computer Networks.*, *148*, 241–261.

Bah, A., Balasundram, S. K., & Husni, M. H. A. (2012). Sensor technologies for precision soil nutrient management and monitoring. *American Journal of Agricultural and Biological Sciences*, *7*(1), 43–49.

Berry, J. K., Delgado, J. A., Khosla, R., & Pierce, F. J. (2003). Precision conservation for environmental sustainability. *Journal of Soil and Water Conservation*, *58*, 332–339.

Brown, D. J., et al. (2006). Global soil characterization with VNIR diffuse reflectance spectroscopy. *Geoderma*, *132*(3), 273–290. Available from https://doi.org/10.1016/j.geoderma.2005.04.025.

Buresh, R.J., Sanchez, P.A., & Calhoun, F. (1997). *Replenishing soil fertility in Africa* (Vol. 51, 252 pp.). SSSA Special Publication, SSSA and ASA.

Cambule, A. H., Rossiter, D. G., & Stoorvogel, J. J. (2013). A methodology for digital soil mapping in poorly-accessible areas. *Geoderma*, *192*, 341–353.

Cavazzi, S., Corstanje, R., Mayr, T., Hannam, J., & Fealy, R. (2013). Are fine resolution digital elevation models always the best choice in digital soil mapping? *Geoderma*, *195–196*, 111–121.

Chang, C. W., Laird, D. A., Mausbach, M. J., & Jr, C. R. H. (2001). Near-infrared reflectance spectroscopy– Principal components regression analyses of soil properties. *Soil Science Society of America Journal. Soil Science Society of America*, *65*, 480–490, 10113/24746/1/IND23216887.

Corwin, D., & Lesch, S. (2003). Application of soil electrical conductivity to precision agriculture. *Agronomy Journal*, *95*(3), 455–471. Available from https://doi.org/10.2134/agronj2003.0455.

Cozzolino, D., & Moron, A. (2003). The potential of near-infrared reflectance spectroscopy to analyse soil chemical and physical characteristics. *Journal of Agricultural Science*, *140*, 65–71.

Dabas, M., & Tabbagh, A. (2003). A comparison of EMI and DC methods used in soil mapping—theoretical considerations for precision agriculture. In J. Stafford, & A. Werner, (Eds.), *Precision agriculture* (Papers from the 4th European Conference on

Precision Agriculture, Berlin, Germany, 15-19 June 2003). Wageningen Academic Publishers.

Dale., et al. (1986). Classification of Reflectance on Colour Infrared Aerial Photographs and Sub-Tropical Salt-Marsh Vegetation Types. *International Journal of Remote Sensing*, *7* (12), 1783–1788. Available from https://doi.org/10.1080/01431168608948968.

Dalia, F. (2008). Endocalcari-Epihypogleyic Cambisol plough layer quality in long-term soil management systems. *Zemes Ukio Mokslai T.*, *15*(4), 60–69.

Davey, B. G. (1990). The chemical properties of soils. In K. O. Campbell, & J. W. Bowyer (Eds.), *Scientific Basis of Modern Agriculture. Sydney*. Sydney, Australia: University Press.

De Jong, E., Pennock, D. J., & Nestor, P. A. (2000). Magnetic susceptibility of soils in different slope positions in Saskatchewan. *Canada. Catena*, *40*, 291–305.

De Menezes, M. D., Silva, S. H. G., de Mello, C. R., Owens, P. R., & Curi, N. (2016). Spatial prediction of soil properties in two contrasting physiographic regions in Brazil. *Sci. Agric.*, *73*, 274–285.

Dehaan, R., & Taylor, G. R. (2003). Image-Derived Spectral Endmembers as Indicators of Salinisation. *International Journal of Remote Sensing*, *24*(4), 775–794.

Dewitte, O., Jones, A., Elbelrhiti, H., Horion, S., & Montanarella, L. (2012). Satellite remote sensing for soil mapping in Africa: An overview. *Progress in Physical Geography*, *36*(4), 514–538, 2012.

Ding, J., Wu, M. C., & Tiyip, T. (2011). Study on soil salinization information in arid region using remote sensing technique. *Agricultural Sciences in China*, *10*(3), 404–411.

Dutkiewicz, A. (2006). *Evaluating hyperspectral imagery for mapping the surface symptoms of dryland salinity*. The University of Adelaide.

Dwivedi, R. S., Kothapalli, R. V., & Amarendra Narayana Singh, A. N. (2008). Generation of farm-level information on salt-affected soils using IKONOS-II multispectral data. In G. Metternicht, & J. Zinck (Eds.), *Remote sensing of soil salinization: Impact on land management*. CRC Press.

Dwivedi, R. S. (2001). Soil resources mapping: A remote sensing perspective. *Remote Sensing Reviews*, *20*(2), 89–122. Available from https://doi.org/10.1080/02757250109532430.

Eldeiry, A., & Garcia., L. A. (2008). Detecting soil salinity in alfalfa fields using spatial modelling and remote sensing. *Soil Science Society of America Journal*, *72*(1), 201–211.

Elhaddad, A., & Garcia, L. (2006). Detecting soil salinity levels in agricultural lands using satellite imagery. In *Proceedings of the American Society for photogrammetry and remote sensing annual conference*, Reno.

Elnaggar, A. A., & Noller, J. S. (2009). Application of remote sensing data and decision-tree analysis to mapping salt affected soils over large areas. *Remote Sensing*, *2*(1), 151–165.

Farifteh, J., Farshad, A., & George, R. J. (2006). Assessing salt-affected soils using remote sensing, solute modelling and geophysics. *Geoderma*, *130*(3-4), 191–206.

Fernandez-Buces, N., Siebe, C., Cram, S., & Palacio, J. L. (2006). Mapping soil salinity using a combined spectral response index for bare soil and vegetation: A case study in the former Lake Texcoco, Mexico. *Journal of Arid Environments*, *65*(4), 644–667. Available from https://doi.org/10.1016/j.jaridenv.2005.08.005.

Fraser, D., & Joseph, S. (1998). Mapping soil salinity in the murray valley (NSW) using satellite imagery. In *Proceeding of the 9th Australaian remote sensing and photogrammetry conference*, Australia Causal Productions Sydney, Sydney.

Fleming, K. L., Westfall, D. G., Wiens, D. W., Rothe, L. E., Cipra, J. E., & Heermann, D. F. (1999). Evaluating farmer developed management zone maps for precision farming. In P. C. Robert, et al. (Eds.), *Precision agriculture. Proc. Int. Conf., 4th, Bloomington, MN. 19–22 July 1998* (pp. 335–343). Madison, WI: ASA, CSSA, and SSSA.

Franzen, D. W., Halvorson, A. D., & Hofman, V. L. (2000). Management zones for soil N and P levels in the northern Great Plains. In P. C. Robert, et al. (Eds.), *Precision agriculture. Proc. Int. Conf., 5th, Bloomington, MN. 16–19 July 2000 [CD-ROM]*. Madison, WI: ASA, CSSA, and SSSA.

Furby, S. et al. (1995). *Detecting and monitoring salt-affected land: A report from the LWRRDC project detecting and monitoring changes in land condition through time using remotely sensed data*. CSIRO, Division of Mathematics & Statistics.

Frogbrook, Z. L., Oliver, M. A., & Derricourt, K. E. (2003). Exploring the spatial relations between soil properties and electro-magnetic induction (EMI) and the implications for management. In J. Stafford, & A. Werner (Eds.), *Precision agriculture* (Precision agriculture: Papers from the 4th European Conference on Precision Agriculture, Berlin, Germany, 15-19 June 2003). Wageningen Academic Publishers.

Gebbers, R., & Adamchuk, V. I. (2010). Precision agriculture and food security. *Science (New York, N.Y.), 327*, 828–831. Available from https://doi.org/10.1126/science.1183899.

Goossens, R., El Badawi, M., Ghabour, T., & De Dapper, M. (1993). A simulation model to monitor the soil salinity in irrigated arable land in arid areas based upon remote sensing and GIS, EARSeL. *Advances in Remote Sensing*, *2*(3), 165–171.

Goulding, K., Jarvis, S., & Whitmore, A. (2008). Optimizing nutrient management for farm systems. *Philosophical Transactions of the Royal Society B.*, *363*, 667–680. Available from https://doi.org/10.1098/rstb.2007.2177.

Grift, T. E., Tekeste, M. Z., & Raper, R. L. (2002). Acoustic compaction layer detection. *Transactions of the ASAE*, *48*, 1–8.

Gupta, R. P. (2003). *Remote sensing geology*. Springer Verlag.

Guzmán, G., Barrón, V., & Gómez, J. A. (2010). Evaluation of magnetic iron oxides as sediment tracers in water erosion experiments. *Catena*, *82*, 126–133.

Hamzeh, S., Naseri, A.A., Alavi Panah, S.K., Mojaradi, B., Bartholomeus, H.M., & Herold, M. (2012). Mapping salinity stress in sugarcane fields with hyperspectral satellite imagery. In *Proceedings SPIE 8531, remote sensing for agriculture, ecosystems, and hydrology XIV*, Edinburgh, September 24, 2012, p. 85312B-1.

Hanesch, M., Rantitsch, G., Hemetsberger, S., & Scholger, R. (2007). Lithological and pedological influences on the magnetic susceptibility of soil: Their consideration in magnetic pollution mapping. *The Science of the Total Environment*, *382*, 351–363.

Hanesch, M., & Scholger, R. (2005). The influence of soil type on the magnetic susceptibility measured throughout soil profiles. *Geophysical Journal International*, *161*, 50–56.

Hick, P., & Russell, W. (1987). *Remote sensing of agricultural salinity*. CSIRO, Division of Exploration Geoscience.

Hillel, D. (2000). *Salinity management for sustainable irrigation: Integrating science, environment, and economics*. World Bank Publications. Available from https://doi.org/10.1596/0-8213-4773-X.

Hornung, A., Khosla, R., Reich, R., & Westfall, D. G. (2003). Evaluation of site-specific management zones: Grain yield and nitrogen use efficiency. In J. Stafford, & A. Werner (Eds.), *Precision agriculture*. Wageningen Academic Publishers. Available from https://doi.org/10.3920/978-90-8686-514-7.

Horta, A., Malone, B., Stockmann, U., Minasny, B., Bishop, T. F. A., McBratney, A. B., Pallasser, R., & Pozza, L. (2015). Potential of integrated field spectroscopy and spatial analysis for enhanced assessment of soil contamination: A prospective review. *Geoderma*, *241–242*, 180–209.

Hou, X., He, Y., & Jones, B. T. (2004). Recent advances in portable X-ray fluorescence spectrometry. *Applied Spectroscopy Reviews.*, *39*, 1–25.

Howari, F. (2003). The use of remote sensing data to extract information from agricultural land with emphasis on soil salinity. *Australian Journal of Soil Research*, *41*(7), 1243–1253.

Inman, D. J., Freeland, R. S., Ammons, J. T., & Yoder, R. E. (2002). Soil investigations using electromagnetic induction and ground-penetrating radar in Southwest Tennessee. *Soil Science Society of America Journal*, *66*(1), 206–211.

Jabro, J. D. (1992). Estimation of saturated hydraulic conductivity of soils from particle size distribution and bulk density data. *Transactions of the ASAE. American Society of Agricultural Engineers*, *35*, 557–560.

Jasiewicz, J., & Stepinski, T. F. (2013). Geomorphons—A pattern recognition approach to classification and mapping of landforms. *Geomorphology*, *182*, 147–156.

Jordanova, D., Jordanova, N., & Petrov, P. (2014). Pattern of cumulative soil erosion and redistribution pinpointed through magnetic signature of chernozem soils. *Catena*, *120*, 46–56.

Karimi, R., Ayoubi, S., Jalalian, A., Sheikh-Hosseini, A. R., & Afyuni, M. (2011). Relationships between magnetic susceptibility and heavy metals in urban topsoils in the arid region of Isfahan, central Iran. *Journal of Applied Geophysics*, *74*, 1–7.

Katawatin, R., & Kotrapat, W. (2005). *Use of LANDSAT-7 ETM + with ancillary data for soil salinity mapping in Northeast Thailand*. In 3rd international conference on experimental mechanics and third conference of the Asian 2004, Sevilla, June 3, 2005, pp. 708–716.

Khosla, R. (2008). *Active remote sensing: An innovative technique for precision fertilization* (pp. 1–4). http://www.extsoilcrop.colostate.edu/newsletter.html.

Khosla, R., Fleming, K., Delgado, J. A., & Shaver, T. M. (2002). Use of site-specific management zones to improve nitrogen management for precision agriculture. *Journal of Soil and Water Conservation*, *57*(6), 513–518.

Kim, H. J., Sudduth, K. A., & Hummel, J. W. (2009). Soil macronutrient sensing for precision agriculture. *Journal of Environmental Monitoring: JEM*, *11*, 1810–1824. Available from https://doi.org/10.1039/b906634a.

King, J. A., Dampney, P. M. R., Lark, R. M., Wheeler, H. C., Bradley, R. I., et al. (2005). Mapping potential crop management zones within fields: Use of yield-map series and patterns of soil physical properties identified by electromagnetic induction sensing. *Precision Agriculture*, *6*, 167–181. Available from https://doi.org/10.1007/s11119-005-1033-4.

King, J.A., Dampney, P.M. R., Lark, M., Mayr, T.R., & Bardley, R.I. (2001). *Sensing soil spatial variability by electro-magnetic induction (EMI): Its potential in precision farming*. In G. Grenier & S. Blackmore (Eds.), Proceedings of the 3rd European conference on precision agriculture (Vol. I, pp. 419–424), Agro, Montpellier.

Kome, G. K., Enang, R. K., Tabi, F. O., & Yerima, B. P. K. (2019). Influence of clay minerals on some soil fertility attributes: A review. *Open Journal of Soil Science*, *9*, 155–188.

Koohafkan, P. (2012). *Water and cereals in drylands*. The Food and Agriculture Organization of the United Nations and Earthscan.

Kravchenko, A. N., Bullock, D. G., Reetz, H. F. (2000). Correlation of corn and soybean grain yield with topography and soil properties. *Agronomy Journal*, 92, 75–83.

Krishna, K. R. (2013). *Soil fertility and productivity aspects. In precision farming*. Apple Academic Press.

Liu, D., Ma, J., Sun, Y., & Li, Y. (2016). Spatial distribution of soil magnetic susceptibility and correlation with heavy metal pollution in Kaifeng, China. *Catena*, *139*, 53–60.

Ludwig, B., Khanna, P. K., Bauhus, P., & Hopmans, P. (2002). Near infrared spectroscopy of forest soils to determine chemical and biological properties related to soil sustainability. *Forest Ecology and Management*, *171*, 121–132.

Maher, B. A. (1986). Characterisation of soils by mineral magnetic measurements. *Physics of the Earth and Planetary Interiors.*, *42*, 76–92.

Manor, G., & Clark, R. L. (2001). *Development of an instrumented subsoiler to map soil hard-pans and real-time control of subsoiler depth* (1st ed., p. 12)ASAE.

Marek, W., Wójtowicz, A., & Piekarczyk, J. (2016). Application of remote sensing methods in agriculture. *Communications in Biometry and Crop Science*, *11*, 31–50.

Matinfar, H. R., et al. (2013). Detection of soil salinity changes and mapping land cover types based upon remotely sensed data. *Arabian Journal of Geosciences*, *6*(3), 913–919.P.

McBratney, A.B., Minasny, B., & Whelan, B. (2011). *Defining proximal soil sensing*. In Proceedings of the the second global workshop on proximal soil sensing, Montreal, PQ, Canada, 15–18 May.

Metternicht, G., & Zinck, J. A. (1997). Spatial discrimination of salt- and sodium-affected soil surfaces. *International Journal of Remote Sensing*, *18*(12), 2571–2586.

Metternicht, G., & Zinck, A. (2008). *Remote sensing of soil salinization: Impact on land management*. CRC Press.

Mishra, A., Pant, P. K., Bhatt, P., Singh, P., & Gangola, P. (2019). Management of soil system using precision agriculture technology. *Journal of Plant Development Sciences*, *11*(2), 73–78.

Mondal, B. P. (2008). Hyper-spectral analysis of soil properties for soil management. In Advances in agriculture for sustainable development, (pp. 59–65). Srijan Publishers P. Ltd, ISBN: 978-93-83583-61-4.

Morgan, M. T., & Ess, D. R. (1997). *The precision-farming guide for agriculturists. An agriculture primer*. Moline, IL: John Deere Publishing.

Moron, A., & Cozzolino, D. (2002). Application of near infrared reflectance spectroscopy for the analysis of organic C, total N and pH in soils of Uruguay. *Journal of Near Infrared Spectroscopy*, *10*, 215–221.

Mosleh, Z., Salehi, M. H., Jafari, A., Borujeni, I. E., & Mehnatkesh, A. (2016). The effectiveness of digital soil mapping to predict soil properties over low-relief areas. *Environmental Monitoring and Assessment*, *188*, 195.

Mulla, D. J. (2013). Twenty five years of remote sensing in precision agriculture: Key advances andremaining knowledge gaps. *Biosystems Engineering*, *114*(4), 358–371. Available from https://doi.org/10.1016/j.biosystemseng.2012.08.009.

PPI. (1999). *Site-specific management guidelines* (1st Edn.). Atlanta: Potash and Phosphate Institute.

Rao, B. R. M., Sharma, R. C., Ravi Sankar, T., Das, S. N., Dwivedi, R. S., Thammappa, S. S., & Venkataratnam, L. (1995). Spectral behaviour of salt-affected soils. *International Journal of Remote Sensing*, *16*(12), 2125–2136.

Reeves, J., McCarty, G., & Mimmo, T. (2002). The potential of diffuse reflectance spectroscopy for the determination of carbon inventories in soils. *Environmental Pollution*, *116*, 277–284.

Rossel, R. A. V., Walvoort, D. J. J., McBratney, A. B., Janik, L. J., & Skjemstad, J. O. (2006). Visible, near infrared, mid infrared or combined diffuse reflectance spectroscopy for simultaneous assessment of various soil properties. *Geoderma*, *131*, 59–75. Available from https://doi.org/10.1016/j.geoderma.2005.03.007.

Roy, R. N., Finck, A., Blair, G. J., & Tandon, H. L. S. (2006). *Plant nutrition for food security. A guide for integrated nutrient management* (p. 348) Food and Agricultural Organization of United Nations, FAO Fertilizer and Plant Nutrition Bulletin 16.

Russell, E. W. (1973). *Soil conditions and plant growth* (10th ed). Longman Group Limited.

Sarkar, S., & Jha, P. K. (2020). Is precision agriculture worth it? Yes, may be. *Journal of Biotechnology and Crop Science*, *9*(14), 4–9.

Sarkar, S., Ramsey, A. F., Cazenave, A.-B., & Balota, M. (2021). Peanut leaf wilting estimation from RGBcolor indices and logistic models. *Frontiers in Plant Science*. Available from https://doi.org/10.3389/fpls.2021.658621.

Schirrmann, M., & Domsch, H. (2011). Sampling procedure simulating on-the-go sensing for soil nutrients. *Journal of Plant Nutrition and Soil Science*, *174*, 333–343. Available from https://doi.org/10.1002/jpln.200900367.

Schmid, T., Koch, M., & Gumuzzio, J. (2008). Application of hyperspectral imagery to soil salinity mapping. In G. Metternicht, & J. Zinck (Eds.), *Remote sensing of soil salinization: Impact on land management* (pp. 113–137). CRC Press.

Scull, P., Franklin, J., Chadwick, O. A., & McArthur, D. (2003). Predictive soil mapping: A review. *Progress in Physical Geography*, *27*, 171–197.

Sethi, M., Bundela, D. S., Sethi, M., Lal, K., & Kamra, S. K. (2010). Remote sensing and geographic information system for appraisal of salt-affected soils in India. *Journal of Environmental Quality*, *39*(1), 5–15.

Sheets, K. R., & Hendrickx, J. M. H. (1995). Noninvasive soil water content measurement using electromagnetic induction. *Water Resources Research*, *31*(10), 2401–2409.

Shepherd, D. K., & Walsh, G. M. (2002). Development of reflectance spectral libraries for characterization of soil properties. *Soil Science Society of America Journal*, *66*, 988–998.

Shi, X., Long, R., Dekett, R., & Philippe, J. (2009). Integrating different types of knowledge for digital soil mapping. *Soil Science Society of America Journal. Soil Science Society of America*, *73*, 1682–1692.

Shrestha, D., Margateb, D. E., van der Meera, F., & Anhc, H. V. (2005). Analysis and classification of hyperspectral data for mapping land degradation: An application in Southern Spain. *International Journal of Applied Earth Observation and Geoinformation*, *7*(2), 85–96. Available from https://doi.org/10.1016/j.jag.2005.01.001.

Silva, S. H. G., Poggere, G. C., De Menezes, M. D., Carvalho, G. S., Guilherme, L. R. G., & Curi, N. (2016). Proximal sensing and digital terrain models applied to digital soil mapping and modeling of Brazilian latosols (oxisols. *Remote Sensing*, *8*, 614. Available from https://doi.org/10.3390/rs8080614.

Singh, R. P., & Sirohi, A. (1994). Spectral reflectance properties of different types of soil surfaces. *ISPRS Journal of Photogrammetry and Remote Sensing*, *49*(4), 34–40.

Siqueira, D. S., Marques, J., Pereira, G. T., Teixeira, D. B., Vasconcelos, V., Carvalho Júnior, O. A., & Martins, E. S. (2015). Detailed mapping unit design based on soil-landscape relation and spatial variability of magnetic susceptibility and soil color. *Catena*, *135*, 149–162.

Sonka, S. T., Baeur, M. E., & Cherry, E. T. (1998). *Precision agriculture in the 21st century: Geospatial and information technologies in crop management* (p. 168) National Academy Press.

Spies, B., & Woodgate, P. (2005). *Salinity mapping methods in the Australian context.* Department of the Environment and Heritage.

Srinivasan, A. (2006). *Handbook of precision agriculture: principles and applications*, (1st Edn., p. 683)(ISBN: 1560229543, p. 683). New York: Routledge.

Stafford, J., & Werner, A. (2003). *Precision agriculture* (1st ed., p. 783)Wageningen: Wageningen Academic Publishers.

Sudduth, K. A., Kitchen, N. R., Bollero, G. A., Bullock, D. G., & Wiebold, W. J. (2003). Comparison of electromagnetic induction and direct sensing of soil electrical conductivity. *Agronomy Journal*, *95*, 472–482.

Sudduth, K. A., Drummond, S. T., & Kitchen, N. R. (2001). Accuracy issues in electromagnetic induction sensing of soil electrical conductivity for precision agriculture. *Computers and Electronics in Agriculture*, *31*, 239–264.

Sun, B., Zhou, S. L., & Zhao, Q. G. (2003). Evaluation of spatial and temporal changes of soil quality based on geostatistical analysis in the hill region of subtropical China. *Geoderma*, *115*, 85–99. Available from https://doi.org/10.1016/S0016-7061(03)00078.

Thomas, D. S. G. (2011). *Arid zone geomorphology: Process, form and change in drylands*. John Wiley & Sons.

Toth, G., Stolbovoy, V., & Montanarella, L. (2007). *Soil quality and sustainability evaluation – an integrated approach to support soil-related policies of the European Union.* EUR 22721 EN [PhD thesis]. Ohio State University, Uganda, Office for Official Publications of the European Communities.

Thomas Sims, J., & McGrath, J. (2012). Soil fertility evaluation. In P. M. Huang, Y. Li, & M. E. Sumner (Eds.), Hand book of soil sciences properties and processes (2nd ed.). Boca Raton: CRC Press, Taylor & Francis Group. Available from https://doi.org/10.1201/b11267.

Triantafilis, J., Huckel, I. A., & Odeh, I. O. A. (2001). Comparison of statistical prediction methods for estimating field-scale clay content using different combinations of ancillary variables. *Soil Science*, *166*, 415–427.

Triantafilis, J., Laslett, G. M., & McBratney, A. B. (2000). Calibrating and electromagnetic induction instrument to measure salinity in soil under irrigated cotton. *Soil Science Society of America Journal*, *64*(3), 1009–1017.

Triantafilis, J., & Lesch, S. M. (2005). Mapping clay content variation using electromagnetic induction techniques. *Computers and Electronics in Agriculture*, *2005*(46), 203–237.

Triantafilis, J., & Lesch, S. M. (2005). Mapping clay content variation using electromagnetic induction techniques. *Computers and Electronics in Agriculture*, *46*, 203–237.

Urselmans, T. T., Vagen, G. T., Spaargaren, O., & Shepherd, D. K. (2010). Prediction of soil fertility properties from a globally distributed soil mid-infrared spectral library. *Soil Science Society of America Journal*, *74*, 1792–1799.

Verma, K., Saxena, R. K., Barthwal, A. K., & Deshmukh, S. N. (1994). Remote sensing technique for mapping salt affected soils. *International Journal of Remote Sensing*, *15* (9), 1901–1914. Available from https://doi.org/10.1080/01431169408954215.

Wang, X. S. (2013). Assessment of heavy metal pollution in Xuzhou urban topsoils by magnetic susceptibility measurements. *Journal of Applied Geophysics*, *92*, 76–83.

Wienhold, B. J., Pikul, J. L. Jr, Liebig, M. A., Mikha, M. M., Varvel, G. E., Doran, J. W., & Andrews, S. S. (2006). Cropping system effects on soil quality in the Great Plains: Synthesis from a regional project. *Renewable Agriculture and Food Systems*, *21*, 49–59.

Yin, Z., Lei, T. W., Yan, Q. H., Chen, Z. P., & Dong, Y. Q. (2003). A near-infrared reflectance sensor for soil surface moisture measurement. *Computers and Electronics in Agriculture*, *99*, 101–107.

Yu, L., Tao, S., Ren, Y., Gao, W., Liu, X., Hu, Y., & Shamshiri, R. R. (2021). Comprehensive evaluation of soil moisture sensing technology applications based on analytic hierarchy process and delphi. *Agriculture*, *11*, 1116. Available from https://doi.org/10.3390/agriculture11111116.

Zhu, A. X., Hudson, B., Burt, J., Lubich, K., & Simonson, D. (2001). Soil mapping using GIS, expert knowledge, and fuzzy logic. *Soil Science Society of America Journal. Soil Science Society of America*, *65*, 1463.

Zibayiwa, M. (2020). Remote sensing full explained. https://www.researchgate.net/publication/340716114_Remote_Sensing_Full_Explained_DEDICATION/link/5e99b9e592851c2f52aa3232/download.

CHAPTER

8 Role of precision agriculture in soil fertility and its application to farmers

R. Karthika

PG & Research Department of Biotechnology, K.S. Rangasamy college of Arts and Science, Tiruchengode, Namakkal, Tamil Nadu, India

8.1 Introduction

The pursuit of increased production means modernizing agriculture, by implementing contemporary methods designed to increase the productivity of land and labor. To maximize agricultural production for sustainability and food security, there is also a larger requirement for consumer demand for food and agricultural products (Wu, 2011). The interaction of seed, soil, water, and agrochemicals (including fertilizers) results in the agriculture production system. Our focus must shift to a new technology to revolutionize our agricultural productivity through careful control of all inputs, which is essential for the long-term viability of such a complex system. To preserve crop productivity, it is crucial to create eco-friendly technology. The scientific community is motivated by this desire to use all the inputs precisely in a particular land area rather than try to expand the land area, which is already a permanent constraint. To measure field conditions and micromanage the elements or resources in small or big land areas, precision management involves the right use of high-tech equipment and software. Precision management's objectives are quite obvious. The farmers' priority is to find ways to increase their harvests and/or income. The family's requirements are met by income, which also benefits the neighborhood and the farming enterprise. Input costs and environmental impacts are typically reduced when inputs are decreased. Additionally, our society looks to farmers to manage natural resources and be responsible environmental stewards.

The practice of using information-technology-based agricultural operations to address spatial variability is known as precision agriculture (PA). It can be seen as a cycle that starts with data collection, then moves on to data analysis and interpretation, recommendation creation, field application, and results from the evaluation (Gebbers & Adamchuk, 2010). In 1990, German scientists used a canola crop to create the first productivity map based on a global positioning system (GPS)-connected yield monitor (Schnug, 1991). Since then, growers have

Remote Sensing in Precision Agriculture. DOI: https://doi.org/10.1016/B978-0-323-91068-2.00017-5

gained access to PA thanks to the development of numerous tools and technologies.

The management of agriculture and livestock production can greatly benefit from the set of geoprocessing, or geotechnologies, which includes tools for gathering, processing, analyzing, and delivering information with spatial context (Filippini Alba et al., 2017; Lu et al., 2011). These layers of data can be combined through modeling using the geographic information system (GIS), which increases the capacity for data interpretation and aids in management decision-making for the production system.

PA was previously only available to large farmers due to its high cost and lack of technical expertise. In contrast, because of current research and technological advancements such as GPS and GIS, small- and medium-sized farmers now have full access to PA. Moreover, given that the training is easy to understand, inexpensive, and frequently offered for free by the same company that provides the equipment, it is related to both economic and technical difficulties. One of the tools most frequently used in Brazil to assess soil fertility is the examination of the earth's chemical composition. The degree of elemental adequacy or deficit in the soil is attempted to be ascertained using chemical extractors, along with the quantification of unfavorable conditions that can hinder plant development (Bertalot et al., 2014). It has been confirmed that the vast majority (93%) of producers who adopt the PA uses this tool.

8.2 Elements/components of precision agriculture

The goal of PA is the development of innovative techniques and technological tools to enhance the process of soil fertility monitoring to equalize it through differential fertilizer application and achieve the maximum crop yields feasibly. Measurement and comprehension of variability are essential to precision farming (PF), and the variability must be addressed by the system's core components. GPS, GIS, remote sensing (RS), variable-rate technologies, PF, real-time nitrogen management, biointensive farming, diagnosis and recommendation integrated system (DRIS) approach, yield monitors, soil sensors, etc., are some examples of the components (the enabling technologies) of PF technology.

8.2.1 Global position systems

GPS technologies give farmers unmatched accuracy, positioning freedom for their tractors, navigation, and data collection about a specific location, field, or farm. Through GPS satellites, the spatial variations are quickly interpreted and recorded. A total of 2431 satellites, located in a high-altitude orbit above the planet, are known to be used by the GPS. Their primary objective is to examine the crops, water, and soil on the surface of the planet. These satellites continuously send out signals that are picked up by specialized receivers. Three components make up the GPS: a control system in Colorado, in the center of the United States, and linked stations, a space segment made up of 21 operational satellites orbiting the globe at 20,000 km, and a user segment made up of receivers that provide positioning, velocity, and time information to the user. The coordinates and location of a GPS receiver are claimed to be determined via at least four satellite contacts. The GPS signals must be precise and crisp. Unrefined and foggy signals are useless for pinpointing a field position. The use of GPS-referenced signals for seed planting, soil fertility mapping, yield fluctuations, soil moisture distribution, and other purposes is common. A network of military satellites is used by the GPS to pinpoint a user's location. There are benefits to properly determining a point in a field or the placement of farm vehicles (Kingsley et al., 2013). The GPS facilitates the geographical attribution of soil properties. Based on soil maps, the GPS aids in precisely dispensing fertilizer at each location. Through yield monitors, the GPS aids in precisely mapping the productivity of crop growth. The GPS is useful for locating locations with high or low yields.

8.2.2 Global information system

GIS generally improves our capacity for planning the distribution of pesticides, water, and fertilizer, as well as decision support tools. The actual or acquired collection of computers aids in the collection, archiving, retrieval, updating, manipulation, and analysis of data. We may integrate geographic data with a variety of factors, such as field history, soil fertility, crop husbandry, and crop production estimates, with its assistance. It enables us to mimic fertilizer delivery and crop response in a location, allowing for the making of the necessary decisions. We can examine data gathered over time at a specific location using the GIS that is pertinent to crop production. The geographical impact of soil management and agronomic parameters on crop productivity can be studied using GIS. For instance, spatial data could be used to determine the effects of soil type, its texture or fertility status, pH, organic matter, or moisture on grain or forage yield. Regression analysis of data on the effects of the three main nutrients (N, P, and K) across the entire field might be kept and used to inform decisions for the upcoming season. It is possible to employ soil nutrient maps as well as maps that show regional variations in many other important soil characteristics, such as texture, pH, electrical conductivity, and others (Zhang et al., 2010). Based on geographical data and agricultural production trends, GIS assists in creating prescription maps for each field or smaller management zones (Pick, 2005). GIS records for larger cropping zones and stretches that cover a state or a small nation are also available, in addition to data on soil fertility and nutrient distribution within particular fields. They assist us in more precisely prescribing crop patterns and fertilizer channels within a given cropping belt. For instance, the United States and several European countries have access to comprehensive data on the pH fluctuations and soil nutrient status in many states and sub-regions. GIS can be used to swiftly access and examine fertilizer recommendations that have been made after taking into account a variety of natural elements, crop production systems, yield targets, and most significantly, the economic advantages (Andraski & Bundy, 2005).

8.2.3 Remote sensing

The science of RS involves gathering details about distant objects or regions, usually via planes or satellites. The data are gathered by a remote sensor. Aerial or satellite sensors are within the broad category of remote sensors. They may show differences in the color of the field that correlate to adjustments in the kind of soil, crop development, field boundaries, roads, water, etc. Vegetative indices, which represent the health of the plant, can be produced by processing aerial and satellite photos (Xue & Su, 2017).

8.2.4 Variable-rate technology

The VRT entails applying seeds, fertilizer, irrigation, and pesticides in amounts tailored to each location and variable. A field map and a decision support system determine the inputs (Debeljak et al., 2019). For instance, when it comes to soil fertility, computer-based decision support takes into account crop species and genotypes, yield targets, and profitability. The amount of input, such as fertilizers or water, channeled at each location depends on the guidance a GPS-guided vehicle receives from decision support systems. An individual field's soil fertility is uniform thanks to variable-rate fertilizer input. Similar to this, variable-rate irrigation based on a soil moisture map produces uniform soil moisture availability to the crop and modifies the water delivery volume for various field parts.

8.2.5 Precision farming

For a map and/or sensor-based technology, it has been suggested that at least two sets of methodologies could be used during PF. The first is known as map-based, and the second is known as sensor-based. Grid sampling, laboratory analysis of soil samples taken at least twice in-depth, creation of maps showing soil fertility and nutrient distribution, or maps that show soil physicochemical properties such as pH, cation exchange capacity (CEC), electrical conductivity, and so on, are all part of map-based methods. A variable applicator is then guided by the map created using soil data. We require a positioning device called a differential global positioning system (DGPS) to locate each point during both soil sample and variable-rate inputs. The second approach, sometimes referred to as "sensor-based PF," makes substantial use of real-time sensors along with a feedback control system to quickly measure the soil parameters "on the fly." Use these signals right away to tell a variable-rate applicator where to go next. A sensor is frequently installed on tractors, either in the front or in a separate truck that follows the tractor with a variable applicator. This gives enough time for data collection through sensors, followed by processing it using computer models and direct variable applicators. If therapies are administered right away, a GPS is not required for the sensor-based approach. However, data can be loaded and stored for the future on traction equipment equipped with GPS. Let us consider an example. "Soil DoctorR" produced and marketed by Crop Technology Inc, Houston, TX, USA, is an example that examines soil type, organic matter content, CEC, soil moisture, and NO_3-N using a rolling electrode as the tractor moves through the field. The need for GPS is eliminated. Yet another example is a sensor-based technology developed by researchers at Purdue University, West Lafayette, IN, USA. It is a sensor (photodiode) surrounded by a light emitting diode. To determine the amount of soil organic matter, the reflected light is measured.

8.2.6 Real-time nitrogen management

It means synchronization between crop N_2 demand and supply for improving nitrogen use efficiency and crop yield. Instrumentation used for real-time N_2 management is chlorophyll (SPAD) meter, leaf color chart (For rice crop), and through green seeker.

8.2.7 Biointensive farming

It is an organic agricultural system. It focuses on maximum yield from a minimum area of land while simultaneously maintaining and improving the fertility of the soil. It aims to produce maximum biomass per unit area. The concept and practices of biointensive farming were introduced by Alan Chadwick in the United States. Components of biointensive farming are: raised bed, BBF, FIRB, intensive planting, intercropping, companion planting, and whole system energy.

8.2.8 Diagnosis and recommendation integrated system approach

This technique is useful for nutrient analysis in plants. The full form of DRIS is diagnosis and recommendation integrated system. It was given by Beaufils, in 1973. This technique considers nutrient concentration ratio in the plant rather than individual element concentration.

8.2.9 Soil sensors

To optimize crop development conditions, combat biotic and abiotic challenges, and increase crop yields, soil sensors play a crucial role in smart and PA by monitoring real-time physical and chemical signals in the soil, such as temperature, moisture, pH, and pollutants. Crop productivity is largely influenced by the soil's moisture content, temperature, pH change, nutritional components, pests and insects, and soil contaminants. Based on the information about the soil condition gathered by various types of soil sensors, agriculture management decisions, such as the necessity and amount of irritants, fertilizers, and pesticides, will be made. To assess the health of the soil and plants, soil sensors are used. The market is anticipated to expand significantly over the coming few years due to the rising demand for the use and development of soil sensors in digital farming (Yin et al., 2021).

8.2.10 Yield sensors

The construction of sensors is based on many physicochemical phenomena. Electromagnetic induction, EC, ion-selective field transistors, optoelectronic sensors, ultrasonic displacement, and visual systems are the technologies used in sensors the most frequently. To make the sensor best suited for various scenarios and to improve accuracy, a variety of strategies are frequently used (Adamchuk & Rossel, 2010). For instance, in a combination harvester, the quantity of grain harvested is often measured using touch sensors, ultrasonic sensors, pressure, linear displacement, and optoelectronic sensors.

8.3 Precision agriculture in soil fertility

The requirement for farmers to produce more and more food will increase as the world's population rises. However, the amount of arable land cannot keep up, and the imminent danger to food security could easily lead to regional or perhaps global instability. Large farms are increasingly using PF to adapt to raise yields, cut waste, and lessen the dangers to the economy and security that inevitability comes along with agricultural uncertainty. Traditionally, decisions on when to plant, harvest, irrigate, and use pesticides and fertilizers are made for entire fields based on local conditions and experience. Contrarily, PF uses a combination of sensors, robots, GPS, mapping tools, and data-analytics software to adapt plant care without adding to labor costs. Drones with cameras and stationary sensors deliver pictures and data about specific plants, such as their stem size, leaf shape, and soil moisture, wirelessly to a computer, which scans them for symptoms of stress and health (Osroosh et al., 2018). Farmers use real-time feedback to provide water, pesticides, or fertilizer in precisely measured dosages to exactly the areas that require them. Farmers can use technology to assist in timing the planting and harvest of their crops. As a result, PF can enhance time management, lower water, and chemical use, and produce healthier crops and higher yields, all of which boost farmers' profits and resource conservation while lowering chemical runoff. The use of PA technologies could enhance the management of soil fertility and on-farm experiments or demonstrations.

8.3.1 Soil sampling for precision agriculture

The findings of soil sampling from numerous fields demonstrate the complexity of the spatial variability of P and K as well as other nutrients in soils and the stark differences in variability patterns between fields. The factors that contribute to variability on a large scale differ from those that contribute to variability on a smaller one. Over a scale of many acres, variation is typically caused by different soil types, terrain features, prior crops, or proximity to feeding lots. On a scale of a few feet or even inches, techniques such as tillage, fertilizer, and manure application also produce a lot of diversity. While in some fields the spatial variations of P, K, and pH tend to follow the distribution of soil types or other landscape characteristics, the variability of P or K (and occasionally of pH) typically does not follow the distribution of soil types, and the patterns change between fields (Trotter et al., 2014). The variability over many acres in various fields is frequently compared with that over a few hundred square feet.

The search for the best soil sampling plan that will work in every field has mainly proven fruitless. Sampling by soil type was as useful and should be more practical financially. Because the variation within those areas is as great as the variation across the entire field and because cell borders typically do not follow soil mapping units or landscape, data from most fields indicate that sampling large cells three to four acres in size does not accurately represent soil nutrient levels in many fields. A survey of a sizable area frequently overestimates the pH and nutrient levels of sizable portions of the fields. This issue cannot be resolved by collecting more soil cores for each composite sample. It could not be profitable to decrease the cell area or increase the number of points sampled. The most cost-effective option is probably a tailored grid sampling plan that takes landscape characteristics or other field data into account. This process is adaptable enough to accommodate various field conditions and sample intensity. An effective sample plan can be created using digital soil maps, soil test results, yield maps, and aerial photos (of bare soil and crop canopy). It has long been known that technically possible soil sample methods might not provide agronomists with the level of precision they desire when describing variations in soil nutrients. Even conventionally advised soil sample techniques have consistently sacrificed detail for economic viability (Krishan et al., 2014). However, soil testing for P and K has worked well as a way to establish fertilizer recommendations despite deficits. Grid soil sampling is not much different than this.

8.3.2 Soil sampling and variable-rate fertilization

Several variables affect how grid sampling and variable-rate fertilizer affect soil fertility management and crop production profitability. Some of these are the nutrient levels discovered concerning crop needs, the patterns of nutrient variability, the fertilizer recommendations used, the anticipated effects of fertilization, and the additional expenses. Four trials with maize and soybeans in fields that tested optimal or high on average indicated no significant yield advantage of variable-rate P fertilization (or uniform fertilization) in the majority of fields because there was minimal response to P (Wittry, 2004). This should not come as a surprise as surveys indicate that more than half of Iowa fields test at or above the optimal level for P and K. Because the probability of yield responses in soils testing optimal or above is limited and the number of low-testing areas is likewise small, significant variability in a field with a preponderance of optimum or high values is likely to be unimportant. The most likely benefit of thorough grid sampling and variable-rate fertilization will be accomplished through savings in maintenance fertilizing given the likelihood of modest responsive regions in many Iowa fields. Producers who understand that most of their fields test above optimal and do not require maintenance fertilizing until levels decline to the optimum range will also benefit from these techniques (Bodner et al., 2015). From the standpoint of sustainability and the environment, these results will be advantageous. Lack of grid sampling typically forces farmers to use a uniform high maintenance rate across the field even when it may not be necessary.

8.3.3 Interpretation of soil test and yield maps

Numerous farmers hold the opinion that differences in nutrient levels can be used to explain a large portion of the yield variability within a field and that there is much to be learned by comparing multiple layers of data. Statistics and visual observations of maps of soil test results, soil types, and yields revealed that the fertility measures could only partially account for the yield variations in each area (Earl et al., 2003). Given that some variables affect crop yields, this outcome may be to be anticipated. High nutritional variance did not always account for extremely variable yields, according to the field data that were connected to yields. The use of tests and data management techniques that cannot typically be used by producers revealed that apparent correlations were frequently deceptive and would result in incorrect conclusions, even though in a few instances low yields could be reasonably explained by the soil test or soil type maps. It is impossible to know for sure which nutrients affect yields without comparing fertilization- and non-fertilization-based treatments. This idea motivates efforts to provide better techniques for contrasting management strategies on a field level.

8.3.4 On-farm comparison of management practices

Numerous on-farm comparisons have produced results demonstrating that PA technologies can be successfully applied to on-farm, field-scale analyses of various management strategies. Large quantities of grain are weighed, and treatments are applied to lengthy strips that are duplicated across the field as a popular technique for on-farm research. Grid sampling, DGPSs, yield monitors, and data management with GISs computer software enable a more thorough assessment of treatment differences for various parts of a field and allow for the estimation of interactions between the response to fertilization and other growth factors (Andreo Verónica, 2013).

By taking into account the spatial variance of yields in commonly used statistical studies, the type of data handling that may be performed by grid sampling, DGPS, and yield monitors enables better evaluations of practices. Additionally, it enables the analysis of treatment variations for various field parameters, such as different soil types, soil test results, and other field characteristics. To minimize errors brought on by DGPS issues, the usage of yield monitors, and border effects of the treatments being compared, a minimal set of quality control measures must be followed. The yield monitor should be calibrated properly. Yield points and field strips should be correctly georeferenced; errors caused by field boundaries, streams, or grass strips; changes in the width of the harvested swath (a frequent problem with soybeans) (Lyle et al., 2013). It must be underlined that regardless of the length or width of the strip used, valid conclusions on differences between treatments applied to strips can only be drawn when treatments are reproduced throughout the field.

8.3.5 General recommendations

The current findings demonstrate that there is no single general advice that applies to all sectors. Because soil nutrient variability varies depending on the field, it is desirable to sample and fertilize each field differently. Knowledge about each field is necessary to make accurate assessments of the advantages of variable-rate fertilizer and grid soil sampling. When variance within a field is such that considerable areas of the field test are both below and above optimal, intensive grid soil sampling and variable-rate fertilizer will likely produce significant economic gains. Unfortunately, only after carrying out a time-consuming, expensive preliminary sampling can the benefits of variable-rate fertilization be reliably calculated. An acceptable substitute is to enhance the conventional "sample by soil map unit" method using the existing field data and carry out a more thorough but still affordable sampling (Nanni et al., 2011). To select specific sites for intense sampling, previous soil test results, yield maps, aerial photos, and field histories might be used. This strategy will probably boost producers' financial gains while using the potential of new PF technologies.

Using on-farm strip trials, PA technologies can be used to assess and illustrate different fertilization or management approaches. This methodology can be used to conduct new research, modify suggestions for regional conditions, and gain a better understanding of the factors that influence yield variability. Maps of soil test results, soil types, other soil or plant parameters, and grain yields should all be interpreted cautiously by farmers. Without fertilization treatments, questions about which nutrients limit yields cannot be answered with precision (Yousaf et al., 2017). To discover nutrient deficiencies and investigate how fertilization interacts with other growth factors, it is helpful to use the variation that is typically seen within fields (in soils, diseases, yields, and other factors) in conjunction with duplicated applications of treatments. This possibility does not negate the requirement to repeat the comparisons across different fields and periods. The first findings indicate that to eliminate errors and produce precise protocols for on-farm comparisons, a minimal set of quality control procedures must be followed.

8.4 Conclusion

A map of the field's soil nutrients is necessary for precision forming processes. It offers a general implementation guide for agronomic techniques, such as precision location-specific fertilizer administration, and practically describes differences in soil fertility. A soil nutrient map is also a solid starting point for managing roughly nutrient dynamics, increasing crop production, and using less fertilizer. The creation of soil nutrient status and treatment maps for large-scale paddy development was made easier by the use of information and communication technology in agriculture. This technology has been used to create better methods for estimating and mapping soil nutrients. With the use of geospatial technology, the decision support system has been created to calculate the number of nutrients needed for paddy cultivation. A soil nutrient sensor is used in conjunction with this method to quickly check the soil's nutrient condition for large-scale rice farming.

References

Adamchuk, V. I., & Rossel, R. A. V. (2010). *Development of on-the-go proximal soil sensor systems. Proximal Soil Sensing* (pp. 15–28). Springer Science + Business Media B. V.

Andraski, T. W., & Bundy, L. G. (2005). Cover crop effects on corn yield response to nitrogen on an irrigated sandy soil. *Agronomy Journal*, *97*(4), 1239.

Andreo, V. (2013). *Remote sensing and geographic information systems in precision farming*. Technical report.

Bertalot, M., Guerrini, I., Mendoza, E., & Pinto, M. (2014). Productivity, leaf nutrient content and soil carbon stocked in agroforestry and traditional management of maize (*Zea mays* L.). *American Journal of Plant Sciences*, *5*, 884–898.

Bodner, G., Nakhforoosh, A., & Kaul, H. P. (2015). Management of crop water under drought: A review. *Agronomy for Sustainable Development*, *35*, 401–442.

Debeljak, M., Trajanov, A., Kuzmanovski, V., Schröder, J., Sandén, T., Spiegel, H., & Henriksen, C. B. (2019). A field-scale decision support system for assessment and management of soil functions. *Frontiers in Environmental Science*, *7*.

Earl, R., Taylor, J., Wood, G., Bradley, I., James, I., Waine, T., & Knight, S. (2003). Soil factors and their influence on within-field crop variability, Part I: Field observation of soil variation. *Biosystems Engineering*, *84*(4), 425–440.

Filippini Alba, J. M., Flores, C. A., & Miele, A. (2017). Geotechnologies and soil mapping for delimitation of management zones as an approach to precision viticulture. *Applied and Environmental Soil Science*, *2017*, 1–10.

Gebbers, R., & Adamchuk, V. I. (2010). Precision agriculture and food security. *Science (New York, N.Y.)*, *327*(5967), 828–831.

Kingsley, J., Bouslihim, Y., Isong, I. A., Hssaini, L., Razouk, R., & Krishna, K. R. (2013). *Precision forming soil fertility and productivity aspects* (p. 48) CRC Press.

Krishan, G., Saha, S. K., Kumar, S., & Patel, N. R. (2014). Remote sensing in soil fertility evaluation and management. In R. Chandra, & K. P. Raverkar (Eds.), *Bioresources for sustainable plant nutrient management* (pp. 509–533). Serial Publishing House.

Lu, D., Li, G., Moran, E., Dutra, L., & Batistella, M. (2011). A comparison of multisensor integration methods for land cover classification in the Brazilian Amazon. *GIScience & Remote Sensing*, *48*(3), 345–370.

Lyle, G., Bryan, B. A., & Ostendorf, B. (2013). Post-processing methods to eliminate erroneous grain yield measurements: Review and directions for future development. *Precision Agriculture*, *15*(4), 377–402.

Nanni, M. R., Povh, F. P., Demattê, J. A. M., Oliveira, R. B., de, Chicati, M. L., & Cezar, E. (2011). Optimum size in grid soil sampling for variable rate application in site-specific management. *Scientia Agricola*, *68*(3), 386–392.

Osroosh, Y., Khot, L. R., & Peters, R. T. (2018). Economical thermal-RGB imaging system for monitoring agricultural crops. *Computers and Electronics in Agriculture*, *147*, 34–43.

Pick, J. B. (2005). *Geographic information systems in business* (p. 324) Idea Group Publishing.

Schnug, E. (1991). Sulphur nutritional status of European crops and consequences for agriculture. *Sulphur in Agriculture*, *15*.

Trotter, M., Guppy, C., Haling, R., Trotter, T., Edwards, C., & Lamb, D. (2014). Spatial variability in pH and key soil nutrients: Is this an opportunity to increase fertiliser and lime-use efficiency in grazing systems? *Crop and Pasture Science*, *65*(8), 817.

Wittry, D. J. (2004). Comparison of uniform and variable-rate phosphorus fertilization for corn–soybean rotations. *Agronomy Journal*, *96*(1), 26–33.

Wu, Z. I. (2011). Research on harmony between agricultural modernization and regional economy development in China. *Asian Agricultural Research*, *3*(3), 6–10.

Xue, J., & Su, B. (2017). Significant remote sensing vegetation indices: A review of developments and applications. *Journal of Sensors*, *2017*, 1–17.

Yin, H., Cao, Y., Marelli, B., Zeng, X., Mason, A. J., & Cao, C. (2021). Soil sensors and plant wearables for smart and precision agriculture. *Advanced Materials*, *33*(20), 2007764.

Yousaf, M., Li, J., Lu, J., Ren, T., Cong, R., Fahad, S., & Li, X. (2017). Effects of fertilization on crop production and nutrient-supplying capacity under rice-oilseed rape rotation system. *Scientific Reports*, *7*(1).

Zhang, Q., Yang, Z., Li, Y., Chen, D., Zhang, J., & Chen, M. (2010). Spatial variability of soil nutrients and GIS-based nutrient management in Yongji County, China. *International Journal of Geographical Information Science*, *24*(7), 965–981.

SECTION

3

Precision agriculture practices for irrigation water management

CHAPTER

9

Remote sensing and geographic information system–based land suitability analysis for precision agriculture: a case of paddy cultivation in East Siang district of Arunachal Pradesh (India)

Ranjit Mahato[1], Dhoni Bushi[1], Gibji Nimasow[1] and Oyi Dai Nimasow[2]

[1]*Department of Geography, Rajiv Gandhi University, Itanagar, Arunachal Pradesh, India*

[2]*Department of Botany, Rajiv Gandhi University, Itanagar, Arunachal Pradesh, India*

9.1 Introduction

Land use planning is essential to increase agricultural productivity, which is mostly shaped by the soil, topography, climate, vegetation cover, and hydrological characteristics (Sarkar et al., 2021). The pressure of food security on shrinking agricultural land is attributed to either an increase in the population or decrease in land or both (Xiang et al., 2020). According to Pham et al. (2013), the available agricultural land in developing countries tends to decrease to an extent that food production can be significantly affected for a long time. Moreover, the crop yields/production and the fundamental natural resources primarily land and water are largely affected by the changing climate (Aryal et al., 2020). Therefore, rational land management systems for sustainability of arable land and precision agriculture are a prerequisite for the development of any country (Uisso & Tanrıvermiş, 2021). Paddy is an important species of grain for the nourishment of humanity (Özkan et al., 2019). China and India have the largest rice planting areas and are the most grain producing and rice consuming nations in the world. Paddy cultivation in these two countries plays a pivotal role for both national and global food security (Zhang et al., 2017).

India possesses the 10th largest arable land resource in the world, which serves as a primary source of livelihood for about 58% of the country's

Remote Sensing in Precision Agriculture. DOI: https://doi.org/10.1016/B978-0-323-91068-2.00023-0

population. There are two major agricultural seasons in India, namely *kharif* (summer), which lasts from April to September, and *rabi* (winter), which lasts from October to March. Rice (Paddy) is the main crop of the *kharif* season, whereas wheat is the main crop of the *rabi* season (India Brand Equity Foundation, 2020). Despite having large arable land, the nation is facing problems in maintaining productivity and sustainability of rice-based agriculture system. The major drawbacks include insufficient application of resources (water, labor, and fertilizers), growing scarcity of resources (water and labor), climate change, changing land use (cropping practices and cropping systems), socioeconomic changes such as urbanization, labor migration, changing attitude of people, lack of concerns on farm-related pollutions, rising fuel prices, and emerging energy crisis and reducing farm profitability (Ladha et al., 2009; Mahajan et al., 2017). Furthermore, the continuous practice of traditional mode of rice cultivation is also imposing serious threats to the natural resources and agricultural sustainability (Kumar et al., 2021).

Land suitability analysis (LSA) is the process of determining the suitability of a given land area for a certain type of use (agriculture, forest, recreation, etc.) and the level of suitability (Akıncı et al., 2013). The application of geographic information system (GIS) can handle and simulate different geo-environmental parameters acquired from various sources (Al-Hanbali et al., 2021). The combination of GIS with multicriteria decision-making (MCDM) techniques is popular and mostly applied in land suitability analysis (Pramanik, 2016). Among the different MCDM techniques, analytical hierarchy process (AHP) is commonly used to resolve complex decision-making processes, which include multiple criteria, scenarios, and factors (Bozdağ et al., 2016). AHP was first developed by Saaty (1980) where the decision-maker performs a pair-wise comparison judgment to develop final priorities for ranking the alternatives without any inconsistency (Saaty & Vargas, 2012). The combination of GIS and AHP methods has been largely used in land suitability analysis of agricultural crops in recent years (Ennaji et al., 2018; Everest et al., 2021; Tashayo et al., 2020; Yalew et al., 2016). Therefore, this study attempted to delineate the land suitability of paddy cultivation in the East Siang district of Arunachal Pradesh, India. The district is mostly plain and possesses high potential for paddy cultivation. The land suitability analysis could be fruitful in identifying the suitable areas for paddy cultivation, increased rice productivity, efficient land use planning, and thereby precision agriculture to ensure rural food security.

9.2 Materials and methods

9.2.1 Study area

The study area (East Siang district) is located approximately between 27°43′ to 29°20′ North latitudes and 94°42′ to 95°35′ East longitudes (Fig. 9.1).

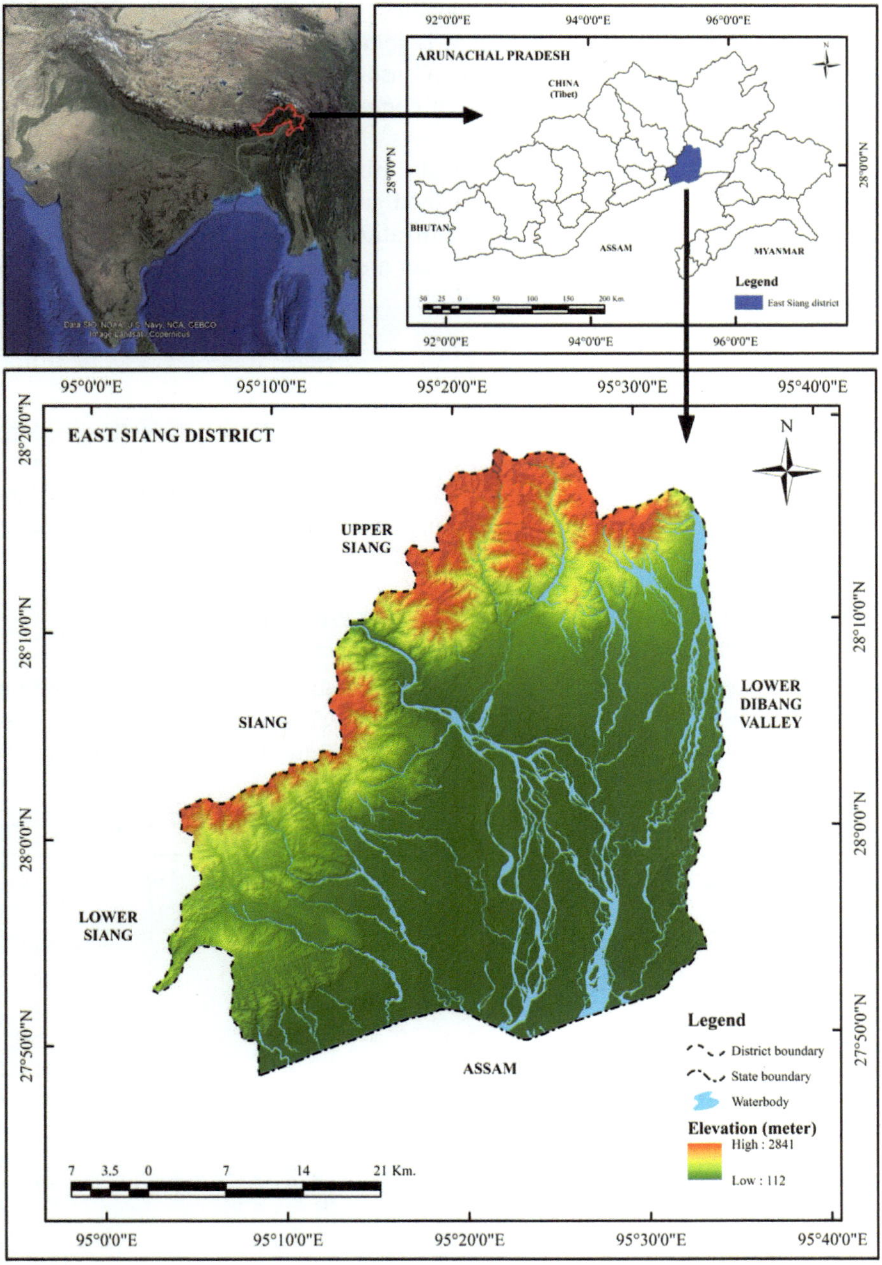

FIGURE 9.1

Location of the study area (East Siang district, Arunachal Pradesh, India).

The Mighty River Siang, which originates from Tibet (China), where it is called Tsangpo, is the major river that is joined by many tributaries. The total population of the district was 81,882 (Census of India, 2011). The district is mostly inhabited by the Adi tribe, which comprises a large number of sub-tribes such as the Minyongs, Padams, Shimongs, Milangs, Pasis, Karkos, Ashings, Pangis, Tangams, and Boris, and the Galo tribe. They are mostly dependent on agriculture and allied activities such as food gathering, hunting, and fishing. The villagers practice traditional shifting cultivation in the hilly mountainous areas and wet rice cultivation (WRC) in the plain areas. Paddy is an important crop of both shifting and WRC agricultural fields.

9.2.2 Methodology

The study consists of five main stages, namely data collection, selection of parameters and creation of thematic maps, standardization of criteria, collection of expert opinion, pair-wise comparison matrix using AHP, and derivation of land suitability map. The schematic representation of the land suitability analysis is shown in Fig. 9.2.

9.2.2.1 Data collection, selection of parameters, and creation of thematic maps

A detailed literature review on land suitability analysis of paddy was carried out and 16 parameters were selected for this study. The parameters include three topographic (elevation, slope, and aspect), two climatic (temperature and rainfall),

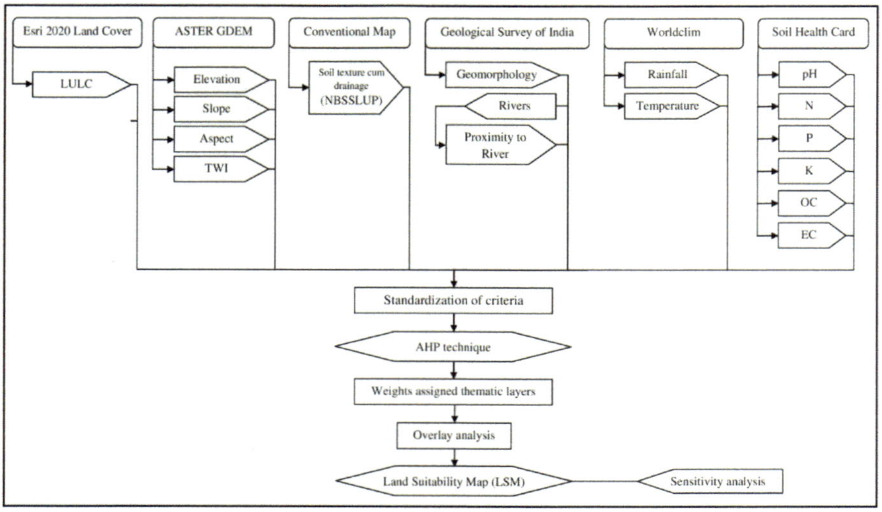

FIGURE 9.2

Schematic representation of the whole work.

two hydrological (topographic wetness index [TWI] and proximity to river), soil type, geomorphology, land use/land cover, and six layers of soil chemical properties (potential of hydrogen [pH], nitrogen [N], phosphorus [P], potassium [K], organic carbon [OC], and electrical conductivity [EC]). The data of each parameter have been collected from different sources (Table 9.1). The elevation, slope, aspect, and TWI were generated using ASTER GDEM with 30 m spatial resolution downloaded from https://search.earthdata.nasa.gov/search. Rainfall and temperature map in raster format with 1 km resolution was downloaded from https://www.worldclim.org. The geomorphology map in shape file was downloaded from https://bhukosh.gsi.gov.in. Proximity to river was generated in GIS environment using extracted river layer associated with geomorphology map. Soil type map was extracted from the soil map of Arunachal Pradesh published by ICAR-National

Bureau of Soil Survey & Land Use Planning, Nagpur, land use/land cover (LULC) map of the year 2020 was extracted from 10 class global LULC map with 10 m resolution (Karra et al., 2021). Lastly, the thematic maps of soil chemical properties were prepared using the datasets available in Soil Health Card (https://www.soilhealth.dac.gov.in) with the help of inverse distance weighting interpolation technique.

Elevation plays a fundamental role in the agricultural productivity (Everest et al., 2021) as well as land suitability. The vertical temperature fluctuation and

Table 9.1 Data sources of the selected parameters.

Criteria	Sources
Temperature	WorldClim (https://www.worldclim.org/)
Rainfall	WorldClim (https://www.worldclim.org/)
Geomorphology	Geological Survey of India (https://bhukosh.gsi.gov.in/Bhukosh/Public)
Proximity to River	Geological Survey of India (https://bhukosh.gsi.gov.in/Bhukosh/Public)
LULC	ESRI 2020 Global Land Use Land Cover from Sentinel-2 (Karra et al., 2021)
Elevation	ASTER GDEM with 30 m spatial resolution (https://search.earthdata.nasa.gov/search)
Slope	ASTER GDEM with 30 m spatial resolution (https://search.earthdata.nasa.gov/search)
Aspect	ASTER GDEM with 30 meter spatial resolution (https://search.earthdata.nasa.gov/search)
TWI	ASTER GDEM with 30 m spatial resolution (https://search.earthdata.nasa.gov/search)
Soil texture cum drainage	The National Bureau of Soil Survey and Land Use planning (NBSS&LUP)
Soil chemical properties (pH, OC, EC, N, P, and K)	Soil Health Card (https://www.soilhealth.dac.gov.in/)

resultant variation in plant distribution particularly in high lands are largely determined by the altitudinal effects (Akıncı et al., 2013; Bozdağ et al., 2016). The elevation of the study area ranges in between 112 and 2841 m, which was classified into five classes using natural break (Jenks) method (Fig. 9.3d). The plain or gently sloping areas have been assigned the highest weight as they are ideal for growing paddy due to its high requirement of stagnant water (Al-Hanbali et al., 2021). The slope map was classified into 0–5, 5–15, 15–25, 25–35, and above 35 degrees (Fig. 9.3c). Aspect also plays a significant role in the crop production (Everest et al., 2021) because sun exposure at certain intervals is important for the plant to maintain physiological activities. Generally, most cultigens exhibit optimum growth in the southern and western aspects that receive sunlight for a substantial portion of the day (Akıncı et al., 2013). The TWI, which was first introduced by Beven and Kirkby (1979), is a widely used compound topographic attribute for precision agriculture (Qin et al., 2011). TWI is a useful criterion that provides an estimate of the water accumulation (Al-Hanbali et al., 2021). A high index value represents relatively higher water availability than a small index value (Sørensen & Seibert, 2007). The TWI of the study area was calculated using the Eq. (9.1) and classified into five classes using natural break (Jenks) method:

$$\mathrm{TWI} = \ln\left(\frac{As}{\tan\beta}\right) \tag{9.1}$$

where As represents cumulative upslope area draining through a point and $\tan\beta$ slope angle at the point.

The climatic parameters such as temperature and rainfall have significant impact on crop productivity all over the world (Abbas & Mayo, 2021). Temperature and rainfall map of the study area was classified into five classes using natural break (Jenks) method, and the highest ratings were given to the highest values (Fig. 9.3a and b). Proximity to river is another important factor for paddy cultivation because river sedimentation is the main process in alluvial soil formation. The areas far away from the river networks have low potential of groundwater. Paddy cultivation is water-intensive that requires a specific distance from the river network. The proximity to river was classified into five categories, and highest ratings were given to the areas closest to the river (Fig. 9.3e). The agricultural crops also strongly depend on the soil type (texture and drainage). The soil drainage should prevent water from quickly percolating into the subsurface during paddy field cultivation (Al-Hanbali et al., 2021). The soil map was classified into 13 categories, and the highest rating was assigned to the class characterized by imperfectly drained coarse-silty soils on active flood plain (Fig. 9.3f). Land use data enables the estimation of an area's coverage under vegetation, settlements, forests, and water bodies (Binte Mostafiz et al., 2021). Rivers and water bodies, forest cover, and built-up areas were considered restricted area and thus assigned lowest weights for this study (Fig. 9.3n). The geomorphology plays an important role in the suitability of agricultural crops. The older flood

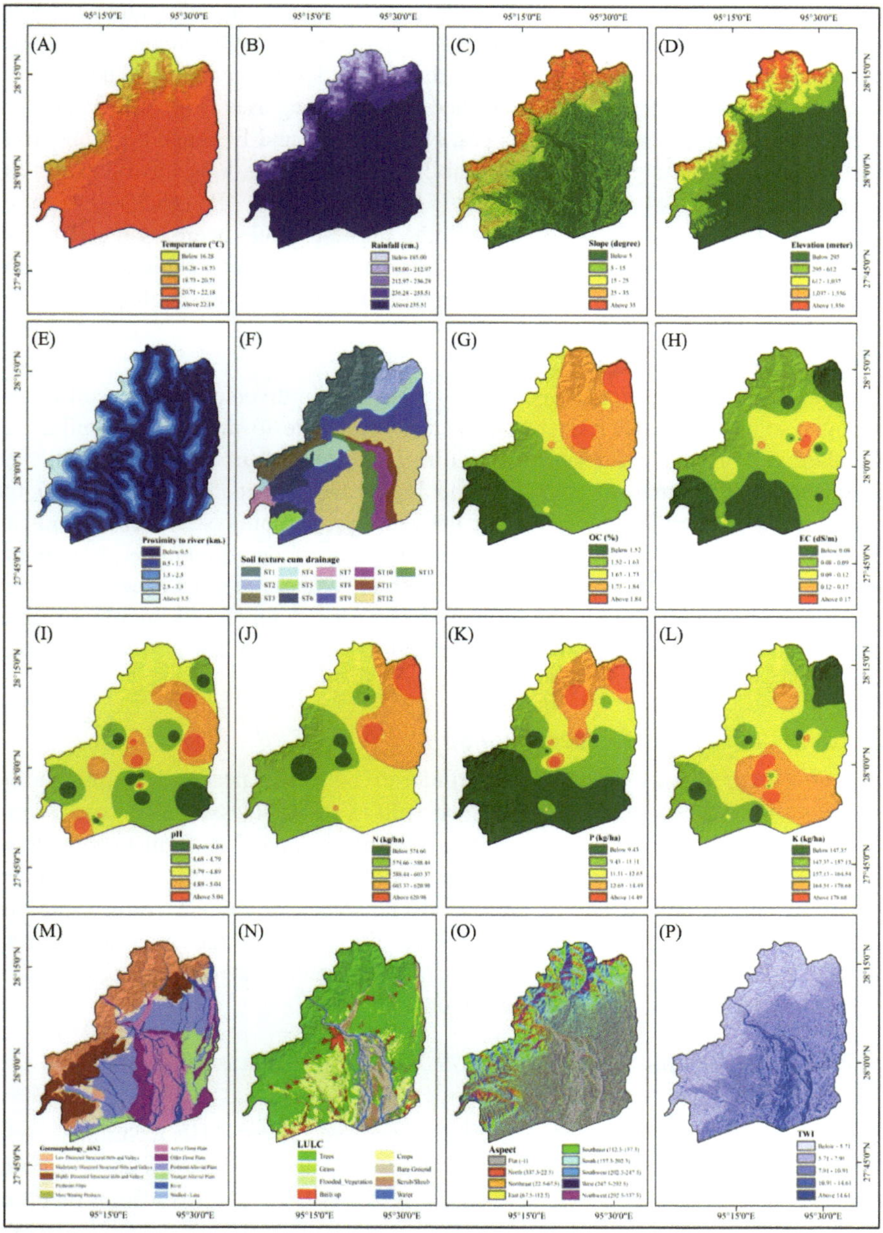

FIGURE 9.3

Thematic maps used in the LSA for paddy cultivation.

plain and piedmont alluvial plain have high potential of cultivation, whereas structural hills and valleys and active floodplain have low potential. Similarly, younger alluvial plain has less potential than the older alluvial plain. The geomorphology map was prepared and classified into 11 categories (Fig. 9.3m). Further, the soil chemical properties such as pH, N, P, K, OC, and EC have been prepared, and weights were assigned for the suitability analysis (Fig. 9.3g–l). Based on the ranges of pH (4.42–5.43), N (549.00–646.64), P (6.35–9.43), K (125.15–210.66), and OC (1.33–2.02), the highest values were assigned higher ratings. In terms of EC (0.04–0.30), which has negative correlation with paddy cultivation, the lowest ratings were assigned to the areas having highest EC.

9.2.2.2 Standardization of criteria

All the selected parameters were initially measured in diverse units as well as different resolution, format (raster and vector), and scale to maintain a similar and standard data format with geographic dimension before proceeding to AHP and weighted overlay analysis (WOA). The layers have been transformed into the same geographic environment having the same projection system. Thereafter, all the layers were projected into raster format with similar dimension and spatial resolution. Further, the layers were reclassified, and ratings have been assigned on 1–5 scale, where 1 represents not suitable and 5 represents highly suitable. The standardization of parameters is essential to maintain equal pixel size for perfect calculation of different parameters in WOA.

9.2.2.3 Collection of expert opinion

We developed a well-structured schedule to collect expert judgments on the weights and rankings of the parameters as well as the subclasses. The experts consist of two professors, three agricultural scientists, and five farmers who have been engaged in paddy cultivation since last 20 years. The judgments of professors and agricultural scientists were collected through Google form while the feedbacks of farmers have been collected through personal interview during October–December, 2020. Finally, the expert judgments have been aggregated to determine the ranking of influential factors.

9.2.2.4 Preparation of decision matrix using analytical hierarchy process

The weightage of different parameters was acquired using AHP technique. AHP (Saaty, 1980) is a specified decision-aiding tool (Saaty, 1989), which is used to derive priorities in MCDM problem (Saaty & Vargas, 1998). It is a semi-qualities method, which analyzes the different factors based on the pair-wise comparison matrix (Roy & Saha, 2018). During pair-wise comparison, the numerical judgments use the fundamental scale of absolute numbers ranging from 1 to 9 in order to prioritize parameters relative to one another (Saaty, 2004). The pair-wise comparison matrix and the normalized weights of selected parameters are given in Table 9.2.

Table 9.2 Pair-wise comparison and normalized weights of the parameters.

	TM	RN	SL	EL	PR	ST	OC	EC	pH	N	P	K	GM	LU	AS	TWI	Weights
TM	1.00	0.17	0.20	0.25	0.50	0.33	3.00	4.00	2.00	2.00	2.00	2.00	0.25	0.50	0.50	0.33	0.0355
RN	6.00	1.00	2.00	3.00	6.00	4.00	7.00	7.00	5.00	6.00	6.00	6.00	3.00	4.00	4.00	5.00	0.1913
SL	5.00	0.50	1.00	2.00	5.00	4.00	7.00	7.00	6.00	5.00	5.00	5.00	2.00	3.00	3.00	4.00	0.1481
EL	4.00	0.33	0.50	1.00	4.00	3.00	6.00	6.00	5.00	5.00	5.00	5.00	3.00	4.00	4.00	5.00	0.1358
PR	2.00	0.17	0.20	0.25	1.00	0.33	2.00	2.00	3.00	3.00	3.00	3.00	0.33	0.50	0.50	0.50	0.0412
ST	3.00	0.25	0.25	0.33	3.00	1.00	3.00	3.00	2.00	2.00	2.00	2.00	0.50	2.00	2.00	2.00	0.0610
OC	0.33	0.14	0.14	0.17	0.50	0.33	1.00	2.00	0.50	0.50	0.50	0.50	0.25	0.33	0.33	0.50	0.0183
EC	0.25	0.14	0.14	0.17	0.50	0.33	0.50	1.00	0.50	0.50	0.50	0.50	0.33	0.50	0.50	0.50	0.0176
pH	0.50	0.20	0.17	0.20	0.33	0.50	2.00	2.00	1.00	2.00	2.00	2.00	0.20	0.25	0.25	0.33	0.0277
N	0.50	0.17	0.20	0.20	0.33	0.50	2.00	2.00	0.50	1.00	1.00	1.00	0.25	0.33	0.33	0.33	0.0230
P	0.50	0.17	0.20	0.20	0.33	0.50	2.00	2.00	0.50	1.00	1.00	1.00	0.25	0.33	0.33	0.33	0.0230
K	0.50	0.17	0.20	0.20	0.33	0.50	2.00	2.00	0.50	1.00	1.00	1.00	0.25	0.33	0.33	0.33	0.0230
GM	4.00	0.33	0.50	0.33	3.00	2.00	4.00	3.00	5.00	4.00	4.00	4.00	1.00	3.00	3.00	2.00	0.0922
LU	2.00	0.25	0.33	0.25	2.00	0.50	3.00	2.00	4.00	3.00	3.00	3.00	0.33	1.00	3.00	2.00	0.0605
AS	2.00	0.25	0.33	0.25	2.00	0.50	3.00	2.00	4.00	3.00	3.00	3.00	0.33	0.33	1.00	0.50	0.0497
TWI	3.00	0.20	0.25	0.20	2.00	0.50	2.00	2.00	3.00	3.00	3.00	3.00	0.50	0.50	2.00	1.00	0.0521

AS, *aspect;* EC, *electrical conductivity;* EL, *elevation;* GM, *geomorphology;* K, *phosphorus;* LU, *land use/land cover;* N, *nitrogen;* OC, *organic carbon;* P, *potassium;* pH, *potential of hydrogen;* PR, *proximity to river;* RN, *rainfall;* SL, *slope;* ST, *soil texture cum drainage;* TM, *temperature;* TWI, *topographic wetness index.*

The AHP provides mathematical measures to determine the inconsistencies of the judgment matrix. The efficiency criteria of AHP are predicted by consistency ratio (CR), which is measured as Eq. (9.2).

$$CR = \frac{CI}{RI} \tag{9.2}$$

where CI represents consistency index and RI indicates random index based on the size of the matrices. RI is the mean value of the CI depending on the computed matrix order. If, the value of CR is >0.10, then the weight values of the matrix indicate inconsistencies, and the method (AHP) may not give meaningful results (Saaty, 1980). The value of CI was calculated using Eq. (9.3).

$$CI = \frac{\lambda_{max} - n}{n - 1} \tag{9.3}$$

where λ_{max} is the largest Eigen value of matrix and "n" is the order of the same matrix.

9.2.2.5 Generation of land suitability map

The WOA is an effective tool of GIS, which can perform suitability analysis based on multi-thematic layers. AHP is applied to generate the land suitability map of paddy using Eq. (9.4):

$$LSI = \sum_{i=1}^{n} W_i X_i \tag{9.4}$$

where LSI denotes the land suitability index, W_i represents the weight of the parameters, X_i denotes the assigned sub-criteria ratings of ith land suitability criteria, and n denotes the total number of selected parameters. The LSI was classified into four classes, namely highly suitable, moderately suitable, marginally suitable, and not suitable based on the FAO (1976).

9.2.3 Sensitivity analysis

Sensitivity analysis indicates which layer is the most/least significant in determining the output map (Fenta et al., 2015). Both the map-removal (Lodwick et al., 1990) and single-parameter (Napolitano & Fabbri, 1996) sensitivity analyses were carried out to justify the influence of thematic layers on the land suitability map. The map-removal method was determined using Eq. (9.5):

$$S = \frac{\left|\frac{A}{N} - \frac{B}{n}\right|}{A} \times 100 \tag{9.5}$$

where S is the index of sensitivity analysis associated with the removal of one map, A is the LSI computed using all the thematic layers, B is the LSI computed by excluding one thematic layer at a time, and N and n are the numbers of thematic layers used to compute A and B, respectively.

The single-parameter method compares the “effective” or “real” weight with the “empirical” weight assigned to each thematic layer in the LSI map. The effective weight was calculated using Eq. (9.6):

$$W = \frac{P_r P_w}{A} \times 100 \tag{9.6}$$

where W is the effective weight of each thematic layer, P_r and P_w are the rates and weight values of each thematic layer, and A is the land suitability map generated using all the thematic layers.

9.3 Results

9.3.1 Assessment of the selected parameters

The AHP-based decision matrix showed the largest Eigen value (λ_{max}) = 17.2259, RI = 1.5978, CI = 0.0817, CR = 0.0511, and $n = 16$ in this study. Hence, the selected parameters are considered to be essential in the AHP technique for land suitability analysis of paddy. The parameters and subclasses were assigned specific weights based on expert opinions (Table 9.3). The most important factors were rainfall (19.13%), slope (14.81%), elevation (13.58%), geomorphology (9.22%), soil texture cum drainage (6.10%), LULC (6.05%), TWI (5.21%), aspect (4.97%), proximity to river (4.12%), and temperature (3.55%) while organic carbon (1.83%) and electrical conductivity (1.76%) were the least contributing parameters. The southern region is characterized by floodplains with gentle slopes and high annual rainfall, whereas the northern parts are characterized by dissected structural hills and valleys and dense forest. Similarly, the conditions of TWI and proximity to river were better in the southern parts compared to the northern region.

9.3.2 Land suitability map for paddy

The land suitability map of paddy crop has been generated by combining remote sensing (RS), GIS, and MCDM techniques and selecting the most contributing factors for precision agriculture. We classified the land suitability zones into four categories, namely highly suitable, moderately suitable, marginally suitable, and not suitable. The results show an area of 164.92 km^2 (9.79%) as highly suitable, 619.07 km^2 (36.74%) as moderately suitable, 454.98 km^2 (27%) as marginally suitable, and 445.95 km^2 (26.47%) as not suitable for paddy cultivation in the study area (Table 9.4). The highly suitable areas are mostly located in the southwestern and southeastern parts comprising Bilat, Sille-Oyan, Pasighat, Ruksin, and Mebo circles (Fig. 9.4). The suitable areas predicted by the LSA are characterized by high annual rainfall, gentle slopes, alluvial flood plains, and good soil conditions (texture and drainage) essential for paddy cultivation. Similarly, the

Table 9.3 Parameter weights and ratings of subclasses.

Parameters	Subclass	Ratings	Parameter weights
Temperature (°C)	Below 16.28	1	0.0355
	16.28–18.73	2	
	18.73–20.71	3	
	20.71–22.18	4	
	Above 22.18	5	
Rainfall (cm)	Below 185.00	1	0.1913
	185.00–212.97	2	
	212.97–236.28	3	
	236.28–255.51	4	
	Above 255.51	5	
Slope (degree)	Below 5	5	0.1481
	5–15	4	
	15–25	3	
	25–35	2	
	Above 35	1	
Elevation (m)	Below 295	5	0.1358
	295–612	4	
	612–1037	3	
	1037–1556	2	
	Above 1556	1	
Proximity to river (km)	Below 0.5	2	0.0412
	0.5–1.5	5	
	1.5–2.5	4	
	2.5–3.5	3	
	Above 3.5	1	
Soil texture cum drainage	ST1 (excessively drained loamy soil on hill slope)	1	0.0610
	ST2 (well-drained fine-loamy soils on hill slope)	2	
	ST3 (well-drained fine-clayey soils on hill slope)	1	
	ST4 (well-drained fine-loamy soils on steep summits)	1	
	ST5 (excessively drained fine-loamy soils on hill slope)	1	
	ST6 (excessively drained loamy-skeletal soils on hill slope)	2	
	ST7 (well-drained coarse-loamy soils on piedmont/hill slope)	2	
	ST8 (well-drained loamy-skeletal soils on piedmont slope)	3	

(*Continued*)

Table 9.3 Parameter weights and ratings of subclasses. *Continued*

Parameters	Subclass	Ratings	Parameter weights
	ST9 (well-drained coarse-loamy soils on gentle piedmont slope)	4	
	ST10 (moderately well-drained coarse-loamy soils on flood plain)	1	
	ST11 (well-drained coarse-loamy soils on active flood plain)	1	
	ST12 (imperfectly drained coarse-silty soils on active flood plain)	5	
	ST13 (excessively drained sandy soils on bar lands)	1	
Organic carbon	Below 1.52	1	0.0183
	1.52–1.63	2	
	1.63–1.73	3	
	1.73–1.84	4	
	Above 1.84	5	
Electrical conductivity	Below 0.08	5	0.0176
	0.08–0.09	4	
	0.09–0.12	3	
	0.12–0.17	2	
	Above 0.17	1	
pH	Below 4.68	1	0.0277
	4.68–4.79	2	
	4.79–4.89	3	
	4.89–5.04	4	
	Above 5.04	5	
N	Below 574.66	1	0.0230
	574.66–588.44	2	
	588.44–603.37	3	
	603.37–620.98	4	
	Above 620.98	5	
P	Below 9.43	1	0.0230
	9.43–11.11	2	
	11.11–12.65	3	
	12.65–14.49	4	
	Above 14.49	5	
K	Below 147.37	1	0.0230
	147.37–157.13	2	
	157.13–164.54	3	
	164.54–178.68	4	
	Above 178.68	5	

(Continued)

Table 9.3 Parameter weights and ratings of subclasses. *Continued*

Parameters	Subclass	Ratings	Parameter weights
Geomorphology	Active flood plain	1	0.0922
	Highly dissected structural hills and valleys	1	
	Moderately dissected structural hills and valleys	1	
	Low dissected structural hills and valleys	1	
	Mass wasting products	1	
	Older flood plain	5	
	Piedmont alluvial plain	4	
	Piedmont slope	2	
	River	1	
	Other water bodies	1	
	Young alluvial plain	3	
LULC	Water	1	0.0605
	Trees	1	
	Grass	3	
	Flooded vegetation	3	
	Crop land	5	
	Scrub/shrub	1	
	Built up	1	
	Bare ground	1	
Aspect	Flat	5	0.0497
	North	1	
	North east	2	
	East	3	
	South east	4	
	South	5	
	South west	4	
	West	3	
	North west	2	
TWI	Below 5.71	1	0.0521
	5.71–7.91	2	
	7.91–10.91	3	
	10.91–14.61	4	
	Above 14.61	5	

moderately suitable areas are also interspersed along the southwestern and southeastern highly suitable areas. These areas are characterized by piedmont alluvial plain with somewhat higher slopes and lesser annual rainfall. The central part

Table 9.4 Total area and percentage area under suitable classes.

Suitability classes	Area (km^2)	Percentage
Not suitable	445.95	26.47
Marginally suitable	454.98	27.00
Moderately suitable	619.07	36.74
Highly suitable	164.92	9.79
Total	1684.92	100.00

(Daying Ering Wildlife Sanctuary), northwestern and northeastern mid-altitude parts have been categorized as marginally suitable areas. The whole of northern region characterized by dissected structural hills and valleys with lesser annual rainfall, steep slope, and dense forest has been classified as not suitable for paddy crop. Overall, the LSA predicted 46.53% (highly + moderately suitable) of the study area as suitable for paddy cultivation. The findings show significant potential areas for paddy cultivation in East Siang district provided the farmers are prepared to cultivate paddy in the predicted suitable areas. As per the LULC map of the study area, the predicted suitable areas mostly fall in the current agricultural fields under paddy crop. Therefore, a combined application of RS and GIS technology with MCDM techniques can effectively identify the land suitability of important crops and can prove fruitful for precision agriculture.

9.3.3 Sensitivity analysis

9.3.3.1 Map-removal sensitivity analysis

The statistics of the variation index of map-removal sensitivity analysis resulting from the removal of one parameter at a time is shown in Table 9.5. The highest variation index with a mean of 2.16% was found in the removal of the rainfall layer. The other layers showing high variation index were elevation and slope with mean values of 0.70 and 0.68, respectively. The geomorphology, aspect, soil texture cum drainage, and temperature recorded the lowest variation index with mean value ranges of 0.19%–0.12%.

9.3.3.2 Single-parameter sensitivity analysis

The statistics of single-parameter sensitivity analysis are shown in Table 9.6. There are some deviations in the effective weights when compared with the empirical weights. In conformity with the results of the map-removal analysis, the single-parameter analysis also shows rainfall as the most effective layer in mapping of LSI with a mean effective weight of 25.15%. The next higher mean effective weights of 16.77 and 16.49 were recorded by the elevation and slope layers. The geomorphology, aspect, soil texture cum drainage, temperature, and proximity to river exhibit slightly moderate mean effective weights of 5.52%, 5.10%,

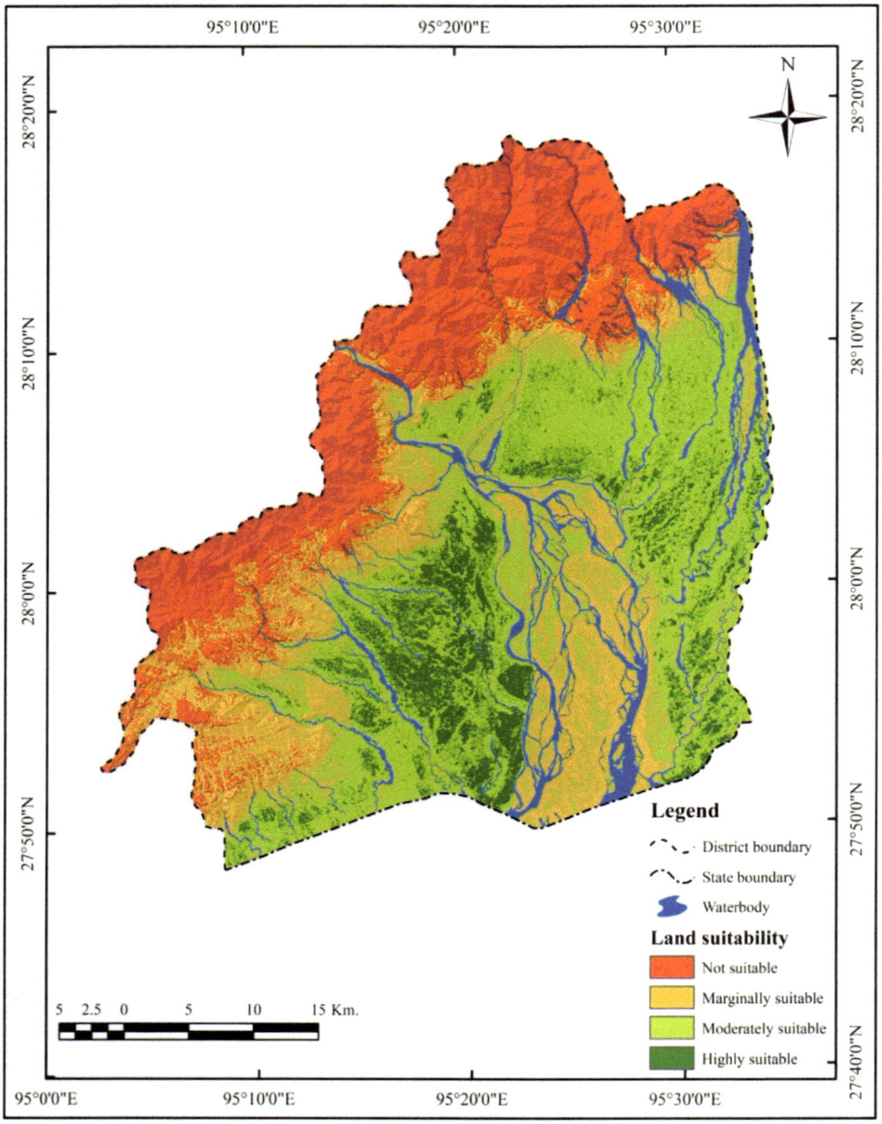

FIGURE 9.4

Land suitability for paddy cultivation in East Siang district.

4.75%, 4.41%, and 4.14%, respectively. The geomorphology, soil texture cum drainage, and LULC with little higher empirical weights of 9.22%, 6.10%, and 6.05% appear to be less effective layers compared with the mean effective weights of 5.52%, 4.75%, and 2.61%. The empirical and effective weights for soil chemical property layers and TWI were low and found close to each other.

Table 9.5 Statistics of map-removal sensitivity analysis.

	Variation index (%)			
Layer removed	**Minimum**	**Maximum**	**Mean**	**Standard deviation**
TM	0.00	0.33	0.12	0.04
RN	0.10	2.25	2.16	0.25
SL	0.00	2.22	0.68	0.25
EL	0.00	1.39	0.70	0.17
PR	0.00	0.53	0.17	0.10
ST	0.00	0.33	0.14	0.10
OC	0.05	0.39	0.31	0.06
EC	0.05	0.39	0.27	0.05
Ph	0.00	0.38	0.25	0.07
N	0.00	0.38	0.28	0.06
P	0.00	0.38	0.31	0.08
K	0.06	0.38	0.29	0.05
GM	0.00	0.75	0.19	0.07
LU	0.00	0.33	0.26	0.08
AS	0.00	0.70	0.14	0.10
TWI	0.00	0.47	0.20	0.09

AS, *aspect;* EC, *electrical conductivity;* EL, *elevation;* GM, *geomorphology;* K, *phosphorus;* LU, *land use/land cover;* N, *nitrogen;* OC, *organic carbon;* P, *potassium;* pH, *potential of hydrogen;* PR, *proximity to river;* RN, *rainfall;* SL, *slope;* ST, *soil texture cum drainage;* TM, *temperature;* TWI, *topographic wetness index.*

9.4 Discussion

Currently, the application of an integrated approach (RS, GIS, and AHP method) in LSA of specific crops for precision agriculture is increasingly becoming popular. A survey of recent literature shows the use of such combined approach in LSA of paddy (Özkan et al., 2019; Rukhsana & Molla, 2021; Sarkar et al., 2021), wheat (Pilevar et al., 2020), maize (Tashayo et al., 2020), apple (Chozom & Nimasow, 2021), etc., across the globe. Generally, LSA considers various topographic, climatic, and edaphic factors that determine the growth conditions of a particular crop (Mohammed et al., 2020). The integration of GIS and AHP method has been reported to be conducive in avoiding the problems associated with subjectivity, uncertainty, and hierarchical properties of LSA (Zhang et al., 2015). Therefore, we used AHP method to calculate the weights and relative importance of the parameters, and the LSA map for paddy cultivation was generated using WOA in GIS.

The rainfall, slope, elevation, and geomorphology were found to be the main factors in the determination of suitable areas for paddy cultivation in this study based on the mean variation index of both map-removal and single-parameter

Table 9.6 Statistics of single-parameter sensitivity analysis.

Layer removed	Empirical weights (%)	Variation index (%)			
		Minimum	Maximum	Mean	Standard deviation
TM	3.55	1.30	6.97	4.41	0.67
RN	19.13	7.70	40.07	25.15	3.76
SL	14.81	4.40	39.49	16.49	3.84
EL	13.58	5.13	27.14	16.77	2.53
PR	4.12	0.96	14.28	4.14	2.11
ST	6.10	1.37	9.78	4.75	2.21
OC	1.83	0.40	5.55	1.57	0.89
EC	1.76	0.38	5.49	2.18	0.78
Ph	2.77	0.58	6.48	2.42	1.04
N	2.30	0.48	6.60	2.03	0.88
P	2.30	0.79	6.54	1.61	1.72
K	2.30	0.49	5.38	1.96	0.79
GM	9.22	2.14	17.50	5.52	2.98
LU	6.05	1.31	9.23	2.61	1.76
AS	4.97	1.08	16.77	5.10	2.34
TWI	5.21	1.17	13.26	3.27	1.43

AS, *aspect;* EC, *electrical conductivity;* EL, *elevation;* GM, *geomorphology;* K, *phosphorus;* LU, *land use/land cover;* N, *nitrogen;* OC, *organic carbon;* P, *potassium;* pH, *potential of hydrogen;* PR, *proximity to river;* RN, *rainfall;* SL, *slope;* ST, *soil texture cum drainage;* TM, *temperature;* TWI, *topographic wetness index.*

sensitivity analysis. Many past studies have also prioritized rainfall, temperature, elevation, slope, soil texture cum drainage, and geomorphology as the highly influential determinants in the delineation of potential suitable areas of paddy cultivation (Ayehu & Besufekad, 2015; Islam et al., 2018; Kihoro et al., 2013; Roy & Saha, 2018; Sarkar et al., 2021; Ujoh et al., 2019; Zolekar & Bhagat, 2015). However, in some studies, the physical and chemical properties of soil have been highly prioritized in LSA for paddy cultivation (Hao et al., 2019; Mandal et al., 2020; Özkan et al., 2019). The study found a high mean annual rainfall (>255.51 cm), gentle slope (<5 degree), and low elevation (<295 m) as highly suitable areas in conformity with the findings of Sarkar et al. (2021). Besides, the highly suitable areas are also characterized by alluvial flood plains (Roy & Saha, 2018) and clayey soil with high water-holding capacity (Özkan et al., 2019; Sarkar et al., 2021). The study indicates that paddy cultivation in the study area requires a high mean annual rainfall, gently sloping area, lower elevation, older alluvial plains/piedmont alluvial plain, and clayey soil with more water retention capacity. East Siang is the only medium-low productivity districts (1500–2000 kg/ha) of paddy in the state (Directorate of Rice Development, 2002). The area under paddy cultivation was 14,060 ha with a total production of

38,800 metric tons (Directorate of Agriculture, 2020). The LSA map shows 9.79% (164.92 km^2) of the total area as highly suitable, but currently 140.60 km^2 of the area is under actual paddy cultivation. Therefore, the findings of the study could be disseminated to the prospective farmers to explore the highly suitable locations for further expansion of paddy cultivation area. Additionally, the moderately suitable areas (619.07 km^2) can be used by the farmers with little inputs for increased paddy productivity. Simple efforts for location-specific agriculture and proper management of the current area under paddy cultivation can ensure food security as rice is the staple food of the people.

9.5 Conclusion

The integration of RS, GIS, and MCDM (AHP) techniques in LSA offers an opportunity for paddy cultivation in the study area by providing the essential information related to precision agriculture. It will also help the farmers in identifying the suitable locations of paddy cultivation for greater production and income. Out of the limited cultivable land, the results show about 10% of the area as highly suitable and around 37% of the area as moderately suitable for paddy cultivation. However, the study found that most of potential areas are currently unexplored or underutilized. Therefore, the findings of the study need to be disseminated to the farmers through the Directorate of Agriculture, Government of Arunachal Pradesh to help the farmers in the selection of suitable areas for paddy cultivation. The state government particularly the Directorate of Agriculture could play a crucial role to increase the total production and area under paddy cultivation in the district. The methods applied in this study can be effectively replicated to other potential areas of the state to maximize paddy production in coming days.

References

Abbas, S., & Mayo, Z. A. (2021). Impact of temperature and rainfall on rice production in Punjab, Pakistan. *Environment, Development and Sustainability*, *23*(2), 1706–1728. Available from https://doi.org/10.1007/s10668-020-00647-8.

Akıncı, H., Özalp, A. Y., & Turgut, B. (2013). Agricultural land use suitability analysis using GIS and AHP technique. *Computers and Electronics in Agriculture*, *97*, 71–82. Available from https://doi.org/10.1016/j.compag.2013.07.006.

Al-Hanbali, A., Shibuta, K., Alsaaideh, B., & Tawara, Y. (2021). Analysis of the land suitability for Paddy fields in Tanzania using a GIS-based analytical hierarchy process. *Geo-spatial Information Science*, 1–17. Available from https://doi.org/10.1080/10095020.2021.2004079.

Aryal, J. P., Sapkota, T. B., Khurana, R., Khatri-Chhetri, A., Rahut, D. B., & Jat, M. L. (2020). Climate change and agriculture in South Asia: Adaptation options in smallholder production systems. *Environment, Development and Sustainability*, *22*(6), 5045–5075. Available from https://doi.org/10.1007/s10668-019-00414-4.

Ayehu, G. T., & Besufekad, S. A. (2015). Land suitability analysis for rice production: A GIS based multi-criteria decision approach. *American Journal of Geographic Information System*, *4*(3), 95–104. Available from https://doi.org/10.5923/j.ajgis.20150403.02.

Beven, K., & Kirkby, N. (1979). A physically based, variable contributing area model of basin hydrology/UN modèle a base physique de zone d'appel variable de l'hydrologie du bassin versant. *Hydrological Sciences Bulletin*, *24*(1), 43–69. Available from https://doi.org/10.1080/02626667909491834.

Binte Mostafiz, R., Noguchi, R., & Ahamed, T. (2021). Agricultural land suitability assessment using satellite remote sensing-derived soil-vegetation indices. *Land*, *10*(2), 223. Available from https://doi.org/10.3390/land10020223.

Bozdağ, A., Yavuz, F., & Günay, A. S. (2016). AHP and GIS based land suitability analysis for Cihanbeyli (Turkey) County. *Environmental Earth Sciences*, *75*(9), 813. Available from https://doi.org/10.1007/s12665-016-5558-9.

Census of India. (2011). *Primary census abstracts*. Registrar General of India, Ministry of Home Affairs, Government of India. Available from https://www.censusindia.gov.in/2011census/PCA/pca_highlights/pe_data.html.

Chozom, K., & Nimasow, G. (2021). GIS- and AHP-based land suitability analysis of *Malus domestica* Borkh. (apple) in West Kameng district of Arunachal Pradesh, India. *Applied Geomatics*, *13*, 349–360. Available from https://doi.org/10.1007/s12518-021-00354-7.

Directorate of Agriculture. (2020). *Crop production scenario – Target and achievements*. Department of Agriculture, Government of Arunachal Pradesh, Itanagar. https://agri.arunachal.gov.in/crop.php.

Directorate of Rice Development. (2002). *Rice productivity analysis in India*. Directorate of Rice Development, Government of India, Ministry of Agriculture & Farmer's Welfare. Available from https://www.drdpat.bih.nic.in/Productivity%20Analysis.

Ennaji, W., Barakat, A., El Baghdadi, M., Oumenskou, H., Aadraoui, M., Karroum, L. A., & Hilali, A. (2018). GIS-based multi-criteria land suitability analysis for sustainable agriculture in the Northeast area of Tadla plain (Morocco). *Journal of Earth System Science*, *127*(6), 1–14. Available from https://doi.org/10.1007/s12040-018-0980-x.

Everest, T., Sungur, A., & Özcan, H. (2021). Determination of agricultural land suitability with a multiple-criteria decision-making method in northwestern Turkey. *International Journal of Environmental Science and Technology*, *18*(5), 1073–1088. Available from https://doi.org/10.1007/s13762-020-02869-9.

FAO. (1976). *A framework for land evaluation, Soil Bulletin 32*. Food and Agriculture Organization of the United Nations. Available from http://www.fao.org/3/x5310e/x5310e00.htm.

Fenta, A. A., Kifle, A., Gebreyohannes, T., & Hailu, G. (2015). Spatial analysis of groundwater potential using remote sensing and GIS-based multi-criteria evaluation in Raya Valley, northern Ethiopia. *Hydrogeology Journal*, *23*, 195–206. Available from https://doi.org/10.1007/s10040-014-1198-x.

Hao, N., Ha, K., & Van, P. (2019). Applying AHP method and GIS to evaluate land suitability for paddy rice crop in Quang Xuong district, Thanh Hoa province. *Can Tho University Journal of Science*, *11*(3), 1–10.

India Brand Equity Foundation. (2020). https://www.ibef.org/download/Agriculture-andAllied-Industries-April-2020.pdf.

Islam, M., Ahamed, T., & Noguchi, R. (2018). Land Suitability and insurance premiums: a gis-based multi-criteria analysis approach for sustainable rice production. *Sustainability*, *10*(6), 1759. Available from https://doi.org/10.3390/su10061759.

Karra, K., Kontgis, C., Statman-Weil, Z., Mazzariello, J.C., Mathis, M., & Brumby, S.P. (2021). Global land use/land cover with Sentinel 2 and deep learning. *2021 IEEE international geoscience and remote sensing symposium (IGARSS)*. https://doi.org/10.1109/igarss47720.2021.9553499.

Kihoro, J., Bosco, N. J., & Murage, H. (2013). Suitability analysis for rice growing sites using a multi-criteria evaluation and GIS approach in great Mwea region, Kenya. *Sprin Plus*, *2*(1), 1–9. Available from https://doi.org/10.1186/2193-1801-2-265.

Kumar, N., Chhokar, R. S., Meena, R. P., Kharub, A. S., Gill, S. C., Tripathi, S. C., Gupta, O. P., Mangrauthia, S. K., Sundaram, R. M., Sawant, C. P., Gupta, A., Naorem, A., Kumar, M., & Singh, G. P. (2021). Challenges and opportunities in productivity and sustainability of rice cultivation system: A critical review in Indian perspective. *Cereal Research Communications*. Available from https://doi.org/10.1007/s42976-021-00214-5.

Ladha, J. K., Kumar, V., Chandna, P., Gathala, M., Sharma, S., Pathak, H., Alam, M. M., Regmi, R. P., Singh, U. P., Hafiz, M., Erenstein, O., & Balasubramanian, V. (2009). Integrating crop and resource management technologies for enhanced productivity, profitability, and sustainability of the rice-wheat system in South Asia. *Integrated Crop and Resource Management in the Rice–Wheat System of South Asia*, 69–108. Available from https://agris.fao.org/agris-search/search.do?recordID = QR2010000043.

Lodwick, W. A., Monson, W., & Svoboda, L. (1990). Attribute error and sensitivity analysis of map operations in geographical information systems: Suitability analysis. *International Journal of Geographical Information System*, *4*(4), 413–428.

Mahajan, G., Kumar, V., & Chauhan, B. S. (2017). Rice production in India. *Rice Production Worldwide*, 53–91. Available from https://doi.org/10.1007/978-3-319-47516-5_3.

Mandal, V. P., Rehman, S., Ahmed, R., Masroor, M., Kumar, P., & Sajjad, H. (2020). Land suitability assessment for optimal cropping sequences in Katihar district of Bihar, India using GIS and AHP. *Spatial Information Research*, *28*(5), 589–599.

Mohammed, S., Alsafadi, K., Ali, H., Mousavi, S. M. N., Kiwan, S., Hennawi, S., Harsanyie, E., Pham, Q. B., Linh, N. T. T., Ali, R., Anh, D. T., & Mohammed, V. N. T. (2020). Assessment of land suitability potentials for winter wheat cultivation by using a multi-criteria decision support-geographic information system (MCDS-GIS) approach in Al-Yarmouk Basin (S. Syria). *Geocarto International*, 1–19. Available from https://doi.org/10.1080/10106049.2020.1790674.

Napolitano, P., & Fabbri, A. G. (1996). Single parameter sensitivity analysis for aquifer vulnerability assessment using DRASTIC and SINTACS. In K. Kovar, & H. P. Nachtnebel (Eds.), *Application of geographical information systems in hydrology and water resources management* (pp. 559–566). IAHS Publ. 235.

Özkan, B., Dengiz, O., & Demirağ Turan, İ. (2019). Site suitability assessment and mapping for rice cultivation using multi-criteria decision analysis based on fuzzy-AHP and TOPSIS approaches under semihumid ecological condition in delta plain. *Paddy and Water Environment*, *17*(4), 665–676. Available from https://doi.org/10.1007/s10333-019-00692-8.

Pham, T. T., Tong, T. H., & Pham, V. C. (2013). Becoming urban: How urbanization influences the loss of arable land in Peri-urban Hanoi. *Lecture Notes in Computer Science*, 238–252. Available from https://doi.org/10.1007/978-3-642-39649-6_17.

Pilevar, A. R., Matinfar, H. R., Sohrabi, A., & Sarmadian, F. (2020). Integrated fuzzy, AHP and GIS techniques for land suitability assessment in semi-arid regions for wheat and maize farming. *Ecological Indicators*, *110*, 105887. Available from https://doi.org/10.1016/j.ecolind.2019.105887.

Pramanik, M. K. (2016). Site suitability analysis for agricultural land use of Darjeeling district using AHP and GIS techniques. *Modeling Earth Systems and Environment*, *2*(2), 56. Available from https://doi.org/10.1007/s40808-016-0116-8.

Qin, C., Zhu, A., Pei, T., Li, B., Scholten, T., Behrens, T., & Zhou, C. (2011). An approach to computing topographic wetness index based on maximum downslope gradient. *Precision Agriculture*, *12*(1), 32–43. Available from https://doi.org/10.1007/s11119-009-9152-y.

Roy, J., & Saha, S. (2018). Assessment of land suitability for the Paddy cultivation using analytical hierarchical process (AHP): A study on Hinglo river basin, Eastern India. *Modeling Earth Systems and Environment*, *4*(2), 601–618. Available from https://doi.org/10.1007/s40808-018-0467-4.

Rukhsana., & Molla, S. H. (2021). Investigating the suitability for rice cultivation using multi-criteria land evaluation in the Sundarban region of south 24 parganas district, West Bengal, India. *Journal of the Indian Society of Remote Sensing*. Available from https://doi.org/10.1007/s12524-021-01441-3.

Saaty, T. L., & Vargas, L. G. (1998). Diagnosis with dependent symptoms: Bayes theorem and the analytic hierarchy process. *Operations Research*, *46*(4), 491–502. Available from https://doi.org/10.1287/opre.46.4.491.

Saaty, T. L., & Vargas, L. G. (2012). *Models, methods, concepts & applications of the analytic hierarchy process. International series in operations research & management science* (2nd ed). Springer. Available from https://doi.org/10.1007/978-1-4614-3597-6.

Saaty, T. L. (1980). *The analytic hierarchy process*. McGraw-Hill.

Saaty, T. L. (1989). Group decision making and the AHP. *The Analytic Hierarchy Process*, 59–67. Available from https://doi.org/10.1007/978-3-642-50244-6_4.

Saaty, T. L. (2004). Decision making – the analytic hierarchy and network processes (AHP/ANP). *Journal of Systems Science Systems Engineering*, *13*, 1–35. Available from https://doi.org/10.1007/s11518-006-0151-5.

Sarkar, B., Das, P., Islam, N., Basak, A., Debnath, M., & Roy, R. (2021). Land suitability analysis for paddy crop using GIS-based Fuzzy-AHP (F-AHP) method in Koch Bihar district, West Bengal. *Geocarto International*, 1–26. Available from https://doi.org/10.1080/10106049.2021.2007299.

Sørensen, R., & Seibert, J. (2007). Effects of DEM resolution on the calculation of topographical indices: TWI and its components. *Journal of Hydrology*, *347*(1–2), 79–89. Available from https://doi.org/10.1016/j.jhydrol.2007.09.001.

Tashayo, B., Honarbakhsh, A., Akbari, M., & Eftekhari, M. (2020). Land suitability assessment for maize farming using a GIS-AHP method for a semi- arid region, Iran. *Journal of the Saudi Society of Agricultural Sciences*, *19*(5), 332–338. Available from https://doi.org/10.1016/j.jssas.2020.03.003.

Uisso, A. M., & Tanrıvermiş, H. (2021). Driving factors and assessment of changes in the use of arable land in Tanzania. *Land Use Policy*, *104*, 105359. Available from https://doi.org/10.1016/j.landusepol.2021.105359.

Ujoh, F., Igbawua, T., & Paul, M. O. (2019). Suitability mapping for rice cultivation in Benue State, Nigeria using satellite data. *Geo-Spatial Information Science*, *22*(4), 332–344.

Xiang, T., Malik, T. H., & Nielsen, K. (2020). The impact of population pressure on global fertiliser use intensity, 1970–2011: An analysis of policy-induced mediation. *Technological Forecasting and Social Change*, *152*119895. Available from https://doi.org/10.1016/j.techfore.2019.119895.

Yalew, S. G., Van Griensven, A., Mul, M. L., & Van der Zaag, P. (2016). Land suitability analysis for agriculture in the Abbay basin using remote sensing, GIS and AHP techniques. *Modeling Earth Systems and Environment*, *2*(2), 1–14. Available from https://doi.org/10.1007/s40808-016-0167-x.

Zhang, J., Su, Y., Wu, J., & Liang, H. (2015). GIS based land suitability assessment for tobacco production using AHP and fuzzy set in Shandong province of China. *Computers and Electronics in Agriculture*, *114*, 202–211, In this issue. Available from https://doi.org/10.1016/j.compag.2015.04.004.

Zhang, G., Xiao, X., Biradar, C. M., Dong, J., Qin, Y., Menarguez, M. A., Zhou, Y., Zhang, Y., Jin, C., Wang, J., Doughty, R. B., Ding, M., & Moore, B. (2017). Spatiotemporal patterns of Paddy rice croplands in China and India from 2000 to 2015. *Science of the Total Environment*, *579*, 82–92. Available from https://doi.org/10.1016/j.scitotenv.2016.10.223.

Zolekar, R. B., & Bhagat, V. S. (2015). Multi-criteria land suitability analysis for agriculture in hilly zone: Remote sensing and GIS approach. *Computers and Electronics in Agriculture*, *118*, 300–321.

CHAPTER

Precision agriculture practices for smart irrigation 10

Md. Saddam Hossain[1], Uttam Kumar Ghosh[1], Md. Nahidul Islam[2] and Md. Arifur Rahman Khan[1]

[1]Department of Agronomy, Bangabandhu Sheikh Mujibur Rahman Agricultural University (BSMRAU), Gazipur, Bangladesh

[2]Department of Agro-Processing, Bangabandhu Sheikh Mujibur Rahman Agricultural University (BSMRAU), Gazipur, Bangladesh

10.1 Introduction

Precision agriculture practices are considered as a "group of technologies which consists of enhanced machinery, sensors, information systems, an up-to-date management to maximize production by accounting for variability and uncertainties within the agricultural systems" (Rodríguez Díaz & Montesinos, 2015). Smart irrigation (SI) includes the correct evaluation of plant water necessities and exact application of the required amount at the required time. Hence, it is also essential to decide the suitable irrigation frequency/scheme to escape leaching in the soil (characteristically very sandy) toward deeper layers and the probability of aquifers pollution. SI technologies could be useful in achieving this surely increasing water use efficiency (WUE) and its environmental sustainability (Morillo et al., 2015).

Agriculture is a vital sector that ensures a healthy diet for the world's rising population. By 2050, the world's population is expected to exceed 10 billion. Global food production therefore must be greatly boosted by 2050, according to the United Nations Food and Agriculture Organization, due to rising population demands. Farmers and agricultural firms will have to drive innovations by repressing present techniques to fulfill demand, as evidence from various researchers shows that agricultural areas and productivity are diminishing day by day. Furthermore, it is predicted that the usage of smart farming instruments and technology in the agriculture industry would increase in the coming years. Many developing countries, in South Asia, are currently confronted with a number of hostile events, including extreme climatic variability, a lack of agricultural labor, water competition from industries and agriculture, and groundwater exhaustion, all of which threaten the viability of irrigated farming (Kumar et al., 2020). Agriculture's environmental footprint has accelerated in recent decades, resulting in a slew of environmental issues such as water scarcity, climatic change, and

Remote Sensing in Precision Agriculture. DOI: https://doi.org/10.1016/B978-0-323-91068-2.00009-6

land degradation (Boursianis et al., 2020; Sun et al., 2016). To this purpose, the fourth agricultural revolution incorporates information and communications technologies into traditional farming practices, taking into account all of the basic principles of sustainable agriculture (Jeschke et al., 2016).

Precision agriculture is an agricultural production technique aimed at quantifying plant (or animal) production through the observation, measurement, and response to farm metrics to achieve less-input, high-output, and sustainable agriculture (Boursianis et al., 2020). One of the most important elements of agricultural productivity is irrigation. Irrigation aids in the preservation of soil moisture and the regulation of soil temperature. For thousands of years, it has been an integral part of agriculture. Surface irrigation, sprinkler irrigation, center pivot irrigation, and drip irrigation are all labor-intensive irrigation systems that create flood due to considerable seepage into the field (Munoth et al., 2016). Over 70% of farmers still use traditional irrigation methods, with irrigation applied at regular intervals. Because water is a finite resource, it is critical to provide a sufficient quantity of water at the appropriate moment to crops based on their water requirements (Garg et al., 2016). It necessitates over 85% of the available freshwater resources (Barkunan et al., 2019).

Climate change and resource overuse have had a significant impact on water resources worldwide in recent decades (García et al., 2020). The scarcity of freshwater, the deterioration of water quality, and the salinity of the water are only a few of the major factors that the agricultural community must consider while planning irrigation operations (García et al., 2020; Lv et al., 2017). The only feasible solution to the current water crisis issues is SI. SI is an agricultural technique that employs irrigation management to reduce agricultural usage of water. Wireless technology (e.g., wireless sensor networks – WSNs) and Internet of things (IoT) technologies are commonly used in SI systems (Farooq et al., 2019; García et al., 2020). Knowledge of crop water utilization and irrigation management strategies are essential components of good agricultural practices (Bencini et al., 2010; Rahim Khan et al., 2013). One strategy to reduce water consumption and enhance WUE is to use irrigation water wisely. In modern precision agriculture, the WSN acts as the backbone. It is used in the field for real-time irrigation to give the consistent and effective volume of water to the crop at each step of its growth cycle (Li et al., 2010; Sakthipriya, 2014).

Global food insecurity, climatic change, and poverty have all been exacerbated as a result of lack of agricultural water resources. In the face of dwindling freshwater resources, agriculture's greatest task is to provide food and fiber for over 8 billion people around the world (Neupane & Guo, 2019). Because rainfall only meets a percentage of the entire demand for crop evapotranspiration, the difficulty is more acute in dry and semiarid countries, where subterranean water sources, particularly aquifers, are exploited for irrigation (Mauget et al., 2017). Many aquifers have been rapidly depleted as a result of heavy irrigation. Many other water resources around the world have been quickly disappearing in recent years with little replenishment, suggesting the necessity for agricultural water conservation strategies (Mauget et al., 2013; Takács et al., 2018). To better preserve water and improve water usage efficiency for sustainable agricultural production through SI, more effective and efficient water management is necessary. In this chapter, we have presented the current status, technological requirements, and future prospect of precision agriculture practices for SI to address water scarcity challenges.

10.2 Components of smart irrigation system

10.2.1 Data acquisition

An irrigation management system must be able to detect and measure the geographical and temporal variation in agricultural soil and crop conditions within a field and between fields. Existing technology can measure a variety of items of the land continuum, many of them in real time, to provide precision and/or real-time irrigation control.

10.2.2 Interpretation

Data must be collected, evaluated, and analyzed on a regular and suitable basis. A key bottleneck for the interpretation of real-time data and implementation of precision agriculture technologies has been identified as the lack of development of decision support systems (McBratney et al., 2005).

10.2.3 Control

An SI system must be capable of optimizing inputs and adjusting irrigation management at suitable temporal and spatial scales. The irrigation system will determine how different depths of water are applied throughout a field. The most dependable and accurate method of controlling irrigation applications should be automatic controllers with real-time data.

10.3 Smart irrigation systems

SI is a type of artificial irrigation that regulates the amount of water used by assessing the requirements. It is the most important component in agriculture, having a substantial impact on the health, cost, and yield of crops. Because most nations in the world are experiencing water scarcity, one of the most important aspects of SI is to reduce water waste (Shafi et al., 2019). Several water management models or platforms that combine diverse technologies to create SI systems in agriculture are available. The wireless sensor connectivity was built on the Global System for Mobile Communications protocol to transport the measured data from the field to the farmer's mobile terminal, and Barkunan et al. (2019)

suggested a smart sensor for automatic drip irrigation that was used for rice farming in India. Similarly, in Spain, Navarro-Hellín et al. (2015) developed a wireless sensor architecture for irrigation water management, with the proposed design based on multiple wireless nodes equipped with the General Packet Radio Services (GPRS) protocol for communication with their database. A more sophisticated approach has been applied by Khan et al. (2017). The authors proposed a novel architecture for smart agriculture by using cognitive radio technology in the IoT-driven system. They used two different data types to model the various conditions of crops, especially in countries with limited resources. Cambra Baseca et al. (2019) proposed a new precision agriculture technology that included SI and fertilization control systems. Their goal was to create an open-source precision agricultural platform based on open communication protocols and open software technologies. Trilles et al. (2020) gave an intriguing solution that described an interconnected ecology. The authors tested their IoT agnostic architecture as an IoT platform in a smart farming environment.

The issue of water management in harsh weather contexts has been addressed, demonstrating the importance of the system. Ogidan et al. (2019) highlighted issues of water management in many parts of Africa, and they demonstrated a SI system based on an Arduino microcontroller for algorithm control configuration to effectively manage water supply. Gelcer et al. (2018) presented another fascinating project in which they developed an irrigation tool to help growers monitor their fields' water supply. To provide optimal water management, a few SI systems for precision agriculture use fuzzy logic in their scheduling algorithms. By incorporating a fuzzy logic method, Viani (2016) demonstrated a scalable SI system for precision agriculture. Their goal was to reduce water usage and improve product quality while optimizing the volumetric water content in the soil. Keswani et al. (2018) also demonstrated an IoT-enabled WSN platform for precision agriculture, which automates numerous agricultural procedures. To maximize water utilization in soil, they used neural network prediction and structure similarity approaches. They used a fuzzy logic-based weather condition modeling system to apply valve management in a SI system based on these two methodologies. Finally, Paucar et al. (2015) exhibited the design of an SI system with monitoring and actuation capabilities, as well as the use of fuzzy logic as a decision support tool.

Several intelligent irrigation systems make use of important enablers including thermal imaging, big data, and machine learning. To develop an SI system for precision agriculture practices, Roopaei et al. (2017) combined the use of thermal images from the field to provide information about soil moisture, an IoT sensor network to assess temperature and humidity, and the cloud to collect and process all the acquired data in a cost-effective manner. Furthermore, Navarro-Hellín et al. (2016) developed and evaluated a decision support system for autonomous SI that pits two machine learning algorithms (partial least square regression and adaptive neuro fuzzy inference system) against cultivator expert recommendations. Similarly, Goap et al. (2018) presented and tested an SI system that predicts the field's irrigation needs using open source technologies and machine learning techniques. Finally, Zhang et al. (2017) incorporated a smart water and fertilizer system based on big data.

Several developing technologies, such as as the Message Queue Telemetry Transport Protocol (MQTT) and the LoRa (Low Range) protocol, have been used in SI systems in recent years. MQTT is a lightweight network messaging protocol for small sensors in WSNs that can be used for a variety of IoT applications, such as agricultural and livestock (Mahbub, 2020). MQTT has been successfully used to transport data between sensors and the central unit in numerous SI systems (Nawandar & Satpute, 2019; Pernapati, 2018).

Akubattin et al. (2016) demonstrated an SI system that included a Raspberry Pi and two sensors: a soil moisture sensor to check the water level in the soil, and a temperature and humidity sensor to monitor the environmental condition. These sensors and the water supply network were linked to the Raspberry Pi. A smartphone application was created for remote monitoring and water flow control, which may be done manually or automatically. Water flow was automatically turned ON/OFF based on the water level of the soil in automatic mode, without the need for human interaction. The user could check the soil moisture level in manual mode. When the water level in the soil fell below a certain threshold, an alert was sent out, and the user could turn it on or off using a mobile application.

Kamienski et al. (2019) developed an IoT-based SI platform for precision agriculture, which was tested in four distinct experimental pilots in Brazil, Spain, and Italy, revealing the systems' specific on-site requirements. Many academics have developed power-efficient systems because power is a major concern in IoT-based platforms. Solar power was used to present a power-efficient water irrigation system (Harishankar et al., 2014), in which the controller was connected to the soil sensor and water supply valve. The water valve was turned on/off based on the moisture sensor's readings of the water level. The solar panel provided the power; therefore, the system was not reliant on any other power source. Another sensor-based IoT water irrigation system was demonstrated (Kansara et al., 2015) in which the controller controlled the opening and closing of a solenoid valve based on the water level of the soil. A series of weather notifications were also delivered to the user via a mobile application to keep them up to speed on the temperature and humidity of the surroundings, which had a direct impact on the soil's water level. A WSN was used to offer an energy-efficient irrigation system for cultivated crops, in which water was successfully managed based on environmental circumstances. Based on humidity, temperature, and wind speed received by sensors, as well as past data, this system predicted the amount of water required for typical irrigation (Nikolidakis et al., 2015).

Rawal (2017) demonstrated an IoT-based irrigation system that included soil moisture sensors controlled by an ATMEGA 328P on an Arduino UNO board and a GPRS module (Fig. 10.1). The sensor data were sent to the cloud, that is, Things speak, where graphs were created to visualize the data trends. An online site was also created so that the farmer could monitor the status of the water,

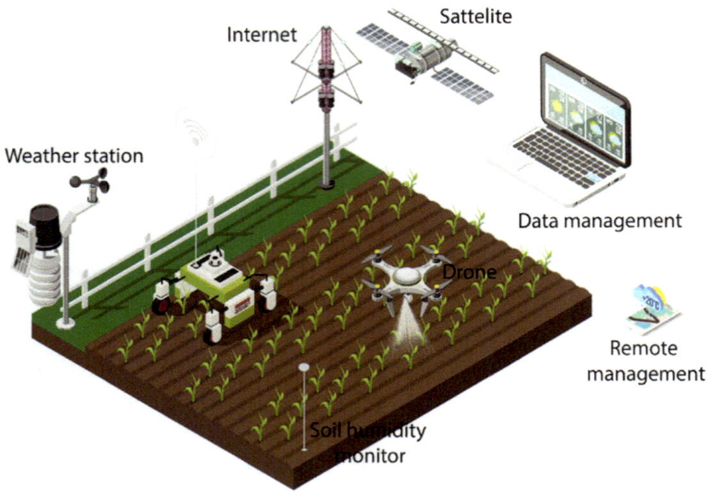

FIGURE 10.1

Smart irrigation systems.

such as whether it was on or off. A real-time prototype of an irrigation system was also shown (Vellidis et al., 2008), in which soil moisture sensors and soil temperature sensors were used to assess the water status of the soil. Radio frequency identification (RFID) technology was used to send data to the cloud so that it could be analyzed further. Using ATMEGA 328 and temperature, humidity, and soil moisture sensors, a water sprinkler system for SI was shown. The water spray was controlled based on the soil moisture level to save water and reduce human work (Kumar et al., 2017). A Raspberry Pi, Arduino, electronic water control valve, and relay were combined to create a cost-effective drip irrigation system for a residence. For communication, ZigBee protocols were employed. The user activated or deactivated the water valve by sending commands to the Raspberry Pi, which were then processed by the Arduino (Agrawal & Singhal, 2015).

Sensor placement is a major issue that has an impact on sensor accuracy. Soulis et al. (2015) offered a full description of soil moisture placement in the field and its accuracy. Complete software and hardware requirements, problems and obstacles, and advantages for real-time irrigation systems were described, along with a broad picture of the entire system (Yousif et al., 2018).

10.4 Different types of irrigation methods

Irrigation is the use of artificial water as an input to an agriculture field for the growth and development of crop plants. It is a critical component that aids in the absorption of micronutrients from the soil. Traditional irrigation methods including boarder, furrow, and flood irrigation all necessitate a large amount of water for agricultural cultivation. Freshwater scarcity became a driving force behind the development of pressurized irrigation technology for SI management.

10.4.1 Surface irrigation methods

Traditional irrigation methods (border, basin, and furrow) deliver water to the upper borders of banks/bunds in accordance with traditional plant-based indications, resulting in ponds rather than irrigation. According to Hamdy et al. (2003), irrigation water was used inefficiently through traditional irrigation technologies, resulting in lower crop intensities and yields. It was also discovered that traditional water application procedures have resulted in water stagnation and soil salinity, reducing overall irrigation efficiency significantly. Furthermore, moisture distribution is unequal in such a scenario, and profound percolation is unavoidable. Initial cost, field size and shape, soil qualities, nature and availability of water supply, climate, cutting patterns, societal preferences and structures, historical experiences, and external influences are only a few of the advantages and disadvantages of each surface irrigation system.

10.4.2 Modern or microirrigation and pressurized irrigation

By using pipes as a water transport medium, pressurized irrigation or microirrigation decreases seepage and evaporation, as well as water conveyance losses. The entire area wetting problem is eliminated, and weed development is reduced, thanks to these irrigation strategies (Kumar et al., 2020).

10.4.2.1 Drip irrigation

Drip irrigation uses drip emitters to deliver water to the root zone one drop at a time. A pressurized pipe (laterals) system that runs along agricultural rows is typically used for drip irrigation. Emitters are set at specific distances on the lateral tubes, allowing water to flow to the crop drop by drop. It was discovered that drip irrigation techniques reduced the amount of irrigation water used excessively and saved a significant amount of freshwater (Kumar et al., 2020). This strategy conserved freshwater by minimizing evaporation and water transportation losses associated with conventional surface irrigation. Drip irrigation could reduce the amount of water applied by 50%–100%. Squash, cowpea, faba bean, pea, sorghum sunflower, and other crops were found to benefit from good drip irrigation management practices (Abdel-Mawgoud et al., 2009). The results showed that adjusting applied water improved yield components, all vegetative development variables, and overall yield significantly.

10.4.2.2 Sprinkler irrigation

In the sprinkler irrigation method, water is sprayed into the air and allowed to fall on the ground surface, imitating rainfall. This method is ideally suited for water-scarce places and regions with an uneven ground level where watering with other irrigation systems is impractical. Sprinkler irrigation creates a humid atmosphere for crops that might otherwise perish owing to excessive temperatures. Table 10.1 shows the irrigation efficiency of surface, drip, and drip irrigation techniques (Kumar et al., 2020).

10.4.3 Advanced or automated irrigation

Since the beginning of agriculture, cultivation practices have changed dramatically. In recent years, there has been increased technical innovation. The IoT has become a part of every sector in today's world, and it is a boon for development in a variety of industries, including agriculture and farming. Water conservation and correct use of freshwater are becoming critical issues in irrigation. The primary goal of using an advanced or automated irrigation system is to provide the water that crops require. Precision irrigation refers to the application of water at the correct time, in the correct amount, and at the correct location. Precision irrigation aids in crop productivity and water conservation (WUE). As a result, the cost of energy per irrigation is reduced. As a result, it is possible to manage variability in farmer's fields and experimental fields (Kumar et al., 2020).

10.5 Benefits of smart irrigation

SI has the ability to reduce water consumption while also increasing economic efficiency. According to reports, SI (Drip and Sprinkler) can increase water application efficiency by 80%–90%, compared with 40%–45% for surface irrigation (Michael & Johannes, 2004). Case studies of variable rate irrigation revealed

Table 10.1 Irrigation efficiency of surface and pressurized irrigation methods (Kumar et al., 2020).

Irrigation efficiencies	Methods of irrigation Surface	Sprinkler	Drip
Conveyance efficiency	40–50 (Canal) 60–70 (Well)	100	100
Application efficiency	60–70	70–80	90
Water use efficiency	35–40	80–95	80–95
Surface water moisture evaporation	30–40	30–40	20–25
Overall efficiency	30–35	50–60	80–90

water savings varying from 0% to 50% in various years. SI has the ability to save money by lowering the input costs or increasing the yield for the same inputs.

10.5.1 Water savings

The main purpose of SI is to apply the right amount of water to the right places in the field. Many scholars believe it is the most likely way to save considerable amounts of water (Evans & Sadler, 2008). A crucial or essential component of SI is site-specific or variable-rate irrigation. Most academics predict a decrease in water use on at least some fields, if not an overall decrease in value (Sadler et al., 2005). Variable rate irrigation has been found to save 10%–15% of the water used in traditional irrigation practices. Water reductions of roughly 25% are feasible, according to (Hedley & Yule, 2009), thanks to improvements in application efficiency achieved by spatially diversified irrigation applications.

10.5.2 Yield and profit

King et al. (2006) conducted experimental tests to determine the yield of potatoes under spatially variable irrigation treatments. It was reported that yields were higher for 2 years in a row when irrigation was managed uniformly. Bronson et al. (2006) studied cotton yields and WUE over a 4-year period using spatially variable irrigation. Cotton appears to be unexpected to handle with spatially variable irrigation, according to the researchers. The effort of Bronson et al. (2006) contributes to this result.

10.6 Conclusion and future perspectives

Today's innovations can be combined to reduce costs and increase input utilization. Farmers currently control irrigation methods manually and irrigate their land on a regular basis. Such strategies require a large amount of water and result in water loss; moreover, irrigation in arid locations is difficult due to low rainfall. The SI system ensures increased productivity while conserving water. SI systems are useful not just in open regions such as agricultural fields but also in greenhouses and tunnel farming. In countries where there is a lack of water, water management is crucial. It is essential to automate the irrigation control process by using the detected environmental factors. SI can be automated using the above-mentioned technologies, with the main benefits being increased productivity, reduced water usage, and reduced soil degradation. To address the issue of water scarcity, precision agriculture practices must be developed in agricultural production system. Future advancements in SI systems will lead to a better understanding of how to use irrigation water more efficiently to reduce water usage in agriculture.

Though numerous SI systems have been developed, in the future, a rain gage or rain sensor might be integrated into an existing system to measure rainfall. As a result, the system can be changed to supply less water than the customer specified based on the amount of rain received. These methods should be expanded to accommodate greater irrigation areas. In addition, the device can measure the pH of the soil and advise farmers on how much fertilizer to use. A camera module can also be connected to the Arduino Uno board to monitor crop growth.

Acknowledgment

Authors are thankful to research management wing (37.01.3303.133.34.002.22.1765-003), Bangabandhu Sheikh Mujibur Rahman Agricultural University, Gazipur 1706, Bangladesh.

References

Abdel-Mawgoud, A., Gameh, M., Abd-Elaziz, S., & El-Sayed, M. (2009). Sunflower water relations at various irrigation regimes with modern irrigation systems under climatic conditions of Assiut Governorate, Upper Egypt. In *Paper presented at the 13th international water technology conference (IWTC 2009)*, Hurghada, Egypt.

Agrawal, N., & Singhal, S. (2015). Smart drip irrigation system using Raspberry Pi and Arduino. In *Paper presented at the international conference on computing, communication & automation*. Available from https://doi.org/10.1109/ccaa.2015.7148526.

Akubattin, V., Bansode, A., Ambre, T., Kachroo, A., & SaiPrasad, P. (2016). Smart irrigation system. *International Journal of Scientific Research in Science and Technology*, *2* (5), 343–345.

Barkunan, S. R., Bhanumathi, V., & Sethuram, J. (2019). Smart sensor for automatic drip irrigation system for paddy cultivation. *Computers & Electrical Engineering*, *73*, 180–193. Available from https://doi.org/10.1016/j.compeleceng.2018.11.013.

Bencini, L., Di, D., Collodi, G., Manes, A., & Manes, G. (2010). *Wireless sensor networks for on-field agricultural management process. Wireless sensor networks: Application-centric design*. InTech. Available from https://doi.org/10.5772/13001.

Boursianis, A. D., Papadopoulou, M. S., Diamantoulakis, P., Liopa-Tsakalidi, A., Barouchas, P., Salahas, G., … Goudos, S. K. (2020). Internet of things (IoT) and agricultural unmanned aerial vehicles (UAVs) in smart farming: A comprehensive review. *Internet of Things*, *18*, 100187. Available from https://doi.org/10.1016/j.iot.2020.100187.

Bronson, K. F., Booker, J. D., Bordovsky, J. P., Keeling, J. W., Wheeler, T. A., Boman, R. K., … Nichols, R. L. (2006). Site-specific irrigation and nitrogen management for cotton production in the southern high plains. *Agronomy Journal*, *98*(1), 212–219. Available from https://doi.org/10.2134/agronj2005.0149.

Cambra Baseca, C., Sendra, S., Lloret, J., & Tomas, J. (2019). A smart decision system for digital farming. *Agronomy*, *9*(5), 216.

Evans, R. G., & Sadler, E. J. (2008). Methods and technologies to improve efficiency of water use. *Water Resources Research*, *44*(7). Available from https://doi.org/10.1029/2007wr006200.

Farooq, M. S., Riaz, S., Abid, A., Abid, K., & Naeem, M. A. (2019). A survey on the role of IoT in agriculture for the implementation of smart farming. *IEEE Access*, *7*, 156237–156271. Available from https://doi.org/10.1109/access.2019.2949703.

García, L., Parra, L., Jimenez, J. M., Lloret, J., & Lorenz, P. (2020). IoT-based smart irrigation systems: An overview on the recent trends on sensors and IoT systems for irrigation in precision agriculture. *Sensors (Basel, Switzerland)*, *20*(4), 1042. Available from https://doi.org/10.3390/s20041042.

Garg, A., Munoth, P., & Goyal, R. (2016). Application of soil moisture sensor in agriculture. In *Paper presented at the proceedings of internation conference on hydraulic*.

Gelcer, E., Fraisse, C., Zotarelli, L., Perondi, D., Malia, H., Ecole, C., & Migliaccio, K. (2018). A smart irrigation tool to determine the effects of ENSO on water requirements for tomato production in Mozambique. *Water*, *10*(12), 1820. Available from https://doi.org/10.3390/w10121820.

Goap, A., Sharma, D., Shukla, A. K., & Rama Krishna, C. (2018). An IoT based smart irrigation management system using machine learning and open source technologies. *Computers and Electronics in Agriculture*, *155*, 41–49. Available from https://doi.org/10.1016/j.compag.2018.09.040.

Hamdy, A., Ragab, R., & Scarascia-Mugnozza, E. (2003). Coping with water scarcity: Water saving and increasing water productivity. *Irrigation and Drainage: The Journal of the International Commission on Irrigation and Drainage*, *52*(1), 3–20.

Harishankar, S., Kumar, R. S., Sudharsan, K., Vignesh, U., & Viveknath, T. (2014). Solar powered smart irrigation system. *Advance in Electronic and Electric Engineering*, *4*(4), 341–346.

Hedley, C. B., & Yule, I. J. (2009). Soil water status mapping and two variable-rate irrigation scenarios. *Precision Agriculture*, *10*(4), 342–355. Available from https://doi.org/10.1007/s11119-009-9119-z.

Jeschke, S., Brecher, C., Meisen, T., Özdemir, D., & Eschert, T. (2016). *Industrial Internet of things and cyber manufacturing systems. Industrial Internet of things* (pp. 3–19). Springer International Publishing. Available from https://doi.org/10.1007/978-3-319-42559-7_1.

Kamienski, C., Soininen, J.-P., Taumberger, M., Dantas, R., Toscano, A., Salmon Cinotti, T., ... Torre Neto, A. (2019). Smart water management platform: IoT-based precision irrigation for agriculture. *Sensors (Basel, Switzerland)*, *19*(2), 276. Available from https://doi.org/10.3390/s19020276.

Kansara, K., Zaveri, V., Shah, S., Delwadkar, S., & Jani, K. (2015). Sensor based automated irrigation system with IOT: A technical review. *International Journal of Computer Science and Information Technologies*, *6*(6), 5331–5333.

Keswani, B., Mohapatra, A. G., Mohanty, A., Khanna, A., Rodrigues, J. J. P. C., Gupta, D., & de Albuquerque, V. H. C. (2018). Adapting weather conditions based IoT enabled smart irrigation technique in precision agriculture mechanisms. *Neural Computing and Applications*, *31*(S1), 277–292. Available from https://doi.org/10.1007/s00521-018-3737-1.

Khan, A. A., Shaikh, A. Z., Naqvi, S., & Altaf, T. (2017). A novel cognitive radio enabled IoT system for smart irrigation. *Journal of Informatics and Mathematical Sciences*, *9* (1), 129–136.

King, B. A., Stark, J. C., & Wall, R. W. (2006). Comparison of site-specific and conventional uniform irrigation management for potatoes. *Applied Engineering in Agriculture*, *22*(5), 677–688. Available from https://doi.org/10.13031/2013.22000.

Kumar, B. D., Srivastava, P., Agrawal, R., & Tiwari, V. (2017). Microcontroller based automatic plant irrigation system. *International Research Journal of Engineering and Technology*, *4*(5), 1436–1439.

Kumar, S. V., Singh, C., & Upendar, K. (2020). Review on IoT based precision irrigation system in agriculture. *Current Journal of Applied Science and Technology*, *39*(45), 15–26.

Li, S., Simonian, A., & Chin, B. A. (2010). Sensors for agriculture and the food industry. *The Electrochemical Society Interface*, *19*(4), 41–46. Available from https://doi.org/10.1149/2.f05104if.

Lv, Z., Song, H., Basanta-Val, P., Steed, A., & Jo, M. (2017). Next-generation big data analytics: State of the art, challenges, and future research topics. *IEEE Transactions on Industrial Informatics*, *13*(4), 1891–1899. Available from https://doi.org/10.1109/tii.2017.2650204.

Mahbub, M. (2020). A smart farming concept based on smart embedded electronics, internet of things and wireless sensor network. *Internet of Things*, *9*, 100161. Available from https://doi.org/10.1016/j.iot.2020.100161.

Mauget, S. A., Adhikari, P., Leiker, G., Baumhardt, R. L., Thorp, K. R., & Ale, S. (2017). Modeling the effects of management and elevation on West Texas dryland cotton production. *Agricultural and Forest Meteorology*, *247*, 385–398. Available from https://doi.org/10.1016/j.agrformet.2017.07.009.

Mauget, S., Leiker, G., & Nair, S. (2013). A web application for cotton irrigation management on the U.S. Southern High Plains. Part I: Crop yield modeling and profit analysis. *Computers and Electronics in Agriculture*, *99*, 248–257. Available from https://doi.org/10.1016/j.compag.2013.10.003.

McBratney, A., Whelan, B., Ancev, T., & Bouma, J. (2005). Future directions of precision agriculture. *Precision Agriculture*, *6*(1), 7–23. Available from https://doi.org/10.1007/s11119-005-0681-8.

Michael, D. D., & Johannes, M. S. (2004). *Automated subsurface drip irrigation based on soil moisture*. Paper presented at the 2004, Ottawa, Canada, August 1–4, 2004. Available from https://doi.org/10.13031/2013.16407.

Morillo, J. G., Martín, M., Camacho, E., Díaz, J. A. R., & Montesinos, P. (2015). Toward precision irrigation for intensive strawberry cultivation. *Agricultural Water Management*, *151*, 43–51. Available from https://doi.org/10.1016/j.agwat.2014.09.021.

Munoth, P., Goyal, R., & Tiwari, K. (2016). *Sensor based irrigation system: A review*. NCACE.

Navarro-Hellín, H., Martínez-del-Rincon, J., Domingo-Miguel, R., Soto-Valles, F., & Torres-Sánchez, R. (2016). A decision support system for managing irrigation in agriculture. *Computers and Electronics in Agriculture*, *124*, 121–131. Available from https://doi.org/10.1016/j.compag.2016.04.003.

Navarro-Hellín, H., Torres-Sánchez, R., Soto-Valles, F., Albaladejo-Pérez, C., López-Riquelme, J. A., & Domingo-Miguel, R. (2015). A wireless sensors architecture for efficient irrigation water management. *Agricultural Water Management*, *151*, 64–74. Available from https://doi.org/10.1016/j.agwat.2014.10.022.

Nawandar, N. K., & Satpute, V. R. (2019). IoT based low cost and intelligent module for smart irrigation system. *Computers and Electronics in Agriculture*, *162*, 979–990. Available from https://doi.org/10.1016/j.compag.2019.05.027.

Neupane, J., & Guo, W. (2019). Agronomic basis and strategies for precision water management: A review. *Agronomy*, *9*(2), 87. Available from https://doi.org/10.3390/agronomy9020087.

Nikolidakis, S. A., Kandris, D., Vergados, D. D., & Douligeris, C. (2015). Energy efficient automated control of irrigation in agriculture by using wireless sensor networks. *Computers and Electronics in Agriculture*, *113*, 154–163. Available from https://doi.org/10.1016/j.compag.2015.02.004.

Ogidan, O. K., Onile, A. E., & Adegboro, O. G. (2019). Smart irrigation system: A water management procedure. *Agricultural Sciences*, *10*(1), 25–31. Available from https://doi.org/10.4236/as.2019.101003.

Paucar, L. G., Diaz, A. R., Viani, F., Robol, F., Polo, A., & Massa, A. (2015). Decision support for smart irrigation by means of wireless distributed sensors. In *Paper presented at the 2015 IEEE 15th Mediterranean Microwave Symposium (MMS)*. Available from https://doi.org/10.1109/mms.2015.7375469.

Pernapati, K. (2018). IoT based low cost smart irrigation system. In *Paper presented at the 2018 2nd international conference on inventive communication and computational technologies (ICICCT)*. Available from https://doi.org/10.1109/icicct.2018.8473292.

Rahim Khan, I. A., Suryani, M. A., Ahmad, M., & Zakarya, M. (2013). Wireless sensor network based irrigation management system for container grown crops in Pakistan. *World Applied Sciences Journal*, *24*(8), 1111–1118.

Rawal, S. (2017). IOT based smart irrigation system. *International Journal of Computer Applications*, *159*(8), 7–11. Available from https://doi.org/10.5120/ijca2017913001.

Rodríguez Díaz, J., & Montesinos, P. (2015). The role of precision irrigation in environmentally sensitive areas. *Irrigation & Drainage Systems Engineering*, *1*, e001.

Roopaei, M., Rad, P., & Choo, K.-K. R. (2017). Cloud of things in smart agriculture: Intelligent irrigation monitoring by thermal imaging. *IEEE Cloud Computing*, *4*(1), 10–15. Available from https://doi.org/10.1109/mcc.2017.5.

Sadler, E., Evans, R., Stone, K., & Camp, C. (2005). Opportunities for conservation with precision irrigation. *Journal of soil and water conservation*, *60*(6), 371–378.

Sakthipriya, N. (2014). An effective method for crop monitoring using wireless sensor network. *Middle-East Journal of Scientific Research*, *20*(9), 1127–1132.

Shafi, U., Mumtaz, R., García-Nieto, J., Hassan, S. A., Zaidi, S. A. R., & Iqbal, N. (2019). Precision agriculture techniques and practices: From considerations to applications. *Sensors (Basel, Switzerland)*, *19*(17), 3796. Available from https://doi.org/10.3390/s19173796.

Soulis, K. X., Elmaloglou, S., & Dercas, N. (2015). Investigating the effects of soil moisture sensors positioning and accuracy on soil moisture based drip irrigation scheduling systems. *Agricultural Water Management*, *148*, 258–268. Available from https://doi.org/10.1016/j.agwat.2014.10.015.

Sun, Y., Song, H., Jara, A. J., & Bie, R. (2016). Internet of things and big data analytics for smart and connected communities. *IEEE Access*, *4*, 766–773. Available from https://doi.org/10.1109/access.2016.2529723.

Takács, S., Bíró, T., Helyes, L., & Pék, Z. (2018). Variable rate precision irrigation technology for deficit irrigation of processing tomato. *Irrigation and Drainage*, *68*(2), 234–244. Available from https://doi.org/10.1002/ird.2299.

Trilles, S., González-Pérez, A., & Huerta, J. (2020). An IoT platform based on microservices and serverless paradigms for smart farming purposes. *Sensors (Basel, Switzerland)*, *20*(8), 2418. Available from https://doi.org/10.3390/s20082418.

Vellidis, G., Tucker, M., Perry, C., Kvien, C., & Bednarz, C. (2008). A real-time wireless smart sensor array for scheduling irrigation. *Computers and Electronics in Agriculture*, *61*(1), 44–50. Available from https://doi.org/10.1016/j.compag.2007.05.009.

Viani, F. (2016). Experimental validation of a wireless system for the irrigation management in smart farming applications. *Microwave and Optical Technology Letters*, *58*(9), 2186–2189. Available from https://doi.org/10.1002/mop.30000.

Yousif, M. E.-R., Ghafar, K., Zahari, R., & Lim, T. H. (2018). A rule-based smart automated fertilization and irrigation systems. In *Paper presented at the 9th international conference on graphic and image processing (ICGIP 2017)*.

Zhang, P., Zhang, Q., Liu, F., Li, J., Cao, N., & Song, C. (2017). The construction of the integration of water and fertilizer smart water saving irrigation system based on big data. In *Paper presented at the 2017 IEEE international conference on computational science and engineering (CSE) and IEEE international conference on embedded and ubiquitous computing (EUC)*. Available from https://doi.org/10.1109/cse-euc.2017.258.

CHAPTER

11 Precision irrigation management: a step toward sustainable agriculture

Vishwa Jyoti Baruah[1], Mahima Begum[2,3], Bhaswati Sarmah[3,4], Bipul Deka[5], Raktim Bhagawati[2], Shantonu Paul[6] and Marami Dutta[6]

[1]Centre for Biotechnology and Bioinformatics, Faculty of Biological Sciences, Dibrugarh University, Dibrugarh, Assam, India

[2]Department of Agronomy, Assam Agricultural University, Jorhat, Assam, India

[3]AICRP on Potato, Assam Agricultural University, Jorhat, Assam, India

[4]Department of Plant Breeding and Genetics, Assam Agricultural University, Jorhat, Assam, India

[5]AICRP on Irrigation Water Management, Assam Agricultural University, Jorhat, Assam, India

[6]Department of Soil Science, Assam Agricultural University, Jorhat, Assam, India

11.1 Introduction

The ever-increasing population and global warming have created an unparalleled strive to systematically use natural assets such as water resources to ensure food security and check the water crisis. In the developmental planning of any state, water plays a decisive and multifaceted role. Water management is the core input in any agricultural practice around the globe. Around 70% of all freshwater worldwide is used for farming, divulging that agriculture is the primary sector for global water use. India shares 17% of the global population with only 2.4% of land and 4% of the water resources (Jain et al., 2019). India is expected to be among the most populated countries by 2050, with a population of over 1.6 billion. The food demand is overgrowing, but the per capita land and water availability show a declining trend (Jain et al., 2019). It is estimated that 40% of the freshwater used for agriculture in developing countries is lost, either by evaporation, spills, or absorption by the deeper layers of the soil, beyond the reach of plants' roots (Panchard et al., 2007). The existing agricultural system will not be sufficient to feed the growing population. Competition for resources among various sectors has also increased to meet this demand. The massive increase in population and substantial income growth demand an extra 2.5 million tons of food grains annually in India. Therefore, it is essential to economize water in agriculture and bring more area under irrigation.

Nevertheless, excess irrigation water use can lead to wastage of water, leaching of beneficial plant nutrients, and imbalance of beneficial microbes in the soil.

Remote Sensing in Precision Agriculture. DOI: https://doi.org/10.1016/B978-0-323-91068-2.00021-7

Also, it can increase soil salinity and waterlogging leading to a reduction in yield and making the land unproductive. On the contrary, insufficient water can lead to poor growth and development, reduced yield and quality of produce, or even complete crop failure. As such, there is a need for proper irrigation management. With the advent of information technologies and cost-effective machinery, the new concept of irrigation scheduling is when and how much to irrigate and when, how much, and where to irrigate (Sarma, 2016). Thus the new concept is shifted from "uniform irrigation treatment" of the conventional system to "differential irrigation treatment."

Precision irrigation is "applying water in the right place with the right amount" (Al-Karadsheh et al., 2002). It curtails the water requirement of every individual plant, which helps in better productivity and reduces the cost of production with increasing water use efficiency, minimizes related adverse environmental impacts (Hogo et al., 2019), and improves water use efficiency up to 80%−90% as compared with surface irrigation methods having an efficiency of 40%−45% (Dukes & Scholberg, 2004). The critical factor for the success of any precision irrigation system is to apply an adequate amount of water at the critical time of its requirement with the accurate information of the field production functions for all locations within a field. The components of a precision irrigation system may vary depending on the specific system design and requirements, but generally, they include a water source, a water delivery system, a pumping system, sensors, controllers, drip irrigation emitters, fertigation system, monitoring and control software and weather station (Fig. 11.1).

11.2 Importance of irrigation management

To meet the growing demand, the misuse/excess use of agricultural resources leads to degradation of the environment and reduction in crop yield (Mondal & Tewari, 2007). In traditional irrigation systems, the quantities of water are not explicitly calculated to fit crop requirements as they have to compensate for possible losses such as evaporation loss, which may be over 50% of the applied water (Batchelor et al., 2002). The complex agricultural production system consumes about 70% of the fresh available water, about 1500 billion m^3 out of the 2500 billion m^3 of water worldwide annually (Goodwin & O'Connell, 2008). In developing countries, only 60% of freshwater is efficiently used for crop growth, and the rest, 40%, is lost through different modes (Panchard et al., 2007). In a country like India, 80% of groundwater is consumed by the agriculture sector in India (Harsh, 2017), which leads to scarcity of water or decline in the groundwater table in some parts of the country (Central Groundwater Board, 2017). The declining trend of groundwater level in all parts of the country also indicates that the assured supply will challenge the country's development (Manivannan et al., 2017). The only possible way is to increase the water application efficiency

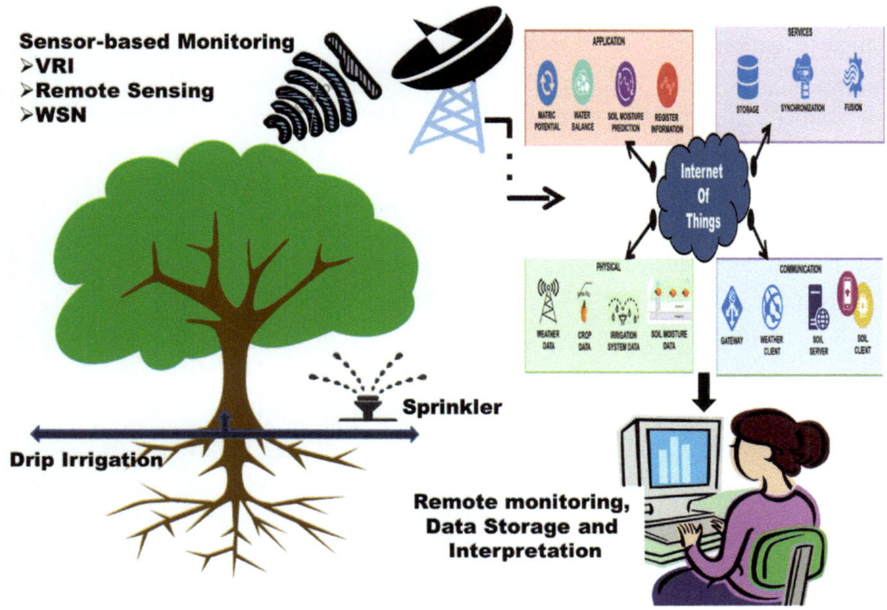

FIGURE 11.1

Components of a precision irrigation system (Campos et al., 2020).

Adapted from Campos N. G. S., Rocha, A. R., Gondim, R., Coelho da Silva, T. L., & Gomes, D. G. (2020). Smart & green: An internet-of-things framework for smart irrigation. Sensors, 20*(1), 190.*

through the irrigation method, as the flood irrigation system has an overall efficiency of only 25%–40% (Amarasinghe et al., 2007).

Therefore, the basic concept of precision irrigation management fulfills the following objectives:

1. To apply sufficient quantity to crops and reduce wastage
2. To ensure proper use of water for improving crop production
3. To use the water efficiently without much conveyance loss
4. To use, store, and manage water resources for future or off-season use.
5. To use the water considering the cost–benefit, that is, economic use of water
6. To dispense water uniformly without any social barriers

11.3 Need and advantages of precision irrigation

The potential of precision irrigation in terms of both water use and economic efficiencies is higher in comparison with the conventional irrigation system (Table 11.1).

Table 11.1 Comparative performances of conventional irrigation with microirrigation.

Performance indicator	Conventional irrigation methods	Precision irrigation	References
Field condition	Consider the field as a homogeneous one without considering the spatial variability	Consider the spatial variability including soil, topography, crop growth conditions	De Caires et al. (2015)
	It does not divide the field; instead, the whole field is considered as one unit one	For précised use of irrigation water divided the field into small management units	McBratney et al. (2005)
Water saving	Lots of water is wasted. It may be due to evaporation percolation or runoff.	Almost 27–55 mm per hectare of irrigation could be saved through VRI. Almost 10%–15% water could be saved than the conventional. However, potential water savings using precision irrigation is 25%	Türker et al. (2015), Sadler et al. (2005), Hedley and Yule (2009)
Water use efficiency	30%–50%	80%–90% depending on the technology adopted; in the case of VRI, it goes up to 90% and goes above 90% in microirrigation	Yari et al. (2017), Evett et al. (2006)
Saving in labor	Labor requirement is high	Labor requirement is periodic in nature	Khatri and Smith (2006)
Weed infestation	Very high	Less due to less wetting of the soil	Sayed and Bedaiwy (2001), Shrestha et al. (2007)
Pest and diseases problems	Disease pressure is relatively high	Relatively rarer because of less atmospheric humidity	Neibling and Gallian (1997), Rogers et al. (1997)
Water control	Inadequate	Very precise and easy	Pierce (2010)
Efficiency of fertilizer use	Low, due to heavy losses through runoff and leaching	Very high. Leaching and runoff water reduced considerably	King et al. (1995), Exner et al. (2014)
Energy saving	Required high energy	Less energy required than the conventional. Saved almost 23–67 kg CO_2-eq/ha/year energy	Carolyn et al. (2010)

(*Continued*)

Table 11.1 Comparative performances of conventional irrigation with microirrigation. *Continued*

Performance indicator	Conventional irrigation methods	Precision irrigation	References
Environmental impact	Increased CO_2e emissions	Minimize any adverse effect on the environmental. Mainly reduced CO_2e emissions by 23%	Raine et al. (2007), El Chami et al. (2019)
Increase in crop yield and economics	Crop yield is reduced due to the nonuniformity of available moisture	Yield can be increased much fold, and economic return is also high compared with conventional irrigation methods	King et al. (2006), Nijbroek et al. (2003)

11.3.1 Increasing water use efficiency

With the increasing scarcity of freshwater, the proper use of each drop of water is the present need of the hour. In India, the water consumption for producing each unit of a major food crop is much higher than that in European countries (Sarma, 2016). This indicates the low water use efficiency in most present irrigation technologies. This necessitates the adoption of a precision irrigation management strategy for better water use efficiency (WUE). Precision irrigation methods such as drip and sprinkler have the potential to increase the application efficiency of water up to 80%–90% compared with surface irrigation, which has only 40%–45% (Dukes & Scholberg, 2004).

11.3.2 Water and energy savings

The prime objective of precision irrigation is to use the optimal amount of water throughout the field. Variable-rate irrigation (VRI) saves around 10%–15% than conventional irrigation (Yule et al., 2008). Hedley and Yule (2009) reported that 25% of water could be saved with necessary improvements in application efficiency, which can be obtained through spatially varied irrigation applications. Through sensor-based precision irrigation, a saving of 52.5% in maize, 34% in finger millets and 36.4% in tomato could be obtained (Thimmegowda, 2020), and that value hiked up to 90% water saving through an automated irrigation system (Joaquin et al., 2014). In an earlier research, Dukes et al. (2003) also reported saving water in bell through sensor-based irrigation. The measurement of soil moisture through an electromagnetic sensor saved almost 53% water in sprinkler irrigation (Blonquist et al., 2006). Yule et al. (2008) also reported that a precise irrigation system could save 10%–15% irrigation water. This benefits the farmer in lowering his/her cost of the application in terms of electrical expenses and labor involved and is more sustainable.

11.3.3 Reduction in disease and pest

Excessive irrigation is expensive and wasteful and reduces soil aeration and aggravates the soil-borne pathogens, putting the crop in a more vulnerable state and getting infected by various diseases (Allmaras et al., 1988; Carter & Johnston, 1989). Precise application of irrigation avoids unnecessary wetting of the aerial parts of the plant, causing less foliar diseases in addition to more WUE (Evans & Sadler, 2008). This type of precision irrigation system has demonstrated that disease occurrences vary in response to irrigation patterns where a significant reduction in the movement of virus vector was observed (Lage et al., 2019; Rur et al., 2018).

The type and method of irrigation system also determine the frequency and intensity of disease incidence and its spread. Although economic and allowing better water availability to crops, it has been observed that sprinkler irrigation systems promote the wetting of the foliage, making it more prone to pathogens, especially fungal and bacterial spores. The time of irrigation also determines the incidence of disease and its proliferation. It was reported that some fungal spores of Phytophthora infestans are dispersed by afternoon irrigations while they can be checked by night irrigation (Rotem & Palti, 1969). The pathogen's success in establishing itself in the aerial parts of the plant is directly affected by irrigation timing. It is exceedingly reliant on the foliar drenching period and other factors (Wang et al., 2015). The leaf wetness duration has been considered the most determinant microclimatic variable for disease establishment and progress, and it is one of the main variables monitored in disease prediction systems (Lee et al., 2016). Therefore, precision irrigation systems can be employed to manage the time and duration of irrigation so that disease incidence and spread can be kept at a minimum.

11.3.4 Reduction in weed infestation

Precise irrigation application can check weeds from flourishing and is also considered a modern production system. Although sprinkler systems are costlier due to the involvement of high-pressure pumps to deliver water, it provides superior application uniformity compared with surface irrigation systems and can operate with significantly less water (Locascio, 2005; Sammis, 1980). Due to the superior application uniformity, sprinkler irrigation is the method of choice when applying herbicides or other agrichemicals through the irrigation system (Ogg, 1986). Through sprinkler irrigation methods, herbicides or other agrichemicals are sprayed that can reduce almost eight times weed infestation (Sayed & Bedaiwy, 2001). An eightfold reduction in weed pressure was noted when applying herbicides through sprinkler irrigation compared with traditional methods (Sayed & Bedaiwy, 2001).

The drip irrigation system applies water to the vicinity of the emitter or drip line facilitating farmers to apply water only in required critical areas of the crop at any critical time. As water is precisely applied to the vicinity of the target area, the area between the rows is devoid of water and nutrients/fertilizers, reducing weed growth. On combining drip irrigation with plastic mulch, weeds can be effectively controlled within rows, leaving only between-row areas to be managed. Subsurface drip irrigation (SDI) is another method that is more capable of controlling weeds compared with furrow irrigation (Shrestha et al., 2007). In several crops grown in the arid zones such as cotton, corn, tomato, and pistachio, weed pressure was seen to be reduced due to the SDI system of water application (Dastorani et al., 2010; Grattan et al., 1988; Sutton et al., 2006). In another study, it was observed that SDI was capable of lowering the weed growth up to fourfold in pistachio orchards of Iran compared with the surface irrigation (Dastorani et al., 2010).

11.3.5 Reduction in problems associated with over/underirrigation

Due to excessive irrigation, leaching of nitrogen beyond the root zone takes place, as well as waterlogging, leaching of nutrients and chemicals (Tilman et al., 2001), groundwater population (Exner et al., 2014), waste of water and energy (Shepherd et al., 1993), depletion of the groundwater table especially in an arid–semiarid region (McBratney et al., 2005), ultimately affecting the yield (Locascio et al., 1989). Moreover, an increase in the problem of soil salinity and sodicity in areas having shallow groundwater tables was also observed (Tanji, 2002).

On the other hand, deficit or underirrigation failed to supply the required quantity of water to fulfill the crop water demand that resulted in yield and economic loss of the growers (Zhang et al., 2021). The water deficit has a negative impact on the growth and development of crops, or even there may be total crop failure due to efficient uptake of nutrients from the soil (Groves & Bailey, 1997). Thereby, the only efficient means of water saving is précised use of water.

11.3.6 Increase in yield and profit

VRI/precession irrigation has the potential to increase yield over other irrigation systems. King et al. (2006) revealed that the yield of potatoes was higher with spatially varied irrigation compared with the uniform irrigation application system. Sensor-based precision irrigation management significantly increased the yield of maize (100 q/ha), finger millets (38 q/ha), and tomato (112 q/ha) reporting a yield improvement of 8%, 17%, and 12% in comparison with conventional irrigation (Thimmegowda, 2020). The precision irrigation management through subsurface increased the productivity and WUE of Alfalfa (Mohammed et al.,

2011). In the case of some horticultural crops such as banana, coconut, and grapes, the return per unit of energy was found to be 92.3 (Rs/kWh), 98 (Rs/kWh), and 378 (Rs/kWh), respectively, which were much higher than that in traditional irrigation method (Kumar & Palanisami, 2010). Therefore, it can be concluded that precision irrigation methods can be applied to increase the yield of crops, which in turn will give better remuneration. Variable rate of irrigation is advantageous over conventional method as it gives an added value of £3.02/m^3 (El Chami et al., 2019).

11.4 Impact of microirrigation in precision agriculture

In traditional irrigation systems, the quantities of water are not explicitly calculated to fit crop requirements as they have to compensate for possible losses such as evaporation loss, which may be over 50% of the applied water (Batchelor et al., 2002). The microirrigation technique is well accepted for plantation crops, small areas of high-value vegetable crops, flowers, etc. This technique is more suitable in arid regions or any region where the water supply is limited or expensive. This technique is also widely used in greenhouse cultivation. However, its application could be much more extended or could even easily replace any traditional irrigation installation. The microirrigation technique has remarkable advantages over conventional irrigation. In this technique, the water is applied at the closest to the root zone, and as such, there is less water loss through deep percolation and evaporation, resulting in higher WUE. Keeping in consideration the water-holding capacity of soil, the irrigation is scheduled in time and interval so that high uniformity in distribution is achieved (Barragan et al., 2010). Owing to high-frequency irrigation, the soil moisture remains optimal and constant (Barragan & Wu, 2001). It facilitates each plant to grow at its best, resulting in improved crop quality and increased crop yield. Proper management of water and nutrients through a microirrigation system results in a sustained economic benefit. Further, as water is prevented from running all over the field, the growth of weed is also restricted. The direct consequence is a reduction in agronomic cost and effort for weed control, fertilization, and tillage (Ayars et al., 2015).

11.4.1 Drip irrigation

In precision agriculture, generally, we try to supply the inputs/resources at the right place at the right time with an object to minimize the loss. Drip irrigation is often regarded as epitomizing precise irrigation because of its ability to control application rate and timing (Sarma, 2016) and can enhance irrigation efficiency because of the slow and recurrent supply of water to fulfill the crop evaporation demands. The irrigation requirements of drip-irrigated maize in a sub-humid climate could be evaluated with the help of Hybrid-Maize mode (Liu et al., 2018).

Drip irrigation also increased both cumulative WUE by 11% and WUE by 15% in maize, mainly due to a reduction in soil evapotranspiration ratio (Wang et al., 2021). The irrigation efficiency of drip irrigation generally ranged from 80% to 91% (Chimonides, 1995), while, in the case of sprinkler, it ranged from 54% to 80% (Zalidis et al., 1997). Similarly, Zhang et al. (2018) reported that in maize, drip irrigation with mulch reduced the evapotranspiration by 2.8%–5.2% by increasing the WUE by 10.7%–13.1%. In comparison, Liao et al. (2019) reported that drip irrigation could reduce evapotranspiration up to 24% and increase WUE by 19% in the cherry crop. This technique also saved almost 50%–80% water in comparison with the surface irrigation method (Kadasiddap & Rao, 2018). Because of this advantage, it is used chiefly in a large proportion of areas under dryland agriculture (Han et al., 2015). Surface drip irrigation was also an efficient technique for managing N and water, even in sandy soils (Liu et al., 2015). Apart from this, drip irrigation also accelerates crop maturity and increases the quality of the final product (Pawar et al., 2013).

Both drip irrigation and sprinkler irrigation were found to be effective in increasing the tuber yield, quality, and nitrogen uptake of potato (Waddell et al., 1999). However, they reported more water productivity in drip irrigation compared with sprinklers. In drip irrigation, about 8% higher yield of capsicum was obtained while saving 40% water compared with the surface irrigation (Paul et al., 2013). A comparison study of different microirrigation was conducted in Alfalfa by Mohammed et al. (2011). They reported that SDI supplied the least water, resulting in more irrigation water use efficiency (IWUE) followed by surface drip and sprinkler irrigation. The reduction in deep percolation and surface evaporation loss under SDI resulted in more IWUE (Alam et al., 2002).

11.4.2 Sprinkler irrigation

Sprinkler irrigation is another technique for water saving and fertigation that helps in the accurate supply of irrigation water at an accurate time (Li & Rao, 2003). In this irrigation method, water is sprayed into the air through a pressurized pipe network to the nozzles of the sprinkler (Ali, 2011). Many researchers reported more crop yield and higher WUE under sprinkler irrigation compared with surface irrigation (Yang et al., 2000). Through this technique, the transpiration rate of the crop could be reduced by 50% (Tolk et al., 1995). Nevertheless, the performance of sprinkler always depends upon the type of sprinkler, frequency of water application (Abdelraouf et al., 2014), type and nozzles of the sprinkler (Keller & Bliesner, 1990), coefficient of uniformity (CU), and distribution uniformity (DU) (Serem et al., 2016). By increasing the frequency of irrigation under the solid set sprinkler, more yield and higher WUE were obtained compared with hand-move laterals. (Abdelraouf et al., 2014). While evaluating the performance of three different types of sprinklers, namely Jalpar, Taiwan, and Naan, Serem et al. (2016) reported that Taiwan sprinkler performed better than others two in terms of increasing crop yield of green gram because of having higher water application rate (8.5 mm/h), DU (75.88%), and CU (83.2%).

11.5 Recent advancements in precision irrigation technologies

11.5.1 Variable-rate irrigation

Spatial variability of soils, topography, nutrient availability, and crop growth are noticed in any agriculture field. Different areas of the same field may have different irrigation needs. VRI can help a farmer address this variability, conserve resources, and improve yield. The main objective of VRI is to reduce water consumption (Lo et al., 2016) by maximizing the crop productivity (Ge et al., 2008) that mainly relies on both soil water content sensors and soil water balance (Li et al., 2018).

The variable-rate technology allows spatially precise irrigation to the crop root zone to meet the crop water demands in a controlled way for specific management zones (MZs) within a field. Drip and sprinkler irrigation systems form the backbone of VRI. Outfitted with a VRI package, drip and sprinkler irrigation can efficiently increase the WUE and site-specific water demands of the crop (Chavez et al., 2010; Han et al., 2009; King & Wall, 1998). In the practical field, farmers should choose low-elevation spray application (Lyle & Bordovsky, 1983) instead of high-impact or mid-elevation spray application to decrease the drift between field MZs (O'Shaughnessy et al., 2013). In addition to this, the system should include a means of remote monitoring and communication for precise spatial and temporal information and operations. According to Evett et al. (2014), the first step for VRI involves dividing the whole field into MZs based on physiological aspects. Thereafter, soil water sensors are installed in each MZs, which are to be delineated through mapping different soil physicochemical parameters in conjunction with soil survey data and topography features (Grisso et al., 2009; Hedley et al., 2013; Landrum et al., 2015). Post delineation of MZs is subjected to data acquisition procedures using geographical information systems, remote sensing, etc. This is very important as variability between the MZs has to be identified and quantified for further analytical and agronomic modification to the irrigation system. Based on the recorded data, farmers can develop irrigation schedules specific to the existing crop in the field. For example, in rice crop, Rowshon and Amin (2010) developed a water balance model based on quantitative estimation of significant components of water balance such as crop evapotranspiration, adequate rainfall, seepage, percolation, cumulative relative water supply, and ponding water index and GIS-based data to complement the operational model for precise irrigation management in rice. VRI can increase yield WUE and reduce the cost of operations. The problems associated with surface irrigation due to untimely irrigation, such as deep percolation leaching of nutrients, could be alleviated by adopting VRI (Xiumei et al., 2021). A study on the assessment of VRI to save water over different weather regimes and across several crops demonstrated that VRI has the capability to save as much as 10%–15% water (Sadler et al., 2005). It is the best suitable for center pivot and lateral sprinkler irrigation systems because of its high application accuracy (Carolyn et al., 2010).

King et al. (2006) revealed that precise irrigation in potatoes out-yielded conventional uniform irrigation management during two consecutive years. Yari et al. (2018) recorded 43% lesser water application under the VRI and a 12% reduction in irrigation pumping cost in the potato crop field. VRI produced tubers of uniform size and glucose content and produced a higher yield compared with regular irrigation. Water savings can be increased by 25% by a spatially VRI system (Hedley & Yule, 2009). Although the VRI helps in managing the spatial and temporal variability of a crop field, the adoption of VRI is very limited and not considered the best irrigation schedule (Sharma & Irmak, 2020) (Fig. 11.2).

11.5.2 Intelligent system using low-cost sensor network

In precision agriculture, it is necessary to analyze the data obtained from sensor devices implemented in different wireless sensor networks (WSNs; Khanna & Kaur, 2019). For decision-making, crop and environmental conditions need to be closely monitored (Popovic et al., 2017). To find out the spatial and temporal variation, soil moisture analysis is the prime factor in precision irrigation management. Among all variables, the soil hydraulic parameters and spatial heterogeneity deeply influence the moisture and crop rooting patterns (Elmaloglou et al., 2013). The WSN helps in identifying the optimal crop

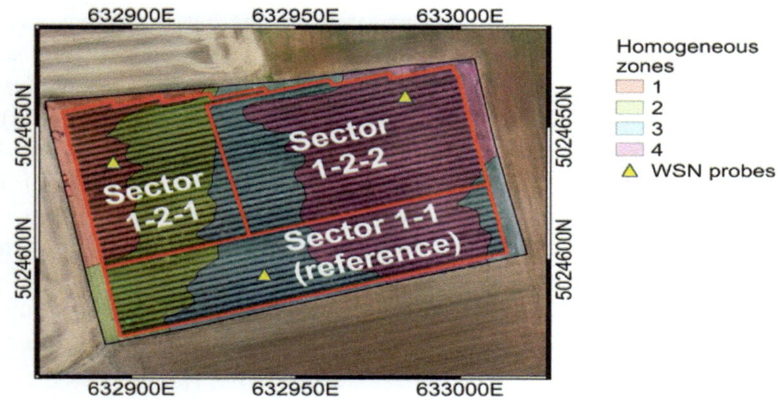

FIGURE 11.2

Layout of variable-rate irrigation (VRI) system showing the three sectors and measurement points of soil water content where zones 1 and 2 comprise coarse soils and zones 3 and 4 comprise heavy soils (Ortuani et al., 2019).

growing status by monitoring and controlling the different environmental parameters such as temperature, humidity, and soil moisture (Srbinovska et al., 2015). Out of different variables, because of its nonlinear manners, the soil moisture variability is typically considered in precision irrigation management (Soulis et al., 2015; Soulis & Elmaloglou, 2018). Similarly, other hydrodynamic soil properties such as water drainage and water-holding capacity are equally important while deciding on precision irrigation management (Kramer & Boyer, 1995). Hence, multiple sensors are needed for efficient irrigation scheduling to the plants (Naor et al., 2001). The soil sensors provide dimensions of soil volumetric water content, water potential, soil salinity/temperature, and many more (Vaz et al., 2013).

In recent years, the significance of soil moisture measurements in WSN is mainly gaining popularity in achieving the efficient and best possible water use technique for irrigating a huge crop area (Goap et al., 2018; Lozoya et al., 2016; Muangprathub et al., 2019). Using a fuzzy-based intelligent irrigation scheduling system with a low-cost WSN results in a 59.61% and 67.35% decrease in water use and electrical energy consumption, respectively, increasing the crop yield by 22.58% (Jamroen et al., 2020). Being a network connecting many powered sensing devices, WSN requires a competent energy running system for implementing it in the agricultural system (Band et al., 2019), which can be alleviated by introducing a sleep/wake-up strategy in wireless devices that can be kept in standby mode (Jawad et al., 2017).

11.5.3 Precision irrigation automating farm irrigation

Instead of a predetermined irrigation schedule in agricultural crops, the irrigation systems can be automated by obtaining information such as volumetric soil water content (Jones, 2004), using dielectric moisture sensors (Nemali & Iersel, 2006) to optimize water use. An automated irrigation system has been developed consisting of a distributed wireless network of soil moisture and temperature sensors placed in the plants' root zone. A gateway unit handles sensor information, triggers actuators, and transmits data to a web application (Joaquin et al., 2014). The automatic irrigation scheduling efficiently increases the cotton yields and WUE compared with traditional/manual irrigation (Shaughnessy & Evett, 2010). It can be recommended as a model method due to low-labor requirements, cost-effectiveness, efficient use of water resources, complete irrigation, and high productivity (Dehghanisanij et al., 2007).

11.5.4 Internet of Things

There is a rising threat to food security due to the combined effects of global warming, depleting food and water resources, and unprecedented rise in the population (Premanandh, 2011). Irrigation is an essential exhaustion of freshwater; wastage of these resources can significantly reduce overall food security (Oborkhale et al., 2015). Precision irrigation system plays a substantial part in delivering considerable impacts to food production by precisely designing an irrigation system that can predict, adapt, and deliver the appropriate amount of water to the crops where and when it is needed in response to the changing dynamics of the soil, weather, and the plant (Abioye et al., 2020; Işık et al., 2017). Therefore, to increase WUE, precision irrigation through controlled and optimized Internet of Things (IoT) strategies comes into play (Benyezza et al., 2018; Koech & Langat, 2018).

The term Internet has been associated with things and is now being identified as IoT, which implies the interconnection of electronic devices through the Internet via Wi-Fi, radio frequency identification device, LoraWan, Zigbee, Bluetooth, Long-term evolution, and other wireless communication technologies (Abdulrahman et al., 2016; Khanna & Kaur, 2019). Irrigation cannot be compromised to compensate the crop yield, but shortage in the workforce can be compensated with IoT (Kumar et al., 2020). Automation of irrigation has been done with sensors, microcontroller, and android, while IoT involves Internet systems connecting, collecting information, and exchanging data with other devices and systems (Hendricks, 2015). Various essential parameters such as soil temperature, moisture, pH, humidity required for proper implantation of precision irrigation can be acquired using the IoT platform (Kumar et al., 2020).

In the current decade, IoT has provided an efficient means in the monitoring system as the user can monitor and control the system anywhere and at any time (Tzounis et al., 2017). In agriculture, the application of IoT has the main aim to connect physical objects (things) such as sensors, cameras, flow meters, actuators, and robots to the Internet using wireless network connectivity to measure variables such as soil moisture, temperature, humidity, images of plant and other weather conditions (Cambra Baseca et al., 2019; Mohanraj et al., 2016; Sreekantha & Kavya, 2017; Tzounis et al., 2017; Zamora-Izquierdo et al., 2019). Therefore, it would be advantageous to leverage an IoT system platform to enhance the monitoring and control process of irrigation operation and observe plant response in terms of growth and water stress (Abioye et al., 2020). In an experimental finding conducted on tomato and eggplant based on IoT platform, automatic irrigation was set up, and precise irrigation was applied. In 1-month time, under IoT platform, total volume of water applied in both tomato and eggplant was 4009.875 and 4543.08 mL, respectively, while in the traditional system, consumption of water by tomato and eggplant was 6650 and 8550 mL, respectively, which indicated a 44% savings in total water consumption in IoT-based precise irrigation system (Palconit et al., 2020).

11.5.5 Remote sensing–based precision irrigation

For site-specific water management, to forecast the soil or plant water status, some indices such as crop water stress index, normalized difference vegetation index, leaf area index (Urban et al., 2018), multispectral images obtained from remote sensing are broadly used for a large area (Masseroni et al., 2017). In addition, by determining the crop canopy temperature with the help of digital infrared thermography, the scheduling of irrigation water can be managed easily (Kimm et al., 2020; O'Shaughnessy et al., 2012). The crop water content of the standing crop is determined with the help of thermal remote sensing, and accordingly, irrigation scheduling is done (Mulla, 2013).

Several researchers reported the précised use of irrigation with remote sensing and soil/plant-based irrigation scheduling technology (Evett et al., 2020; Foster et al., 2019; Pathak et al., 2019) and remote sensing–based ET model (Anderson et al., 2004).

11.6 Fertigation and precision agriculture

Drip irrigation frequently supplied the lower quantity of water with nutrients to the crop root zone as and when required, reducing the quantity of fertilizers required compared with the broadcast method (Yaseen et al., 2014). The advantage of fertigation is precise nutrient supply in time and space. Economical and sustainable crop production can be achieved with optimal fertigation. In fertigation, precise timing and precise rate of fertilizer can be applied to crops as per requirement and at different growth stages, which is not possible in the conventional method of fertilization (FAO & ITPS, 2021). The application of precision reduces nutrient losses and increases the fertilizer uptake efficiency (Gärdenäs et al., 2005).

Fertigation allows the regular supply of nutrients, reduces its fluctuation in soil, and maintains optimal nutrients in irrigated soil, not possible in broadcasting or band application of fertilizers (Kafkafi & Kant, 2005). Precision in nutrient and water application also increases soil organic carbon (SOC) sequestration and carbon retention due to crop production (Lal, 2020).

Some of the advantages of fertigation are listed below (Kafkafi & Kant, 2005):

- Precise application of nutrients to subsurface and wetted soil increases nutrient use efficiency and reduces nutrient losses as crop roots are more active in this zone.
- Crop foliage remains dry, reducing the incidence of pest and diseases and decreasing the chances of foliage burn, which is not possible in sprinkler irrigation.
- Uniform distribution of nutrients around the emitters is possible in fertigation.
- In subsurface fertigation, evaporation and runoff losses can be reduced as well as weed germination can be restricted.

11.6.1 Effect of fertigation on organic carbon in soil

Research findings indicate that fertigation increases crop production and biomass carbon in soil (Lal, 2020). Similar results were observed in the arid region of

India (Singh et al., 2014). Application of nitrogen, phosphorus and potash (NPK) through drip fertigation in aeracanut increased soil (SOC; Bhat & Sujatha, 2009)

11.6.2 Effect of fertigation on soil properties

Fertigation and biochar in alkaline soils were found to improve the soil properties such as water content, phosphorous content, and capacity of microbial population to use the soil compounds and biomass carbon and vanquish the parasitic nematodes of plants (Zhang et al., 2020). The addition of nutritive salts in fertigation was found to increase soil aggregate stability (Moreira et al., 2015). It has also been reported that drip fertigation increased the availability of soil micronutrients (Glonek & Komosa, 2013).

11.6.3 Effect of fertigation on production

Experimental findings in potato crops with drip fertigation resulted in a higher yield compared with furrow irrigation (Janat, 2007). Another research conducted in Italy on tomato crops showed increased yield and N uptake from the soil with fertigation (Farneselli et al., 2020). Yield recorded under drip fertigation with 125% recommended dose of fertilizer (RDF) was found to be equivalent with drip fertigation at 150% RDF (Hadole et al., 2020).

11.6.4 Effect of fertigation in mitigating climate change

Greenhouse gas such as N_2O increases with an increase in irrigation water (Horváth et al., 2010). It primarily depends on the availability of N in the soil; therefore, in precise drip fertigation, the application of N nutrient is less and need-based, which creates less emission of N_2O. Researchers also explained the decrease of methane (CH_4) emission and N losses (Badr et al., 2012; Maris et al., 2015). There was a reduction in N_2O by 19.9% in maize crops due to drip fertigation compared with traditional flooding systems (Zhang et al., 2019).

11.7 Impact of precision irrigation on soil properties

The precision irrigation always considered both spatial and temporal variation for proper decision-making. Hence, an in-depth study of both these factors is the most sensible part of precision irrigation management. The spatial variation of both soil physical and chemical properties influences the field variations, ultimately affecting crop production (Batchelor et al., 2002; Steiner et al., 2018). The water droplets of sprinklers significantly influenced the soil compaction (Toromo et al., 2012), and sprinklers with a high discharge rate caused more soil compaction compared with sprinklers with a lower discharge rate (Serem et al., 2016). In addition to that, sprinklers with larger water droplets created crust formation, reduced infiltration, encouraged more surface runoff (Silva, 2006), and sprinkler droplets energy degraded the soil aggregate stability (Lehrsch & Kincaid, 2006). According to Stylianou and Orphanos (1981), potatoes could be irrigated through sprinkler irrigation, but trickle was not the suitable method of irrigation for potato crop because of many drawbacks such as soil cracking and attack of potato tuber moth.

In addition to improving soil physical properties, drip irrigation was an efficient technique for reclamation of salt-affected soil of China (Tang & Kang, 2009). The supply of water to the crop root zone (Xue et al., 2017) led to high leaching efficiency, which helped remove salts beyond the root zone (Bresler et al., 1982; Jiao et al., 2008; Kang et al., 2008).

11.8 Conclusion

Considering the global population boom, water availability for irrigation will be a significant challenge. Precision irrigation is a sustainable agricultural approach that gives farmers the ability to use irrigation water more effectively and efficiently. It produces more crop yield and quality products without polluting the environment and helps save energy by decreasing the cost of cultivation using low-cost wireless sensors. It also helps the farmers by providing stability over all uncertainties associated with cultivation by predicting the climate and its consequent management.

Due to the scarcity of water and cost associated with irrigation, much less development has been achieved in the high-valued crops of arid and semiarid regions where the crop is primarily rainfed. Therefore, with the assistance of precision irrigation systems, there is a potential scope to improve upon these regions by making cultivation possible and achieving higher yields with minimum water input.

Acknowledgments

VJB and BS thankfully acknowledge the Department of Biotechnology, Government of India, for providing research grants (Grant No. BT/PR25099/NER/95/1014/2017). BS and MB also thank Dr. Bidyut Chandan Deka, Vice Chancellor, AAU, Jorhat, for his constant encouragement and support.

References

Abdelraouf, R. E., Ahmed, M. E., Bakry, A. B., & Elbegawy, M. K. (2014). Effect of sprinkler irrigation systems and irrigation frequency on water use efficiency and economical parameters for wheat production. *International Journal of Scientific Research in Agricultural Sciences*, *1*(4), 56–66.

Abdulrahman, T. A., Isiwekpeni, O. H., Surajudeen-Bakinde, N. T., & Otuoze, A. O. (2016). Design, specification and implementation of a distributed home automation system. *Procedia Computer Science*, *94*, 473–478.

Abioye, E. A., Abidin, M. S. Z., Mahmud, M. S. A., Buyamin, S., Ishak, M. H. I., Abd Rahman, M. K. I., Otuoze, A. O., Onotu, P., & Ramli, M. S. A. (2020). A review on monitoring and advanced control strategies for precision irrigation. *Computers and Electronics in Agriculture*, *173*, 105441.

Alam, M., Trooien, T. P., Rogers, D. H., & Dumler, T. J. (2002). An efficient irrigation technology for alfalfa growers. *Journal of Extension*, *40*(3).

Ali, H. (2011). *Practices of irrigation & on-farm water management* (2). Springer.

Al-Karadsheh, E., Sourell, H., & Krause, R. (2002). *Precision irrigation: New strategy irrigation water management*. In Conf. int. research on food security, natural resource management and rural development, Deutscher Tropentag, Witzenhausen, pp. 9–11.

Allmaras, R. R., Kraft, J. M., & Miller, D. E. (1988). Effects of soil compaction and incorporated crop residue on root health. *Annual Review of Phytopathology*, *26*(1), 219–243.

Amarasinghe, U. A., Shah, T., Turral, H., & Anand, B. K. (2007). *India's water future to 2025–2050: Business asusual scenario and deviations* (Research Report 123). International Water Management Institute, Colombo, Sri Lanka.

Anderson, M. C., Norman, J., Mecikalski, J. R., Torn, R. D., Kustas, W. P., & Basara, J. B. (2004). A multiscale remote sensing model for disaggregating regional fluxes to micrometeorological scales. *Journal of Hydrometeorology*, *5*, 343–363.

Ayars, J. E., Fulton, A., & Taylor, B. (2015). Subsurface drip irrigation in California – Here to stay? *Agricultural Water Management*, *157*, 39–47.

Badr, M. A., El-Tohamy, W. A., & Zaghloul, A. M. (2012). Yield and water use efficiency of potato grown under different irrigation and nitrogen levels in an arid region. *Agricultural Water Management*, *110*, 9–15. Available from https://doi.org/10.1016/j.agwat.2012.03.008.

Band, Đ., Jakšic, B., Bandur, M., & Jovic, S. (2019). An analysis of energy efficiency in wireless sensor networks (WSNs) applied in smart agriculture. *Computers and Electronics in Agriculture*, *156*, 500–507.

Barragan, J., Cots, L. l, Monserrat, J., Lopez, R., & Wu, I. P. (2010). Water distribution uniformity and scheduling in micro-irrigation systems for water saving and environmental protection. *Biosystems Engineering*, *107*, 202–211.

Barragan, J., & Wu, I. P. (2001). Optimum scheduling of a microirrigation system under deficit irrigation. *Journal of Agricultural Engineering Research*, *80*(2), 201–208.

Batchelor, W. D., Basso, B., & Paz, J. O. (2002). Examples of strategies to analyze spatial and temporal yield variability using crop models. *European Journal of Agronomy*, *18*, 141–158.

Benyezza, H., Bouhedda, M., Djellout, K., & Saidi, A. (2018, November). *Smart irrigation system based ThingSpeak and Arduino*. In 2018 international conference on applied smart systems (ICASS) (pp. 1–4). IEEE.

Bhat, R., & Sujatha, S. (2009). Soil fertility and nutrient uptake by arecanut (*Areca catechu* L.) as affected by level and frequency of fertigation in a laterite soil. *Agricultural Water Management*, *96*(3), 445–456. Available from https://doi.org/10.1016/j.agwat.2008.09.007.

Blonquist, J. M., Jones, S. B., Jr., & Robinson, D. A. (2006). Precise irrigation scheduling for turfgrass using a subsurface electromagnetic soil moisture sensor. *Agricultural Water Management*, *84*(1–2), 153–165.

Bresler, E., McNeal, B. L., & Carter, D. L. (1982). *Saline and sodic soils: Principles-dynamics-modeling*. Berlin Heidelberg: Springer-Verlag.

Cambra Baseca, C., Sendra, S., Lloret, J., & Tomas, J. (2019). A smart decision system for digital farming. *Agronomy*, *9*(5), 216.

Campos, N. G. S., Rocha, A. R., Gondim, R., Coelho da Silva, T. L., & Gomes, D. G. (2020). Smart & green: An internet-of-things framework for smart irrigation. *Sensors*, *20*(1), 190.

Carolyn, H., Ian, Y., & Stu, B. (2010). *Analysis of potential benefits of precision irrigation for variable soils at five pastoral and arable production sites in new Zealand*. In 19th world congress of soil science, soil solutions for a changing world (pp. 1–6).

Carter, M. R., & Johnston, H. W. (1989). Association of soil macroporosity and relative saturation with root rot severity of spring cereals. *Plant and soil*, *120*(1), 149–152.

Central Groundwater Board. (2017). *Dynamic ground water resources of India*. Ministry of Water Resources, River Development and Ganga Rejuvenation Government of India, Faridabad.

Chavez, J. L., Pierce, F. J., Elliott, T. V., Evans, R. G., Kim, Y., & Iversen, W. M. (2010). A remote irrigation monitoring and control system (RIMCS) for continuous move systems: Part B. Field testing and results. *Precision Agriculture*, *11*(1), 11–26. Available from https://doi.org/10.1007/s11119-009-9110-8.

Chimonides, S. J. (1995). Irrigation management under water shortage conditions. In N. X. Tsiourtis (Ed.), *Water resources management under drought or water shortage conditions* (pp. 73–78). Balkema.

Dastorani, M. T., Heshmati, M., & Sadeghzadeh, M. A. (2010). Evaluation of the efficiency of surface and subsurface irrigation in dryland environments. *Irrigation and Drainage: The Journal of the International Commission on Irrigation and Drainage*, *59*(2), 129–137.

De Caires, S. A., Wuddivira, M. N., & Bekele, I. (2015). Spatial analysis for management zone delineation in a humid tropic cocoa plantation. *Precision Agriculture*, *16*, 129–147.

Dehghanisanij, H., Naseri, A., Anyoji, H., & Eneji, A. E. (2007). Effects of deficit irrigation and fertilizer use on vegetative growth of drip irrigated cherry trees. *Journal of Plant Nutrition*, *30*(3), 411–425.

Dukes, M. D., Simonne, E. H., Davis, W. E., Studstill, D. W., & Hochmuth, R. (2003). *Effect of sensor-based high frequency irrigation on bell pepper yield and water use*. In Proceedings of 2nd international conference on irrigation and drainage (pp. 12–15), May.

Dukes, M. D., & Scholberg, J. M. (2004). *Automated subsurface drip irrigation based on soil moisture*. ASAE Paper No. 052188.

El Chami, D., Knox J., W., Daccache, A., & Keith Weatherhead, E. (2019). Assessing the financial and environmental impacts of precision irrigation in a humid climate. *Horticultural Science. (Prague)*, *46*, 43–52.

Elmaloglou, S., Soulis, K. X., & Dercas, N. (2013). Simulation of soil water dynamics under surface drip irrigation from equidistant line sources. *Water Resources Management*, *27*, 4131–4148.

Evans, R. G., & Sadler, E. J. (2008). Methods and technologies to improve efficiency of water use. *Water Resources Research*, *44*(7).

Evett, S. R., O'Shaughnessy, S. A., Andrade, M. A., Kustas, W. P., Anderson, M. C., Schomberg, H. H., & Thompson, A. (2020). Precision agriculture and irrigation: Current U.S. perspectives. *Transactions of the ASABE*, *63*, 57–67.

Evett, S. R., Peters, R. T., & Howell, T. A. (2006). *Controlling water use efficiency with irrigation automation: Cases from drip and center pivot irrigation of corn and soybean.* In Southern conservation systems conference (pp. 57–66), Amarillo, TX.

Evett, S. R., O'Shaughnessy, S. A., Colaizzi, P. D., & Schwartz, R. C. (2014). *Soil water sensing: Implications of sensor capabilities for variable-rate irrigation management.* Irrigation Associations Exposition and Technical Conference Proceedings, Fairfax, VA.

Exner, M. E., Hirsh, A. J., & Spalding, R. F. (2014). Nebraska's groundwater legacy: Nitrate contamination beneath irrigated cropland. *Water Resources Research*, *50*, 4474–4489.

FAO, & ITPS. (2021). Recarbonizing global soils: A technical manual of recommended management practices. *Volume 3: Cropland, Grassland, Integrated systems and farming approaches – Practices overview*. Rome. Available from https://doi.org/10.4060/cb6595en.

Farneselli, M., Benincasa, P., Tosti, G., Guiducci, M., & Tei, F. (2020). Combining green manuring and fertigation maximizes tomato crop yield and minimizes nitrogen losses. *Agronomy*, *10*(7), 977. Available from https://doi.org/10.3390/agronomy10070977.

Foster, T., Gonçalves, I. Z., Campos, I., Neale, C. M. U., & Brozović, N. (2019). Assessing landscape scale heterogeneity in irrigation water use with remote sensing and in situ monitoring. *Environmental Research Letters*, *14*, 024004.

Gärdenäs, A. I., Hopmans, J. W., Hanson, B. R., & Šimůnek, J. (2005). Two-dimensional modeling of nitrate leaching for various fertigation scenarios under micro-irrigation. *Agricultural Water Management*, *74*(3), 219–242. Available from https://doi.org/10.1016/j.agwat.2004.11.011.

Ge, Y. F., Thomasson, J. A., Sui, R. X., Morgan, C. L., Searcy, S. W., & Parnell, C. B. (2008). Spatial variation of fiber quality and associated loan rate in a dryland cotton field. *Precision Agriculture*, *9*(4), 181–194.

Glonek, J., & Komosa, A. (2013). fertigation of highbush blueberry (*Vaccinium corymbosum* L.). Part II. The effect on soil nutrient content. *Acta Scientiarum Polonorum, Hortorum Cultus*, *12*(3), 59–68.

Goap, A., Sharma, D., Deepak, A. K., & Krishna, C. (2018). An IoT based smart irrigation management system using machine learning and open source technologies. *Computers and Electronics in Agriculture*, *155*, 41–49.

Goodwin, I., & O'Connell, M. G. (2008). The future of irrigated production horticulture – world and Australian perspective. *Acta Horticulturae*, *792*, 449–458.

Grattan, S., Schwankl, L., & Lanini, W. (1988). Weed control by subsurface drip irrigation. *California Agriculture*, *42*(3), 22–24.

Grisso, R., Alley, M., Holshouser, D., & Thomason, W. (2009). *Precision farming tools: Soil electrical conductivity*. Publication 442-508. Virginia Tech Cooperative Extension. http://pubs.ext.vt.edu/442/442-508/442-508_pdf.

Groves, S. J., & Bailey, R. J. (1997). The influence of sub-optimal irrigation and drought on crop yield, N uptake and risk of N leaching from sugarbeet. *ADAS Soil Use and Management*, *13*(4), 190–195.

Hadole, S. S., Patidar, G., Age, A. B., Sarap, P. A., Satyanaryana, E., & Dhule, D. T. (2020). Effect of fertigation on soil fertility, yield and nutrient uptake by brinjal,. *International Journal of Chemical Studies*, *8*(3), 1013–1017.

Han, M., Zhao, C. Y., Feng, G., Yan, Y. Y., & Sheng, Y. (2015). Evaluating the effects of mulch and irrigation amount on soil water distribution and root zone water balance using Hydrus-2D. *Water*, *7*, 2622–2640.

Han, Y. J., Khalilian, A., Owino, T. W., Farahani, H. J., & Moore, S. (2009). Development of Clemson variable-rate lateral irrigation system. *Computers and Electronics in Agriculture*, *68*(1), 108–113.

Harsh, J. (2017). *Micro-irrigation in India: An assessment of bottlenecks and realities.* http://www.globalwaterforum.org/2017/06/13/micro-irrigation-in-indiaan-assessment-of-bottlenecks-and-realities/, accessed on July 18 through internet.

Hedley, C. B., & Yule, I. J. (2009). Soil water status mapping and two variable-rate irrigation scenarios. *Precision Agriculture*, *10*, 342–355.

Hedley, C. B., Roudier, P., Yule, I. J., Ekanayake, J., & Bradbury, S. (2013). Soil water status and water table depth modelling using electromagnetic surveys for precision irrigation scheduling. *Geoderma*, *199*, 22–29.

Hendricks, D. (2015). *The trouble with the Internet of things.* London Datastore. Greater London Authority.

Hogo, A., Das, J., & Kumar, A. (2019). Precision irrigation the nest avenue of highly irrigated Punjab. *Bulletin of Environment, Pharmacology and Life Sciences*, *8*, S5–S9.

Horváth, L., Grosz, B., Machon, A., Tuba, Z., Nagy, Z., Czóbel, S. Z., Balogh, J., Péli, E., Fóti, S. Z., Weidinger, T., Pintér, K., & Führer, E. (2010). Estimation of nitrous oxide emission from Hungarian semi-arid sandy and loess grasslands: Effect of soil parameters, grazing, irrigation and use of fertilizer. *Agriculture, Ecosystems & Environment*, *139*(1), 255–263. Available from https://doi.org/10.1016/j.agee.2010.08.011.

Işık, M. F., Sönmez, Y., Yılmaz, C., Özdemir, V., & Yılmaz, E. N. (2017). Precision irrigation system (PIS) using sensor network technology integrated with IOS/Android application. *Applied Sciences*, *7*(9), 891.

Jain, R., Kishore, P., & Singh, K. D. (2019). Irrigation in India, challenges and options. *Indian Journal of Soil and Water Conservation*, *18*(4), 1–11.

Jamroen, C., Komkum, P., Fongkerd, C., & Krongpha, W. (2020). Intelligent irrigation scheduling system using low-cost WSN toward sustainable and precision agriculture. *IEEE Access*, *4*(1), 172756–172769.

Janat, M. (2007). Efficiency of nitrogen fertilizer for potato under fertigation utilizing a nitrogen tracer technique. *Communications in Soil Science and Plant Analysis*, *38* (17–18), 2401–2422. Available from https://doi.org/10.1080/00103620701588775.

Jawad, H. M., Nordin, R., Gharghan, S. K., Jawad, A. M., & Ismail, M. (2017). Energy-efficient wireless sensor networks for precision agriculture: A review. *Sensors*, *17*, 1781.

Jiao, Y., Kang, Y., Wan, S., Sun, Z., Liu, W., & Dong, F. (2008). Effect of soil matric potential on the distribution of soil salt under drip irrigation on saline and alkaline land in arid regions. *Transactions of the CSAE*, *24*, 53–58.

Joaquin, G., Juan Francisco, V.-M., Alejandra, N.-G., & Miguel Ángel, P.-G. (2014). Automated irrigation system using a wireless sensor network and GPRS module. *IEEE Transactions on Instrumentation and Measurement*, *63*(1), 166–176.

Jones, H. G. (2004). Irrigation scheduling: advantages and pitfalls of plant-based methods. *Journal of Experimental Botany*, *55*, 2427–2436.

Kadasiddap, K. K., & Rao, V. P. (2018). Irrigation scheduling through drip and surface methods: A critical review on growth, yield, nutrient uptake and water use studies of rabi maize. *Agricultural Research Communication Center*, *39*(4), 300–306.

Kafkafi, U., & Kant, S. (2005). Fertigation. In D. Hillel (Ed.), *Encyclopedia of soils in the environment* (pp. 1–9). Elsevier. Available from https://doi.org/10.1016/B0-12-348530-4/00227-7.

Kang, Y., Wan, S., Jiao, Y., Tan, J., & Sun, Z. (2008). *Saline soil salinity and water management with tensiometer under drip irrigation*. In Symposia on the 5th annual meeting of agricultural land and water engineering of Chinese Society of Agricultural Engineering (pp. 124–131).

Keller, J., & Bliesner, R. D. (1990). *Sprinkle and trickle irrigation AVI book*. Van Nostrand Reinhold.

Khanna, A., & Kaur, S. (2019). Evolution of Internet of things (IoT) and its significant impact in the field of precision agriculture. *Computers and Electronics in Agriculture*, *157*, 218–231.

Khatri, K. L., & Smith, R. J. (2006). Real-time prediction of soil inflitration characteristics for the managment of furrow irrigation. *Irrigation Science*, *25*, 33–43. Available from https://doi.org/10.1007/s00271-006-0032-1.

Kimm, H., et al. (2020). Deriving high-spatiotemporal-resolution leaf area index for agroecosystems in the US Corn Belt using Planet Labs CubeSat and STAIR fusion data. *Remote Sensing of Environment*, *239*, 111615.

King, B. A., & Wall, R. W. (1998). Supervisory control and data acquisition system for site-specific center-pivot irrigation. *Applied Engineering in Agriculture*, *14*(2), 135–144. Available from https://doi.org/10.13031/2013.19374.

King, B. A., Stark, J. C., & Wall, R. W. (2006). Comparison of site-specific and conventional uniform irrigation management for potatoes. *Applied Engineering in Agriculture*, *22*(5), 677–688.

King, B. A., Brady, R. A., McCann, I. R., & Stark, J. C. (1995). Variable rate water application through sprinkler irrigation. In P. C. Robert, R. H. Rust, & W. E. Larson (Eds.), *Site-specific management for agriculture systems* (pp. 485–493). ASA-CSSA-SSSA.

Koech, R., & Langat, P. (2018). Improving irrigation water use efficiency: A review of advances, challenges and opportunities in the Australian context. *Water*, *10*(12), 1771.

Kramer, P. J., & Boyer, J. S. (1995). *Water relations of plants and soils*. Academic Press.

Kumar, D. S., & Palanisami, K. (2010). Impact of drip irrigation on farming system in southern part of India. *Agricultural Economics Research Review*, *23*, 265–272.

Kumar, S. V., Singh, C. D., & Upendar, K. (2020). Review on IoT based precision irrigation system in agriculture. *Current Journal of Applied Science and Technology*, *39*(45), 15–26.

Lage, D. A., Marouelli, W. A., & Cafe-Filho, A. C. (2019). Management of powdery mildew and behaviour of late blight under different irrigation configurations in organic tomato. *Crop Protection*, *125*, 104886.

Lal, R. (2020). Managing soils for negative feedback to climate change and positive impact on food and nutritional security. *Soil Science and Plant Nutrition*, *66*(1), 1–9.

Landrum, C., Castrignano, A., Mueller, T., Zourarakis, D., Zhu, J., & De Benedetto, D. (2015). An approach for delineating homogeneous within-field zones using proximal sensing and multivariate geostatistics. *Agricultural Water Management*, *147*, 144–153. Available from https://doi.org/10.1016/j.agwat.2014.07.013.

Lee, K. J., Kang, J. Y., Lee, D. Y., Jang, S. W., Lee, S., Lee, B. W., & Kim, K. S. (2016). Use of an empirical model to estimate leaf wetness duration for operation of a disease warning system under a shade in a ginseng field. *Plant Disease*, *100*(1), 25–31.

Lehrsch, G. A., & Kincaid, D. C. (2006). Sprinkler droplet energy effects on soil penetration resistance and aggregate stability and size distribution. *Soil Science*, *171*(6), 435–447.

Li, J., & Rao, M. (2003). Field evaluation of crop yield as affected by non-uniformity of sprinkler-applied water and fertilizers. *Agricultural Water Management*, *59*, 1–13.

Li, X. M., Zhao, W. X., Li, J. S., & Li, Y. F. (2018). Crop yield and water use efficiency as affected by different soil-based management methods for variable-rate irrigation in a semi-humid climate. *Transactions of the ASABE*, *61*(6), 1915–1922.

Liao, R. K., Wu, W. Y., Hu, Y. Q., Xu, D., Huang, Q. N., & Wang, S. Y. (2019). Micro-irrigation strategies to improve water-use efficiency of cherry trees in Northern China. *Agricultural Water Management*, *221*, 388–396.

Liu, Y., Hai-shun, Y., Jiu-sheng, L. I., Yan-feng, L. I., & Hai-jun, Y. A. N. (2018). Estimation of irrigation requirements for drip-irrigated maize in a sub-humid climate. *Journal of Integrative Agriculture*, *17*(3), 677–692.

Liu, Y., Li, J., & Li, Y. (2015). Effects of split fertigation rates on the dynamics of nitrate in soil and the yield of mulched drip-irrigated maize in the sub-humid region. *Applied Engineering in Agriculture*, *31*, 103–117.

Lo, T., Heeren, D. M., Martin, D. L., Mateos, L., Luck, J. D., & Eisenhauer, D. E. (2016). Pumpage reduction by using variable rate irrigation to mine undepleted soil water. *Transactions of the ASABE*, *59*(5), 1285–1298.

Locascio, S. J. (2005). Management of irrigation for vegetables: Past, present, and future. *HortTechnology*, *15*(3), 482–485.

Locascio, S. J., Olson, S. M., & Rhoads, F. M. (1989). Water quantity and time of N and K application for trickle irrigated tomatoes. *Journal of the American Society for Horticulture Science*, *114*(2), 265–268.

Lozoya, C., Mendoza, C., Aguilar, A., Roman, A., & Castello, R. (2016). Sensor-based model driven control strategy for precision irrigation. *Journal of Sensors*, *2016*, 9784071.

Lyle, W. M., & Bordovsky, J. P. (1983). LEPA irrigation system evaluation. *Transactions of the ASAE*, *26*(3), 776–781. Available from https://doi.org/10.13031/2013.34022.

Manivannan, S., Thilagam, V. K., & Khola, O. P. S. (2017). Soil and water conservation in India: Strategies and research challenges. *Journal of Soil and Water Conservation*, *16* (4), 312–319.

Maris, S. C., Teira-Esmatges, M. R., Arbonés, A., & Rufat, J. (2015). Effect of irrigation, nitrogen application, and a nitrification inhibitor on nitrous oxide, carbon dioxide and methane emissions from an olive (*Olea europaea* L.) orchard. *Science of the Total Environment*, *538*, 966–978. Available from https://doi.org/10.1016/j.scitotenv.2015.08.040.

Masseroni, D., Ortuani, B., Corti, M., Gallina, P. M., Cocetta, G., Ferrante, A., & Facchi, A. (2017). Assessing the reliability of thermal and optical imaging techniques for detecting crop water status under different nitrogen levels. *Sustainability*, *9*, 1548.

Mc Bratney, A., Whelan, B., Ancev, T., & Bouma, J. (2005). Future directions of precision agriculture. *Precision Agriculture*, *6*, 7–23.

Mohammed, H., Almarshadi., & Ismail, S. M. (2011). Effects of precision irrigation on productivity and water use efficiency of alfalfa under different irrigation methods in arid climates. *Journal of Applied Sciences Research*, *7*(3), 299–308.

Mohanraj, I., Ashokumar, K., & Naren, J. (2016). Field monitoring and automation using IOT in agriculture domain. *Procedia Computer Science*, *93*, 931–939.

Mondal, P., & Tewari, V. K. (2007). Present status of precision farming: A review. *International Journal Agricultural Research*, *2*(1), 1–10.

Moreira, B., Abdelfattah, J. M., Matula, A., & Dolezal, F. (2015). Effect of fertigation on soil salinization and aggregate stability. *Journal of Irrigation and Drainage Engineering*, *141*(4), 05014010. Available from https://doi.org/10.1061/(ASCE)IR.1943-4774.0000806.

Muangprathub, J., Boonnam, N., Kajornkasirat, S., Lekbangpong, N., Wanichsombat, A., & Nillaor, P. (2019). IoT and agriculture data analysis for smart farm. *Computers and Electronics in Agriculture*, *156*, 467–474.

Mulla, D. J. (2013). Tventy five years of remote sensing in precision agriculture: Key advances and reaming knowledge gaps. *Biosystems Engineering Journal*, *114*(4), 358–371.

Naor, A., Hupert, H., Greenblat, Y., Peres, M., Kaufman, A., & Klein, I. (2001). The response of nectarine fruit size and midday stem water potential to irrigation level in stage III and crop load. *Journal of the American Society for Horticultural Science*, *126*, 140–143.

Neibling, H., & Gallian, J. (1997). Sugar beet irrigation: Essential in Idaho. *Agricultural Irrigation*, 22–23.

Nemali, K. S., & Iersel, M. W. V. (2006). An automated system for controlling drought stress and irrigation in potted plants. *Scientia Horticulturae*, *110*(3), 292–297.

Nijbroek, R., Hoogenboom, G., & Jones, J. W. (2003). Optimizing irrigation management for a spatially variable soybean field. *Agricultural Systems*, *76*, 359–377.

O'Shaughnessy, S. A., Evett, S. R., Colaizzi, P. D., & Howell, T. A. (2012). A crop water stress index and time threshold for automatic irrigation scheduling of grain sorghum. *Agricultural Water Management*, *107*, 122–132.

O'Shaughnessy, S. A., Urrego, Y. F., Evett, S. R., Colaizzi, P. D., & Howell, T. A. (2013). Assessing application uniformity of avariable-rate irrigation system in a windy location. *Applied Engineering in Agriculture*, *29*(4), 497–510.

Oborkhale, L., Abioye, A. E., Egonwa, B. O., & Olalekan, T. A. (2015). Design and implementation of automatic irrigation control system. *IOSR Journal of Computer Engineering (IOSR-JCE)*, *17*(4), 99-11.

Ogg, A. G. (1986). Applying herbicides in irrigation water: A review. *Crop Protection*, *5* (1), 53–65.

Ortuani, B., Facchi, A., Mayer, A., Bianchi, D., Bianchi, A., & Brancadoro, L. (2019). Assessing the effectiveness of variable-rate drip irrigation on water use efficiency in a Vineyard in Northern Italy. *Water*, *11*(10), 1964.

Palconit, M.G. B., Macachor, E.B., Notarte, M.P., Molejon, W.L., Visitacion, A.Z., Rosales, M.A., & Dadios, E.P. (2020). *IoT-based precision irrigation system for eggplant and tomato*. The 9th International Symposium on Computational Intelligence and Industrial Applications (pp. 0–6), November.

Panchard, J., Rao, S., Prabhakar, T. V., Hubaux, J. P., & Jamadagni, H. (2007). Common sense net: A wireless sensor network for resource-poor agriculture in the semi-arid areas of developing countries. *Information Technologies and International Development*, *4*(1), 51–67.

Pathak, H. S., Brown, P., & Best, T. (2019). A systematic literature review of the factors affecting the precision agriculture adoption process. *Precision Agriculture*, *20*, 1292–1316.

Paul, J. C., Mishra, J. N., Pradhan, P. L., & Panigrahi, B. (2013). Effect of drip and surface irrigation on yield, water use-efficiency and economics of capsicum (*Capsicum annum* L.)

grown under mulch and non mulch conditions in eastern coastal India. *European Journal of Sustainable Development*, *2*(1), 99–108.

Pawar, D. D., Dingre, S. K., Shinde, M. G., & Kaore, S. V. (2013). *Drip fertigation for higher crop productivity*. MPKV Research Publication: 99:3.

Pierce, F. J. (2010). Precision irrigation. *Landbauforsch SH*, *340*, 45–56.

Popovic, T., Latinovi c, N., Pesi c, A., Zečević, Ž., Krstajić, B., & Djukanović, S. (2017). Architecting an IoT-enabled platform for precision agriculture and ecological monitoring: A case study. *Computers and Electronics in Agriculture*, *140*, 255–265.

Premanandh, J. (2011). Factors affecting food security and contribution of modern technologies in food sustainability. *Journal of the Science of Food and Agriculture*, *91*(15), 2707–2714.

Raine, S. R., Meyer, W. S., Rassam, D. W., Hutson, J. L., & Cook, F. J. (2007). Soil–water and solute movement under precision irrigation: Knowledge gaps for managing sustainable root zones. *Irrigaion Science*, *26*, 91–100. Available from https://doi.org/10.1007/s00271-007-0075-y.

Rogers, D.H., Lamm, F.R., Alam, M., Trooien, T.P., Clark, G.A., Barnes, P.L., Mankin, K. (1997) Efficiencies and water losses of irrigation systems. *Irrigation Management Series MF-2243*. Kansas State University Cooperative Extension Service.

Rotem, J., & Palti, J. (1969). Irrigation and plant diseases. *Annual Review of Phytopathology*, *7*(1), 267–288.

Rowshon, M. K., & Amin, M. S. M. (2010). GIS-based irrigation water management for precision farming of rice. *International Journal of Agricultural and Biological Engineering*, *3*(3), 27.

Rur, M., Rämert, B., Hökeberg, M., Vetukuri, R. R., Grenville-Briggs, L., & Liljeroth, E. (2018). Screening of alternative products for integrated pest management of cucurbit powdery mildew in Sweden. *European Journal of Plant Pathology*, *150*(1), 127–138.

Sadler, E. J., Evans, R. G., Stone, K. C., & Camp, C. R. (2005). Opportunities for conservation with precision irrigation. *Journal of Soil and Water Conservation*, *60*(6), 371–379.

Sammis, T. W. (1980). Comparison of sprinkler, trickle, subsurface, and furrow irrigation methods for row crops 1. *Agronomy Journal*, *72*(5), 701–704.

Sarma, A. (2016). *Precision irrigation: A tool for sustainable management of irrigation water*. In Proceedings of the civil engineering for sustainable development-opportunities and challenges (pp. 19–21), Guwahati, India.

Sayed, M. A., & Bedaiwy, M. N. A. (2001). Effect of controlled sprinkler chemigation on wheat crop in a sandy soil. *Soil and Water Research*, *6*(2), 61–72.

Serem, P. K., Kipkorir, E. C., & Ng'etich, W. K. (2016). Effects of sprinkler irrigation on physical properties of soil and on the yields of green gram. *Journal of Agricultural and Crop Research*, *4*(3), 35–42.

Sharma, V., & Irmak, S. (2020). Economic comparisons of variable rate irrigation and fertigation with fixed (uniform) rate irrigation and fertigation and pre-plant fertilizer management for maize in three soils. *Agricultural Water Management*, 106307. Available from https://doi.org/10.1016/j.agwat.2020.106307.

Shaughnessy, S. A. O., & Evett, S. R. (2010). Canopy temperature based system effectively schedules and controls center pivot irrigation of cotton. *Agricultural Water Management*, *97*(9), 1310–1316.

Shepherd, M. A., Davies, D. B., & Johnson, P. A. (1993). Minimizing nitrate losses from arable soils. *Soil Use and Management*, *9*, 94–99.

Shrestha, A., Mitchell, J. P., & Lanini, W. T. (2007). Subsurface drip irrigation as a weed manage- ment tool for conventional and conservation tillage tomato (*Lycopersicon esculentum* Mill.) production in semi-arid agroecosystems. *Journal of Sustainable Agriculture*, *31*(2), 91−112.

Silva, L. P. (2006). The effect of spray head sprinklers with different deflector plates on irrigation uniformity, runoff and sediment yield in a Mediterranean soil. *Agricultural Water Management*, *85*(3), 243−252.

Singh, A., Gulati, I. J., Chopra, R., Sharma, D., & Gochar, R. (2014). Effect of drip-fertigation with organic manures on soil properties and tomato (*Lycopersicon esculentum* Mill.) yield under arid condition. *Annals of Biology*, *30*, 345−349.

Soulis, K. X., Elmaloglou, S., & Dercas, N. (2015). Investigating the effects of soil moisture sensors positioning and accuracy on soil moisture based drip irrigation scheduling systems. *Agricultural Water Management*, *148*, 258−268.

Soulis, K. X., & Elmaloglou, S. (2018). Optimum soil water content sensors placement for surface drip irrigation scheduling in layered soils. *Computers and Electronics in Agriculture*, *152*, 1−8.

Srbinovska, M., Gavrovski, C., Dimcev, V., Krkoleva, A., & Borozan, V. (2015). Environmental parameters monitoring in precision agriculture using wireless sensor networks. *Journal of Cleaner Production*, *88*, 297−307. Available from https://doi.org/10.1016/j.jclepro.2014.04.036.

Sreekantha, D.K., & Kavya, A.M. (2017, January). *Agricultural crop monitoring using IOT: A study*. In 2017 11th international conference on intelligent systems and control (ISCO) (pp. 134−139). IEEE.

Steiner, J. L., Briske, D. D., Brown, D. P., & Rottler, C. M. (2018). Vulnerability of southern plains agriculture to climate change. *Climate Change*, *146*, 201−218.

Stylianou, Y., & Orphanos, P. I. (1981). Irrigation of potatoes by sprinklers or tricklers on the basis of pan evaporation in a semi-arid region. *Potato Research*, *24*, 159−170.

Sutton, K. F., Lanini, W. T., Mitchell, J. P., Miyao, E. M., & Shrestha, A. (2006). Weed control, yield, and quality of processing tomato production under different irrigation, tillage, and herbicide systems. *Weed Technology*, *20*(4), 831−838.

Tang, J., & Kang, Y. (2009). Changes in soil properties under the influences of cropping and drip irrigation during the reclamation of severe salt-affected soils. *Agricultural Sciences in China*, *8*(10), 1228−1237.

Tanji, K. K. (2002). *Salinity in the soil environment. Salinity: Environment-plants-molecules* (pp. 21−51). Springer.

Thimmegowda, M.N. (2020). *Application of sensors in precision irrigation management*. https://drive.google.com/file/d/1d0iMFGhc5jhCC-73Loa4uKgv8ktqNKbj/view?usp = sharing.

Tilman, D., Fargione, J., Wolff, B., D'Antonio, C., Dobson, A., Howarth, R., Schindler, D., Schlesinger, W. H., Simberloff, D., & Swackhamer, D. (2001). Forecasting agriculturally driven global environmental change. *Science*, *292*(5515), 281−284.

Tolk, J. A., Howell, T. A., Steiner, J. L., & Krieg, D. R. (1995). Role of transpiration suppression by evaporation of intercepted water in improving irrigation efficiency. *Irrigation Science*, *16*, 89−95.

Toromo, A. K., Kipkorir, E. C., Shitote, S. M., & Kibiiy, J. K. (2012). Can shelterbelts improve sprinkler irrigation performance under windy semi arid conditions? *Irrigation and Drainage Journal*, *25*(4), 335−345.

Türker, U., Akdemir, B., Topakci, M., Tekin, B., Aydin, İ.Ü. A., Özoğul, G., & Evrenosoğlu, M. (2015). *Developments in precision agriculture technologies*. In TMMOB Chamber of Agricultural Engineers, Agriculture Engineering. VIII. Technical congress proceedings (Vol. 1, pp. 295–320), January 12–16, 2015, Ankara, Turkey (in Turkish).

Tzounis, A., Katsoulas, N., Bartzanas, T., & Kittas, C. (2017). Internet of things in agriculture, recent advances and future challenges. *Biosystems Engineering*, *164*, 31–48.

Urban, D., Guan, K., & Jain, M. (2018). Estimating sowing dates from satellite data over the US Midwest: A comparison of multiple sensors and metrics. *Remote Sensing of Environment*, *211*, 400–412.

Vaz, C. M., Jones, S., Meding, M., & Tuller, M. (2013). Evaluation of standard calibration functions for eight electromagnetic soil moisture sensors. *Vadose Zone Journal*, *12*, 1–16.

Waddell, J. T., Gupta, S. C., Moncrief, J. F., Rosen, C. J., & Steele, D. D. (1999). Irrigation and nitrogen management effects on potato yield, tuber quality, and nitrogen uptake. *Agronomy Journal*, *91*, 991–997.

Wang, B., Li, B. H., Dong, X. L., Wang, C. X., & Zhang, Z. F. (2015). Effects of temperature, wetness duration, and moisture on the conidial germination, infection, and disease incubation period of Glomerella cingulata. *Plant Disease*, *99*(2), 249–256.

Wang, Y., Li, S., Cui, Y., Qin, S., Guo, H., Yang, D., & Wang, C. (2021). Effect of drip irrigation on soil water balance and water use efficiency of maize in northwest China. *Water*, *13*, 217. Available from https://doi.org/10.3390/w13020217.

Xiumei, L., Weixia, Z., Jiusheng, L., & Yanfeng, L. (2021). Effects of irrigation strategies and soil properties on the characteristics of deep percolation and crop water requirements for a variable rate irrigation system. *Agricultural Water Management*, *257*, 107143.

Xue, D. X., Zhang, H. J., Ba, Y. C., Zhang, M., & Wang, S. J. (2017). Effects of regulated deficit irrigation on soil environment and yield of potato under drip irrigation. *Acta Agriculturae Boreali-Sinica*, *03*(32), 229–238.

Yang, X., Chen, F., Gong, F., & Song, D. (2000). Physiological and ecological characteristics of winter wheat under sprinkler irrigation condition. *Transactions of the Chinese Society of Agricultural Engineering*, *16*(3), 35–37.

Yari, A., Madramootoo, C. A., Woods, S. A., & Adamchuk, V. I. (2017). Performance evaluation of constant versus variable rate irrigation. *Irrigaion and Drainage*, *66*, 501–509.

Yari, A., Madramootoo, C.A., Woods, S.A., Viacheslav I., Adamchuk, & Gilbert, L. (2018). *Application of variable-rate irrigation for potato productivity*. In 14th international conference on precision agriculture, Montreal, Quebec, Canada.

Yaseen, R., Shafi, J., Ahmad, W., Rana, M. S., Salim, M., & Qaisrani, S. A. (2014). Effect of deficit irrigation and mulch on soil physical properties, growth and yield of maize. *Environment and Ecology Research*, *2*, 122–137.

Yule, I.J., Hedley, C.B., & Bradbury, S. (2008). *Variable-rate irrigation*. In 12th annual symposium on precision agriculture research and application in Australasia. Sydney.

Zalidis, G., Dimitriads, X., Antonopoulos, A., & Geraki, A. (1997). Estimation of a network irrigation efficiency to cope with reduced water supply. *Irrigation and Drainage Systems*, *11*, 337–345.

Zamora-Izquierdo, M. A., Santa, J., Martínez, J. A., Martínez, V., & Skarmeta, A. F. (2019). Smart farming IoT platform based on edge and cloud computing. *Biosystems Engineering*, *177*, 4–17.

Zhang, J., Guan, K., Peng, B., Jiang, C., Zhou, W., Yang, Y., Pan, M., Trenton, E., Derek, M., eeren., Daran, R., Olufemi, A., Hyungsuk, K., Kelly, C., Stephen, G., Madhu, K., John, G., & Yaping, C. (2021). Challenges and opportunities in precision irrigation decision-support systems for center pivots. *Environmental Research Letters*, *16*, 053003.

Zhang, X., Meng, F., Li, H., Wang, L., Wu, S., Xiao, G., & Wu, W. (2019). Optimized fertigation maintains high yield and mitigates N_2O and NO emissions in an intensified wheat–maize cropping system. *Agricultural Water Management*, *211*, 26–36. Available from https://doi.org/10.1016/j.agwat.2018.09.045.

Zhang, X., Qu, J., Li, H., La, S., Tian, Y., & Gao, L. (2020). Biochar addition combined with daily fertigation improves overall soil quality and enhances water-fertilizer productivity of cucumber in alkaline soils of a semi-arid region. *Geoderma*, *363*, 114170. Available from https://doi.org/10.1016/j.geoderma.2019.114170.

Zhang, Y. Q., Wang, J. D., Gong, S. H., Xu, D., Sui, J., Wu, Z. D., & Mo, Y. (2018). Effects of film mulching on evapotranspiration, yield and water use efficiency of a maize field with drip irrigation in Northeastern China. *Agricultural Water Management*, *205*, 90–99.

CHAPTER

Applications of remote sensing in water quality assessment

12

Mohit Kumar Srivastava[1], Shishir Gaur[1], Anurag Ohri[1], Prashant K. Srivastava[2] and Nikhilesh Singh[1]

[1]*Department of Civil Engineering, Indian Institute of Technology (BHU), Varanasi, Uttar Pradesh, India*

[2]*Institute of Environment and Sustainable Development (IESD), Banaras Hindu University, Varanasi, Uttar Pradesh, India*

12.1 Introduction

It's an undeniable fact that the world is going through a water crisis today. The matter is not the fact that there is a shortage in the availability of water, but it is the fact that the available water is not being sustainably managed. Modern techniques such as Remote Sensing can be used to develop an effective and integrated approach to managing these water resources. Water quality is the process of determining the chemical, physical, and biological characteristics of various water bodies. It also includes the identification of possible contamination sources that degrade the water quality.

There could be various reasons for water quality getting degraded such as the influence of waste discharges, nutrients, heavy metals, sediments, and microorganisms. However, the real challenge is getting to know the minimum or maximum limit that these external factors are allowed to be present in water bodies without degrading them. That is why there are different water quality standards that have been developed by various monitoring agencies for water.

However, in addition to being time-consuming and expensive, conventional methods of studying water quality frequently fail to adequately represent heterogeneous and patchy areas (Khorram, 1991). The traditional in situ methods of obtaining water quality parameters by obtaining samples are laborious, costly, and time-consuming, bringing many challenges. Remote Sensing technology can be effectively used to ascertain the spatial and temporal changes in water quality even in inaccessible locations. Owing to high-frequency data acquisition and large-scale coverage, various spaceborne sensors with visible, infrared, and microwave wavelengths can be used to monitor water quality.

Remote Sensing in Precision Agriculture. DOI: https://doi.org/10.1016/B978-0-323-91068-2.00019-9

12.2 Basis of remote sensing

Remote sensing is defined as the science and art of obtaining information about an object, area, or phenomenon through the analysis of data acquired by a device that is not in contact with the object, area, or phenomenon under investigation (Lillesand et al., 2015). Remote Sensing techniques measure the spectral signature backscattered from water. The amount of backscattering from the water surface changes with the particular substance present in it. Remote sensing techniques depend on measuring these changes in the spectral signature backscattered from water and relating these measured changes by empirical or analytical models to a water quality parameter. The optimal wavelength used to measure a water quality parameter depends on the substance, its concentration, and the sensor characteristics (Ritchie et al., 2003).

Some of the major factors that play a major role in affecting the quality of water are the presence of algae (i.e., carotenoids, chlorophyll), suspended sediments (i.e., turbidity), dissolved organic matters, some chemicals (such as nutrients, pesticides, and metals), pathogens, aquatic plants, and oils. These factors change the energy spectra of the reflected and emitted radiations coming from the water surfaces of ponds, lakes, and rivers. These radiations are then measured using various satellite sensors. Thus, these remote sensing techniques can easily help make spatial and temporal analyses of surface water quality parameters, which may not be possible to obtain directly by field measurements (Fig. 12.1).

12.3 Spectral reflectance curve for water bodies

The optical properties of pure water are required for remote sensing of surface water. Knowledge of the spectral properties of components in the surface water is required for accurate interpretation of measured reflection and attenuation spectra in terms of their concentrations. One of the most relevant properties of water that helps in remote sensing is its optical properties. The optical properties of water can be categorized as: Inherent Optical Properties (IOPs) and Apparent Optical Properties (AOPs). IOPs include the absorption coefficient, the scattering coefficient, the volume scattering function, the index of refraction, the beam attenuation coefficient, and the single-scattering albedo, while AOPs include diffuse attenuation coefficients and irradiance reflectance. IOPs depend on the medium in which the light travels, while the AOPs are dependent on the medium in which light travels as well as its spatial distribution.

Water has very low spectral reflectance in the visible part of the electromagnetic region, whereas snow or ice has very high spectral reflectance in the visible and near-infrared part of the EMR. Pure water absorbs nearly all incident energy in both the near-infrared and middle infrared wavelengths. The low reflectance of water in visible and NIR band has advantage in Remote Sensing as water

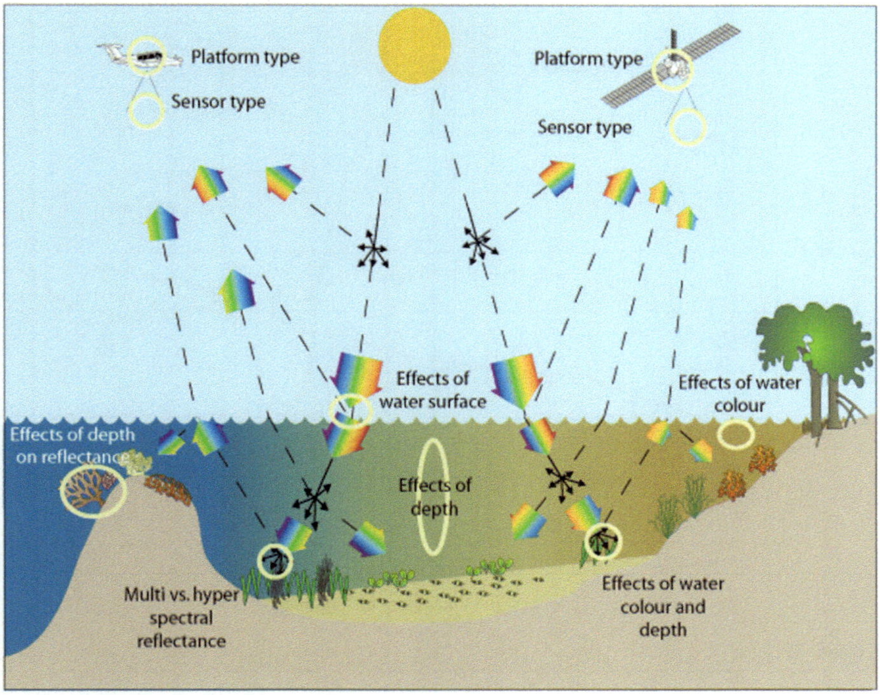

FIGURE 12.1

Representation of remote sensing of a water body.

Courtesy: SEES, RSRC.

becomes clearly distinguishable from either vegetation or soil cover throughout the reflective infrared portion. Thus satellite imageries have been used successfully in determination of various water quality parameters such as total suspended solids, turbidity, chlorophyll content, color, and temperature by using the Visible, Reflected Infrared, and occasionally Thermal Infrared bands of the electromagnetic spectrum.

This is done by measuring the spectral reflectance. The ratio of the energy reflected by the earth's surface to the incident energy, measured as a function of wavelength, is known as the spectral reflectance of the surface. It is also known as the *albedo* of the surface. It may vary from 0% to 100%. The energy that is reflected by features on the earth's surface over a variety of different wavelengths will give their spectral responses in the remote sensing systems (Fig. 12.2). The energy radiance captured by a sensor comprises roughly of two components: scattered energy from the atmosphere and radiant energy reflected from the water body. The energy reflected from the atmosphere is essentially considered noise and is hence removed for accurate remote sensing quantification of water quality parameters. Each type of feature/object has a unique

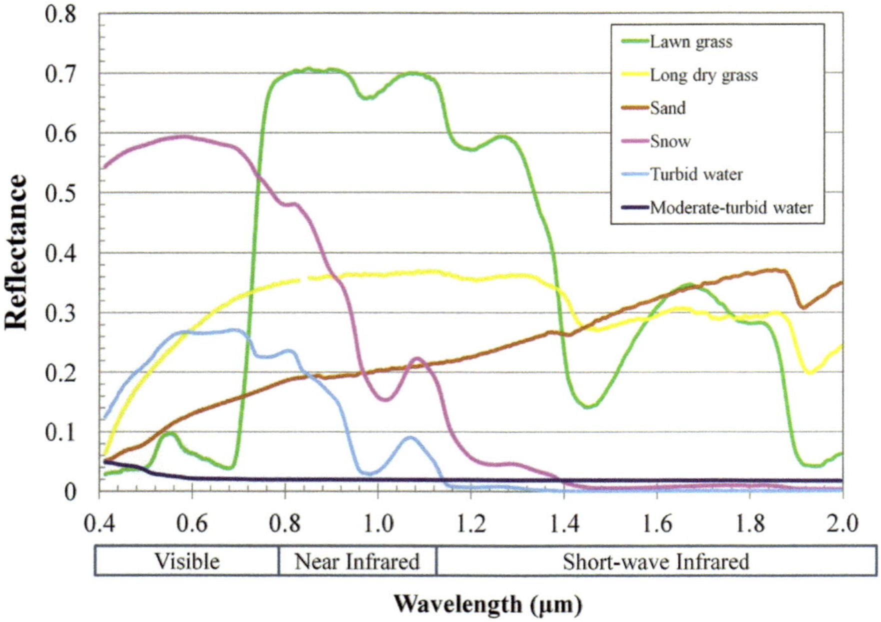

FIGURE 12.2

Reflectance of several typical land cover objects.

Courtesy: https://speclab.cr.usgs.gov/spectral-lib.html.

spectral response/reflectance characteristic, also known as a spectral signature, which can be used to identify the respective surface features and to study their properties. The graphical representation of the spectral response of an object over different wavelengths of the electromagnetic spectrum is termed the spectral reflectance curve. The reflectance characteristics of the surface features are represented using these curves.

The spectral signature of water is dependent on several factors—time of the year, sun elevation angle, aerosol and molecular content of the atmosphere, water vapor content of the atmosphere, specular reflection of the skylight from the water surface, the roughness of water surface, water color and content of dissolved colored material in water, reflectance and absorbance characteristics of suspended particles, multiple reflections and scattering of solar energy in water, depth of water and reflectance of bottom sediment, submerged or emerged vegetation, and turbidity of water.

In the case of a water body, the spectral response also varies with the wavelength of the radiation and the physical and chemical characteristics of the water. In the visible region (0.4 and 0.7 μm) water has a high reflectance. Wavelengths beyond 0.7 μm are completely absorbed. This is why a clear water body will appear in a darker tone in the NIR image. This greatly helps in locating and delineating water bodies with remote sensing data using infrared wavelengths. Notably, when used for water quality monitoring, remote sensing uses the information of water depth and bottom characteristics of the particular water body as well in its analysis. However, the influence of water depth becomes negligible with depths greater than 50 cm (Seyhan, 1974).

12.4 RS techniques in water quality monitoring

In the earlier days, to assess the water quality of a water body, the difference between spectral energies was recorded. Thereafter, an empirical relationship between water quality parameters and their spectral properties was established. These equations were of the form:

$$Y = A + BX \quad \text{or} \quad Y = AB^X \tag{12.1}$$

where X and Y are the respective water quality parameters and their measured reflectance, while A and B are empirically derived factors. Prior information on the parameter's spectral property helped in selecting the best wavelength range for their detection.

But all of this was mainly empirical, that is, confined to the state and condition where the data were obtained. This made these studies limited in their applications. Eutrophication assessment frameworks such as the Australian National Water Quality Management Strategy, Oslo Paris (OSPAR) Commission Common Procedure, Water Framework Directive (WFD) of the European Union, Marine Strategy Framework Directive (MSFD) from the European Commission, Helsinki Commission (HELCOM) Eutrophication Assessment Tool (HEAT), and in the USA, the National Coastal Assessment and National Aquatic Resource Assessment use various approaches for assessing water quality and status of coastal and inland waters (Devlin et al., 2011; Ferreira et al., 2011; Park et al., 2010).

12.5 Shallow water spectral characteristics

The reflected energy from the water body is indicative of the inherent optical properties (scattering, absorption, and fluorescence) of the water column and its various constituents (Morel, 1980; Sturm, 1981). It is also affected by the illumination conditions, water depth, as well as bottom reflectance if the depth is shallow (Liu et al., 2003). That is why, it is recommended to do a simultaneous retrieval of depth, bottom reflectance (bottom type), and water-column absorption and scattering properties during remote sensing of shallow water. Also, such analysis requires a very high spatial resolution (1–10 m), and some previous work has established that in such cases, band-ratio algorithms cannot be used (Gao et al., 2007).

12.6 Groundwater remote sensing

Groundwater is one of the fundamental resources for humans. With increased water extraction across the world due to increasing population and weather changes, this resource is lowering toward the critical mark, resulting in water scarcity in towns and villages. According to recent studies, around 2.1 billion people worldwide do not have access to freshwater (Benameur et al., 2022). The chemistry of groundwater is very intricate and different from that of surface water. Many natural processes such as dissolution, percolation, precipitation, cation exchange, as well as anthropogenic activities, affect the mechanism of groundwater chemistry (Panneerselvam et al., 2022; Vespasiano et al., 2021; Zhou et al., 2020). In the case of the analysis of groundwater, if the field data measurements are accompanied by the remote sensing data and GIS (Geographical Information System), then it may help a lot in the analysis and visualization of data with the help of spatial maps. However, owing to the problems faced during correct groundwater data acquisition from space (considering the hindrances caused by external factors such as cloud coverage), the efficacy of spectral indices for predicting the quality of groundwater (for drinking or irrigation) is limited in the available literature.

Remote Sensing and GIS (Geographic Information System) can help in obtaining thematic map results −such as Land Use and Land Cover (LULC), drainage network, drainage density, the slope of the area, geology, geomorphology, rainfall distribution, groundwater level, and lineament density, which in turn can help in generation of groundwater potential maps. This is done with the help of overlay analysis and assigning weights to each of the influencing factors.

12.7 Water quality parameters

Water quality indicators are categorized as (1) Biological: bacteria, algae, (2) Physical: temperature, turbidity, and clarity, color, salinity, suspended solids, dissolved solids, (3) Chemical: pH, dissolved oxygen, biological oxygen demand, nutrients (including nitrogen and phosphorus), organic and inorganic compounds (including toxicants), and (4) Esthetic: odor, taints, color, and floating matter. However, extracting water quality information from remotely sensed radiance alone is nearly impossible. This is mainly because the boundary conditions of the water−surface interface and the characteristics of the atmospheric aerosol are still unknown (Sturm, 1981). This limits our inferences to be empirical or semiempirical-based. Image data can be a single band, a combination of bands, or their ratio. Their transformation however, would require the acquisition of both remotely sensed data and in situ water quality samples, and the development and validation of empirical or semiempirical models. This ensures taking into account the dynamics of water quality. According to Doerffer et al. (1999), water quality parameters that can be quantified using remote sensing broadly fall into categories as follows:

12.7.1 Suspended sediment concentration

When it comes to freshwater systems, suspended sediments are the most common pollutants. In the past, many attempts have been made to develop a relationship between suspended sediment concentration and incident wavelength. However, most of these studies have been empirical in nature. With the techniques and current spatial resolution available of the satellite data, a detailed mapping of water bodies is sometimes still a tedious task in a few studies, especially those requiring finer resolution images. In contrast, large water bodies for the sediments concentrated on their surface can easily be mapped using available sensors. Ritchie et al. (1990) managed to find a curvilinear relationship between suspended sediments and reflectance, which showed that the amount of reflectance tended to increase as the concentration of the suspended sediments increased in the water body. Peterson et al. (2018) have attempted to use the ELM (Extreme Learning Machine) model along with Feed-Forward Neural Network (FFNN) and Cascade Forward Neural Network (CFNN) to train a reflectance-suspended sediment model using Landsat 4–5, 7, and 8 sensors.

12.7.2 Chlorophyll concentration

One of the very prominent problems that persist with smaller water bodies such as lakes is eutrophication. Eutrophication is the process by which a body of water becomes enriched in dissolved nutrients (such as phosphates), which stimulates the growth of aquatic plant life usually resulting in the depletion of dissolved oxygen. This could be checked by estimating the chlorophyll concentrations, which can be done using Remote Sensing. However, similar to sediment concentration measurement, most remote sensing studies on chlorophyll using remote sensing are based on the empirical relationship between chlorophyll concentration and reflectance. It does require field or in situ data measurements to validate the findings. Chlorophyll is an optically active material present in plants and algae and special types of photosynthesis-capable bacteria (usually referred to as cyanobacteria), and it links nutrients' concentration and algal production and is considered a good indicator of a water body's trophic state (Hussein et al., 2020). Various visual spectral bands and their ratios are widely used to quantify Chl-*a* and Chl-*b* based on amount of light absorbed (Fig. 12.3). Spectral band ratios can reduce irradiance, and atmospheric-water surface influences in the remotely sensed signal (Lillesand et al., 2015).

Studies have also shown that broad-wavelength spectral data are not helpful when it comes to measuring chlorophyll concentrations because this spectrum is mainly dominated by reflectance spectra of suspended sediments. Hence, a narrow-band spectrum is advisable. National Aeronautics and Space Administration's (NASA) Moderate Resolution Imaging Spectroradiometer (MODIS) Aqua satellite is used highly for the estimation of oceanic chlorophyll. The satellite operates by collecting reflected wavelength energy daily from the earth as data, which is then

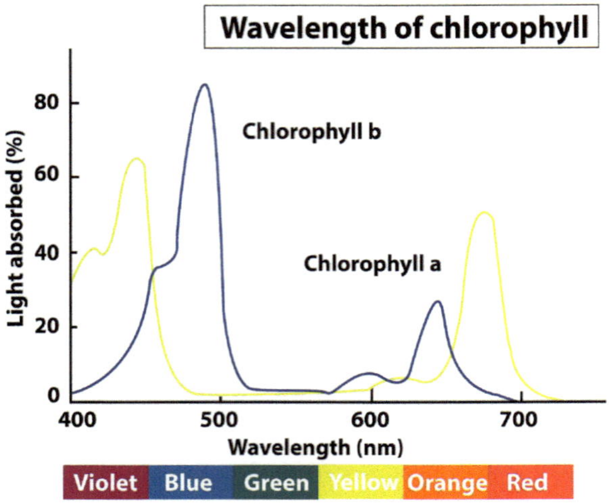

FIGURE 12.3

Absorption spectra of Chlorophyll-*a* and Chlorophyll-*b* pigments.

Courtesy: Vedantu.

converted via a chain of algorithms to chlorophyll concentration (Martin et al., 2010; Svab et al., 2005). Thus, MODIS data not only provides a large amount of data but can also help in determining the spatial and temporal distribution of chlorophyll concentration in a water body. Recent studies have shown that hyperspectral data from some of the recently launched satellite sensors such as SeaWiFS, Modular Optical Scanner, Ocean Color and Temperature Scanner, and Ikonos hold great promise for measuring biological productivity (chlorophyll) in aquatic systems.

12.7.3 Algae

Algae are single-celled organisms containing chlorophyll and other pigments. Algae are a critical component of lake food webs and support primary consumers. Owing to their basal position in lake food webs, algae are responsive both to changes among higher trophic levels, for example, zooplankton and fish, and to lake chemical characteristics (Papenfus et al., 2020). Eutrophication leads to increased ecological stress on the water body and helps in the formation of an overabundance of algae and cyanobacteria. These algal blooms lower the dissolved oxygen and give a foul taste and odor, which further has large socioeconomic consequences. Chlorophyll a and b are indicative of the photosynthetic potential of a plant and also give some indication of the physiological status of the plant (Danks et al., 1983; Gamon & Surfuis, 1999; Young & Britton, 1990). Chlorophyll-*a* is the major indication of the algal population. The 2012 National Lakes Assessment (NLA) estimated that 55% of lakes in the continental USA (CONUS) are either eutrophic or hyper-eutrophic based on chl-*a* measures (Oudejans, 2017).

12.7.4 Temperature

Water temperature is one of the most critical elements when it comes to restoring a freshwater aquatic ecosystem (such as rivers and streams). The stream temperature reflects watershed conditions and directly influences the biology of aquatic organisms (Beschta et al., 1987). Traditional methods of temperature measurement by use of in situ data required continuous temperature monitoring, and although it provided continuous temporal data, it remained highly coherent spatially. Remote measurements of water temperature are made with a sensor that detects thermal radiation (3–5 and 8–14 μm wavebands) emitted from the upper 0.1 mm of the water surface (Atwell et al., 1971; Anderson & Wilson, 1984; Robinson et al., 1984).

As was said earlier, it is not possible to quantify all the water quality variables from the currently available satellite sensors. Some examples of such parameters include total nitrogen, total phosphorous, dissolved oxygen levels, and microorganisms/pathogens. It is so because many of these parameters were either not very optically active or it just requires very fine spatial resolutions in hyperspectral imagery. As the nutrients are very significant factors in the initiation, propagation, and maintenance of HABs (Schindler, 1977), the ability to remotely estimate nutrients and a suite of other water quality parameters associated with toxins is imperative.

12.8 Sensors for water quality

To correctly assess the water quality, one should correctly understand the right kind of sensor that would be required for it. Depending on the platform on which it is situated, sensors are broadly classified into two categories (Tables 12.1 and 12.2).

12.8.1 Airborne sensors

Sensors mounted on a platform that remains within the Earth's atmosphere (such as a boat, balloon, helicopter, or UAVs). Airborne sensors are prone to imaging geometry problems, but they are really flexible and somewhat portable in their capabilities to collect data. However, the biggest drawback with this kind of sensor is how much they are dependent on weather and flying conditions. Such kinds of sensors are susceptible to a sudden change in velocity, this may be due to a sudden gust of wind. To avoid image artifacts or geometric positioning errors due to random variations in the motion of the aircraft, the radar system must use sophisticated navigation/positioning equipment and advanced image processing to compensate for these variations. An advantage of airborne remote sensing is its capability of offering very-high-spatial-resolution images (20 cm or less). At the same time, it has some major disadvantages –such as low coverage area, high cost per unit area of ground coverage, not being cost-effective enough to map a

Table 12.1 Airborne hyperspectral and multispectral sensors used for water quality assessments.

Type of sensor	Sensor name	Acronym	Number of bands	Spectral range (μm)	Spatial resolution (m)	Purpose	Developed by	Data availability	Technical References
HSS	Airborne Visible Infrared Imaging Spectrometer	AVIRIS	224	0.40–2.50	17	To identify, measure, and monitor constituents of the Earth's surface/ atmosphere based on molecular absorption and particle scattering signatures	Jet Propulsion Laboratory (JPL), NASA, USA	1987	Curran and Dungan (1989)
HSS	Compact Airborne Spectrographic Imager	CASI	288	0.38–1.05	0.4–1.05	To be used in Agriculture, Environment, Water Quality, LIDAR, IFSAR, Forestry, and Defense	ITRES Research Limited Calgary, Alberta, Canada	1989	http://www.itres.com
MSS	Multispectral Infrared and Visible Imaging Spectrometer	MIVIS	102 VIS/ NIR (28), MIR (64), TIR (10)	VIS (0.43–0.83), NIR (1.15–1.55), MIR (2.0–2.5) TIR (8.2–12.7)	3–8	Land cover classifications, Water quality, environment monitoring, identifying terrestrial vegetation species	Sensytech	1993	http://www.sensytech.com

HSS	Environmental Protection System	EPS-H	VIS/NIR (76), SWIR1 (32), SWIR2 (32), TIR (12)	VIS/NIR (0.43–1.05), SWIR1 (1.50–1.80), SWIR2 (2.00–2.50), TIR (8–12.50)	upto 1 m	To be used in every kind of environmental and ecological research	GER Corporation	1995	http://www.ger.com
HSS	Hyperspectral Digital Imagery Collection Experiment	HYDICE	210	0.40–2.50	17		Naval Research Laboratory, Hughes Danbury Optical Systems, Environmental Research Institute of Michigan	1995	Resmini et al. (1997)
MSS	ARSF Airborne Thematic Mapper (ATM) - Daedalus	Daedalus	12	0.42–14.00	25	to provide calibrated thermal information for determining radiometric temperature relationships for various remote sensing applications	EG&E Energy Measurements Remote Sensing Laboratory (RSL), Las Vegas, NV & NERC Airborne Research and Survey Facility (ARSF)	1982–1998	http://www.nln.geos.ed.ac.uk

(Continued)

Table 12.1 Airborne hyperspectral and multispectral sensors used for water quality assessments. *Continued*

Type of sensor	Sensor name	Acronym	Number of bands	Spectral range (μm)	Spatial resolution (m)	Purpose	Developed by	Data availability	Technical References
HSS	HyMap	–	126	0.45–2.5	5	to provide affordable, high quality hyperspectral data on a commercial basis	Integrated Spectronics Inc., Sydney, Australia	1996	Cocks et al. (1998)
HSS	Airborne Prism Experiment	APEX	Up to 300 VIS/NIR (114), SWIR (199)	VIS/NIR (0.38–0.97), SWIR1 (0.97–2.50)	2–5	as a simulator and a calibration and validation device for future spaceborne hyperspectral imagers	E.S.A. (European Space Agency)	2005	Schaepman et al. (2015)

VIS stands for Visible Region of the spectra (0.4–0.8 μm).
NIR stands for Near-Infrared Region of the spectra (0.75–1.4 μm).
SWIR stands for Short-Wave Infrared region of spectra (1.4–3.0 μm).
MIR stands for Mid-Infrared Region of spectra (3–8 μm).
HSS, *Hyperspectral Sensor;* MSS, *Multispectral Sensor.*

Table 12.2 Spaceborne sensors in water quality assessments.

Satellite-sensor	Launch year	Number of bands (in nm)	Spatial resolution (m)	Temporal resolution (day)	Orbit altitude (km)	Orbit type	Swath width (km)	Whether operational
Worldview-1	2007	Panchromatic (pan)	0.5	1.7	496	Sun-synchronous	17.6	Yes
Worldview-2	2009	8 (400–1040)-1 Pan (450–800)	1.85–0.46	1.1	770	Sun-synchronous	16.4	Yes
Worldview-3	2014	8 (400–1040)-1 Pan (450–800)-8 SWIR (1195–2365)	0.3 (pan), 1.24–1.38 (MSS), 3.7–4.1 (SWIR),	1–4.5	617	Sun-synchronous	13.1	Yes
Quickbird	2001	4 (430–918)-1 Pan (450–900)	0.65–0.73 (pan), 2.62–2.9 (MSS)	1–3.5	480–482	Sun-synchronous	16.8–18	No (Decommissioned in 2015)
Geoeye-1	2010	4 (450–920)-1 Pan (450–800)	0.5 (pan), 1.84 (MSS)	2.1–8.3	770	Sun-synchronous	15.2	Yes
IKONOS	1999	4 (445–853)-1 Pan (526–929)	1 (pan), 4 (MSS)	3	681	Sun-synchronous	11.3–13.8	No (decommissioned in 2015)
SPOT-7	2014	1-pan, 4 – Visible (455–695), NIR (760–890)	1.5 (pan), 6 (MSS)	26	694	Sun-synchronous	60	Yes
CARTOSAT-1	2005	Pan (500–850)	2.5	5	617.99	Polar Sun-synchronous	30	No (Decommissioned in 2019)
CARTOSAT-3	2019	Pan, MSS, HSS	0.25 - Pan, 1.13 - MSS, 12 - HSS	5	509	Polar Sun-synchronous	16	Yes
Landsat-9	2021	Pan (503–676), Visible (452–673), NIR (851–879), SWIR1 (1566–1651), SWIR2 (2107–2294)	15 (pan), 30 (MSS)	16	705	near-polar, sun-synchronous	185	Yes

(Continued)

Table 12.2 Spaceborne sensors in water quality assessments. *Continued*

Satellite-sensor	Launch year	Number of bands (in nm)	Spatial resolution (m)	Temporal resolution (day)	Orbit altitude (km)	Orbit type	Swath width (km)	Whether operational
Landsat-8 (OLI/ TIRS)	2013	5 (430–880)-1 Pan (500–680)-2 SWIR (1570–2290)-1 cirrus cloud detection (1360–1380)-2 TIRS (10,600–12,510)	15 (pan), 30 (MSS)	16	705	near-polar, sun-synchronous	185	Yes
Landsat-7 (ETM +)	1999	6 (450–1750)-1 Pan (520–900)-1 (2090–2350)-1 (1040–1250)	15 (pan), 30 (VNIR/SWIR), 60 (TIR)	16	705	Polar sun-synchronous	185	No (Decommissioned in 2021)
Landsat 5 (MSS)	1984	4 (450–1750)-1 Pan (1040–1250)	60 (MSS), 30 (Thematic Mapper)	16	705	Near polar, sun-synchronous	185	No (Decommissioned in 2013)
EO-1 Hyperion	2000	242 (350–2570)	30	16	705	Sun-synchronous	7.75	No (Decommissioned in 2017)
PROBA-1 CHRIS	2001	19 in the VNIR range (400–1050)	18	17	556	Sun-synchronous	14	Yes
Terra MODIS	1999	2 (620–876)-5 (459–2155)-29 (405–877 and thermal)	250 (2 bands), 500 (5 bands), 1000 (29 bands)	1–2	705	Sun-synchronous	2330 (across), 10 (along)	Yes
OrbView-2 SeaWiFS	1997	8 (402–885)	1000	16	705	Sun-synchronous	2800	No (Decommissioned in 2010)
ENVISAT AATSR	2002	3 VIS-NIR (555–865), 1 SWIR (1600), 1 MWIR (3700), TIR (10,850–12,000)	1000	1–3	800	Sun-synchronous	512	No (Decommissioned in 2012)
NOAA-16 AVHRR	2000	6 (650–1230)	1100	2	850	Sun-Synchronous	3000	No (Decommissioned in 2014)

SPOT, *Satellite Pour l'Observation de la Terre (Satellite for observation of Earth)*; CARTOSAT, *Cartography Satellite;* OLI, *Operational Land Imager;* TIRS, *Thermal Infrared Sensor;* EO, *Earth Orbital;* MODIS, *Moderate Resolution Imaging Spectroradiometer;* ENVISAT AATSR, *Environment Advanced Along-Track Scanning Radiometer;* NOAA AVHRR, *National Oceanic and Atmospheric Administration Advanced Very High-Resolution Radiometer.*

large area, and inability to provide continuous data. Also, if platforms such as aircraft are used for data acquisition, then they require permission to intrude into a foreign space, which in itself is a lengthy and frustrating task.

12.8.2 Space-borne sensors

The sensors that are mounted on-board a spacecraft (space shuttle or satellite) orbiting the earth are called space-borne sensors. Their altitudes vary from 100 to 36,000 km above the earth's surface. Today's space-borne sensors provide data in a wide range of scales (or pixel resolutions), radiometry, band numbers, and bandwidths. Much of the data are free and accessible online. The resolution of the images offered by space-borne sensors is relatively lower than those offered by airborne sensors. However, there are also some very-high-resolution satellites available that can offer images with spatial resolution upto even 1 m or less (IKONOS-2). But the period of acquisition of high-resolution images is larger than those with coarser resolution. In spite of that, space-borne remote sensing has proven advantageous in many aspects, such as large area coverage, repetitive coverage, provide a wide range of applications (modeling, forecasting, mapping, monitoring, and analysis), and has relatively lower cost per unit area of coverage.

12.9 Hyperspectral sensors

Hyperspectral imaging is the capturing and processing of an image at a very large number of wavelengths. On the other hand, multispectral imaging generally evaluates an image in three or four colors (red, green, blue, and near-infrared (NIR), for example). Hyperspectral imaging breaks the image down into tens or hundreds of colors. In other words, hyperspectral imaging (HSI) uses continuous and contiguous ranges of wavelengths (e.g., 400–1100 nm in steps of 1 nm), while multiband imaging (MSI) uses a subset of targeted wavelengths at chosen locations (e.g., 400–1100 nm in steps of 20 nm). Flores-Anderson et al. (2020) used hyperspectral Hyperion imageries to develop a semiempirical algorithm for Lake Atitlan, Guatemala. They used a Blue-to-Green band ratio for the modeling of Chlorophyll-*a* concentrations in the lake. Adam Goodenough (2001) used imageries from AVIRIS and MISI to model the reflectance spectra of Chlorophyll, total suspended solids, and Colored Dissolved Organic Matter (CDOM), and their atmospheric conditions are modeled using MODTRAN. Sometimes in situ water quality data is used along with hyperspectral imageries to quantitatively predict various water quality parameters e.g., phosphorus, nitrogen, biochemical oxygen demand (BOD), chemical oxygen demand (COD), and Chlorophyll-*a* (Zhang et al., 2020).

12.10 Multispectral sensors

Water quality in different sectors of any given water body can be assessed using narrow wavebands of the electromagnetic spectrum of multispectral sensors (Nieuwoudt & Backeberg, 2011; Soticha et al., 2014). Multispectral data basically divide the electromagnetic wave of a ground object into a narrower spectral section and obtain the information of different bands of the same target in the way of a full picture or line sweep. Working mostly with 3–10 bands, multispectral sensors include MSS, TM, ETM + , OLI, Sentinel-2, MODIS, SPOT, and AVHRR. However, Landsat series data are mostly used to assess the water quality monitoring owing to their large accessibility and having considerably better spatial, spectral, and temporal resolution. The data from this are used for the estimation of Suspended Matter, Chemical Oxygen Demand, and Total Phosphorous.

However, as multispectral data are limited in their resolution extent, their application is particularly limited to empirical usage (suitable only for a certain area or specified period). Hussein et al. (2020) used multispectral imageries (from Sentinel-2 L1C MSI data) to monitor Chlorophyll-*a* (Chl-*a*) concentrations and develop a model from the samples collected from three dam sites of Jordan: King Talal Dam, Mujib Dam, and Wadi Al-Arab Dam. Later the developed model was used to predict Chl-*a* concentration from surface reflectance (SR) data. Similar to this, Tong et al. (2005) developed a model to measure Chlorophyll-*a* and suspended sediments from the multispectral data obtained from Landsat-5 TM images and the in situ measurements for Lake Tai, China. However, using multispectral data also comes with some disadvantages such as radiation calibration, atmospheric correction, geometric corrections, and the necessity of noise elimination from the data.

12.11 Advantages of using remote sensing in water quality assessment

As has been specified earlier, estimating water quality using remote sensing comes with many advantages. Using the data and results obtained from a remotely sensed image, one can plan for sustainable management of the available water resource. On account of its lower reflectance, water bodies become clearly distinguishable from soil cover or vegetation cover. Using various bands in different wavelengths such as visible region, NIR, or sometimes TIR of the EM spectrum helps in the determination of various water quality parameters. This would in turn help in hydrological modeling, flood management, irrigation management, drought and flood management, city or town planning, etc. In the times when remote sensing was not available, in situ measurements were taken for measuring the water quality of a water body. It was done by collecting samples from the sites and then testing them in the laboratories for individual parameters.

This procedure was not only tiresome but was also time-consuming and prone to many kinds of errors that could get included in the measurements. Also, such an approach limited the obtained result to only the location (or time) from (or at) which the sample was obtained. Remote Sensing data, however, can easily provide spatial as well as temporal information about the results. It also helps in correlating various parameters and their characteristics and thus helps in the development of nearly accurate models. These models can encompass the temporal variation of the data and can thus indicate seasonal and inter-annual variability. In some cases, they can also help in predicting a future scenario. In conclusion, remote sensing and GIS technologies coupled with computer modeling are useful tools in providing a solution for future water resources planning and management to the government, especially in formulating policies related to water quality. Satellite imageries obtained from Landsat Thematic Mapper (TM), Linear Imaging and Self Scanning (LISS), and Wide Field Sensor (WiFS) have been used extensively by various workers for calculation of drainage basin area, drainage density, Normalized Difference Vegetation Index (NDVI), and Leaf Water Content Index (LWCI). Application of Remote Sensing in flood management, drought management, and Irrigation Command Area management was demonstrated in India by a number of researchers from time to time.

12.12 Barriers in RS measurements of WQ

Even though the capabilities of remote sensing to assess water quality are undeniable, this technique alone is not sufficiently precise. It must be used in conjunction with traditional sampling methods and field surveying. Thus, integrated use of remote sensing, in situ measurements, and computer water quality modeling must be used for better insight. Also, most of the remote sensing measurements that are aimed at measuring the water quality must be made in the absence of clouds. Not only that, the accuracy of data that are obtained (even in the absence of clouds) is debatable, as much information gets lost in performing atmospheric corrections and during processing.

Certain important parameters such as water discharge and vertical quality profile cannot be obtained using remote sensing alone. Handling of such data also requires skill and knowledge, which may not be available. Many of the studies conducted so far are concentrated on optically active variables (such as Chlorophyll-*a*, CDOM, TSS, and Turbidity), but other quality parameters such as pH, total nitrogen (TN), ammonia nitrogen (NH_3-N), nitrate nitrogen (NO_3-N), and dissolved phosphorus (DP) remain unexplored due to their weak optical characteristics. Also, there are many natural processes that affect these parameters' distribution such as frontal formation, waves, wind re-suspension, and biological growths over the surface (Schaeffer et al., 2013).

If one looks closely at the available literature, not much work has been done to separate the spectral signatures for Chl-*a*, CDOM, and inorganic suspended matters too. They often get confused with each other, thus making in situ measurements even more necessary. The atmospheric interference also restricts the optical signals coming from water bodies (Gholizadeh et al., 2016). Also, dealing with satellite data comes with many disadvantages too as the accuracy of the obtained result also depends on the precision and accuracy by which a particular model is calibrated, which itself depends on many factors.

References

Anderson, J. M., & Wilson, S. B. (1984). The physical basis of current infrared remote-sensing techniques and the interpretation of data from aerial surveys. *International Journal of Remote Sensing*, *5*(1), 1–18.

Atwell, B. H., McDonald, R. B., Bartolucci, L. A. (1971). Thermal mapping of streams from airborne radiometric scanning. *Water Resources Bulletin*, *7*, 228–243.

Benameur, T., Benameur, N., Saidi, N., Tartag, S., Sayad, H., & Agouni, A. (2022). Predicting factors of public awareness and perception about the quality, safety of drinking water, and pollution incidents. *Environmental Monitoring and Assessment*, *194*, 22.

Beschta, R. L., Bilby, R. E., Brown, G. W., Holtby, L. B., & Hofstra, T. D. (1987). Stream temperature and aquatic habitat: Fisheries and forestry interactions. In E. O. Salo, & T. W. Cundy (Eds.), *Streamside management: Forestry and fishery interactions* (pp. 191–232). Seattle, WA: University of Washington, Institute of Forest Resources. (1987).

Cocks, T., et al. (1998). The HyMap airborne hyperspectral sensor: The system, calibration and performance. In Proc. 1st EARSeL workshop on imaging spectroscopy (M. Schaepman, D. Schläpfer, & K. I. Itten, Eds.), 6–8 October 1998, Zurich, EARSeL, Paris, pp. 37–42.

Curran, P. J., & Dungan, J. L. (1989). Estimation of signal-to-noise: A new procedure applied to AVIRIS data. *IEEE Transactions on Geoscience and Remote sensing*, *27*(5), 620–628.

Danks, S. M., Evans, E. H., & Whittaker, P. A (1983). *Photosynthetic systems: structure, function, and assembly*. John Wiley & Sons.

Devlin, M., Bricker, S., & Painting, S. (2011). Comparison of five methods for assessing impacts of nutrient enrichment using estuarine case studies. *Biogeochemistry*, *106*, 177–205.

Doerffer, R., Sorensen, K., & Aiken, J. (1999). MERIS potential for coastal zone applications. *International Journal of Remote Sensing*, *20*(9), 1809–1818.

Ferreira, J. G., Andersen, J. H., Borja, A., Bricker, S. B., Camp, J., Silva, M. C. D., Garcés, E., Heiskanen, A.-S., Humborg, C., Ignatiades, L., Lancelot, C., Menesguen, A., Tett, P., Hoepffner, N., & Claussen, U. (2011). Overview of eutrophication indicators to assess environmental status within the European Marine strategy framework directive. *Estuaries, Coastal, and Shelf Science*, *93*, 117–131.

Flores-Anderson, A. I., Griffin, R., Dix, M., Romero-Oliva, C. S., Ochaeta, G., Skinner-Alvarado, J., Ramirez Moran, M. V., Hernandez, B., Cherrington, E., Page, B., & Barreno, F. (2020). Hyperspectral satellite remote sensing of water quality in lake Atitlán, Guatemala. *Frontiers in Environmental Science*, *8*, 7.

Gamon, J. A., & Surfuis, J. S. (1999). Assessing leaf pigment content and activity with a reflectometer. *The New Phytologist*, *143*(1), 105–117.

Gao, B. C., Montes, M. J., Li, R.-R., Dierssen, H. M., & Davis, C. O. (2007). An atmospheric correction algorithm for remote sensing of bright coastal waters using MODIS land and ocean channels in the solar spectral region. *IEEE Transactions on Geoscience and Remote Sensing*, *45*(6), 1835–1843.

Gholizadeh, M. H., Melesse, A. M., & Lakshmi, R. (2016). A comprehensive review on water quality parameters estimation using remote sensing techniques. *Sensors*, *16*(8), 1298.

Goodenough, A. (2001). Evaluating water quality monitoring with hyperspectral imagery. Thesis. Rochester Institute of Technology.

Hussein, N. M., & Assaf, M. N. (2020). Multispectral remote sensing utilization for monitoring chlorophyll-a levels in inland water bodies in Jordan. *The Scientific World Journal*, *14*, 2020, Article ID 5060969.

Khorram, S., et al. (1991). Water quality mapping of Augusta Bay, Italy from Landsat-TM data. *International Journal of Remote Sensing*, *12*(4), 803–808.

Lillesand, T., Kiefer, R. W., & Chipman, J. (2015). *Remote sensing and image interpretation*. John Wiley & Sons.

Liu, Y., Islam, M. A., & Gao, J. (2003). Quantification of shallow water quality parameters by means of remote sensing. *Progress in Physical Geography: Earth and Environment*, *27*(1), 24–43.

Martin, J., Tremblay, J. E., Gagnon, J., Tremblay, G., Lapoussière, A., Jose, C., Poulin, M., Gosselin, M., Gratton, Y., & Michel, C. (2010). Prevalence, structure and properties of subsurface chlorophyll maxima in Canadian Arctic waters. *Marine Ecology Progress Series*, *412*, 69–84.

Morel, A. Y. (1980). In-water and remote measurement of ocean colour. *Boundary Layer Meteorology*, *18*, 177–201.

Nieuwoudt, W. L., & Backeberg, G. R. (2011). A review of the modelling of water values in different use sectors in South Africa. *Water SA*, *37*(12), 703–710.

Oudejans, L. (2017) Report on the 2016 U.S. Environmental Protection Agency (EPA) International Decontamination Research and Development Conference. U.S. Environmental Protection Agency, Washington, DC, EPA/600/R-17/174.

Panneerselvam, B., Muniraj, K., Duraisamy, K., Pande, C., & Karuppannan, S. (2022). An integrated approach to explore the suitability of nitrate-contaminated groundwater for drinking purposes in a semiarid region of India. *Environmental Geochemistry and Health*, *3*, 1–17.

Papenfus, M., Schaeffer, B., Pollard, A. I., et al. (2020). Exploring the potential value of satellite remote sensing to monitor chlorophyll-a for US lakes and reservoirs. *Environmental Monitoring and Assessment*, *192*, 808.

Park, Y.-J., Ruddick, K., & Lacroix, G. (2010). Detection of algal blooms in European waters based on satellite chlorophyll data from MERIS and MODIS. *International Journal of Remote Sensing*, *31*, 6567–6583.

Peterson, K. T., Sagan, V., Sidike, P., Cox, A. L., & Martinez, M. (2018). Suspended sediment concentration estimation from landsat imagery along the lower missouri and middle Mississippi Rivers using an extreme learning machine. *Remote Sensing*, *10*(10), 1503.

Resmini, R. G., Kappus, M. E., Aldrich, W. S., Harsanyi, J. C., & Anderson, M. (1997). Mineral mapping with hyperspectral digital imagery collection experiment (HYDICE) sensor data at Cuprite, Nevada, USA. *International Journal of Remote Sensing*, *18*(7), 1553–1570.

Ritchie, J. C., Cooper, C. M., & Schiebe, F. R. (1990). The relationship of MSS and TM digital data with suspended sediments, chlorophyll, and temperature in Moon Lake, Mississippi. *Remote Sensing of Environment*, *33*(2), 137–148.

Ritchie, J. C., Zimba, P. V., & Everitt, J. H. (2003). Remote sensing techniques to assess water quality. *Photogrammetric Engineering & Remote Sensing*, *69*(6), 695–704.

Robinson, I. S., Wells, N. C., & Charnock, H. (1984). The sea surface thermal boundary layer and its relevance to the measurement of sea surface temperature by airborne and spaceborne radiometers. *International Journal of Remote Sensing*, *5*, 19–45.

Schaeffer, B. A., Schaeffer, K. G., Keith, D., Lunetta, R. S., Conmy, R., & Gould, R. W. (2013). Barriers to adopting satellite remote sensing for water quality management. *International Journal of Remote Sensing*, *34*(21), 7534–7544.

Schaepman, M. E., Jehle, M., hueni, A., D'Odorico, P., Damm, A., Weyermann, J., Schneider, F. D., Laurent, V., Popp, C., Seidel, F. C., & Lenhard, K. (2015). Advanced radiometry measurements and Earth science applications with the Airborne Prism Experiment (APEX). *Remote Sensing of Environment*, *158*, 207–219.

Schindler, D. W. (1977). Evolution of phosphorus limitation in lakes. *Science (New York, N.Y.)*, *195*, 260–262.

Seyhan, E. (1974). *Measurement of spectral signatures for water quality monitoring. NIWARS internal report*. Delft.

Soticha, K., Jareeya, Y., Sudjit, K., & Prapat, P. (2014). Assessing water quality of rural water supply in Thailand. *Journal of Clean Energy Technologies*, *2*(3), 226–228.

Sturm, B. (1981). The atmospheric correction of remotely sensed data and the quantitative determination of suspended matter in marine water surface layers. In A. P. Cracknell (Ed.), *Remote sensing in meteorology, oceanography, and hydrology* (pp. 163–197). Chichester: Ellis Horwood.

Sturm, B. (1981). Ocean colour remote sensing and quantitative retrieval of surface chlorophyll in coastal waters using Nimbus CZCS data. *Oceanography from Space*, 267–279.

Svab, E., Tyler, A. N., Preston, T., Presing, M., & Balogh, K. V. (2005). Characterizing the spectral reflectance of algae in lake waters with high suspended sediment concentrations. *International Journal of Remote Sensing*, *26*(5), 919–928.

Tong, X., Xie, H., Zhang, J., Zhang, Y., Zhao, J., & Qiu, Y. (2005). Multi-spectral remote sensing based water quality monitoring for Lake Tai. *MIPPR 2005: SAR and Multispectral Image Processing*, *6043*, 196–205.

Vespasiano, G., Muto, F., & Apollaro, C. (2021). Geochemical, geological and groundwater quality characterization of a complex geological framework: The case study of the Coreca area (Calabria, South Italy). *Geosciences*, *11*, 121.

Young, A., & Britton, G. (1990). Photobleaching in the unicellular green alga Dunaliella parva 19/9. *Photosynthesis Research*, *25*, 129–136.

Zhang, Q., Xu, P., & Qian, H. (2020). Groundwater quality assessment using improved water quality index (WQI) and human health risk (HHR) evaluation in a semi-arid region of northwest China. *Exposure and Health*, *12*, 487–500.

Zhou, Y., Li, P., Xue, L., Dong, Z., & Li, D. (2020). Solute geochemistry and groundwater quality for drinking and irrigation purposes: A case study in Xinle City. *North China. Geochemistry*, *80*, 125609.

SECTION

4

Precision agriculture practices for crop yield management

CHAPTER

13 Combined use of uncrewed aerial vehicle and satellite remote sensing data to gain crop insights within Colombia

Samantha Lavender[1], Michael Selvaraj[2] and Carlos Bojaca[3]

[1]*Pixalytics Ltd., Plymouth Science Park, Plymouth, Devon, United Kingdom*

[2]*The Alliance of Bioversity International and International Center for Tropical Agriculture (CIAT), Cali, Colombia*

[3]*Colombian Oil Palm Research Center—Cenipalma, Oil Palm Agronomy Research Program, Geomatics Section, Bogotá, Colombia*

13.1 Introduction

The use of uncrewed aerial vehicles (UAVs or drones) has become an expanding area of Remote Sensing (RS), which may be used on its own or with satellite RS. As concluded by Yao et al. (2019), the "new" aspects brought by UAV-based RS lie in its: ultra-high resolution; high availability of geometric data, for example, digital elevation model (DEM) calculated from the captured UAV data; and flexibility in multisensor integration. The Ecological Production Management Information System (EcoProMIS) project aimed to support Colombian rice and oil palm farming using a combination of UAV and satellite RS alongside crop modeling and ground-based measurements. The long-term aim was to improve productivity and stabilize incomes that would, in turn, allow the farmers to compete globally while responding to climate change and producing responsibly.

This chapter focuses on preliminary results from the UAV and satellite RS activities, which included the derivation of vegetation products such as Canopy Cover (CC), Leaf Area Index (LAI), Normalized Difference Red Edge Index (NDRE), and Normalized Difference Vegetation Index (NDVI). CC (fraction of pixel surface area covered by the canopy) and LAI (the projected area of leaves over a unit of land, m^2/m^2) detect the morphological characteristics that define the characteristics of plant canopies and are used to describe the structure and density. Meanwhile, the vegetation indices are used to detect the physiological characteristics linked to the chlorophyll content. NDRE is derived as the ratio of the difference and sum of the Red Edge (RE) to Near-InfraRed (NIR) reflectance

Remote Sensing in Precision Agriculture. DOI: https://doi.org/10.1016/B978-0-323-91068-2.00006-0

(Barnes et al., 2000) while NDVI uses the "classical" red reflectance band (Tucker, 1979):

$$\text{Normalized Difference Index} = (\text{NIR} - \text{Band})/(\text{NIR} + \text{Band})$$

The RE at around 700 nm is a unique feature of green vegetation because it results from two optical properties of plant tissue: high internal leaf scattering, causing a high NIR reflectance and chlorophyll absorption giving low red reflectance (Horler et al., 1983).

RS has been used to estimate yield for various annual crops, e.g., Guan et al. (2017) reviewed regional and global cropland monitoring, alongside fruit trees and vegetables. With the increased use of harvester-mounted yield monitors, intensive yield data can be collected from a field and used to calibrate RS approaches. For this research, the extraction of RS-derived products was considered for rice fields and palm plots in Colombia.

Rice is one of the world's most important crops in terms of economic value and is a prime source of highly nutritional food, with approximately 90% of the world's rice being grown and consumed (Dawe, 2000). In Colombia, rice occupies first place in economic value among short-cycle crops. The main rice seasons are from March–April to July–August, and the second season is from August–October to January–February. As of 2004, Colombia was the second-largest rice producer in Latin America and the Caribbean, and rice was a primary source of calories and protein for around 20% of the country's population (FAO, 2005).

Rice yields are subject to significant year-to-year variations caused by climate variability, including the El Nino Southern Oscillation (ENSO) that affects the frequency of heavy rains, flooding, drought, and heatwaves across much of South America (Barrios-Perez et al., 2021). The global average rice yield variability (standard deviation) is $\sim$0.5 tons/ha/year or $\sim$13% of average rice yields (Ray et al., 2015). Therefore forecasting rice crop growth and crop yields in advance of maturity is essential for farmers and the broader public and private sectors such as governments and the insurance industry.

Short-cycle crops such as rice have specific requirements and sensitivity to environmental factors in each growth stage that are directly related to the physiological processes occurring in the plant (Delerce et al., 2016). Therefore in-season crop yield forecasting supports farmers to adjust agronomic management to obtain improved yields sustainably, for example, focused fertilizer usage and reduction of greenhouse gas emissions, with governments able to develop timely strategies to cope with food market fluctuations. LAI has been identified as one of the main crop features derived from RS that can be used to improve a crop model (Chen & Tao, 2022). As a result, a calibrated LAI RS model is designed to correct, in Near Real-Time (NRT), a crop model's simulated LAI and contribute to improving the estimated yield.

Oil palm is a tropical tree crop mainly grown for the industrial production of vegetable oil with a sequence of production cycles over at least 25 years, with oil palm fruits taking more than three years from flower initiation to harvest maturity (Hoffmann et al., 2017). In recent decades, plantations have expanded rapidly, particularly in the main oil palm producing countries such as Indonesia and Malaysia (Taheripour et al., 2019); the resulting deforestation has drawn public attention due to its negative impacts on biodiversity due to losses of tropical rainforest habitats. The cultivation of palm oil in Colombia dates to the 1940s, when the first plantation was established in the Magdalena department by United Fruit Company (Wight, 2018). Nowadays, Colombia produces more palm oil than any other country in Latin America and is the fourth-largest producer globally (Wallace et al., 2016).

The management of oil palm plantations is conducive to the improvement of palm oil yield and benefits achieving production goals without expanding the area planted. However, climate change and diseases can cause large yield fluctuations and significantly impact the smaller growers with less financial means (Wight, 2018). Therefore, RS has been undertaken to support applications such as plantation mapping, including tree detection, alongside the determination of age, carbon and yield estimation (Zheng et al., 2021). This collection of data is being mirrored by the development of machine learning techniques, such as Bayesian networks and artificial neural networks (Chapman et al., 2018).

13.2 Methodology

The following sections summarize the different approaches undertaken for the rice field and palm plantations.

13.2.1 Uncrewed aerial vehicle data collection and processing

The UAV captured aerial multispectral time-series images using a MicaSense RedEdge 3 camera. The camera has five spectral bands: Blue, Green, Red, NIR, and RE with the wavelengths of 455–495 nm, 540–580 nm, 658–678 nm, 800–880 nm, and 707–727 nm, respectively.

As described in (Selvaraj et al., 2020), a fully automated drone image acquisition protocol was followed. Six PhotoScan coded targets (concentric rings) were printed on 0.5 × 0.5 m plastic sheets and evenly distributed within each rice field as Ground Control Points (GCPs); maintained until all the UAV images were acquired. These GCPs were then georeferenced using the highly accurate Real-Time Kinematic (RTK)-Global Positioning System (GPS) (from South Galaxy G1, China) with a horizontal accuracy of 0.25 m and a vertical accuracy of 0.5 m.

The flight plan was controlled using DJI Ground Station Pro Application (DJI GS Pro, China). Before each image acquisition, one image was taken of the MicaSense reflectance panel for radiometric calibration. To achieve an overlap of successive flightlines, 75% vertically and horizontally, the camera was triggered using the UAV DJI A3 flight controller with an Arduino Nano interface configured by the DJI GS app. The image acquisitions occurred at 10:00 local time at an altitude of between 50 and 100 m above ground level, resulting in a spatial resolution of between 0.04 and 0.1 m.

UAV images were captured at different phenological stages for the rice fields. The generation of orthomosaics and DEMs was undertaken using the Agisoft MetaShape Pro software (Version 1.2.2, Agisoft LLC, http://www.agisoft.com). Products such as canopy height, CC, and vegetation indices were extracted using the International Center for Tropical Agriculture (CIAT) Pheno-i software; detailed in (Selvaraj et al., 2020). For the palm plantations, UAV imagery was captured on several dates throughout the year and orthomosaics generated.

13.2.2 Satellite data download and processing

The satellite data used for the analysis came from the Copernicus Sentinel-2 missions that carry the Multispectral Instrument (MSI) and USGS/NASA Landsat-8 mission that carries the Operational Land Imager (OLI). The first Sentinel-2 mission (A) was launched in June 2015, with Sentinel-2B joining the constellation in March 2017, which resulted in a significant increase in the amount of available data with the repeat coverage reducing to five days assuming cloud-free conditions—the equator crossing time is 10:30. Landsat-8 was launched in February 2013 and has a repeat cycle of 16 days with a 10:00 equator crossing time. When the crop cycle season is cloudy, this repeat coverage for usable data is reduced.

The data for both missions were processed from Level 1 to Level 2 using the iCOR atmospheric correction tool (Doxani et al., 2018), which processes Landsat-8 and Sentinel-2 data and exhibited the most consistent results when multiple atmospheric correction approaches were tested (comparison not shown).

The Sentinel-2 tiles were filtered such that only those with less than 95% cloud cover (as defined by the Level 1 file cloud coverage statistics) were downloaded from the Copernicus Open Access Hub (https://scihub.copernicus.eu/); 95% was choosen as scenes with a higher cloud coverage tended to fail the atmospheric correction that needed some cloud free land pixels. For Landsat-8, Collection 1 scenes were downloaded from Google Cloud (https://cloud.google.com/storage/docs/public-datasets/landsat). The downloaded tiles/scenes are shown in Table 13.1.

Table 13.1 Rice field and oil palm Sentinel-2 tiles and Landsat-8 scenes.

Site	Sentinel-2 tiles	Landsat-8 scenes
Casanare rice field	18NYL	007/057
Magdalena palm plantation	18PWS	009/052 and 009/053
Tolima rice field	18NVK	008/057

Then, iCOR version 2.0 was run with the default parameters such that the atmospheric parameters were:

1. Aerosol Optical Thickness (AOT) estimated using the presence of land pixels and sufficient spectral variability in the scene.
2. Fixed AOT (0.1 m) when the first approach failed.

The output of the atmospheric correction was the Bottom-of-Atmospheric (BOA) reflectance and a cloud mask, which was used to remove cloud-affected pixels. In addition, temporal filtering was used to identify spurious pixels that were not removed by the cloud mask, for example, those affected by partial cloud pixels or cloud shadows.

13.2.3 Uncrewed aerial vehicle and satellite data extraction

13.2.3.1 Rice fields

The two fields under investigation were in the departments of Casanare and Tolima.

- Casanare is in the Orinoquía region, also called the Eastern Plains region, at about 500 m above sea level with a mixture of savannas and forests. The rice is rain-fed and cropped once a year, with a rainy season between April and November and a dry season between December and March.
- Tolima is in the Andean region, extending from the Andean Cordillera (mountains) across the Magdalena River valley to the Cordillera Oriental (branch of the Colombian Andes). The department has a tropical rainforest climate situated around 410 m above sea level. The rice is irrigated and cropped twice per year, with rainy seasons in April–May and September–November.

The UAV and satellite data were extracted from frames (small specific areas within the fields that were the focus for the phenology data collection) represented as circular polygons in Fig. 13.1. There was no change in the frame's position for Tolima, but for Casanare, the positions varied from one crop cycle to another.

In addition, for the satellite data, the extraction was run for a further two options:

- Extraction of all the pixels within the red polygons; shown in Fig. 13.1. This approach will be referred to as JSON as the in-field statistics were stored in a GeoJSON format file.
- Random points were distributed across the field, covering the area imaged by both the UAV and satellite RS data.

The number of pixels included within a frame depended on the satellite's spatial resolution for the satellite data. As Landsat-8 OLI pixels have a 30 m spatial resolution, the number of pixels contained is significantly less than that for Sentinel-2 MSI, which is 10 m resolution. Each pixel that intersects a frame polygon was included in the analysis for that frame.

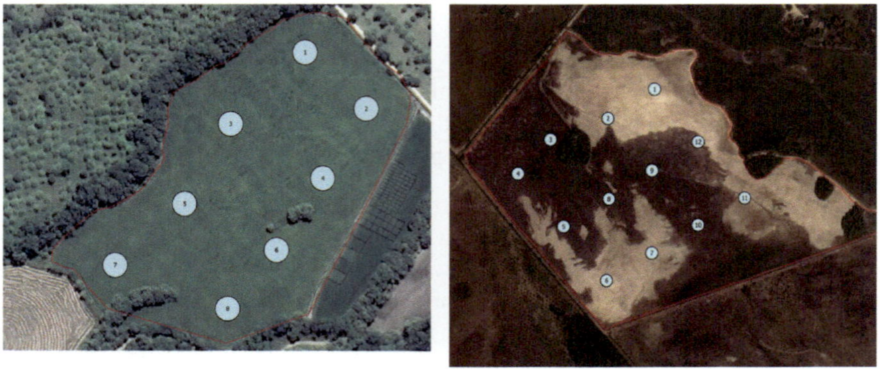

FIGURE 13.1

Rice frames for (left) Tolima and (right) Casanare overlaid on Google Maps (Satellite view) with the red line showing the whole field polygon.

FIGURE 13.2

Zoomed-out view of Palo Alto Plantations (left) and zoomed-in to show the identification of individual trees using circular blue-filled polygons (right) overlaid on the Google Maps (Satellite view) with the red and blue lines showing the whole plot polygons (red for EcoProMIS focused plots).

13.2.3.2 Palm plantations

UAV and satellite data were analyzed for the Palo Alto Plantations in the Magdalena region. The Magdalena department is in the north of Colombia, by the Caribbean Sea. It mainly has hot temperatures with high humidity, a rainy season from April and November, and lower rainfall between June and August.

For the palm plots, the circular polygons identify the location of each palm tree with these data supplied by Cenipalma; see Fig. 13.2. Palm oil trees are grown in a regularly spaced pattern, allowing palm plantations to be recognizable

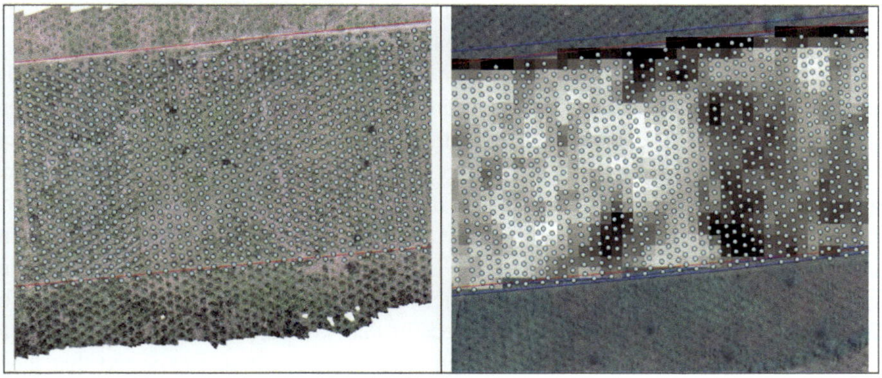

FIGURE 13.3

Comparison of the palm polygons for the RGB UAV data (left) and Sentinel-2 NDVI satellite data (right) as dark-to-light values for low-to-high NDVI values.

from natural forests within satellite imagery. In Fig. 13.2, the plots with the blue polygon outlines are palm plantation plots. With the trees identified inside, the red polygons show the palm plots of specific interest to EcoProMIS, where ground-based measurements and UAV flights were undertaken.

Fig. 13.3 shows the red palm polygons and filled circle blue palm tree locations compared with the UAV RGB (668 nm as Red, 560 nm as Green, and 475 nm as Blue) and NDVI (dark to light as low to high) Sentinel-2 satellite data. As can be seen, for the free-to-access satellite imagery, knowing the oil palm locations does not help with individual palm tree matchups as 10 m resolution data cannot resolve the differences between the palms and in-between palm areas. However, the places with a lower vegetation fraction show a lower NDVI value.

13.2.4 Rice ground-based phenology Leaf Area Index approach

The first step in building an LAI model from RS is to explore the relationship between the RS LAI product and the ground-based LAI data. In this case, ground-based LAI data were collected with the Licor LAI-2200 plant canopy analyzer at times the rice crop reached a particular development stage throughout the cropping cycle. However, the initial comparisons between RS and ground-based LAI data yielded unexpected low correlations.

After checking the data collection and processing protocols for both data sources (RS and ground-based), LAI data collected by the canopy analyzer was not following the expected trend throughout the cropping cycle. It is likely that once the crop reaches the reproductive stages, the LAI should stop increasing, and by the end of the crop cycle, it should decrease due to leaf senescence. In the case of the canopy analyzer data, the LAI continued to increase for most of the cropping cycle. Based on these results, it was concluded that the data collection protocol with the canopy analyzer was not adjusted to correct the panicle's (flower head form's) presence in the measurements.

If the canopy analyzer measurements are not corrected for the presence of the reproductive structures, the LAI is overestimated, yielding erroneous results. Therefore the dry matter measurements were used to estimate the LAI to overcome this situation. This new dataset was then used to explore the relationship between RS and ground-based data. Along with the LAI data collected by the canopy analyzer, the field data collection team also sampled the plant dry matter accumulation and distribution by doing destructive measurements. The plant was divided into its main aboveground structures (stems, leaves and panicles), and the samples were oven-dried for 72 hours, then the dry matter was weighed. In this way, dry matter was allocated to the leaves at each crop development stage as determined. It was expected that these new estimated LAI values should closely resemble the actual LAI as we extracted the information directly from the leaves without any other plant structure interference.

Fig. 13.4 shows the LAI coming from the two data sources, that is, the plant canopy analyzer and the destructive biomass measurements for all the rice cycles from Casanare and Tolima. As expected, the LAI from the biomass measurements indicated that the LAI reduction started after the reproductive stage. The averaged data for each sampling date, for the same cycle, is presented in Fig. 13.5, which reduces the sampling noise so that this drop in the biomass LAI is more evident.

To further check which data source gave the best indication of the LAI trend over the crop cycle, a crop model was used to conduct simulations for some of

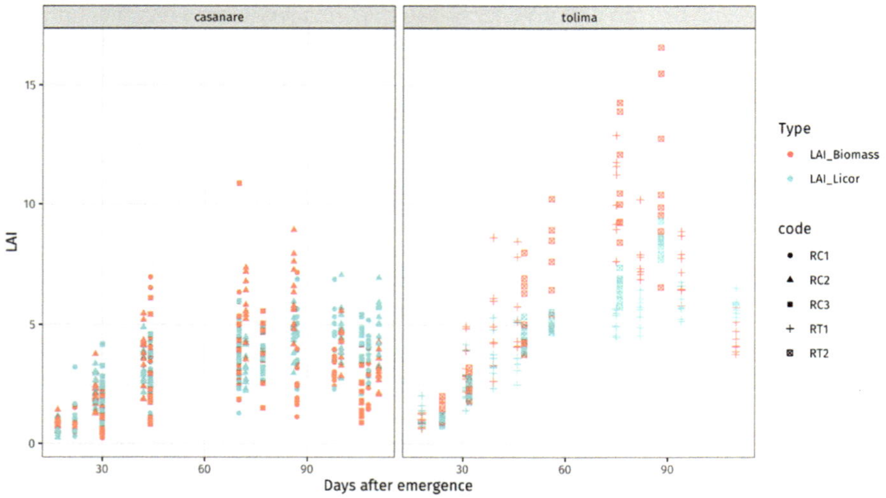

FIGURE 13.4

All individual estimated LAI data points for the plant canopy analyzer (LAI_Licor) and through the destructive measurements (LAI_Biomass) for the experimental rice cycles; the symbols indicate the different rice crop cycles from Casanare and Tolima.

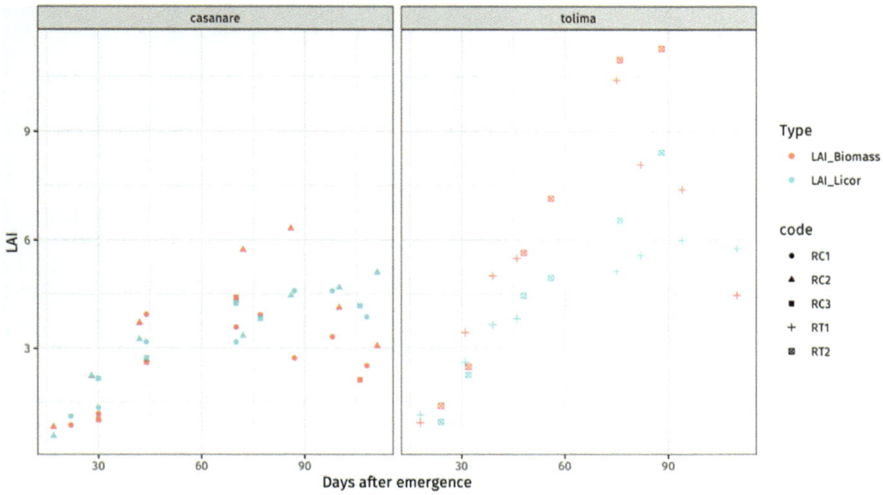

FIGURE 13.5

Averaged data for each sampling date for the estimated LAI by the plant canopy analyzer (LAI_Licor) and the destructive measurements (LAI_Biomass) for the experimental rice cycles; the symbols indicate the different rice crop cycles from Casanare and Tolima.

the experimental cycles (not shown). In this way, the estimated LAI was compared with the simulated (potential) LAI that was overestimated compared with the actual crop status as the model did not consider any growth constraint due to water, nutrients, pests, or diseases. Therefore, these simulated values may not be accurate enough to use for yield prediction but were able to depict the expected LAI trend for a rice crop and so could be used to support consistency checking.

13.3 Results

13.3.1 Rice

13.3.1.1 Normalized Difference Vegetation Index comparisons

Within frame variability was quantified through the extracted data being plotted as the mean and standard deviation of the included pixels values with NDVI values shown in Fig. 13.6.

Looking in further detail at Casanare Frames 5, 6, 7, and 8 for the third crop cycle, Fig. 13.7 shows the two frames from the perspective of the RGB UAV image and satellite NDVI product for the Maturity (MAT) stage of the crop cycle. Although the Sentinel-2 satellite data capture some of the in-field variability, its spatial resolution (10 m) means it cannot capture the highly spatially varying nature of the field. Also, there were clouds on this closest date to the UAV data,

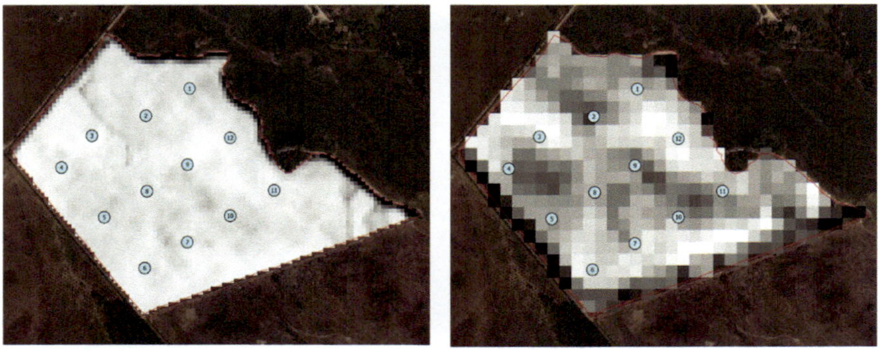

FIGURE 13.6

Rice frames for Casanare overlaid on (left) Sentinel-2 and (right) Landsat-8 NDVI products, overlaid on Google Maps (Satellite view).

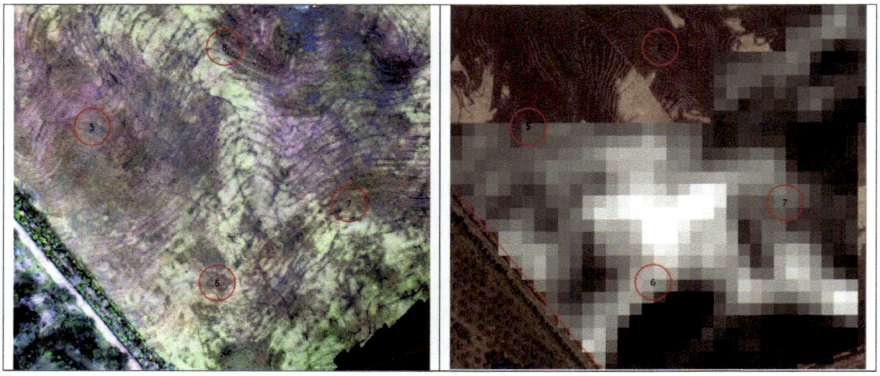

FIGURE 13.7

View of several frames as both the UAV RGB (left) and Sentinel-2 satellite NDVI overlaid on Google Maps (Satellite view) (right) products for the crop cycle 3 maturity development stage.

and therefore large areas of the field have been masked, with Frame 5 only having partial coverage. The remaining satellite pixels may have artifacts from the high aerosol load between the clouds.

The comparison of UAV with satellite data is shown in Fig. 13.8 for the Casanare, with crop cycles 1 and 3, and Fig. 13.9 for the Tolima crop cycles 1 and 2.

For both, the latest UAV crop cycle data are annotated with the development stages: Early Tillering (ET, a tiller is a grain-bearing branch), Medium Tillering (MT), Panicle Initialization (PI, start of the reproductive phase when the flower head forms), Booting (B, bulging of the leaf stem that conceals the developing panicle), Flowering (F), Dough (D, milky part of the grain becomes soft and then reaches a hard paste), Maturity (MAT).

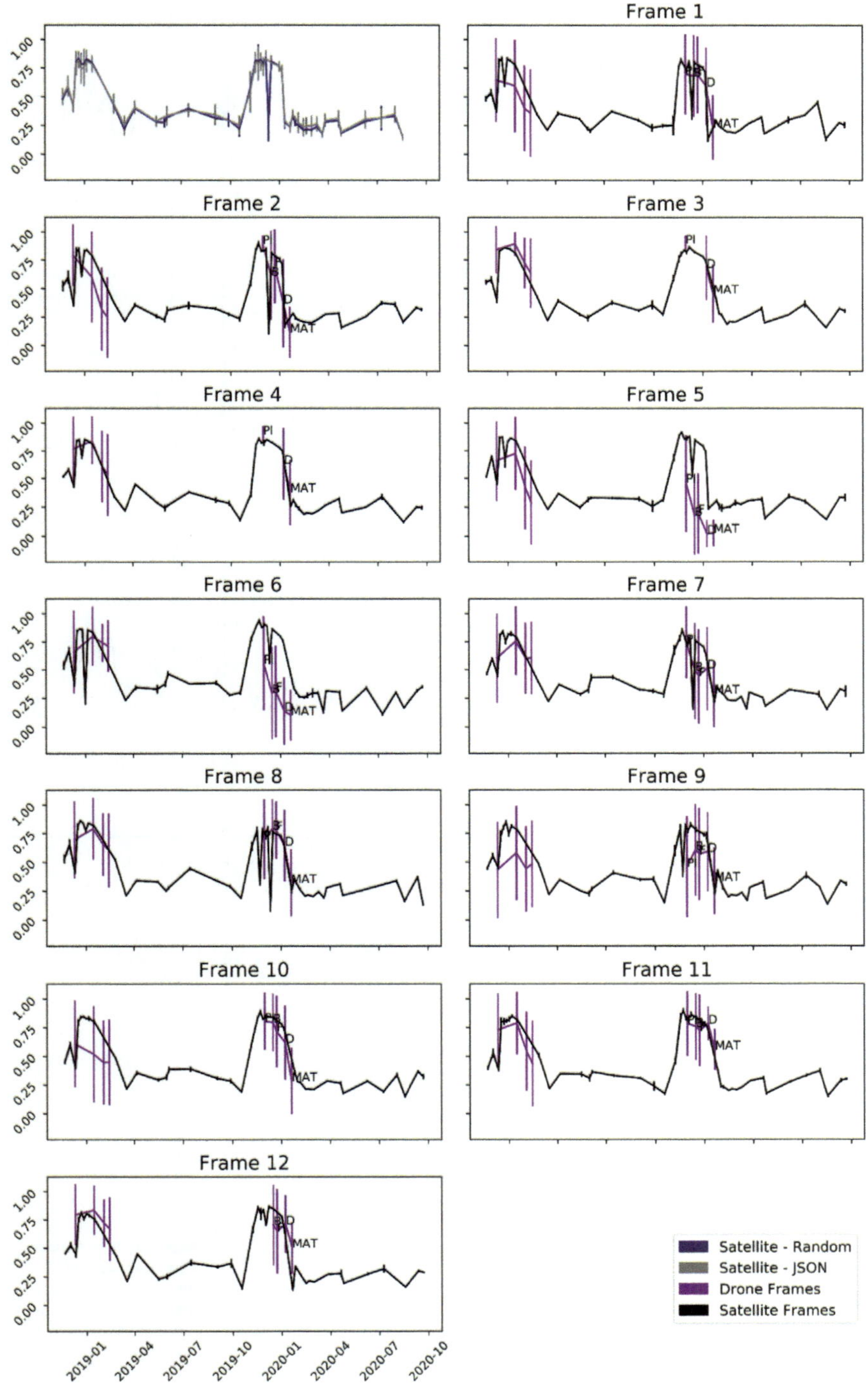

FIGURE 13.8

Time-series plots of Casanare NDVI satellite whole field extraction (top left) and satellite and UAV for each frame (remainder of the subplots).

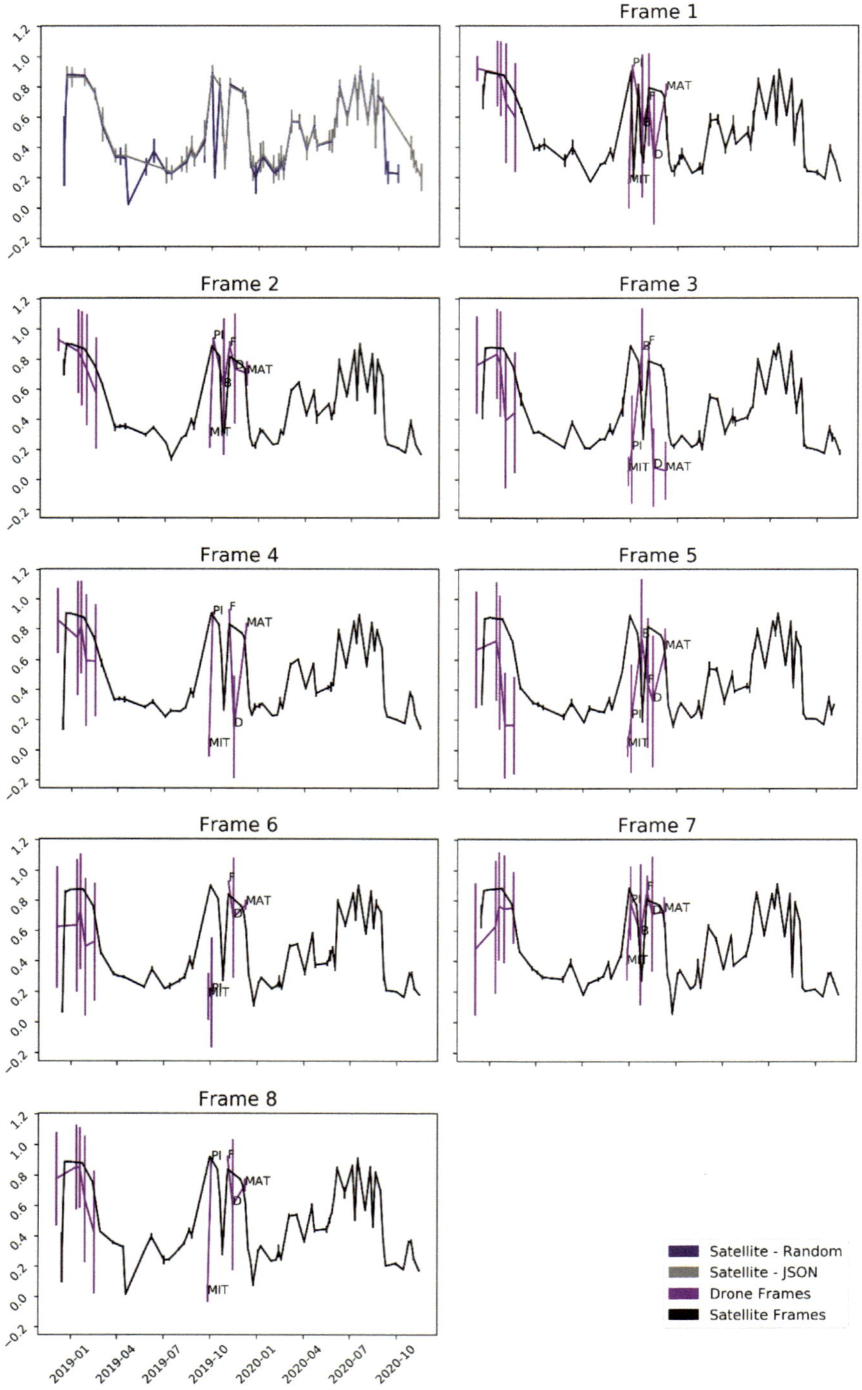

FIGURE 13.9

Time-series plots of Tolima NDVI satellite whole field extraction (top left) and satellite and UAV for each frame (remainder of the subplots).

For the booting development stage, Frames 5 and 6 were not covered by the UAV mosaic. Therefore Fig. 13.10 shows Frames 2, 3, and 4 that had both UAV and satellite coverage. The areas with more bare ground had a lower satellite NDVI value as the vegetation fraction within the satellite pixel was lower. The natural vegetation seen as an "island" within the middle of the field by the UAV data was not visible in the satellite NDVI product as it had an NDVI comparable with the rice crop.

13.3.1.2 Leaf Area Index comparisons

The LAI_Biomass (described in Section 13.2.4) was plotted against the satellite data as a time series for Casanare and Tolima; see Fig. 13.11. For Casanare, there

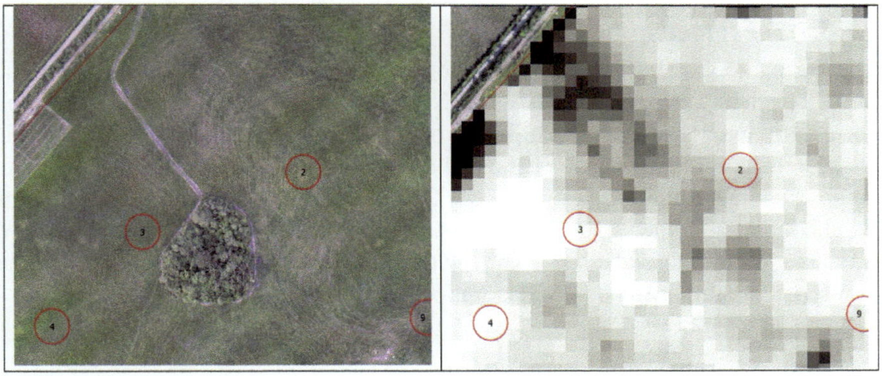

FIGURE 13.10

View of several frames as both the UAV RGB (left) and Sentinel-2 satellite NDVI overlaid on Google Maps (Satellite view) (right) products for the crop cycle 3 booting development stage.

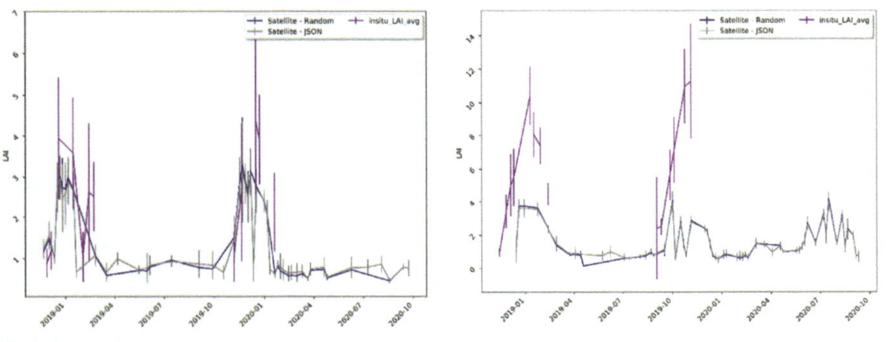

FIGURE 13.11

Time-series plots of the (left) Casanare and (right) Tolima LAI satellite whole field (Random and JSON) data against LAI_Biomass.

was a close agreement but with the phenological data being higher than the satellite data. There was a more significant difference for Tolima, with the phenological values being much higher than that estimated by the satellite data. The Tolima plot shows satellite data with a more irregular appearance (alternating low and high values), especially for the 2020 fourth crop cycle. This irregular appearance indicates that the satellite data are being affected by cloud artifacts that have not been entirely removed during the cloud/temporal filtering.

The results for the Casanare frames were plotted, in Fig. 13.12, to aid an understanding of the infield spatial variability. The result was similar to that for the satellite UAV comparisons in that sometimes there was a close relationship. In contrast, for other frames/cycles, the satellite and phenological data appear to be seeing something different.

13.3.1.3 Estimating rice yield

The correlation analysis for the phenological growth stages revealed that panicle initiation to flowering was consistent; NDVI, NDRE, and CC have a reasonable correlation with final grain yield based on Tolima and Casanare data. Among the tested vegetation indices, NDVI showed a highly significant correlation with grain yield at panicle initiation ($r = 0.87$) for 2019 crop cycle 3; Fig. 13.13 (left). This trend continued in the 2020 crop cycle 4 data, see Fig. 13.13 (right), where NDVI was also highly correlated ($r = 0.65$) with grain yield again at the panicle initiation stage. A similar trend was also noticed in the Tolima 2018 crop cycle 1.

13.3.2 Palm

For the Paolo Alto palm plot, shown in Fig. 13.3, the time-series plot for LAI is shown in Fig. 13.14 with the ERA5-Land (Muñoz-Sabater et al., 2021) reanalysis monthly data for surface solar radiation [J/m^2] and total precipitation [m]; hourly data extracted for 12:00 local time shown. It appears that during the 2019–20 and 2020–21 dry seasons (December to March), the LAI reduces.

13.4 Discussion and conclusions

When comparing UAV and satellite RS products, the high in-field rice spatial variability could explain the high standard deviation seen for individual frames for the UAV NDVI data and the difficulty in achieving a consistent match between the UAV and satellite data. The UAV data were processed such that only the pixels containing vegetation were used in the frame-based calculation. At the same time, the satellite data calculated a lower NDVI when there was an increased fraction of bare ground. These differences might suggest that the satellite would calculate lower values than the UAV data where there was divergence. Still, the UAV data had lower values when a more significant variation from the

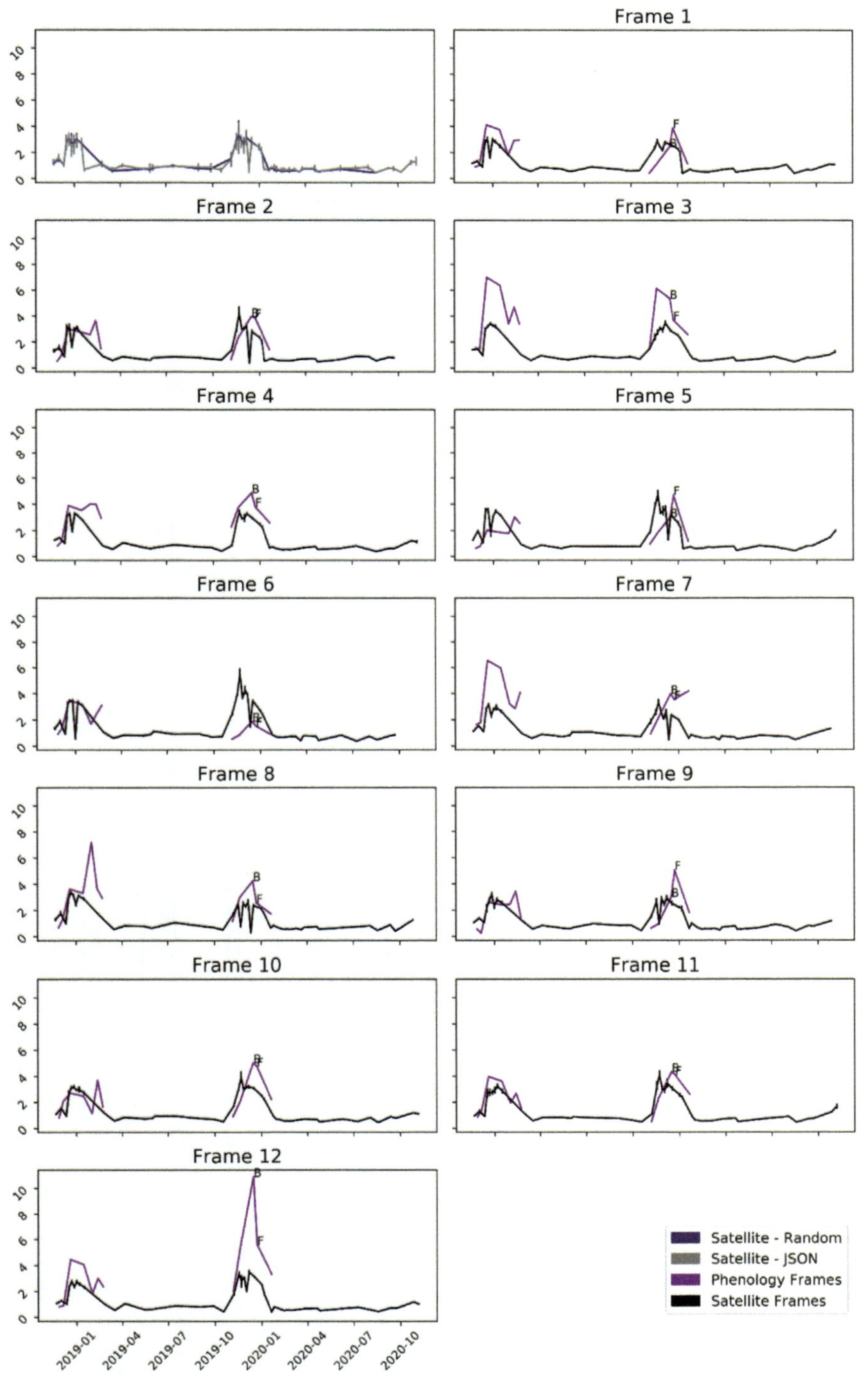

FIGURE 13.12

Time-series plots of Casanare LAI satellite whole field extraction (top left) alongside frame-based satellite and phenological (LAI_Biomass) LAI for each frame (remainder of the subplots).

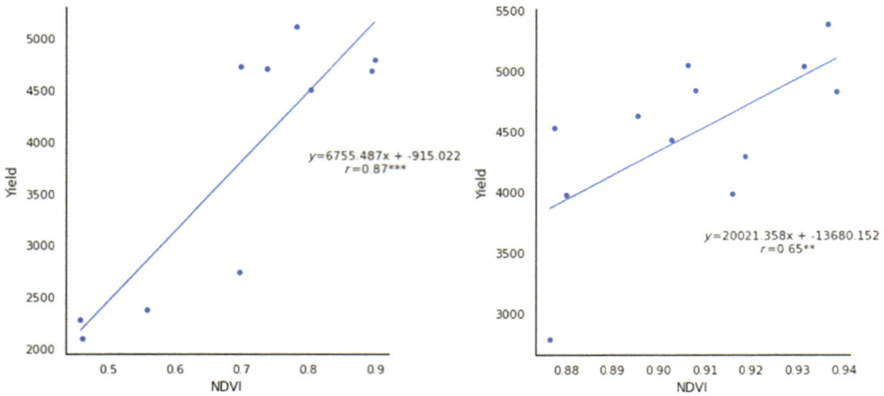

FIGURE 13.13

Pearson correlation coefficient (r) between UAV Casanare NDVI at panicle initiation stage and the grain yield parameter for the (left) 2019 cycle 3 and (right) 2020 cycle 4.

Asterisks indicate the strength of the correlation.

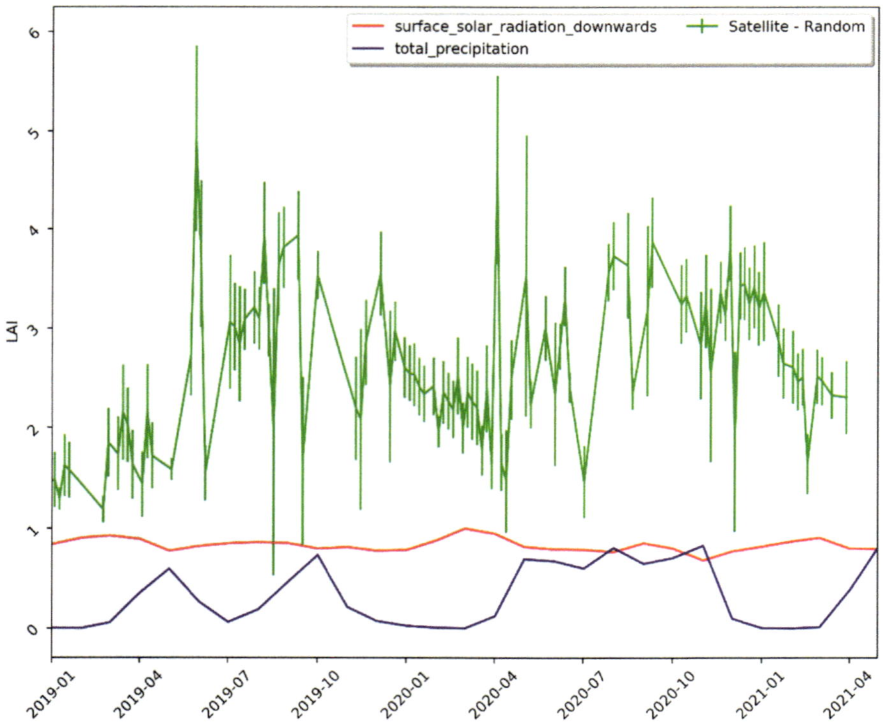

FIGURE 13.14

Time-series plots of Paolo Alto satellite-derived LAI with the normalized monthly averaged midday ERA5-Land modeled surface solar radiation and total precipitation (Muñoz Sabater, 2021).

satellite data was observed. Therefore further work is needed to understand the causes of the differences between the UAV and satellite RS data.

Both NDVI and LAI vary according to the phenological growth stage, with the NDVI increasing until the panicle initialization stage when it plateaus and then drops as maturity is achieved. For some frame/stage sets, the shape and magnitude of the satellite and UAV data are in close agreement, while for others, the plots show a different temporal evolution. For LAI, the satellite data show a similar pattern to NDVI, but the in situ phenological LAI continues to rise after panicle initialization until it peaks at booting or flowering.

Compared with Tolima, Casanare showed more consistent results for rice yield estimation as there were some issues with the collection of the Tolima UAV images. Predicting grain yield around the time of panicle initiation to flowering was consistent with NDVI, NDRE, and CC, providing a reasonable and consistent correlation with final grain yield based on the Casanare data, shown in Fig. 13.13 for the NDVI results.

The difference in spatial variability for the palm plot determined whether individual palm trees were visible or just the overall impact on the RS products. The advantage of the satellite dataset time series was that it was available for multi-annual analysis, with Fig. 13.14 showing a potential relationship between the meteorological conditions and LAI. Palm trees thrive under evenly distributed heavy, 2000–2500 mm/year, rainfall (Farmanta et al., 2021); therefore a lack of rain or uneven rainfall requires water supply through irrigation. Chapman et al. (2018) concluded that yield showed a clear positive response to rainfall in the year prior to harvest, but above a maximum there was a diminishing response. Whereas Monzon et al. (2022) showed that meteorological anomalies up to three years before harvest markedly influence yield, and a high yield year will negatively influence subsequent years due to the palms becoming "exhausted".

The preliminary results from the UAV and satellite RS have shown the potential to detect factors that influence rice and palm oil yield. However, more complex RS/crop modelling is needed to capture all the significant contributors to yield, which is where machine learning can play a role if sufficient data sources exist. The data that is collected needs to be carefully quality controlled before it is included in such approaches.

Acknowledgments

The Global Challenges Research Fund funded the EcoProMIS project through the UK Space Agency International Partnership Programme. Partners included Agricompas (United Kingdom and Colombia), CIAT (International Centre for Tropical Agriculture, Colombia), Cenipalma (Colombia), Fedearroz (Colombia), Pixalytics (United Kingdom), Rothamsted Research (United Kingdom), and Solidaridad (Colombia). The research presented in this chapter builds on the work undertaken by all partners with Cenipalma and Fedearroz conducting the ground-based data collection in the palm plantations and rice fields, respectively.

References

Barnes, E. M., Clarke, T. R., Richards, S. E., Colaizzi, P. D., Haberland, J., Kostrzewski, M., Waller, P., Choi C., R. E., Thompson, T., Lascano, R. J., Li, H., & Moran, M. S. (2000, January). Coincident detection of crop water stress, nitrogen status and canopy density using ground based multispectral data. In: *Proceedings of the 5th international conference*.

Barrios-Perez, C., Okada, K., Varón, G. G., Ramirez-Villegas, J., Rebolledo, M. C., & Prager, S. D. (2021). How does El Niño Southern Oscillation affect rice-producing environments in central Colombia? *Agricultural and Forest Meteorology*, *306*. Available from https://doi.org/10.1016/j.agrformet.2021.108443, May.

Chapman, R., Cook, S., Donough, C., Lim, Y. L., Ho, P. V. V., Lo, K. W, & Oberthür, T. (2018). Using Bayesian networks to predict future yield functions with data from commercial oil palm plantations: A proof of concept analysis. *Computers and Electronics in Agriculture*, *151*, 338–348. Available from https://doi.org/10.1016/j.compag.2018.06.006.

Chen, Y., & Tao, F. (2022). Potential of remote sensing data-crop model assimilation and seasonal weather forecasts for early-season crop yield forecasting over a large area. *Field Crops Research*, *276*, 108398. Available from https://doi.org/10.1016/j.fcr.2021.108398, December 2021.

Dawe, D. (2000). The contribution of rice research to poverty alleviation. *Studies in Plant Science*, *7*(Issue C). Available from https://doi.org/10.1016/S0928-3420(00)80003-8, Elsevier Masson SAS.

Delerce, S., Dorado, H., Grillon, A., Rebolledo, M. C., Prager, S. D., Patiño, V. H., Varón, G. G., & Jiménez, D. (2016). Assessing weather-yield relationships in rice at local scale using data mining approaches. *PLoS One*, *11*(8). Available from https://doi.org/10.1371/journal.pone.0161620.

Doxani, G., Vermote, E., Roger, J. C., Gascon, F., Adriaensen, S., Frantz, D., Hagolle, O., Hollstein, A., Kirches, G., Li, F., Louis, J., Mangin, A., Pahlevan, N., Pflug, B., & Vanhellemont, Q. (2018). Atmospheric correction inter-comparison exercise. *Remote Sensing*, *10*(2). Available from https://doi.org/10.3390/rs10020352.

FAO. (2005). *Proceedings of the FAO Rice Conference 2004: Rice in Global Markets: Rome, 12–13 February 2004*; FAO Rice Conference, Sarris, A., Food and Agriculture Organization of the United Nations, Eds.; FAO commodities and trade proceedings; Food and Agriculture Organization of the United Nations: Rome, 2005; ISBN 978-92-5-105350-8. Available from https://www.fao.org/3/y5682e/y5682e.pdf. (2005) Accessed 18.04.23.

Farmanta, Y., Sung, C. T. B., Giap, S. E. G., Paing, T. N., Ko, H., & Impron. (2021). Net rainfall components under various ages of the oil palm. *Malaysian Journal of Soil Science*, *25*(January), 159–169.

Guan, K., Wu, J., Kimball, J., Anderson, M., Frolking, S., Hain, C., & Lobell, D. (2017). The shared and unique values of optical, fluorescence, thermal and microwave satellite data for estimating large-scale crop yields. *Remote Sensing of Environment*, *199*, 333–349. Available from https://doi.org/10.1016/j.rse.2017.06.043.

Hoffmann, M. P., Donough, C. R., Cook, S. E., Fisher, M. J., Lim, C. H., Lim, Y. L., Cock, J., Kam, S. P., Mohanaraj, S. N., Indrasuara, K., Tittinutchanon, P., & Oberthür, T. (2017). Yield gap analysis in oil palm: Framework development and application in commercial operations in Southeast Asia. *Agricultural Systems*, *151*, 12–19. Available from https://doi.org/10.1016/j.agsy.2016.11.005.

Horler, D. N. H., Dockray, M., Barber, J., & Barringer, A. R. (1983). Red edge measurements for remotely sensing plant chlorophyll content. *Advances in Space Research*, *3* (2), 273–277. Available from https://doi.org/10.1016/0273-1177(83)90130-8.

Monzon, J. P., Jabloun, M., Cock, J., Caliman, J. P., Couëdel, A., Donough, C. R., Vui, P. H. V., Lim, Y. L., Mathews, J., Oberthür, T., Prabowo, N. E., Edreira, J. I. R., Sidhu, M., Slingerland, M. A., Sugianto, H., & Grassini, P. (2022). Influence of weather and endogenous cycles on spatiotemporal yield variation in oil palm. *Agricultural and Forest Meteorology*, *314*. Available from https://doi.org/10.1016/j.agrformet.2021.108789.

Muñoz-Sabater, J., Dutra, E., Agustí-Panareda, A., Albergel, C., Arduini, G., Balsamo, G., Boussetta, S., Choulga, M., Harrigan, S., Hersbach, H., Martens, B., Miralles, D. G., Piles, M., Rodríguez-Fernández, N. J., Zsoter, E., Buontempo, C., & Thépaut, J. N. (2021). ERA5-Land: A state-of-the-art global reanalysis dataset for land applications. *Earth System Science Data*, *13*(9), 4349–4383. Available from https://doi.org/10.5194/essd-13-4349-2021.

Muñoz Sabater, J. (2021). ERA5-Land monthly averaged data from 1950 to 1980. *Copernicus Climate Change Service (C3S) Climate Data Store (CDS)*. Available from https://doi.org/10.24381/cds.68d2bb3.

Ray, D. K., Gerber, J. S., Macdonald, G. K., & West, P. C. (2015). Climate variation explains a third of global crop yield variability. *Nature Communications*, *6*, 1–9. Available from https://doi.org/10.1038/ncomms6989.

Selvaraj, M. G., Valderrama, M., Guzman, D., Valencia, M., Ruiz, H., Acharjee, A., Acharjee, A., & Acharjee, A. (2020). Machine learning for high-throughput field phenotyping and image processing provides insight into the association of above and below-ground traits in cassava (Manihot esculenta Crantz. *Plant Methods*, *16*(1), 1–19. Available from https://doi.org/10.1186/s13007-020-00625-1.

Taheripour, F., Hertel, T., & Ramankutty, N. (2019). Market-mediated responses confound policies to limit deforestation from oil palm expansion in Malaysia and Indonesia. *Proceedings of the National Academy of Sciences*, *116*, 19193–19199. Available from https://doi.org/10.1073/pnas.1903476116.

Tucker, C. J. (1979). Red and photographic infrared linear combinations for monitoring vegetation. *Remote Sensing of Environment*, *8*(2), 127–150. Available from https://doi.org/10.1016/0034-4257(79)90013-0.

Wallace, B. Y. T. C., Velasco, A., Lay, T., Zhang, J., Tromp, J., Tape, C., Liu, Q., Thompson, E. M., Wald, D. J., Thio, H. K., Kanamori, H., ΤΖΕΦΕΡΗΣ, Π., Razafindrakoto, H. N. T., Martin Mai, P., Mai, P. M., Thingbaijam, K. K. S., Jordan, T. H., Juarez, A., Ji, C., & Lavallée, D. (2016). How Colombia became Latin America's palm oil powerhouse. *Bulletin of the Seismological Society of America*, *106* (1), 6465–6489. Available from http://www.bssaonline.org/content/95/6/2373%5Cn, http://www.bssaonline.org/content/95/6/2373.short%0A, http://www.bssaonline.org/cgi/, https://doi.org/10.1785/0120110286%0A, http://gji.oxfordjournals.org/cgi/, https://doi.org/10.1093/gji/ggv142%0A, http://link.springer.com/10.1007/s00024-01.

Wight, A. J. (2018). Using space tech to improve palm oil transparency in Colombia. *Mongabay Environmental News*. Available from https://news.mongabay.com/2018/09/using-space-tech-to-improve-palm-oil-transparency-in-colombia/.

Yao, H., Qin, R., & Chen, X. (2019). Unmanned aerial vehicle for remote sensing applications – A review. *Remote Sensing*, *11*(12), 1–22. Available from https://doi.org/10.3390/rs11121443.

Zheng, J., Fu, H., Li, W., Wu, W., Yu, L., Yuan, S., Tao, W. Y. W., Pang, T. K., & Kanniah, K. D. (2021). Growing status observation for oil palm trees using Unmanned Aerial Vehicle (UAV) images. *ISPRS Journal of Photogrammetry and Remote Sensing*, *173*, 95–121. Available from https://doi.org/10.1016/j.isprsjprs.2021.01.008.

CHAPTER

Prediction of plant pigments for phytosanitary and yield estimation 14

Ayşenur Özcan[1], Uğur Murat Leloğlu[1] and Mehmet Lütfi Süzen[1,2]

[1]*Graduate School of Natural and Applied Sciences, Geodetic and Geographic Information Technologies Department, Middle East Technical University, Ankara, Turkey*

[2]*Geological Engineering Department, Middle East Technical University, Ankara, Turkey*

14.1 Introduction

Agriculture, having a long history dating back thousands of years, played a crucial role in the evolution of human civilization, as tamed species farming resulted in food surpluses that allowed humans to dwell in cities. Agriculture is critical to a country's economic and social well-being, regardless of its degree of development. There appears to be no alternative supply of raw materials and food for this industry, which is essential for human nourishment. As a result, it is common for the agricultural sector to account for a significant portion of overall employment. Consumption per capita grows as incomes rise and commerce expands throughout the world.

Climate change is diminishing the capacity of agriculture to meet the world's growing population's food demands. As a result, agricultural yield, plant health, and global warming are interconnected issues that must be tackled concurrently. The primary activities for addressing these concerns are increasing resource efficiency in agriculture and creating climate resilience. The increase in the world population, which is anticipated to reach 10 billion people by 2050 (Suzuki, 2019), necessitates an increase in healthy crop production. Crop yields prediction is one of the most prominent topics in literature in remote sensing and agriculture economics.

Before harvesting it is essential to estimate the yield over large areas, because government policies and plans can be implemented based on these data. Agriculture also has a load on transportation because the goods are usually transported by road or rail. A country's stable agriculture leads to stable food safety. An early estimate of yield not only assists the private sector to plan the storage, importation, and export of crops and associated goods but also helps international food producing organizations and farmers in planning their next crops and in ordering the proper quantity of seed in advance as well. Hence, accurate and timely yield estimations improve the agricultural system's overall efficiency.

Remote Sensing in Precision Agriculture. DOI: https://doi.org/10.1016/B978-0-323-91068-2.00002-3

As academics began using satellite images, numerous studies have been conducted and numerous models have been proposed to reliably estimate agricultural yields before harvest. However, there is still room to do more in this area. The absence of information from satellite images due to a large percentage of cloud cover during the growing season, the low temporal resolution of most freely available multispectral (MS) images, and the technology to be used in these studies not having been developed quickly enough are the main causes for this regrettable finding. These difficulties make existing agricultural production estimation methodologies less effective.

Phytosanitary conditions and climatic circumstances are crucial in agricultural settings. Diseases and pests are the most common causes of phytosanitary condition of the crop damage, as are drought, excessive precipitation, frost, and freezing, all of which are linked to climate change. Although wheat varieties have shown to be quite adaptable to many agro-ecological zones, the change of conditions within the same region is likely to make people suffer in terms of obtaining sufficient yield. As a result, crop management and strategy design require real-time and intelligent monitoring of plant health and development, as well as wheat quality assessment. Therefore, today's technology must have early warning systems that can estimate the yield and warn of plant health problems before the harvest.

Researchers are willing to find accurate crop estimates so that countries can plan ahead of their economies in accordance with those crops, that is, money for the import of goods not being produced within the country or how much can they export, under the agreements between them and other countries, products such as wheat, rice, tea, and sugar. Agriculture is the primary source of livelihoods, especially in developing countries and, although agricultural work cannot be carried out all year round, is a source of income that contributes to the household economy all year long, even if only seasonally.

Agriculture products are the most common source of humanitarian assistance. Humanitarian organizations, therefore, invest in GIS and remote sensing technology and human power solely to get close enough insight into the crop's situation and information on production, so as to be of timely service to those who are in need. Time is essential in life and in all fields, which affect human lives directly. It is therefore very important that we evaluate the return sooner rather than later that the parties involved can take all precautionary measures and decide on their roadmap for their near future planning. For sustainable agriculture, two related topics, namely, crop health estimation and crop yield estimation problems, should be addressed, among others.

Combining remote sensing and machine learning is seen to have great potential and a wide range of applications in these fields (Zheng et al., 2021). Therefore, to address the aforementioned issues, researchers have attempted to use high-temporal-resolution and low-spatial-resolution photos, as well as agrometeorological data and different indices acquired from remote sensing sensors, but yield predictions have remained inaccurate. Soft computer algorithms such as Artificial Neural Networks (ANN), Fuzzy Logic (FL), Genetic Algorithm (GA), and Random Forests (RF) have also been used to improve the success of traditional computing techniques by achieving more accurate estimation results. Although these algorithms assisted in making significant progress, a new point of view appeared to be required to address the issue of timeliness and accuracy when estimating yields.

Photosynthetic pigments, namely, chlorophylls, xanthophylls, and anthocyanins, can be explored to find a solution to the accurate yield estimation problem. However, because they are deeply intertwined inside a crop, delicate experiments are required to distinguish one from the other. Some researchers studied the impact of chlorophyll indices on yield, whereas others focused on carotenoids. Photochemical Reflectance Index (PRI), which will be discussed in detail in the following section, is one of the most well-known indices in addition to the other pigment-based approaches. Anthocyanins are the least explored pigment, maybe due to the difficulty of its structure and differentiation. Nonetheless, they are included in this chapter due to their impact and hidden significance on yields. Many researchers have investigated the way to predict yields before harvest using different indices, and they will be reviewed in this chapter together with an algorithm for extracting endmembers from satellite images that may or may not match the photosynthetic pigments in plants at the maximum NDVI value when wheat crop canopy closure is considered to be 100%. The abundances of the endmembers, which most likely represent the three main pigments within the crops, as well as some indices generated from them, were calculated and used as inputs for the generalized linear model (GLM), ANN, and RF to estimate yields at least 1 month before harvest, despite the fact that pigment mixing is most likely nonlinear. Some abundance interactions have also been shown to be good predictors, which may partially address the nonlinearity inherent in intimate mixtures. Ground truth data collected by harvesters are used to test the algorithms' performance.

This chapter reviews the studies in the literature regarding estimation of photosynthetic pigments via remote sensing for the purpose of assessing crop health and estimating crop yields. Wheat yield estimation via remote sensing by improving the timeline of yield estimation in such a way that estimates the yield at least a month before harvest; introducing the different point of views in approaching the yield by taking into account the three photosynthetic pigments, namely, chlorophyll a, chlorophyll b, and carotenoids; and by focusing on the sources and parameters that are easily available or can be computed at no additional expense are introduced.

14.2 State of the art

14.2.1 Photosynthetic pigments and their roles in yield estimation

In the structure of plants, there are several pigments. Chlorophylls, carotenoids, and flavonoids (primarily anthocyanins) are the major pigments, which are also called the photosynthetic pigments found in all forms of plants (Blackburn, 2006; Lachman et al., 2017).

Chlorophylls serve a major part in photosynthesis and are key pigments in the life cycle of all living objects. During photosynthesis, chlorophyll absorbs the visible light and reflects green light more than the others. Photosynthesis is possible by the energy absorbed from blue, red, and green light. Chlorophyll concentrations have an important role in agricultural primary production as they regulate the amount of solar radiation absorbed by the leaves, which leads to photosynthetic potential. Furthermore, chlorophylls make up a large percentage of the leaf nitrogen content, which is a nutritional status indicator for plants. Chlorophylls are light-dependent pigments, and their quantity declines under stress, during senescence, or in low or no light conditions.

One of the other significant pigments in plant chloroplasts is the carotenoids. There are around 600 distinct carotenoids classified as xanthophylls, the most significant of which being lutein (which is oxygen-containing) and carotenes (hydrocarbon-containing but oxygen-free). The carotenoid concentration of durum wheat influences its quality by giving the pasta its yellow color (Lachman et al., 2017). Carotenoids have a dual role of collecting photosynthetic energy (a.k.a. incident radiation) and contributing it to assist chlorophylls as well as photo-protection of chlorophyll when exposed to excessive light (Bartley & Scolnik, 1995). The first is because carotenoids (like chlorophyll b) are accessory pigments for chlorophyll a, which absorbs energy from the antenna (accessory pigments). If chlorophyll a concentration is high, it will accept more energy from chlorophyll b and carotenoids, resulting in increased photosynthetic activity and hence primary production (Chappelle et al., 1992). Carotenoids disperse energy in the xanthophyll cycle in the event of excessive radiation, making the xanthophyll pigment directly connected to photosynthetic light usage efficiency (LUE).

The third category of significant pigments in plants is anthocyanins. Flavonoids are a kind of anthocyanins; however, they are the least studied pigment group in the realm of remote sensing. Anthocyanins have been reported to have a variety of functions, including being an antioxidant (Yamasaki et al., 1997), reducing the excitation pressure, and preventing oxidative damage in the case of oxidative stress (Yamasaki et al., 1997). Apart from that, Steyn et al. (2002) discovered that anthocyanins had photo-protective light-screening properties. Anthocyanins have been shown to modify the light environment within a leaf and regulate photosynthesis in such a manner that photo-inhibition and photo-bleaching are limited (Barker et al., 1997).

The spectral, spatial, and temporal dynamics of these vital pigments in plants can give not only scientific insight, but also practical assistance in agricultural and/or environmental management. Pigment concentrations are also essential phytosanitary markers. Finding their quantities using merely conventional remote sensing devices such as MS satellite images and/or drones would be ideal because they can be acquired quickly and easily, in the name of time and frequency. According to literature, multiple regression, stepwise regression, nonlinear methods, PCA, and ANN approaches are the potential spectral discrimination methods of these pigments. However, using solely linear techniques is not enough to discriminate them, as they have a structure of spectra signatures that may be observed in a plant as can be seen in the absorption spectra of the most important plant pigments (Blackburn, 2006).

From simple band ratios to sophisticated unmixing models, various approaches have been explored in the literature, many of which are explained in this chapter.

PRI is the most well-known carotenoid index (Gamon et al., 1992). It's a narrow-band index that measures variations in xanthophyll pigment epoxidation, which may also represent changes in photosynthetic light response efficiency (Jones & Vaughan, 2010). In other words, it displays photosynthetic LUE and functions as a stress indicator (Gamon, 2010). The formula for PRI is as follows:

$$PRI = (R_{531} - R_{ref})/(R_{531} + R_{ref}) \quad (14.1)$$

where R_{531} and R_{ref} are the leaf reflectance values at 531 nm and the reference wavelength. Although 531 nm is usually regarded literally in the scientific world, it has also been used in the range of 505–535 nm (Bilger et al., 1989), 531 nm, and 535 nm (Bilger et al., 1989). The reference reflectance is generally 570 nm, although it can also be 550, 560, or other wavelengths. The xanthophyll pigments were discovered to absorb the least at 531 nm, and this may be used as the reference wavelength. PRI is more sensitive to environmental changes, parallel to xanthophyll epoxidation status and its impacts on the crop in the shortest timescale than NDVI, as it is designed to reduce the effect of daily sun angle variations (Gamon et al., 1992; Penuelas et al., 1994). These calculations, which require narrow band measurements, can only be done using a limited number of satellites, mostly hyperspectral ones. This is another issue that will hopefully be overcome in the coming years.

The Carotenoid Index (CARI) is another nondestructive method for estimating leaf carotenoid concentration using remote sensing methods (Huang et al., 2018). Within the context of CARI/Red-Edge Chlorophyll Index, a Carotenoid/Chlorophyll (car/chl) Ratio Index (CCRI) was recently suggested. CCRI calculated the car/chl concentration with 54% accuracy using winter wheat leaf level data after calibration and validation (Zhou et al., 2018).

Due to the overlap between the chlorophyll and carotenoid absorption peaks (Fig. 14.1) and the higher concentration of chlorophyll compared with carotenoid in most leaves, estimation of leaf carotenoid content from reflectance was also investigated by Lichtenthaler (1987) and Sims and Gamon (2002), which is much more difficult than estimation of chlorophyll.

Therefore, for the carotenoid–chlorophyll ratio estimation indices have been proven more successful than for the estimate of the absolute carotenoid content (Merzlyak et al., 1999; Penuelas et al., 1995).

Multiple indices were created using band ratios close to (not on) absorption of some pigments, generally chlorophyll wavelengths (Zhang, 2011). Radiative Transfer models, for example, PROSPECT and LEAFMOD, have shown successful performance in numerical inversions in the prediction of chlorophyll content in leaf (Demarez, 1999; Ganapol et al., 1998; Jacquemoud & Baret, 1990; Maier et al., 1999; Renzullo et al., 2006). Numerical inversion techniques have the

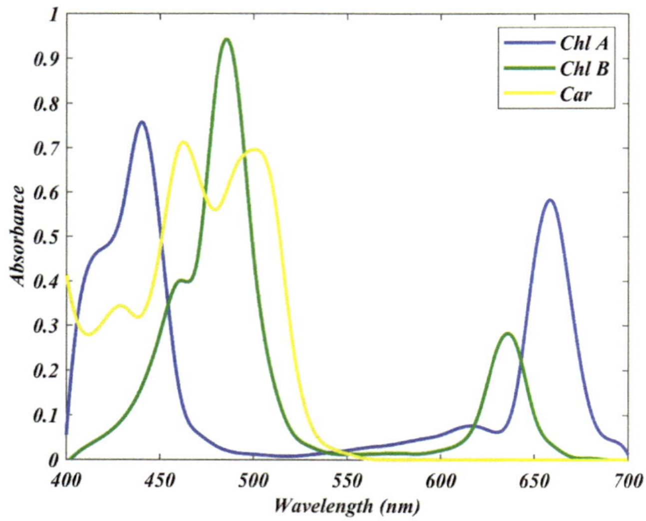

FIGURE 14.1

Absorption spectra of chlorophyll A, chlorophyll B, and carotenoids, interpolated from data samples from plots published in Lichtenthaler (1987).

ability to estimate leaf chlorophyll content from hyperspectral data generally better than spectral indices and other approaches based on empirical calibrations.

There has been little study on the separation of anthocyanins from the entire plant pigment spectrum. The red area where anthocyanins were absorbent in light was investigated and a red/green index was established by (Gamon & Surfus, 1999). However, there was no link to anthocyanins in this index testing. Similar investigations of Gamon, Neill and Gould (2000) have identified the issue as chlorophyll obstructing the discrimination of anthocyanin reflectance spectra. By establishing a 550 and 700 nm narrowband index, Gitelson resolved this difficulty (Gitelson et al., 2001). This new index was found to be effective across a wide range of plant species.

The reflectance spectrum of plants is modified by leaf layers (LAI), proportion of plant covering the ground, regions under shade, and other factors when viewed from canopy level. Using remote sensing methods, a number of studies have investigated pigment discrimination. Single pigments have been attempted to break down at particular wavelengths using hyperspectral data in order to obtain unambiguous spectral separation. The use of reflectance spectra of multiple narrow bands and the construction of indices primarily for the identification of chlorophyll by splitting the values in generally two narrow bands were two of the most popular research studies. There have even been four band indices produced (Thenkabail et al., 2002). Broadband data are insufficient for generating indices, according to Thenkabail et al. (2002), whereas narrow band data have a lot of autocorrelation, resulting in redundancy.

The researchers used their absorption spectra to investigate the function of chlorophyll a, chlorophyll b, and carotenoids in plant photosynthetic activity, after discovering the significance of xanthophylls in photosynthetic activity (Gamon et al., 1992; Penuelas et al., 1994). Sims and Gamon (2002) looked at whether they might extract the chlorophyll content of a leaf despite the presence of all the other pigments. The total chlorophyll content overlaps the absorbance of the carotenoids, suggesting that the total chlorophyll content cannot be used to determine the chlorophyll concentration.

Sims and Gamon (2002) also looked into estimating the content of leaf carotenoids from reflectance, which is more difficult to calculate than the content of chlorophyll due to overlapping chlorophyll and carotenoid absorption peaks and the higher chlorophyll concentration in most leaves compared with carotenoid concentration (Blackburn, 2006). As a result, reflection indices have been more successful in measuring the ratio of carotenoid to chlorophyll than in determining carotenoid absolute concentration (Merzlyak et al., 1999; Penuelas et al., 1995).

Ustin et al. (2009) underlined the relevance of new approaches for identifying overlapping pigment absorptions, claiming that they will lead to substantial advancements in our understanding of plant functions and other plant characteristics. They emphasized that new approaches for evaluating pigment concentration and composition using remote sensing data would give a more improved knowledge of photosynthetic processes.

Little investigation has been performed on separating anthocyanins from the whole range of plant pigments. J. A. Gamon and Surfus (1999) studied the red region, where anthocyanins absorb light and produce a red/green index. The index tests, on the other hand, revealed no link to anthocyanins. In a research comparable with that of Gamon and Surfus, Neill and Gould (2000) identified the issue as chlorophyll obscuring anthocyanin discrimination in the reflectance spectrum. This difficulty was solved by constructing a narrow band index at 550 and 700 nm (Gitelson et al., 2001). This new indicator has been proved to work in a variety of plant species.

These investigations show that, while spectral discrimination of specific pigments is achievable to some extent, separating all pigments at the canopy level from satellite data without the use of sophisticated nonlinear models remains a challenge.

We offer a technique for linearly unmixing the observed spectrum into end-members, which presumably correspond to the pigments and can even be used instead of indices in this chapter (Ozcan et al., 2021).

14.2.2 Remote sensing in agricultural applications and yield estimation

Crop monitoring, which began in the 1970s in industrialized nations, started with the initialization of satellite monitoring. At the end of the 1980s, the United

States (US) began monitoring its own wheat output and subsequently expanded its research to include many other nations' primary crop production. Following in the footsteps of the United States, the European Union developed its own prediction and monitoring system, MARS, at the Joint Research Center (JRC) (EU Science Hub., 2020). Other countries, including France, Germany, Russia, Canada, Japan, and India, and the EU-JRC have developed their own monitoring and forecasting systems. Typically, the NOAA/AVHRR satellite system was used, followed by Landsat. Due to the limited ground resolution of these satellites at the time, a significant quantity of reliable data collected from the field (ground truth) was required.

In agriculture, remote sensing has traditionally focused on the plant reflecting solar radiation and being detected by passive sensors. However, there have been research studies on plant transmittance, absorbance, and emittance. Fluorescence sensing refers to plants that release energy for both photosynthetic and biochemical component functions (Apostol et al., 2003). Thermal remote sensing is focused on fluctuations in evaporation rate based on the plant's temperature response to radiation emission, which provides information on water stress (Cohen et al., 2005). Plant absorption is the polar opposite of plant reflection, and as a result, it also changes with incoming wavelength. Plant chlorophyll absorbs radiation in the visible part of the electromagnetic spectrum (EMS) between 400 and 700 nm, whereas reflectance is strong in the near-infrared range (700–1300 nm) (Pinter et al., 2003). The stark difference between the red and NIR regions of the spectrum prompted the creation of several spectral indices (Mulla, 2013). Simple and complex spectral indices can detect changes in the LAI as well as changes in crop status such as chlorophyll and nitrogen content (Wang et al., 2014).

The Normalized Difference Vegetation Index (NDVI) has been proven to be helpful in forecasting yields (Benedetti & Rossini, 1993; Groten, 1993). NDVI has certain limits, such as the interference of soil at low crop densities and saturation in mature crops with LAI larger than two or three, because the red light absorption peak of the leaves is achieved at these LAI values (Thenkabail et al., 2000). To address these issues, some new indices have been developed, including the Soil Adjusted Vegetation Index (SAVI), the Modified Soil Adjusted Vegetation Index (MSAVI), and the Modified Triangular Vegetation Index (MTVI). However, there will always be a demand for indices that can be used with fewer limitations (Mulla, 2013).

Only Landsat ETM+ pictures were used to estimate wheat yield in the studies of Fortin et al. (2010); Jiang and Thelen (2004); Lobell et al. (2003). They combined their knowledge of crop phenology with multitemporal imagery, adjusting a wheat growth model calibrated locally with instantaneous estimates of canopy light absorption (fraction of Absorbed Photosynthetically Active Radiation—fAPAR) from satellite images to estimate yield at each pixel.

Franch et al. (2019) suggested a novel model in agricultural yield called Differential Vegetation Index (DVI). Landsat 8 and MODIS time series data were used to unmix wheat signals from the signal of other surfaces. Following analysis of the unmixed wheat series, regression equations have been adopted for each administrative unit as yield estimation models, and an R^2 of 0.66 was identified at the country level, while 0.70 at subnational level was seen in the US and Ukraine.

In addition, some research enhances and uses preexisting estimated yield models. Wang et al. (2019) used HJ-1A/B and MODIS data to enhance the CASA NPP prediction model, which was based on absorbed photosynthetically active radiation (APAR) and LUE (ε) absorbed by vegetation, to find 56% accuracy in estimating the production in selected areas of China.

The indices are commonly used to estimate the yields of different crops. Crop yield estimation, particularly before harvest, is critical for agricultural policies and decision-making. Yield estimation is commonly done in the literature using various models. The statistical models, the first category of these models, are typically used when there is information on a vast region, and they can estimate in a wide range of values. For accurate or near-accurate calculations, statistical models are not suggested. The basic principle in crop yield forecasting is using statistical regression, which is considered a common and simple method (Lobell et al., 2010). To use it, a simple matrix is formed with some agrometeorological parameters (not too many) and previous yields, and then a regression equation between yields and the parameters is derived. Multiple linear regression (MLR) is the most used regression model; however, it produces incorrect and unstable results, especially when a high number of parameters are included, and there are correlations between these variables (Lobell et al., 2005). Magney et al. (2017) intended to see if RapidEye spectral VIs could be used to estimate cumulative Nitrogen (N) absorption in wheat and if remotely sensed N uptake maps could be used for precision agriculture (PA) applications. The Normalized Difference Red-Edge (NDRE) index was shown to be the best-performing Vegetation Index (VI). They reported that VIs from RapidEye images may be used to estimate wheat N intake using 17 frequently used spectral VIs. The highest R^2 of the polynomial fit was 0.81.

The mechanistic model is the second frequently used model. These models are far more comprehensive than statistical models. They rely on fundamental soil and plant systems (Dourado-Neto et al., 1998).

The third model is the functional model, which is a more complicated model that can simulate models using daily updated data. Its usefulness is derived from the simplification of complicated procedures. However, if not properly constructed, it may produce findings that are less accurate than mechanistic models (Watt, 2013).

Models can be characterized as deterministic or stochastic in a broader sense. The assumption in deterministic models is that all plants and soils are uniform across the space. And because the parameters are changing, stochastic models use a more realistic approach, recognizing that the outcomes may contain some uncertainty due to soil characteristics, weather conditions, and biotic and abiotic influences. These characteristics make the model valid only in limited regions rather than big areas. As most components of the agro-ecosystem are diverse, the crop growth system is more stochastic than deterministic (Basso et al., 2014).

Remote sensing is a particularly effective technique to sample a large number of plants at the plant size, for purposes such as plant breeding and identifying specific physiological features of cultivars (Jones & Vaughan, 2010). It is also worth noting that while research on remote sensing for PA has been done to evaluate changes of parameters inside fields (Plant, 2001), there do not appear to be many on between-field differences throughout the terrain. The three benefits of yield remote sensing over ground-based methods, according to (Lobell et al., 2005), are:

1. Once yield measurements in the field are eliminated, sample sizes can be increased.
2. Field measurements are gathered using sensors and a restricted number of plots inside fields, resulting in sampling errors limited to within-field variability, whereas remote sensing yields estimates at a considerably broader range of geographic scales.
3. There is a large library of remote sensing pictures that may be used to analyze earlier surveys that did not quantify yield.

Traditional approaches may not be cost-effective, analytical, or give a comprehensive answer when dealing with complex systems. Thankfully, several "inexact" approaches for modeling and analyzing the complicated issues of these complex systems have been established. The term "soft computing" refers to these "inexact" computer approaches (Huang et al., 2010).

Agrometeorological factors, as well as other indices generated from remote sensing sensors, have been used as predictors in the past to estimate yields. Photosynthetic pigments, chlorophylls, xanthophylls, and to a lesser extent, anthocyanins were studied when greenness and soil-driven indicators failed to attain adequate precision in yield estimations. Soft computing methods such as ANNs, FL, GA, and RFs have also been used to improve the success of traditional computer approaches by achieving more accurate estimation results.

14.2.3 Soft computing and its applications in yield estimation

When dealing with complicated systems, traditional approaches may not be cost-effective, analytical, or give a comprehensive answer. Fortunately, several "inexact" approaches for modeling and analyzing these complicated issues have been established. Soft computing refers to these imprecise computer methods (Huang et al., 2010).

Some of the most significant soft computing technologies include FL, ANNs, GAs, Bayesian inference (BI), and decision trees (DTs). The traditional hard computing approaches are stochastic and statistical. Soft computing approaches are flexible and open to inaccuracy, ambiguity, partial truth, and approximation as they are inspired by nature. These approaches can be used alone or in combination to improve the system and outcomes. RF has received a lot of attention in recent years, in addition to these traditional techniques.

RFs, Neural Networks (NNs), and the GLM are the soft computing approaches used in this chapter. RF is a supervised learning technique for classification and regression that uses an ensemble learning approach. To enhance predictions, an ensemble method integrates guesses from various machine learning algorithms. The RF, as a metaestimator, constructs many DTs during training and generates the mean prediction (regression) of the individual trees. Individual DTs have a proclivity toward overfitting.

The most well-known form of machine learning algorithm is the NN, which is modeled after the human brain and allows the computer to learn by integrating new data.

GLMs are a type of nonlinear model that describes a nonlinear connection between a response and its predictors. The patterns of interactions and linkages are defined by the model's structural structure. Measurements of association intensity are included in the model parameters.

Linear and nonlinear models were created and assessed to estimate agricultural production using satellite data (Sayago & Bocco, 2018). To obtain soybean and maize yield in the central area of Córdoba, these models were suggested and implemented by using Landsat and SPOT data (Argentina). This study found that Landsat 8 and SPOT 5 photos may be used to accurately estimate maize and soybean yields from early to mid-season crop growth. To make the spatial resolutions of Landsat 8 and SPOT 5 comparable, the pixel size from Landsat 8 was modified to SPOT 5 (each Landsat 8 pixel was divided into nine parts with the same attribute value). The MLR findings were nearly identical to the ANN results (Soybean: $R^2_{NN} = 0.9$ $R^2_{MLR} = 0.82$, Corn: $R^2_{NN} = 0.92$, $R^2_{MLR} = 0.88$).

Agricultural meteorological and remote sensing data are combined in integrated yield models (Basso & T.R. McVicar, 2007; Basso et al., 2001; Dorigo et al., 2007). An example of an integrated method is the use of Principle Component Analysis (PCA) and Factor Analysis (FA) in conjunction with multiple regressions. Various soil nutrients were attempted to be predicted using integrated models.

A model for predicting wheat yield was created and tested (Pantazi et al., 2014). Three ANNs for yield prediction were fed into fusion vectors including the values of eight soil characteristics, historical yield data from the previous 2 years gathered with an online soil sensor, and NDVI values derived from satellite images. In the given method, they used Self-Organizing Map Models, such as Counter-Propagation Artificial Neural Networks, XY-Fused Networks (XY-Fs), and Supervised Kohonen Networks, which incorporate the component that limits the yield in a multilayer fusion model. (Wu et al., 2015) created two inversion models and compared them using the regression model and the Back-Propagation Neural Network (BPNN) model to estimate the LAI of a temperate meadow steppe in China. The BPNN technique (accuracy: 82.2%) beat the Statistical Regression model, according to the comparative findings (accuracy: 78.8%).

In Maryland, the creation of ANN models was presented as a real approach for forecasting maize and soybean yields in nutrient management planning (Kaul et al., 2005). The findings revealed that the ANN yield forecast was more accurate than the MLR yield model. The accuracy of ANN-based corn prediction ranged from 77% to 90%, whereas MLR findings were only 42% accurate. (Kang & Özdoğan, 2019) used an Ensemble Kalman Filter to disaggregate one country-level (US) maize yield data into a 30 m Landsat resolution yield map using LAI time series data calculated from Landsat images, EVI, meteorological, and soil texture data. When they compared their findings to farmer-reported production statistics, they discovered that the correlation coefficient (R) was at its maximum of 0.82.

In recent years, RF has received a lot of attention for its resilience to overfitting issues and dataset noise. Because it can ignore spatial autocorrelation, it is virtually unaffected by the multicollinearity problem. It may also be used to boost the effectiveness of other approaches such as regression and kriging. As shown in the following experiments, RFs outperform ANN in many situations and are unaffected by dataset size.

Cai et al. (2019) compared the LASSO regression model, SVM, and RF performances using climate, satellite (MODIS EVI), and chlorophyll fluorescence (SIF from GOME-2 and SCIAMACHY) data. With an R^2 of 0.80, SVM outperformed the others. To predict maize yield in their research region, Leroux et al. (2019) integrated MODIS NDVI, MODIS LST, and SMOS SSM with outputs of the SARRA-O crop model. When the 10-fold cross-validated GLM and RF performances were evaluated, it was discovered that RF outperformed GLM and predicted 46% of the reported end-of-season yields 2 months before harvest.

Heremans et al. (2015) used NDVI from the SPOT-VEGETATION sensor, climatic factors, and fertilization levels to evaluate two regression tree techniques, namely, Boosted Regression Trees (BRT) and RF, to assess the accuracy of winter wheat yield. The R^2 for both techniques was more than 0.80, indicating that BRT was sensitive to noise, prone to overfitting, and significantly slower than bagging. In terms of precision, RFs were equivalent to boosting, but without the drawbacks. Using different regression techniques and hybrid geostatistical methods, (Li, Wang, et al., 2016) achieved reliable and timely forecasts of grassland LAI. The findings indicated that for regression models including Partial Least Squares Regression (PLSR), ANNs, RFs, and Regression Kriging, the RF model produced the best accurate predictions (RK). All of the beneficial aspects of RF were demonstrated by Li, Wang, et al. (2016). The R^2 was estimated as 0.77 for PLSR, 0.81 for ANN, 0.89 for RF, 0.92 for RK, and 0.91 for RFRK for various techniques. Guo et al. (2015) evaluated two methods for predicting and mapping the geographical distribution of soil organic matter in a rubber plantation, namely, Stepwise Linear Regression (SLR) and Random Forest Residual Kriging (RFRK). The RFRK model was shown to not require any assumptions about the relationships between the target and predictor variables. These connections might be nonlinear or hierarchical in nature. The R^2 values for SLR, RF, and RFRK were determined to be 0.43, 0.65, and 0.86, respectively. Above-Ground Biomass

(AGB) is calculated using 54 vegetation indicators and eight statistical regression approaches by Yue et al. (2018). Their findings revealed that, among the eight methods studied, PLSR and GLM perform the best in terms of stability and are the most appropriate when high-accuracy and stable estimates are required from a small number of samples. RF has also been found to be very noise resistant, making it suitable for dealing with recurrent observations involving distant sensing data. The results indicated that when using multicollinearity data to predict AGB, GLM performed badly. The models that were least influenced by the problem of multicollinearity were ANN, BBRT, and RF. Their findings revealed that PLSR, GLM, and RF are suitable for work that requires high-precision estimate models.

Hunt et al. (2019) used RF regression to predict within-field yield variability in 39 wheat fields in the UK using Sentinel-2 and environmental data such as meteorological and soil characteristics. They used data from harvester yield monitors as ground truth and discovered that Sentinel-2 data can estimate within-field production variability; however, integrating satellite data with environmental data improved accuracy. They also found that RF beats basic linear regression based on VI.

Mulla (2013) identified a number of future requirements in remote sensing using soft and hard computing approaches. He stated that additional research is needed in precision agricultural applications using chemometric or spectral decomposition/derivative techniques of analysis, as well as the creation of novel sensors to assess nutrient shortages without the need of reference strips. He also emphasized the necessity for new spectrum indices to analyze different crop properties (e.g., LAI and biomass) and highlighted the need of developing more spectral indices (e.g., water and N; weeds and insects). Historical collections of satellite remote sensing data with moderate-to-high spatial resolution and traditional spectral resolution should be integrated with high-spatial and -spectral-resolution real-time remote sensing data to improve precision agricultural decision-making. Soft computing has also been used in the literature to calculate yields. Huang et al. (2010), on the other hand, claimed in their study that there were no applications for the merger of soft and hard computing approaches, but that this might be a worthwhile research issue. Huang et al. (2010) offer a system that addresses this need by monitoring crop output during the growing phase and alerting the producer or decision-maker before harvesting so that precautions may be taken to enhance yield or alter pricing.

The research described in this work seeks to address the first component of Mulla's future demands, as well as the use of soft computing for early warnings indicated by Huang et al. (2010).

14.3 Materials and methods

Researchers have investigated the plant health based on pigments with regard to time of sowing, the insect effects, and the climate.

Ma et al. (2018) studied the effect of powdery mildew, one of the dominant diseases in winter wheat, by designing a powdery mildew detection method based on multitemporal satellite data for regional disease mapping. They collected Lansat-8 scenes covering six winter wheat growth periods (numbered 1–6) to calculate typical vegetation indices (VIs), such as the disease water stress index (DSWI), optimized soil adjusted vegetation index (OSAVI), shortwave infrared water stress index (SIWSI), and triangular vegetation index (TVI). The regional disease distribution was then created using a multitemporal VIs-based k-nearest neighbors (KNNs) technique. Meanwhile, the optimal multitemporal combination for the KNN monitoring model was confirmed using a backward stepwise elimination strategy. The initial results were compared and validated using the categorization and regression tree (CART) and BPNN techniques. They obtained 84.6% accuracy, which was higher by 11.5% and 3.8% than the multitemporal image-based CART and BPNN models, suggesting that using satellite images during the early stages of a major disease infection is vital to increasing the accuracy of monitoring models.

Roumenina et al. (2020) presented a method for evaluating and mapping the condition of winter wheat crops using a set of crop variables derived from MS imagery, namely, Sentinel-2, such as the LAI, fAPAR, fraction of vegetation cover (fCover), fresh above-ground biomass (AGBf), and nitrogen uptake. To begin, they graded the crop condition on a qualitative three-grade scale for each variable. In a second stage, these individual assessments were integrated to create an assessment map of the crop's overall status, which distinguishes between three possible states: Good, Fair, and Poor. In two agricultural years—2016/2017 and 2017/2018—the approach was tested on winter wheat fields in Bulgaria at several phenological growth stages. In a number of elementary sampling units (ESUs), the remotely sensed crop status (phenological growth, crop height, phytosanitary, etc) was compared with independent ground-based evaluations. The proposed method was successful, with crop condition being reliably determined in 87%–94% of ESUs, depending on the FGS/agricultural year.

Traditional methods (Rawashdeh, 2015; Sala et al., 2016; Shewry, 2009), along with thorough processing of the results (Krzyśko et al., 2013) and the development of prediction models (Sala & Boldea, 2011), were used to assess the state of vegetation and nutrition of plants, production, and agricultural productivity in the wheat crop, with good results and high precision in estimation.

Cristian et al. (2018) characterized 18 different grain varieties (wheat, barley, and triticale), as well as to test a technique for estimating chlorophyll content and fresh biomass from aerial pictures. They retrieved the spectral data in the RGB spectrum from digital photographs coming from a drone. The normalized values of RGB and the values of the HSB system were calculated on their basis, and the indices Intensity Index (INT), Normalized Difference Index (NDI), and Dark Green Color Index (DGCI) were found using the RGB and HSBs. The DGCI index ($r = 0.846$, p0.01, F = 40.21) and fresh biomass ($r = 0.772$, p0.01, $F = 23.60$) were both positively linked with chlorophyll (Chl). When the correlation coefficient values for several types of regression functions that characterized the link between DGCI and Chl were analyzed, the correlation coefficient values were nearly similar (0.84), and the Sigf values were all less than 0.001.

Messina et al. (2020) monitored the crop growth, guide localized fertilization, phytosanitary treatments, and harvest of onion crops to support PA techniques implementation by using MS data gathered at different spatial resolutions by an unmanned aerial vehicle (UAV) and satellite platforms PlanetScope and Sentinel-2 to track the deployment of onion crops. The study field's vigor was measured using the SAVI. Following that, statistical analysis was used to compare the vigor maps from the two satellite platforms with those obtained from the UAV to assess the contribution of each platform to onion crop monitoring. They discovered that the images provided by UAVs were crucial in guiding agronomic localized operations including fertilization and phytosanitary treatments.

The wheat crop as well as other crops has been investigated in terms of the complex interactions of genetic, physiological, environmental, pedo-climatic, and economic components to describe and model the wheat biological system (Casadebaig et al., 2016) throughout the years. Some researchers have evaluated economic indicators for wheat harvests in terms of the use of restricted fertilizer supplies (Boldea & Sala, 2010; Sala et al., 2016) to optimize crop technology under technological constraints.

Nevertheless, due to the complexity of the pigment structure within the crops, the discrimination of the pigments was mostly done in laboratory conditions or in small geographical areas, although it is known that it may be much more useful if they could be unmixed and used via multitemporal, MS satellite images that are publicly available.

The upcoming sections of this chapter will be concentrating on one method that was developed by the authors to unmix the pigments within wheat and use AI techniques to obtain the yield from these unmixed structures together with some crucial agri-environmental parameters generally used to estimate the yields.

14.3.1 The data and the study area

The yield map of the harvester combines with yield mapping technology, and barley and wheat products were generated in the framework of a "Smart Agriculture" project carried out by the Turkish Ministry of Agriculture and Forestry (Sönmez et al., 2015). These cartographic maps have been transmitted to GIS for usage with electrooptical and SAR imagery. The combined harvesters equipped with precise measuring, scaling, and recording equipment obtained immediately georeferenced yield figures. The combined capacity of the harvester was around 6 tons, and its width was 5 m. With the combined harvesting system moving at a speed of approximately 7 km/h, the system moves 2 m in 1 second and hence 2×5 m in grids. The raw data for harvested results were used for the preparation of coordinated yield distribution maps in the yield program.

In combination with a DGPS receiver, the efficiency system in the display was built. During the procedure, data including location information were captured at 1 second intervals.

14.3.1.1 Yield data

As part of a project spanning fields from different areas of Turkey, a harvester system for 2015 was employed. The harvester with GPS equipment is integrated with high-tech sensors that weigh the product automatically. The Ministry of Agriculture and Forestry has disseminated these records to enable researchers to access the specific locations and yields of the harvester's areas. In 31 towns in six Turkish provinces, 142 wheat fields were selected as fields of research. All these fields were rainfed fields with diverse climatic conditions in different regions of Turkey.

Fig. 14.2, showing the picture of a harvested field, illustrates the field direction for harvesting the stripes shown in the figure. In the popup, the speed of the harvester is shown during the procedure.

The exact boundaries and coordinates of the agricultural field polygon were retrieved from Google Maps to match the fields on satellite imagery to harvester yield statistics. Fig. 14.3 illustrates the places where JRC collected the agrometeorological data in 142 chosen fields in 31 provinces of Turkey and the station sites.

14.3.1.2 Meteorological data

Agrometeorological data, which are either meteorological data or data derived from meteorological data, and used in agricultural research to obtain qualitative and quantitative improvements in agricultural production, are typically required for yield estimation (WMO, 2010). If the area of interest (AOI), such as a crop field, is near a meteorological station, data obtained from that station can be immediately used. It is ideal to interpolate the agrometeorological data if the AOI

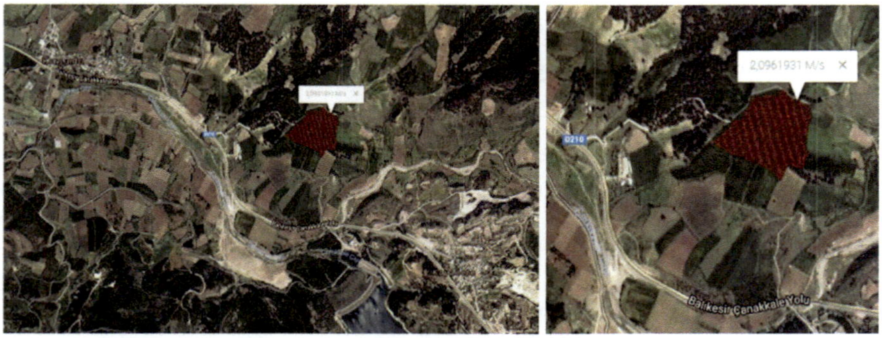

FIGURE 14.2

Google Earth Base picture overlays GPS data of harvester generating yield grids. The right picture is the magnified left image for seeing the harvester speed information.

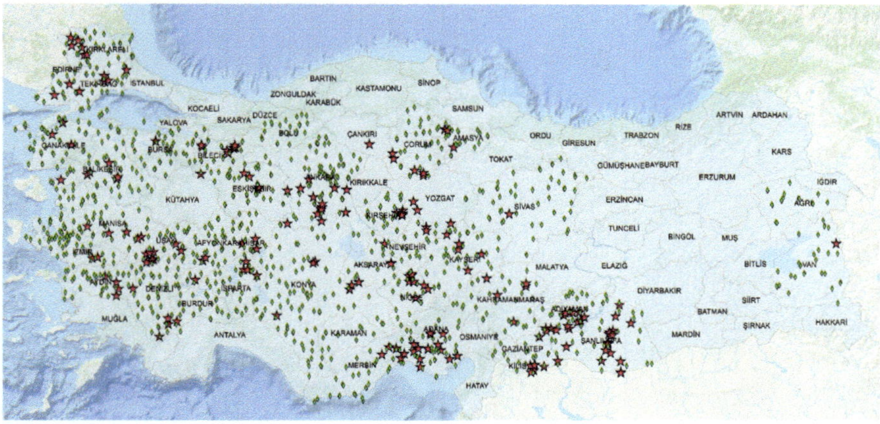

FIGURE 14.3

The 142 fields scattered throughout 31 provinces in Turkey. The green diamonds are the meteorological stations, and the red stars represent the fields.

is located between numerous meteorological stations. The goal of this research is to provide a generic method to yield estimate so that individuals who work in this field may come up with more general solutions. As a result, the JRC data (Andrea, 2014) were used instead of meteorological station data as in the research of Fernandes et al. (2011) and Salvador et al. (2020).

The JRC gathers data from 117 meteorological stations in Turkey on air temperature, precipitation, radiation, air humidity, and wind speed. The data are verified for inconsistencies, inaccuracies, and duplicates, and only then are the results transformed to daily values that fit within the station's uniform weather database. Some factors, such as solar radiation or evapotranspiration, are also added to the database to generate the measured data. These data, on the other hand, are gathered from stations that are located at certain points, resulting in an uneven distribution and density. To transmit these data to places between stations, a conversion is necessary. JRC uses interpolation, aggregation, and analysis, as well as side-by-side grids of 25 km by 25 km that span the whole AOI (Weather Monitoring – Agri4castWiki). The entire number of agrometeorological parameters acquired from JRC and computed, as well as vegetation indices used in the literature (MSAVI, MTVI, etc.) and abundances, as well as their interactions, totals 24. (Table 14.1). It may not be essential to use all of the factors as some are irrelevant or significantly linked with others.

14.3.1.3 Satellite data

Landsat 8 satellite images were used in this investigation because of their worldwide coverage and temporal resolution. The Landsat 8 satellite has a 16-day temporal resolution and 11 bands; however, this study only employed the first

Table 14.1 Names and abbreviations of all the parameters and interactions.

Abbreviation	Parameter
A1	Abundance 1
A2	Abundance 2
A3	Abundance 3
A4	Abundance 4
A1A2	Interaction of abundances 1 and 2
A1A3	Interaction of abundances 1 and 3
A1A4	Interaction of abundances 1 and 4
A2A3	Interaction of abundances 2 and 3
A2A4	Interaction of abundances 2 and 4
A3A4	Interaction of abundances 3 and 4
T_MAX	Maximum temperature (°C)
T_MIN	Minimum temperature (°C)
VPD	Average vapor pressure deficit (hPa)
E0	Potential evapotranspiration of open water (mm/day)
SOILFREEZE	Total number of days the soil temperature was below or equal to 0°C (day)
NoPRECIPITATIONdays	Number of days there was no precipitation until harvest (day)
RADIATION	Average radiation ($KJ/m^2/day$)
WINDSPEED	Average speed of the wind at 10 m (m/s)
Cons_noPrec	Consecutive no precipitation days until harvest
PTU	Photo thermal unit
NDVI	Normalized difference vegetation index
MTVI	Modified triangular vegetation index
MSAVI	Modified soil-adjusted vegetation index
EVI	Enhanced vegetation index

five bands. Table 14.2 lists the characteristics of these bands. Landsat 8 pictures of the specified locations were obtained and processed from the https://earthexplorer.usgs.gov/ website between April and June 2015. At the 0% cloud coverage requirement, at least two distinct pictures were discovered for each field. For each field, the image with the greatest NDVI value was chosen as the input image to be used in the subsequent procedures. It is assumed that the plant cover is about 100% at this stage of crop development, and that soil contribution to the spectrum is minimal.

14.3.2 Methodology

The production of satellite pictures and extraction of specified fields from the appropriate satellite images are the initial stage. The endmembers are then

Table 14.2 Characteristics of Landsat 8 satellite bands used in this study and their properties.

Bands	Wavelength (μm)	Spatial Resolution (m)
Band 1—Coastal aerosol	0.43–0.45	30
Band 2—Blue	0.45–0.51	30
Band 3—Green	0.53–0.59	30
Band 4—Red	0.64–0.67	30
Band 5—NIR	0.85–0.88	30

NIR, *Near infrared.*

identified and analyzed at the peak of the NDVI before being found in abundance in the fields. To forecast yields, GLM and RF algorithms are trained on abundances, NDVI, and chosen agrometeorological factors. The specifics can be found in the subsections.

14.3.2.1 Transformation of absorption to reflection spectra

The mean reflectance spectra of photosynthetic pigments (Fig. 14.4) are calculated from Lichtenthaler (1987) absorbance values using Gitelson and Solovchenko's conversion factors (Gitelson & Solovchenko, 2018).

The absorption spectra of photosynthetic pigments, among other things, are provided by Lichtenthaler (1987). However, the reflectance spectra of pigments are required for this investigation, as well as maybe some other remote sensing studies, to analyze the spectra of pigments and other materials in satellite pictures. Even yet, converting absorbance to reflectance is not an easy task. Although there are complex mathematical methods for estimating reflectance from absorbance (Dawson et al., 1998; Jacquemoud & Baret, 1990), in this work, we choose to use empirical data. As a result, Gitelson and Solovchenko (2018)'s absorbance and reflectance values for Virginia creeper leaf were used.

The values were included in a MATLAB® function that uses these values in the transformation formula for each wavelength where the photosynthetic pigments are efficient (400–700 nm):

$$R_{\text{pigment}}^{-1}(\lambda) = \frac{R_{\text{leaf}}^{-1}(\lambda)}{c \quad A_{\text{leaf}}(\lambda)} A_{\text{pigment}}(\lambda) \tag{14.2}$$

where λ is the wavelength, R_{pigment} is the reflectance of the pigment, A_{pigment} is the absorbance of the pigment, R_{leaf} is the reflectance of the leaf, A_{leaf} is the absorbance of the leaf at λ (Table 14.3), and c is a constant that aligns the range of the two absorbance values, as the absorbance units are different in the two sources.

The bandwidths of the first four bands of the Landsat 8 satellite were used to analyze the newly discovered reflectance spectra of each pigment. In the

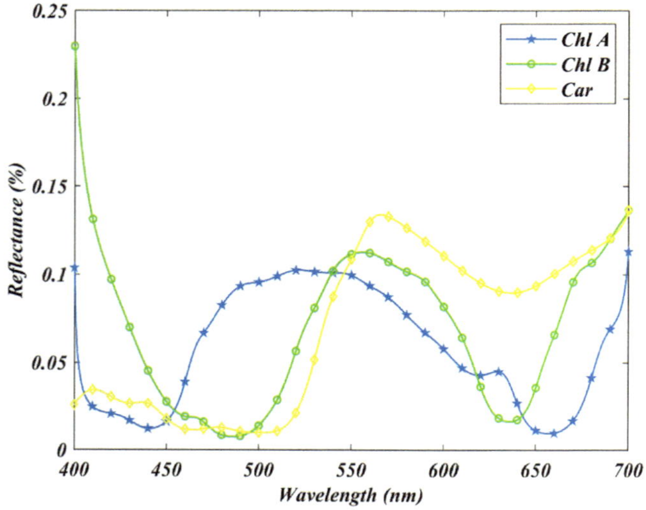

FIGURE 14.4

The absorption spectra in Fig. 14.1 were used to generate the reflection spectra of the main pigments.

Table 14.3 The estimated absorbance and reflectance values of Virginia creeper leaf as graphs in Gitelson & Solovchenko (2018).

Wavelength (nm)	Absorbance values A_{leaf} (λ)	Reflectance values R_{leaf} (λ)
400	2.25	11.5
425	2.1	12
450	1.9	12.5
500	1.5	13
550	0.8	6
600	1	9
650	1.3	14
680	1.75	15.5
700	0.8	6
750	0.35	2

passbands, each band's transmissivity is considered to be unity, and the reflectance spectra are simply combined in each band to produce the reflectance values observed by Landsat 8. Fig. 14.5 depicts the flowchart for this procedure.

The final reflectance spectra of the pigments produced using the methods described earlier are shown in Fig. 14.4. The general work flow is shown in Fig. 14.6.

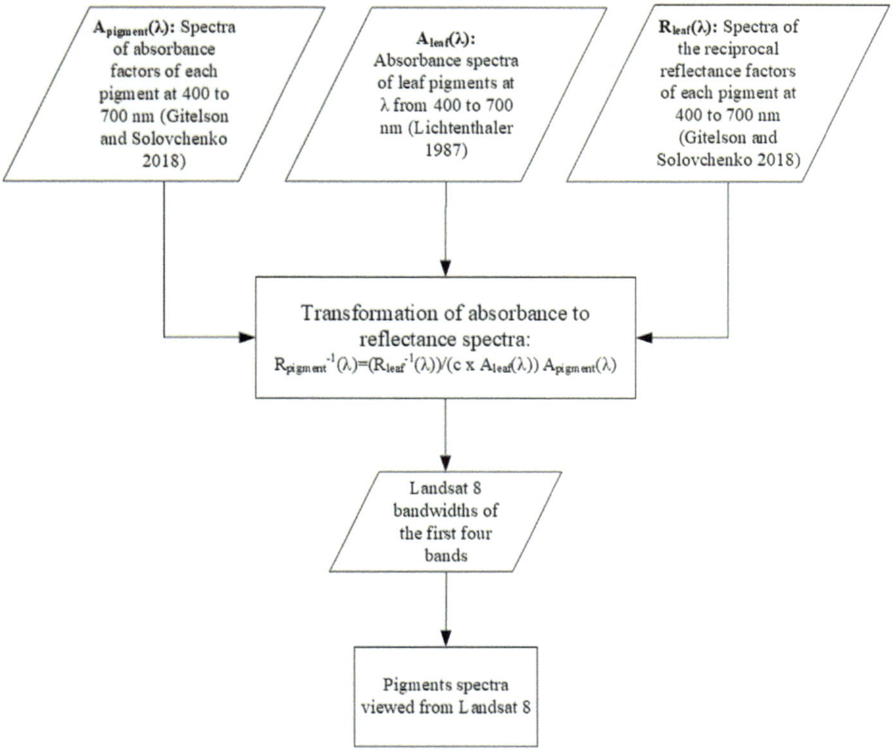

FIGURE 14.5

The transformation of photosynthetic pigment absorbance spectra to reflectance spectra (Özcan, 2020).

14.3.2.2 Preparation of the data

FLAASH in ENVI 5.3 was used to correct all Landsat 8 images of the region of interest for radiometric and atmospheric effects. After being geometrically and radiometrically adjusted, 141 fields were extracted from Landsat 8 images. For each field, the image with the greatest average NDVI value was chosen and used as the dataset for endmember extraction and abundance computation.

In this study, two agrometeorological indicators, *NoPRECIPITATIONdays* and *Cons_noPrec*, were used as indices to see if they had any influence on the assessment of wheat yields. *NoPRECIPITATIONdays* days is the total number of days without rain in the field area from sowing to harvest, and *Cons_noPrec* is the consecutive number of days without precipitation from sowing to harvest, which helps to predict the accurate yield in the event of drought. It is also a reliable indicator for determining whether there are several days without rain in a row and whether or not it has an impact on the yield during that time period.

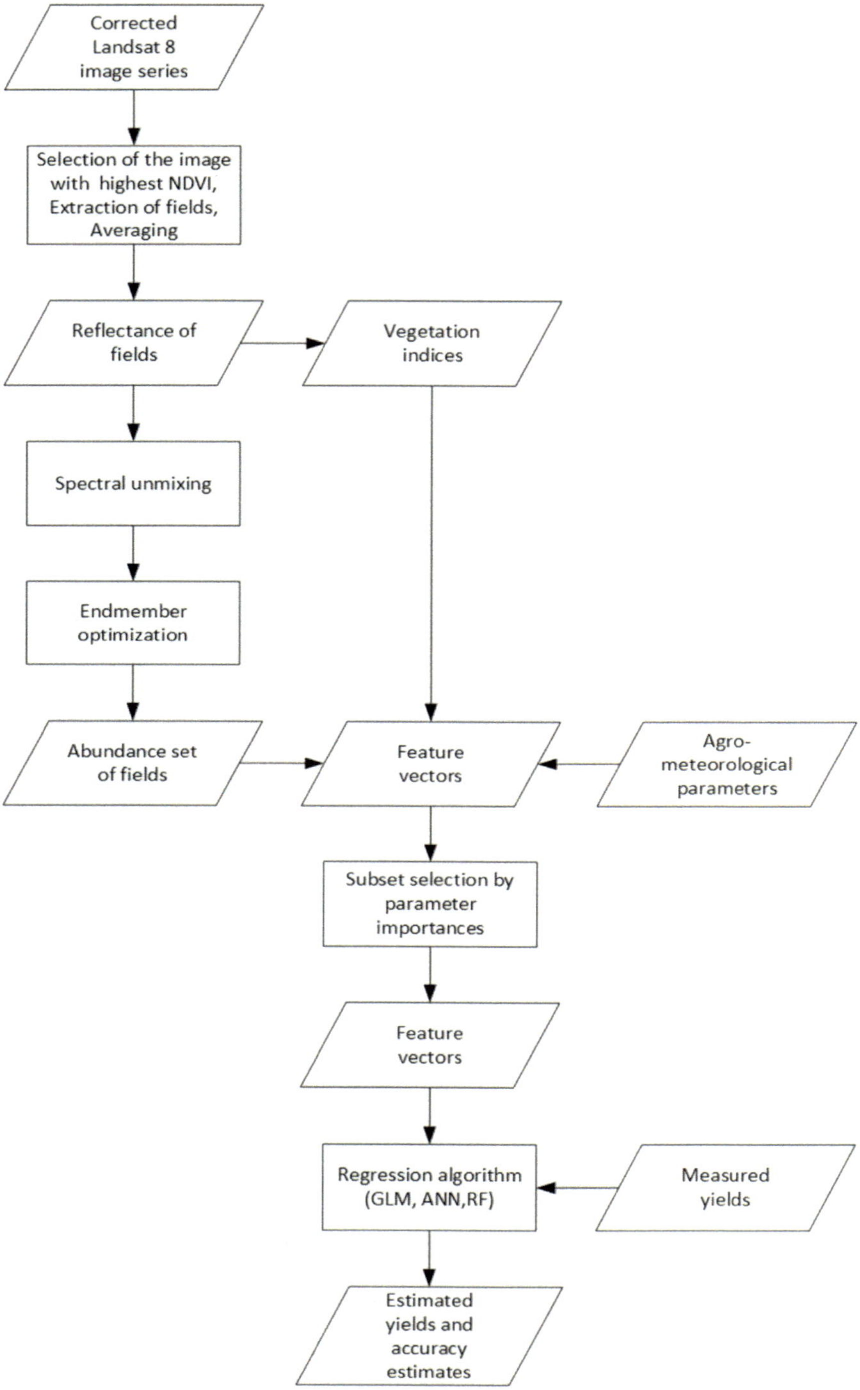

FIGURE 14.6

The flowchart of the proposed algorithm in Özcan (2020).

14.3.2.3 Spectral unmixing

The pixel sizes of MS sensors are large enough to hold a variety of materials, causing the extraction of required material ratios to become crucial for research reasons. The Spectral Unmixing Algorithm is the typical extraction algorithm (Keshava & Mustard, 2002). The mixed pixel is considered to be made up of a collection of constituent spectral signatures (endmembers) weighted by the subpixel fractional cover (abundances) in linear spectral unmixing. In the literature, the unmixing algorithms are created and applied to hyperspectral pictures, and it should be noted that all the scholars listed in the following have applied their algorithms to hyperspectral data.

Only photosynthetic pigments, which are intimately mixed within the plant, are expected to be present in crop fields. When working with intimate mixtures rather than broad scenes of satellite images, endmember extraction usually necessitates the use of sophisticated nonlinear methods. As a result, extracting the endmembers from an intimate combination of photosynthetic pigments necessitates approaches other than linear ones, as well as being user-friendly, which nonlinear techniques may not be. The endmembers are used as inputs in the following phase of the process, which is to generate abundances, which are used as inputs in the yield prediction technique.

The extraction of required components has become crucial for research reasons as the pixel sizes of MS sensors are too big to hold diverse components. Spectral unmixing is used in linear spectral unmixing (Keshava & Mustard, 2002; Somers et al., 2011), in which it is assumed that a mixed pixel consists of a collection of component spectral signatures (endmembers) weighted by a fractional subpixel cover (abundance), as indicated in Eq. (14.3).

$$Y = EA + \varepsilon \tag{14.3}$$

where $Y_{d,n}$ is composed of observed n pixels with d bands, $E_{d,m}$ is the matrix where each column is an endmember, $A_{m,n}$ is the abundance matrix with each column representing the fractional cover occupied by the endmembers for the corresponding pixel, and ε is the error, which is the portion of the spectrum that cannot be modeled using the endmembers (including sensor noise, endmember variability, and other model inadequacies).

Normally, the LMM is constrained by two factors:

1. Nonnegativity constraint: All abundances must be nonnegative. $A \geq 0$ and
2. Full-additivity constraint: $1_m^T A = 1_n^T$.

Crops should ideally be made up entirely of photosynthetic pigments that are intimately mixed throughout the plant. To extract the endmembers from these close mixtures, advanced nonlinear methods must be used rather than linear ones. The endmembers are the inputs for the following processing phase, which creates the abundances, which are the yield prediction method's inputs.

The initial assumption in a multivariate system is that the variables are linearly linked. However, it is theoretically plausible that in certain cases, a second predictor variable Z is actually regulating the impact of a predictor variable *X* on a criterion variable *Y*. Aside from the linear effects 1 and 2, an interaction effect 3 is introduced into the model structure here. To analyze the interaction effect in conjunction with the linear effects in the regression equation, a new variable must be created, namely, the product XZ between the predictors X and Z, which will be included as the third term in the multiple regression equation (Dimitruk et al., 2007).

$$Y = \beta_0 + \beta_1 X + \beta_2 Z + \beta_3 XZ + \varepsilon \tag{14.4}$$

In Eq. (14.4) (Kenny & Judd, 1984) Y is the criterion variable, *X* and *Z* are the predictor variables, XZ is the interaction term, $\beta 0$ is the intercept, $\beta 1$ and $\beta 2$ are the linear effect terms, β3 is the interaction effect, and ε is the error.

Interaction terms proved to be rather useful in the solution of multivariate nonlinear problems. Nonlinear interactions in the mixing phase are rarely driven by material distortions, but rather by nonlinear interactions between them (Klein & Moosbrugger, 2000; Suzuki et al., 2009).

The importance of spectral unmixing has lately been recognized in hyperspectral research. However, unmixing methods have only seen limited use in MS investigations to date, and researchers generally do the unmixing across a broad picture region with a wide range of land cover types. Only plant pigments are predicted to be the endmembers if the scene is an agricultural field with complete covering leaves. The typical LMM performance would be restricted since these pigments create intimate combinations.

Standard unmixing methods in the literature conduct unmixing over a large region of an image. PCOMMEND is a novel method to linear unmixing that is not the same as traditional linear unmixing (Zare et al., 2013). The approach was designed to discover several sets of endmembers, demonstrating that it is a superior representation of hyperspectral images. In contrast to traditional models, PCOMMEND may do iterative fuzzy clustering while also performing spectral unmixing to split a mixture (pixel) into several areas of the space specified by the endmembers. A different collection of endmembers that characterize a simplex exists in each of these areas. This represents all of the pixels in the image as a union of all the simplices. As a result, the pigments may be able to be separated linearly within these tiny areas. Despite the fact that a complicated algorithm was operating in the background, PCOMMEND could be completed quickly and with satisfactory results. The endmember signatures were recovered as two groups of three unique endmembers, for a total of six endmembers. They also demonstrated that the technique was useful when applied to a Landsat TM picture with a restricted number of wavebands.

Another linear technique called Robust Collaborative Nonnegative Matrix Factorization (R-CoNMF) is used as an initial estimate of the endmembers (Li, Bioucas-Dias, et al., 2016). R-CoNMF calculates the endmember abundances, which are subsequently used to estimate the yields. R-CoNMF conducts linear unmixing to the pure crop pixel, which likely comprises closely mixed photosynthetic pigments, while looking for endmembers. Because there are no pixels made entirely of pure photosynthetic pigments, one of the most important properties of R-CoNMF that we used in our work is that the endmembers do not always correspond to pure pixels. Although a pure pixel method determines the initial endmembers, they are iteratively modified so that the final endmembers do not always show as pure pixels in the image.

Nonetheless, the R-CoNMF algorithm's endmembers are not optimal. To that aim, we offer an optimization technique that improves the endmembers' predictive capability. The optimized endmembers are calculated as follows:

$$E_{\mathrm{opt}} = \mathrm{argmin}_E(\mathrm{Rsquared\_glm}(\mathrm{SUnSAL}(E, Y), y)) \tag{14.5}$$

where E represents the endmembers, Y, the field average pixels as earlier, while SUnSAL, an abundance estimation method proposed in (Li, Bioucas-Dias, et al., 2016), returns the abundance matrix A (both constraints are used). In this study, y are yields and Rsquared_glm is the coefficient of determination supplied by the MATLAB function "fitglm" with interactions to partially model the nonlinearity. The function is minimized using unconstrained multivariable minimization as implemented in MATLAB's "fminsearch" function, and the endmembers calculated by R-CoNMF are used as the start values. The R-CoNMF solution is used as the initial value of the optimization algorithm. In other words, the endmembers are modified to maximize the R^2 value between the yields and the abundances.

14.3.2.4 Linear and nonlinear regression

The abundances were used as regression algorithm inputs to produce early yield estimations, which was the end goal. GLM, ANN, and RF were the approaches chosen to achieve the ultimate aim. MATLAB was used to run all GLM, ANN, and RF algorithms.

It is critical to determine which characteristics are truly relevant in calculating yields. The significance of chosen factors varies depending on the parameter configuration. In this work, three techniques were used to choose factors and estimate yield using different datasets, including:

- Only the abundances,
- Abundances with their interactions with each other (to partially address the nonlinear mixing),
- Only the agrometeorological parameters,
- The agrometeorological parameters with NDVI and
- The abundances together with all the other parameters.

The selection and use of suitable parameters are critical in all techniques to obtain the intended results. The RF technique was used to estimate the parameter importance values using out of bag samples. After selecting the suitable parameters in each technique, yield was estimated using RF, ANN, and GLM.

14.4 Results and discussion

This section describes the calculation of the endmembers and abundances by following the proposed methodology's steps. In the discussion part, the results of machine learning algorithms using all available data as well as a selected fraction of all available data that are regarded to be the most important are compared.

14.4.1 The endmembers

The endmembers can be thought of as the spectral signatures of the most prevalent textures in the study area. Because we are looking at wheat fields that are almost completely covered and green (full closure, maximum NDVI), the plant pigments should be found as endmembers. For the 142 fields in 31 cities where Landsat 8 photos existed immediately before the harvesters collected yield data, the endmembers were computed using R-CoNMF (Li, Bioucas-Dias, et al., 2016) in MATLAB. Four endmembers were obtained using the first four MS bands of Landsat 8 images, where the photosynthetic pigments were mainly absorbent.

The newly computed reflectance spectra of each pigment were then processed using the bandwidths of the first four Landsat 8 satellite bands, as shown in Fig. 14.4. In the passbands, each band's transmissivity is considered to be unity, and the reflectance spectra are simply added in each band to obtain the reflectance values observed by Landsat 8.

The final reflectance spectra of the pigments produced using the methods described earlier are shown in Fig. 14.4. For comparison with the endmembers, Fig. 14.7A illustrates how the pigments' spectra might appear if seen directly from the first four bands of Landsat 8.

Fig. 14.7B shows the endmembers produced via the R-CoNMF and optimization. The absorbance and reflectance spectra of photosynthetic pigments are typically generated in a narrow-band format using spectrometers under laboratory circumstances. The wavelengths between 400 and 700 nm are where these pigments are most active. We can only get a highly generalized picture of the real signatures because Landsat 8 only contains four bands in that range.

In this research, R-CoNMF is chosen because it produces endmembers that are comparable with the spectrum signatures of photosynthetic pigments (compare Fig. 14.7B with A). This demonstrates that photosynthetic pigments may be obtained in a completely green wheat field by performing unmixing. This is critical as yield is a critical function of pigments, as evidenced by several previous research studies. The endmember optimization improved the endmembers' resemblance to the real pigment endmembers seen in Fig 14.7A.

As a result, the endmembers in Fig. 14.7B can be read as follows: Endmember 1 is associated with chlorophyll b, Endmember 2 with chlorophyll a, and Endmember 3 with carotenoids. The fourth endmember depicts any nonmodeled components and explains their presence as a result of the nonlinearities that are invariably present in plant structures.

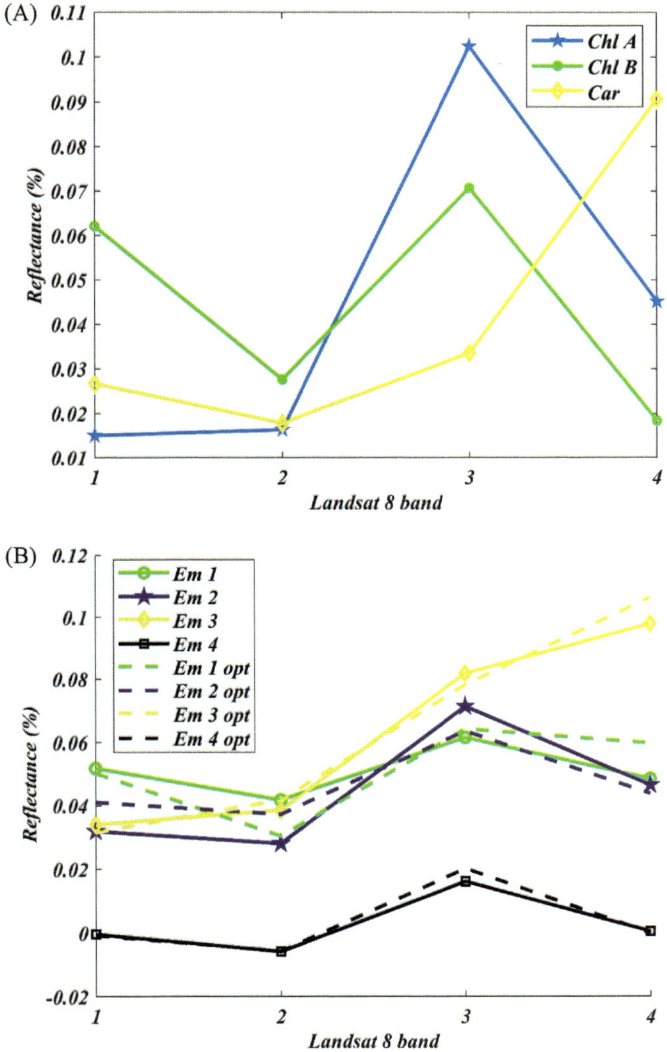

FIGURE 14.7

(A) The pigment reflectance computed for Landsat 8's first four bands; (B) the endmembers identified using the R-CoNMF method and optimization.

14.4.2 The abundances

The four abundances, which were estimated using R-CoNMF as well, are shown against the fields' known yields. Fig. 14.12 shows the connections between each abundance and the yields. The endmembers linked to chlorophyll b and carotenoids (Abundance 1 and 3, respectively) show a substantial connection with yield, but the endmember connected to chlorophyll b (Abundance 1) has a negative correlation. The information from the fourth endmember, which represents all the remaining nonmodeled components, is carried by the fourth abundance. Chlorophylls and carotenoids have been found to be positively associated with wheat dry mass (Sabo et al., 2002), although chlorophyll b shows the least connection. The sum-to-one restriction means that an increase in one quantity will result in a reduction in others; therefore chlorophyll b might have a negative correlation with yield.

14.4.3 Parameter selection and interactions

Table 14.4 contains a list of all the metrics that have been gathered or computed from our dataset, including abundances, agrometeorological factors, and vegetation indices. The table also includes all potential interactions between the abundances. The predictor importance characteristic of RF is used to assess the relevance of the factors in Table 14.4.

We obtained a categorical soil map of Turkey and included the categorical parameters as binary into the dataset to study the influence of the field soil. Their separate contributions, as well as their relationships with the other factors, were examined. Because the influence of any of these characteristics was so small, all soil parameters were excluded from the analysis. As shown in Fig. 14.8, the

Table 14.4 All of the selected parameters' and interactions' names and abbreviations.

Abbreviation	Name of the parameter
A1	Abundance 1
A2	Abundance 2
A3	Abundance 3
A4	Abundance 4
A1A3	Interaction of abundances 1 and 3
A2A3	Interaction of abundances 2 and 3
A2A4	Interaction of abundances 2 and 4
T_MAX	Maximum temperature (°C)
T_MIN	Minimum temperature (°C)
VPD	Average vapor pressure deficit (hPa)
E0	Potential evapotranspiration of open water (mm/day)
SOILFREEZE	Total number of days the soil temperature was below or equal to 0°C (day)
NoPRECIPITATIONdays	Number of days there was no precipitation until harvest (day)
RADIATION	Average radiation (KJ/m^2/day)
Cons_noPrec	Consecutive no precipitation days until harvest
EVI	Enhanced vegetation index

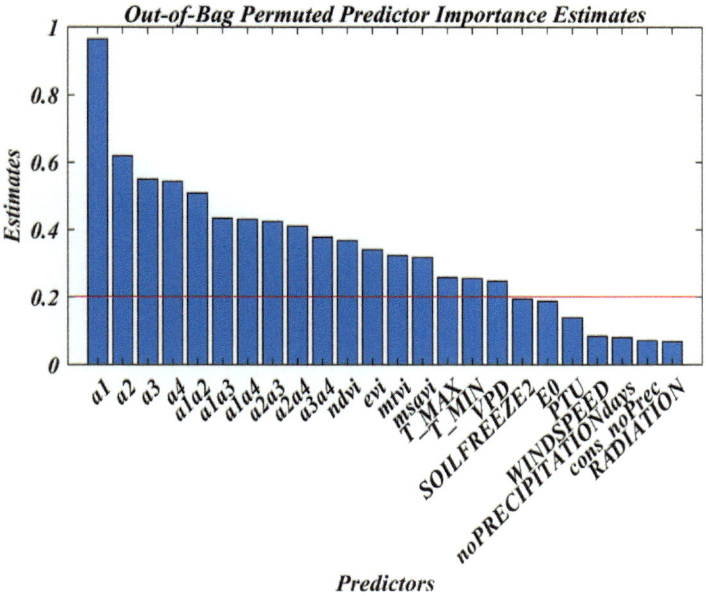

FIGURE 14.8

All parameters' out-of-the-bag significance.

relative significance of all 25 factors used in estimating wheat yields was computed using MATLAB's RF algorithm's "Out of Bag Predictor Importance Estimates." When all of the factors were added together, the VPD, T MIN, T MAX, RADIATION, and A2A4 were shown to be the most essential. Elevation was removed from the predictor list after the predictor association test revealed a strong connection between VPD and Elevation. Given the high relevance of VPD relative to all other characteristics, Elevation was removed from the predictor list. EVI and MTVI were significantly linked as well, and MTVI was eliminated as EVI was more important. PTU, A1A4, A2A4, A3A4, NDVI, and MSAVI had little impact on yield prediction and were thus eliminated from the list. Windspeed was removed from the parameter list due to its minimal relevance when compared with the other parameters. According to the "Predictor Importance Estimates" algorithm, all abundances and most of their interactions appeared to have a substantial contribution to yield, thus they stayed with the most relevant agrometeorological parameters and EVI, which appeared to be more important than NDVI in this study.

Fig. 14.9 shows the contrast of predictor significance estimates produced by permuting out-of-bag data with those obtained by accumulating gains in the mean squared error owing to splits on each predictor.

Surrogate splits were also used to estimate predictor association measures (Fig. 14.10). Predictor association is a 24×24 matrix comprising all the

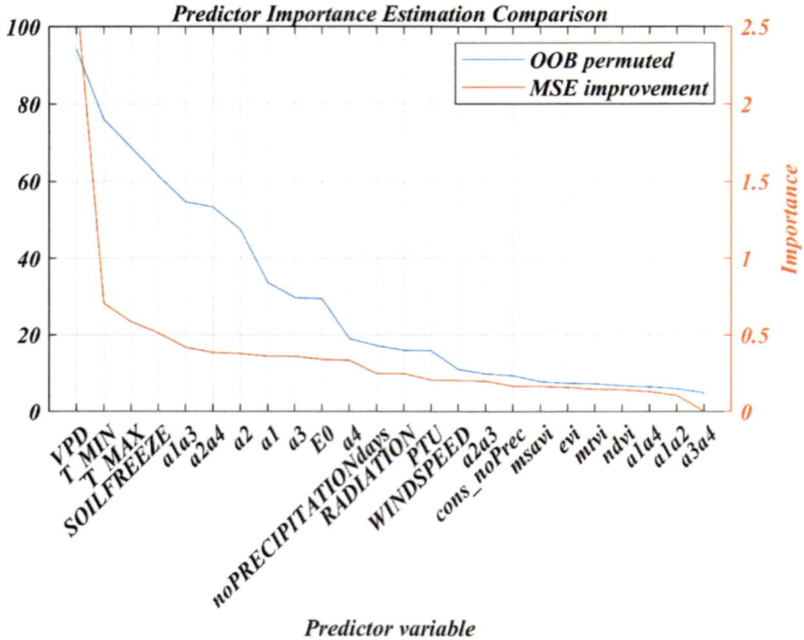

FIGURE 14.9

Comparison of predictor importance estimates.

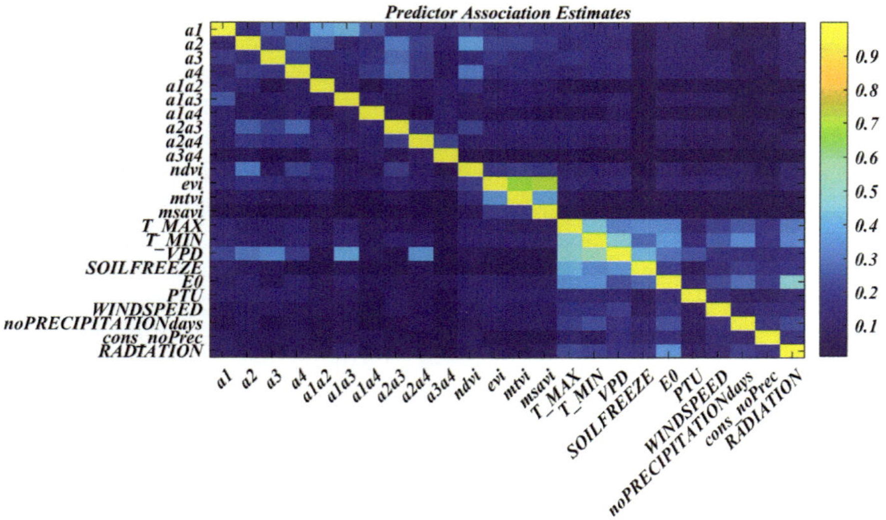

FIGURE 14.10

Estimates of predictor association for all parameters.

characteristics that might be used to estimate wheat yields. The components of the predictor association can be used to infer the strength of the link between pairs of predictors. More closely associated pairs of predictors suggest larger values. With a 78.2% relationship, MTVI and EVI had the strongest link in our dataset; nevertheless, this value was not high enough to suggest a significant relationship between the two variables, and one of them should be eliminated from the dataset.

Following an examination of the findings of the essential parameter selection algorithms, the remaining abundances and agrometeorological parameters below a particular threshold that contributed the most to yield estimation were chosen. Table 14.4 shows the final parameter list.

To explain the selected parameters, the same unbiased and out-of-bag parameter selection methods, as well as the predictor association algorithm, were used, and the importance of the selected parameters can be found in Fig. 14.11.

Fig. 14.12 shows a comparison of predictor significance estimates produced by permuting out-of-bag observations with estimates generated by accumulating gains in the mean squared error owing to splits on each of the selected predictors.

For the selected parameters, the predictor association measures calculated using surrogate splits were also observed (Fig. 14.13). The 16×16 matrix of selected predictor association measures of all the characteristics that can be used to estimate wheat yields is now known as predictor association. T MAX and

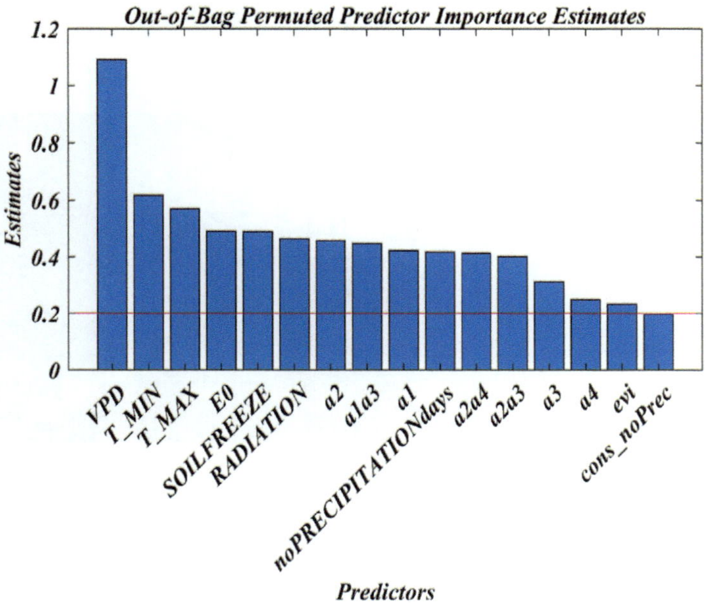

FIGURE 14.11

Out-of-the-box significance of specified parameters.

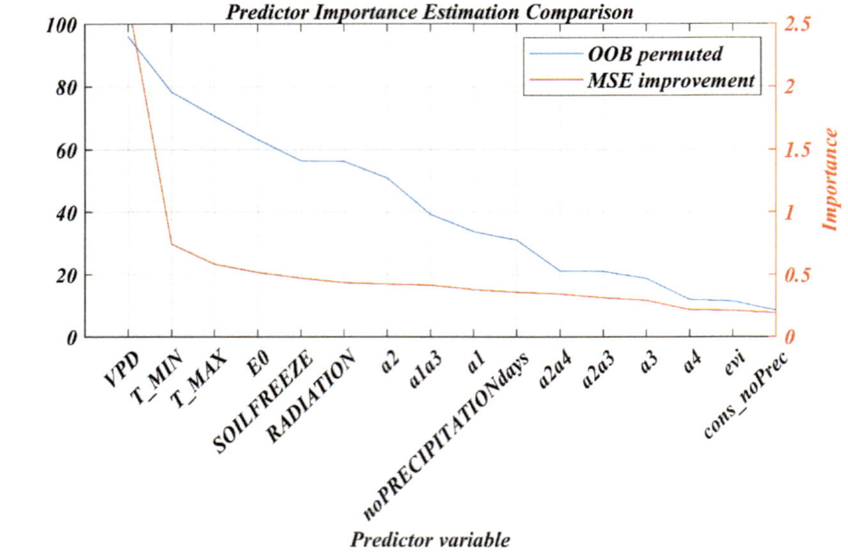

FIGURE 14.12

Comparison of predictor importance estimates.

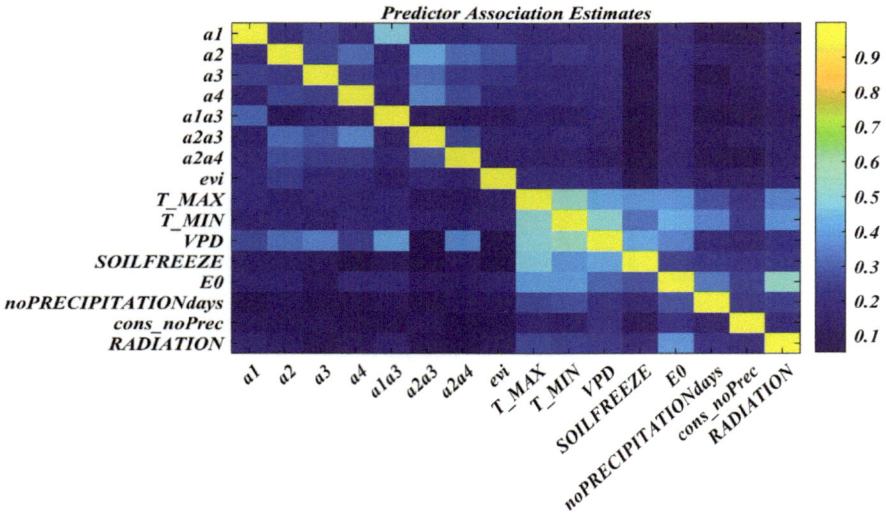

FIGURE 14.13

Estimates of all parameter predictor associations.

T MIN had the strongest link in the selected parameters dataset, with a 53.4% relationship, indicating that all of the parameters were independently contributing to the assessment of wheat yields.

14.4.4 Yield estimation using three different machine learning approaches

This research was carried out using Landsat 8, a medium-resolution satellite dataset, and the techniques that produce the best results when combined with hyperspectral images. The objective of this study, however, is to achieve the best possible results using free and easily accessible data for repeatable research. The dataset was subjected to three distinct machine learning techniques to compare their results. The GLM, ANN, and RF techniques were all run in MATLAB. The validation method was carried out using cross-validation techniques due to the restricted quantity of data that could be used in this study. Because 10-fold cross-validation is often used in the literature, we used it in all of our models (GLM, ANN, and RF) to ensure consistency.

14.4.4.1 The generalized linear model approach

The linear correlation between the response and the predicted parameters is described by linear regression models. Most of the time, though, there may be a nonlinear connection between the parameters. Nonlinear regression is the term used to characterize these nonlinear models in general. The GLM is a type of nonlinear model that uses linear approaches for regression fitting (Models, 2019).

The GLM model is set to linear, with a Poisson distribution. When just the abundances and their chosen interactions are used as input parameters, the GLM yields a coefficient of determination (R^2) of 0.64 for the true yield and an RMSE of 31.53 for the projected yield. The R^2 is 0.60, and the RMSE is 33.53 when the selected agrometeorological parameters and NDVI are used as the model's input set. When all of the specified parameters in Table 14.4 were used in the GLM, the R^2 achieved its maximum value of 0.67 with an RMSE of 31.98.

14.4.4.2 The neural network approach

The input set of the Levenberg–Marquardt backpropagation algorithm, which is the ANN method used in this work, is the matrix of all abundances and their interactions. The yield matrix is the target, and the hidden layer size varies between 5 and 16 depending on the number of parameters according to the 2/3 rule of thumb.

When abundances and their interactions are used as the input set, the R^2 is 0.63. The number of hidden layers is five, and the RMSE is 31.86. The R^2 is 0.75 with an RMSE of 26.52 when the input set of the specified agrometeorological parameters and NDVI is used as the input set. When all of the settings from Table 14.4 are applied, resulting in a hidden layer size of 11, the R^2 reaches its maximum value of 0.78 with an RMSE of 25.00.

14.4.4.3 The random forests approach

The settings chosen are used in the Treebagger Algorithm, which grows 500 trees in the forest. When the abundances and their chosen interactions are used as the input set of six parameters, the R^2 is 0.63 with an RMSE of 32.45. When agrometeorological parameters and NDVI are used, the R^2 improves substantially to 0.78 with RMSE equal to 24.72, and running the specified parameters in Table 14.4 boosts the R^2 to 0.82 (RMSE = 22.51), the best result of all experiments.

Heremans et al. (2015) used 262 input variables during a 12-year period, including total fertilizer usage, 27 meteorological factors, and 234 cumulative NDVI values. Their results indicated that the RF had R^2 values greater than 0.80. The dataset used in this study was small in comparison with theirs, and it can be observed that the RF produced identical findings when just agrometeorological factors and NDVI were used as inputs ($R^2 = 0.77$).

Fig. 14.14 depicts the geographical distribution of real yields vs. estimated yields. The yellow fields have an error margin of 10%. The inaccuracy in the red areas can be as high as 21%, while the error in the blue fields can be as low as 17% lower than the actual yield.

Table 14.5 displays the results of the machine learning methods for various datasets comprising different combinations of parameters using only the training sets. The results of the identical processes applied to test sets generated using the 10-fold cross-validation methodology are shown in Table 14.6. The findings of the training set indicated that RF was capable of estimating yields even when only the abundances were used ($R^2 = 0.84$), and that when all of the chosen factors were included, the R^2 reached 0.93. Table 14.6 compares the results of the three different estimating techniques using the test sets. The best results are achieved if all the parameters of Table 14.4, picked from Table 14.1, are used

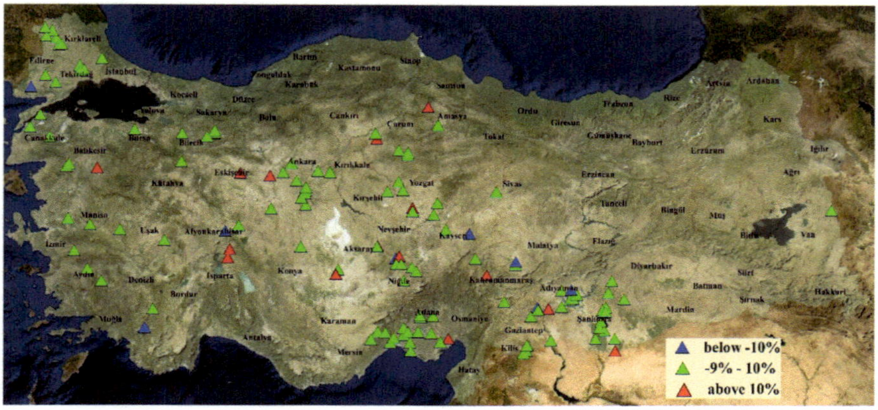

FIGURE 14.14

The spatial distributions of field yields, along with their differences from real yields.

Table 14.5 The real versuss estimated yield accuracies and RMSE of the techniques used: GLM, ANN, and RF for different parameter combinations for the training sets.

Parameters	**GLM**		**ANN**			**RF**	
	R^2	**RMSE**	**Hidden layer size**	R^2	**RMSE**	R^2	**RMSE**
Four abundances	0.59	33.71	3	0.73	27.73	**0.84**	22.32
Four abundances and their selected interactions	0.68	29.85	5	0.73	27.4	**0.84**	22.1
Agrometeorological parameters	0.68	29.73	5	0.83	22.05	**0.89**	18.11
Agrometeorological parameters and NDVI	0.68	29.61	6	0.88	18.64	**0.89**	17.78
All selected parameters	0.82	22.09	11	0.85	20.58	**0.93**	15.21
All parameters	0.83	21.40	16	0.78	24.65	**0.93**	15.11

The bold values represent the highest R^2s obtained when all models are compared.

Table 14.6 The real versus predicted yield accuracies and RMSE of the models used: GLM, ANN, and RF according to different parameter combinations for the test sets.

Parameters	**GLM**		**ANN**			**RF**	
	R^2	**RMSE**	**Hidden layer size**	R^2	**RMSE**	R^2	**RMSE**
Four abundances	0.56	34.89	3	**0.62**	32.55	**0.62**	32.45
Four abundances and their selected interactions	**0.64**	31.53	5	0.63	31.86	0.63	32.32
Agrometeorological parameters	0.63	32.05	5	0.76	25.92	**0.78**	24.96
Agrometeorological parameters and NDVI	0.6	33.53	6	0.75	26.52	**0.78**	24.72
All selected parameters	0.67	31.98	11	0.78	25	**0.82**	23.51
All parameters	0.61	36.21	16	0.61	55.59	**0.80**	24.15

The bold values represent the highest R^2s obtained when all models are compared.

and are the most important factors in the prediction of the returns for artificial intelligence models. When all factors are applied, RF can predict 82% of yields. The findings also show the significance of picking the most critical factors. When specified parameters were used, accuracy for all estimating techniques rose considerably, with the least improvement in RF being seen, as it is resistant to overfitting by nature. This finding demonstrated that, in addition to serving RF itself, parameters determined via an RF algorithm serve the other techniques.

ANN and GLM often need more predictors to enhance their preciseness, whereas RF improves model accuracy by altering each DT's predictor and training data randomly. RF is resistant to noise and overfitting within data. Despite this fact, when combined with agricultural weather factors to find the returns, NDVI did not seem to make a substantial addition to the accuracy. Agrometeorological parameters are indeed able to estimate the yield by themselves using RF (78% accuracy). ANN can also predict above 76% with sufficient precise, but GLM cannot forecast agrometeorological yields as well as other models (63%).

In the majority of situations, RMSE is preferable for RF, even if RF and ANN have done near enough, except when the parameters specified are used to for the output. When just chosen parameters are used, RMSE is computed as low as 22.5. These findings indicate that RF is the best approach of predicting wheat production when using all the key factors and their interactions.

This research demonstrates the significance and strength of the extraction of closely intermingled limbs, probably pigments for photosynthesis. This is achieved despite the fact that the photosynthetic pigments in the wheat crop are intimately mixed linearly with R-CoNMF. In addition to identifying endmembers using R-CoNMF, after optimizing the endmembers with the GLM method, the yield estimation performance rises substantially. The R^2 abundance interactions with the RF model could only achieve the maximum of 0.55 and the RMSE of 35.10. Before optimization was included in the calculation stages, they reached 0.63 with the RMSE of 32.32 following optimization.

The endmembers are linked to pigments because they similarly appear in Landsat 8 bands to the spectral fingerprints of the pigments. Including interactions may be argued that a bilinear technique is employed for a nonlinear mixture (Heylen et al., 2014) and predictions improve. Interactions can contribute more to R^2 growth than some of the abundances themselves (Figs. 14.8 and 14.11). The interactive abundance of these may estimate almost 65% of the yield all in the test set generated with 10x cross-validation by using Landsat 8 data of medium resolution and over 82% by using all the key factors. The estimated yields are likely to be improved if a sophisticated decomposition algorithm of the crops' internal structure with increasing technology is available. For 12 years, 262 input variables consisting of total fertilizer utilization, 27 weather parameters, and 234 cumulative NDVI values were used by Heremans et al. (2015). Their findings indicated RF values of more than 0.80 R^2. Compared with their set, RF exhibits comparable findings when only agrometeorological factors and NDVI were used as inputs ($R^2 = 0.77$). The dataset used in this investigation is rather limited. We are doing our analysis using agrometeorological parameters and with data from only 1 year (2005) with 17 selected parameters that demonstrate their efficiency and easy computation of the data collecting procedure in contrast with Heremans et al. (2015).

What is most essential to notice is that if hyperspectral satellite data were to be used, the results of this study, particularly the relevance of abundance in yield estimate, would definitely rise quickly. This is firstly because additional bands would be used and, secondly, the algorithms used for hyperspectral data were really created.

14.5 Conclusion

This chapter summarized the applications of monitoring and improving the phytosanitary conditions of crops in the literature while also introducing a novel approach of a calculation of wheat yield.

The literature contains many research studies on the phytosanitary of plants and vegetation, mostly on the observation of the effects of insects, nitrogen intake, crops' sowing time, and climate. In the recent years, people have investigated the use of the satellite images in observing the crop phytosanitary and then linked them to the estimation of biomass. However, these were mostly done by using traditional methods of using various indices such as NDVI and SAVI. So far, not many artificial intelligence techniques have been used to find different linkages of the crops' phytosanitary state and yield performance. To make a contribution to the literature to close this gap, the writers proposed a novel approach, which was difficult to implement several years ago. With the improvement of the computing technology and AI applications, a first step was taken to unmix the intimately engaged pigments of a crop, namely, wheat within this study.

The application of the algorithm was done within field using the first four bands of pure Landsat 8 pixels of wheat crop, the performance of which is assessed based on data from 142 fields in 31 provinces in various climatic areas. The harvester statistics collected from the Agricultural Ministry are used as the basis of these fields. We are now able to unmix the intimate pixel of pure wheat crop containing exclusively photosynthetic pigments, but not always have a pure pixel in the unmixing process, with a linear unmixing method called R-CoNMF. The endmembers are further tuned to anticipate their yields and their interactions more precisely. The technique calculated endmembers that exhibited a pattern comparable with the spectral signatures of chlorophylls and carotenoids that process their spectral signatures in Landsat 8 to acquire a view of their look from a satellite viewpoint with medium resolution. Abundances found from the endmembers using the same technique as new indices, and nonlinearity has been addressed by incorporating in the three machine learning algorithms (GLM, ANN, and RF) parameter list interactions that are used to forecast yields.

By using only the abundances and the most significant interactions, GLM forecasted more than 64% of yield. The addition of agrometeorological data and the VIs in the image allows RF to obtain an R^2 of 0.82, a major achievement that may be regarded as a MS satellite.

This study contributed to the literature by demonstrating a new point of view in evaluating yields, by means of soft-computing methods for unmixing pigments that are responsible for the phytosanitary of the crops by introducing a new method in unmixing intimate pigments without the use of nonlinear approaches. The estimated yield could be obtained at least 1 month before the harvest with a small number of sources and criteria for estimating yield. This allows focusing on easily available characteristics or can be determined without any further costs.

New approaches with the usage of remote sensing and AI sources are the future of detecting and watching the phytosanitary conditions of the crop, while estimating its yield.

Sentinel-2 data can be used for future trials as they are accessible publicly just like Landsat 8 and have additional bands and a temporary resolution of 5 days, which is also better than the 16-day time Landsat 8.

To do the analysis and discover the yields, it would be preferable to use hyperspectral satellite data collected from future hyperspectral space missions such as HyspIRI, as each increase in the number of bands will result in a finer endmember computation, resulting in more accurate yields.

The most essential next study would be to perform ground truth measurements using spectroradiometers and laboratory work to see how well the endmembers represent the photosynthetic pigments in the crops. Taking absorption and reflection measurements of wheat leaves in the field would also be a useful future project to determine the real values for the transition from pigment absorbance spectra to wheat reflectance spectra. Implementing the technique for crops other than wheat would also be a worthwhile research project. If Sentinel-2 data are to be used, the most helpful future research would be to integrate the process in Google Earth Engine and record the yield estimation of the wheat fields on a weekly basis. This can be made much easier and more effective if the endmember signatures have been proved not to alter over time. However, for the method to be useful, it must be combined with agricultural parcel segmentation and crop-type detection algorithms.

References

Andrea, T. (2014). *Gridded agro-meteorological data in Europe*. European Commission, Joint Research Centre (JRC) [Dataset] PID. Available from http://data.europa.eu/89h/jrc-marsop4-7-weather_obs_grid_2019.

Apostol, S., Viau, A. A., Tremblay, N., Briantais, J.-M., Prasher, S., Parent, L.-E., & Moya, I. (2003). Laser-induced fluorescence signatures as a tool for remote monitoring of water and nitrogen stresses in plants. *Canadian Journal of Remote Sensing*, *29*(1), 57–65. Available from https://doi.org/10.5589/m02-076.

Barker, D. H., Seaton, G. G. R., & Robinson, S. A. (1997). Internal and external photoprotection in developing leaves of the CAM plant Cotyledon orbiculata. *Plant, Cell and Environment*, *20*(5), 617–624. Available from https://doi.org/10.1111/j.1365-3040.1997.00078.x.

Bartley, G. E., & Scolnik, P. A. (1995). Plant carotenoids: Pigments for photoprotection, visual attraction, and human health. *The Plant Cell*, *7*(7), 1027–1038. Available from https://doi.org/10.1105/tpc.7.7.1027.

Basso, B., Cammarano, D., & Carfagna, E. (2014). *Review of crop yield forecasting methods and early warning systems*. Available from http://www.fao.org/fileadmin/templates/ess/documents/meetings_and_workshops/GS_SAC_2013/Improving_methods_for_crops_estimates/Crop_Yield_Forecasting_Methods_and_Early_Warning_Systems_Lit_review.pdf.

Basso, B., Ritchie, J. T., Pierce, F. J., Braga, R. P., & Jones, J. W. (2001). Spatial validation of crop models for precision agriculture. *Agricultural Systems*, *68*(2), 97–112. Available from https://doi.org/10.1016/S0308-521X(00)00063-9.

Basso, B., T.R. McVicar, B.L. (2007). Remote sensing and GIS applications in agrometeorology. In Remote sensing and GIS applications in agrometeorology. Available from https://www.researchgate.net/publication/241197929.

Benedetti, R., & Rossini, P. (1993). On the use of NDVI profiles as a tool for agricultural statistics: The case study of wheat yield estimate and forecast in Emilia Romagna. *Remote Sensing of Environment*, *45*(3), 311–326. Available from https://doi.org/10.1016/0034-4257(93)90113-C.

Bilger, W., Björkman, O., & Thayer, S. S. (1989). Light-induced spectral absorbance changes in relation to photosynthesis and the epoxidation state of xanthophyll cycle components in cotton leaves. *Plant Physiology*, *91*(2), 542–551. Available from http://www.ncbi.nlm.nih.gov/pubmed/16667067.

Blackburn, G. A. (2006). Hyperspectral remote sensing of plant pigments. *Journal of Experimental Botany*, *58*(4), 855–867. Available from https://doi.org/10.1093/jxb/erl123.

Boldea, M., & Sala, F. (2010). Optimizing economic indicators in the case of using two types of state-subsidized chemical fertilizers for agricultural production. *AIP Conference Proceedings*, *1281*(1), 1390. Available from https://doi.org/10.1063/1.3497988.

Cai, Y., Guan, K., Lobell, D., Potgieter, A. B., Wang, S., Peng, J., Xu, T., Asseng, S., Zhang, Y., You, L., & Peng, B. (2019). Integrating satellite and climate data to predict wheat yield in Australia using machine learning approaches. *Agricultural and Forest Meteorology*. Available from https://doi.org/10.1016/j.agrformet.2019.03.010.

Casadebaig, P., Zheng, B., Chapman, S., Huth, N., Faivre, R., & Chenu, K. (2016). Assessment of the potential impacts of wheat plant traits across environments by combining crop modeling and global sensitivity analysis. *PLoS One*, *11*(1)e0146385. Available from https://doi.org/10.1371/JOURNAL.PONE.0146385.

Chappelle, E. W., Kim, M. S., & McMurtrey, J. E. (1992). Ratio analysis of reflectance spectra (RARS): An algorithm for the remote estimation of the concentrations of chlorophyll A, chlorophyll B, and carotenoids in soybean leaves. *Remote Sensing of Environment*, *39* (3), 239–247. Available from https://doi.org/10.1016/0034-4257(92)90089-3.

Cohen, Y., Alchanatis, V., Meron, M., Saranga, Y., & Tsipris, J. (2005). Estimation of leaf water potential by thermal imagery and spatial analysis. *Journal of Experimental Botany*, *56*(417), 1843–1852. Available from https://doi.org/10.1093/jxb/eri174.

Cristian, C., Mihai, H., Ciprian, R., & Florin, S. (2018). Model prediction of chlorophyll and fresh biomass in cereal grasses based on aerial images. *AIP Conference Proceedings*, 1978. Available from https://doi.org/10.1063/1.5043987.

Dawson, T. P., Curran, P. J., & Plummer, S. E. (1998). LIBERTY – Modeling the effects of leaf biochemical concentration on reflectance spectra. *Remote Sensing of Environment*, *65*(1), 50–60. Available from https://doi.org/10.1016/S0034-4257(98)00007-8.

Demarez, V. (1999). Seasonal variation of leaf chlorophyll content of a temperate forest. Inversion of the PROSPECT model. *International Journal of Remote Sensing*, *20*(5), 879–894. Available from https://doi.org/10.1080/014311699212975.

Dimitruk, P., Schermelleh-Engel, K., Kelava, A., & Moosbrugger, H. (2007). Challenges in nonlinear structural equation modeling. *Methodology*, *3*(3), 100–114. Available from https://doi.org/10.1027/1614-2241.3.3.100.

Dorigo, W. A., Zurita-Milla, R., de Wit, A. J. W., Brazile, J., Singh, R., & Schaepman, M. E. (2007). A review on reflective remote sensing and data assimilation techniques for enhanced agroecosystem modeling. *International Journal of Applied Earth Observation and Geoinformation*, *9*(2), 165–193. Available from https://doi.org/10.1016/j.jag.2006.05.003.

Dourado-Neto, D., Teruel, D. A., Reichardt, K., Nielsen, D. R., Frizzone, J. A., & Bacchi, O. O. S. (1998). Principles of crop modeling and simulation: I. uses of mathematical models in agricultural science. *Scientia Agricola*, *55*(Special Issue), 46–50. Available from https://doi.org/10.1590/S0103-90161998000500008.

EU Science Hub. (2020). *Monitoring Agricultural ResourceS (MARS)*. <https://ec.europa.eu/jrc/en/mars> Accessed 20.03.20.

Fernandes, J. L., Jansle Rocha, V., & Rubens Lamparelli, A. C. (2011). Sugarcane yield estimates using time series analysis of spot vegetation images. *Scientia Agricola*, *68*(2), 139–146.

Fortin, J. G., Anctil, F., Parent, L.-É., & Bolinder, M. A. (2010). A neural network experiment on the site-specific simulation of potato tuber growth in Eastern Canada. *Computers and Electronics in Agriculture*, *73*(2), 126–132. Available from https://doi.org/10.1016/J.COMPAG.2010.05.011.

Franch, B., Vermote, E. F., Skakun, S., Roger, J. C., Becker-Reshef, I., Murphy, E., & Justice, C. (2019). Remote sensing based yield monitoring: Application to winter wheat in United States and Ukraine. *International Journal of Applied Earth Observation and Geoinformation*. Available from https://doi.org/10.1016/j.jag.2018.11.012.

Gamon, J. (2010). The Photochemical Reflectance Index (PRI) – A measure of photosynthetic light-use efficiency. *International Journal of Remote Sensing*. Available from http://www.tandfonline.com/doi/abs/10.1080/01431160210163100.

Gamon, J. A., & Surfus, J. S. (1999). Assessing leaf pigment content and activity with a reflectometer. *The New Phytologist*, *143*(Issue 1). Available from https://www.jstor.org/stable/pdf/2588519.pdf?refreqid = excelsior%3A224101dbba7b7f0cdfedc692316db3d3.

Gamon, J. A., Peñuelas, J., & Field, C. B. (1992). A narrow-waveband spectral index that tracks diurnal changes in photosynthetic efficiency. *Remote Sensing of Environment*, *41*(1), 35–44. Available from https://doi.org/10.1016/0034-4257(92)90059-S.

Ganapol, B. D., Johnson, L. F., Hammer, P. D., Hlavka, C. A., & Peterson, D. L. (1998). LEAFMOD: A new within-leaf radiative transfer model. *Remote Sensing of Environment*, *63*(2), 182–193. Available from https://doi.org/10.1016/S0034-4257(97)00134-X.

Gitelson, A. A., Merzlyak, M. N., & Chivkunova, O. B. (2001). Optical properties and nondestructive estimation of anthocyanin content in plant leaves. *Photochemistry and Photobiology*, *74*(Issue 1). Available from https://pdfs.semanticscholar.org/fb59/6d29c31950b5d370ab5d17508d59e8cebc09.pdf.

Gitelson, A., & Solovchenko, A. (2018). Non-invasive quantification of foliar pigments: Possibilities and limitations of reflectance- and absorbance-based approaches. *Journal of Photochemistry and Photobiology B: Biology*, *178*(September 2017), 537–544. Available from https://doi.org/10.1016/j.jphotobiol.2017.11.023.

Groten, S. M. E. (1993). NDVI—Crop monitoring and early yield assessment of Burkina Faso. *International Journal of Remote Sensing*, *14*(8), 1495–1515. Available from https://doi.org/10.1080/01431169308953983.

Guo, P.-T., Li, M.-F., Luo, W., Tang, Q.-F., Liu, Z.-W., & Lin, Z.-M. (2015). Digital mapping of soil organic matter for rubber plantation at regional scale: An application of random forest plus residuals kriging approach. *Geoderma*, *237–238*, 49–59. Available from https://doi.org/10.1016/J.GEODERMA.2014.08.009.

Heremans, S., Dong, Q., Zhang, B., Bydekerke, L., & Van Orshoven, J. (2015). Potential of ensemble tree methods for early-season prediction of winter wheat yield from short time series of remotely sensed normalized difference vegetation index and *in situ* meteorological data. *Journal of Applied Remote Sensing*, *9*(1)097095. Available from https://doi.org/10.1117/1.JRS.9.097095.

Heylen, R., Parente, M., & Gader, P. (2014). A review of nonlinear hyperspectral unmixing methods. *IEEE Journal of Selected Topics in Applied Earth Observations and Remote Sensing*, *7*(6), 1844–1868. Available from https://doi.org/10.1109/JSTARS.2014.2320576.

Huang, W., Zhou, X., Kong, W., & Ye, H. (2018). *Monitoring crop carotenoids concentration by remote sensing. Progress in carotenoid research.* InTech. Available from https://doi.org/10.5772/intechopen.78239.

Huang, Y., Lan, Y., Thomson, S. J., Fang, A., Hoffmann, W. C., & Lacey, R. E. (2010). Development of soft computing and applications in agricultural and biological engineering. *Computers and Electronics in Agriculture*, *71*, 107–127. Available from https://doi.org/10.1016/j.compag.2010.01.001.

Hunt, M. L., Blackburn, G. A., Carrasco, L., Redhead, J. W., & Rowland, C. S. (2019). High resolution wheat yield mapping using Sentinel-2. *Remote Sensing of Environment*, *233*(December 2018)111410. Available from https://doi.org/10.1016/j.rse.2019.111410.

Jacquemoud, S., & Baret, F. (1990). PROSPECT: A model of leaf optical properties spectra. *Remote Sensing of Environment*, *34*(2), 75–91. Available from https://doi.org/10.1016/0034-4257(90)90100-Z.

Jiang, P., & Thelen, K. (2004). Effect of soil and topographic properties on crop yield in a north-central corn-soybean cropping system. *Agronomy Journal*, *96*, 252–258. Available from https://www.tib.eu/en/search/id/BLSE%3ARN144502824/Effect-of-Soil-and-Topographic-Properties-on-Crop/.

Jones, H. G., & Vaughan, R. A. (2010). Robin A.*Remote sensing of vegetation : Principles, techniques, and applications*. Oxford University Press. Available from https://global.oup.com/academic/product/remote-sensing-of-vegetation-9780199207794?cc = tr&lang = en&.

Kang, Y., & Özdoğan, M. (2019). Field-level crop yield mapping with Landsat using a hierarchical data assimilation approach. *Remote Sensing of Environment*, *228*, 144–163. Available from https://doi.org/10.1016/j.rse.2019.04.005.

Kaul, M., Hill, R. L., & Walthall, C. (2005). Artificial neural networks for corn and soybean yield prediction. *Agricultural Systems*, *85*(1), 1–18. Available from https://doi.org/10.1016/j.agsy.2004.07.009.

Kenny, D., & Judd, C. (1984). Estimating the nonlinear and interactive effects of latent variables. *Psychological Bulletin*, *96*(1), 201–210.

Keshava, N., & Mustard, J. F. (2002). Spectral unmixing. *IEEE Signal Processing Magazine*, *19*(1), 44–57. Available from https://doi.org/10.1109/79.974727.

Klein, A., & Moosbrugger, H. (2000). Maximum likelihood estimation of latent interaction effects with the LMS method. *Psychometrika*, *65*(4), 457−474. Available from https://doi.org/10.1007/BF02296338.

Krzyśko, M., Derejko, A., Górecki, T., & Gacek, E. (2013). Principal component analysis for functional data on grain yield of winter wheat cultivars. *Biometrical Letters*, *50*(2), 81−94. Available from https://doi.org/10.2478/bile-2013-0019.

Lachman, J., Martinek, P., Kotíková, Z., Orsák, M., & Šulc, M. (2017). Genetics and chemistry of pigments in wheat grain − A review. *Journal of Cereal Science*, *74*, 145−154. Available from https://doi.org/10.1016/J.JCS.2017.02.007.

Leroux, L., Castets, M., Baron, C., Escorihuela, M.-J., Bégué, A., & Lo Seen, D. (2019). Maize yield estimation in West Africa from crop process-induced combinations of multi-domain remote sensing indices. *European Journal of Agronomy*, *108*, 11−26. Available from https://pdf.sciencedirectassets.com/271285/1-s2.0-S1161030119X00052/1-s2.0-S116103011830354X/main.pdf?X-Amz-Security-Token = AgoJb3JpZ2luX2VjEEQaCXVzLWVhc3QtMSJHMEUCIQDMqJwourWfwoEjvMB9dk8jOAee2H8LeKGvOLRPEGqt%2BQIgf3xTqCXa%2FT%2FCjvaWeElB5l%2Fpx7u2I%2FUnZn.

Li, J., Bioucas-Dias, J. M., Plaza, A., & Liu, L. (2016). Robust collaborative nonnegative matrix factorization for hyperspectral unmixing. *IEEE Transactions on Geoscience and Remote Sensing*, *54*(10), 6076−6090. Available from https://doi.org/10.1109/TGRS.2016.2580702.

Li, Z., Wang, J., Tang, H., Huang, C., Yang, F., Chen, B., Wang, X., Xin, X., Ge, Y., Li, Z., Wang, J., Tang, H., Huang, C., Yang, F., Chen, B., Wang, X., Xin, X., & Ge, Y. (2016). Predicting grassland leaf area index in the meadow steppes of Northern China: A comparative study of regression approaches and hybrid geostatistical methods. *Remote Sensing*, *8*(8), 632. Available from https://doi.org/10.3390/rs8080632.

Lichtenthaler, H. K. (1987). Chlorophylls and carotenoids: Pigments of photosynthetic biomembranes. *Methods in Enzymology*, *148*(C), 350−382. Available from https://doi.org/10.1016/0076-6879(87)48036-1.

Lobell, D. B., Asner, G. P., Ortiz-Monasterio, J. I., & Benning, T. L. (2003). Remote sensing of regional crop production in the Yaqui Valley, Mexico: Estimates and uncertainties. *Agriculture, Ecosystems and Environment*, *94*(2), 205−220. Available from https://doi.org/10.1016/S0167-8809(02)00021-X.

Lobell, D. B., Cassman, K. G., & Field, C. (2010). Crop yield gaps: Their importance, magnitudes, and causes. *SSRN*. Available from https://doi.org/10.1146/annurev.environ.041008.093740.

Lobell, D. B., Ortiz-Monasterio, J. I., Asner, G. P., Naylor, R. L., & Falcon, W. P. (2005). Combining field surveys, remote sensing, and regression trees to understand yield variations in an irrigated wheat landscape. *Agronomy Journal*, *97*(1), 241−249. Available from https://doi.org/10.2134/agronj2005.0241.

Ma, H., Jing, Y., Huang, W., Shi, Y., Dong, Y., Zhang, J., & Liu, L. (2018). Integrating early growth information to monitor winter wheat powdery mildew using multi-temporal landsat-8 imagery. *Sensors (Switzerland)*, *18*(10). Available from https://doi.org/10.3390/s18103290.

Magney, T. S., Eitel, J. U. H., & Vierling, L. A. (2017). Mapping wheat nitrogen uptake from RapidEye vegetation indices. *Precision Agriculture*, *18*(4), 429−451. Available from https://doi.org/10.1007/s11119-016-9463-8.

Maier, S. W., Lüdeker, W., & Günther, K. P. (1999). SLOP: A revised version of the stochastic model for leaf optical properties. *Remote Sensing of Environment*, *68*(3), 273−280. Available from https://doi.org/10.1016/S0034-4257(98)00118-7.

Merzlyak, M. N., Gitelson, A. A., Chivkunova, O. B., & Rakitin, V. Y. (1999). Non-destructive optical detection of pigment changes during leaf senescence and fruit ripening. *Physiologia Plantarum*, *106*(1), 135–141. Available from https://doi.org/10.1034/j.1399-3054.1999.106119.x.

Messina, G., Peña, J. M., Vizzari, M., & Modica, G. (2020). A comparison of UAV and satellites multispectral imagery in monitoring onion crop. An application in the 'Cipolla Rossa di Tropea' (Italy). *Remote Sensing*, *12*(20), 1–27. Available from https://doi.org/10.3390/rs12203424.

Mulla, D. J. (2013). Twenty five years of remote sensing in precision agriculture: Key advances and remaining knowledge gaps. *Biosystems Engineering*, *114*(4), 358–371. Available from https://doi.org/10.1016/j.biosystemseng.2012.08.009.

Models, G.L. (2019). Generalized Linear Models - MATLAB. Available from http://www.mathworks.com/help/stats/generalized-linear-regression.html [Accessed 19 April 2023].

Neill, S., & Gould, K. S. (2000). Optical properties of leaves in relation to anthocyanin concentration and distribution. *Canadian Journal of Botany*, *77*(12), 1777–1782. Available from https://doi.org/10.1139/b99-153.

Özcan, A. (2020). *Early yield estimation by photosynthetic pigment abundances using Landsat 8 image series*. Middle East Technical University. Available from https://open.metu.edu.tr/handle/11511/69144.

Ozcan, A., Leloglu, U.M., & Suzen, M.L. (2021). *Early wheat yield estimation at field-level by photosynthetic pigment unmixing using Landsat 8 image series*. Available from https://doi.org/10.1080/10106049.2021.1903577.

Pantazi, X. E., Moshou, D., Mouazen, A. M., Kuang, B., & Alexandridis, T. (2014). Application of supervised self organising models for wheat yield prediction. *International Federation for Information Processing*, 556–565. Available from https://doi.org/10.1007/978-3-662-44654-6_55.

Penuelas, J., Baret, F., & Filella, I. (1995). Semi-empirical indices to assess carotenoids/chlorophyll a ratio from leaf spectral reflectance. *Photosynthetica*, *31*(Issue 2), 221–230.

Penuelas, J., Gamon, J. A., Fredeen, A. L., Merino, J., & Field, C. B. (1994). Reflectance indices associated with physiological changes in nitrogen and water-limited sunflower leaves.pdf. *Remote Sensing of Environment*, *48*, 135–146. Available from https://www.creaf.uab.es/global-ecology/Pdfs_UEG/RemSensEnv1994.pdf.

Pinter, P.J., Hatfield, J.L., Schepers, J.S., Barnes, E.M., Moran, M.S., Pinter, P.J., Hatfield, J.L., Schepers, J.S., Barnes, E.M., Moran, M., Susan, Daughtry, C.S. T., Upchurch, D.R., & Daughtry, C.S. T. (2003). Remote sensing for crop management. Photogrammetric Engineering & Remote Sensing. http://digitalcommons.unl.edu/usdaarsfacpubhttp://digitalcommons.unl.edu/usdaarsfacpub/1372http://digitalcommons.unl.edu/usdaarsfacpub/1372.

Plant, R. E. (2001). Site-specific management: The application of information technology to crop production. *Computers and Electronics in Agriculture*, *30*(1–3), 9–29. Available from https://doi.org/10.1016/S0168-1699(00)00152-6.

Rawashdeh, H. (2015). Foliar application with iron as a vital factor of wheat crop growth, yield quantity and quality: A review. *International Journal of Agricultural Policy and Research*. *3*(9), 368–376. Available from https://doi.org/10.15739/IJAPR.062.

Renzullo, L. J., Blanchfield, A. L., Guillermin, R., Powell, K. S., & Held, A. A. (2006). Comparison of PROSPECT and HPLC estimates of leaf chlorophyll contents in a grapevine stress study. *International Journal of Remote Sensing*, *27*(4), 817–823. Available from https://doi.org/10.1080/01431160500239164.

Roumenina, E., Jelev, G., Dimitrov, P., Filchev, L., Kamenova, I., Gikov, A., Banov, M., Krasteva, V., Kercheva, M., & Kolchakov, V. (2020). Qualitative evaluation and within-field mapping of winter wheat crop condition using multispectral remote sensing data. *Bulgarian Journal of Agricultural Science*, *26*(6), 1129–1142.

Sabo, M., Teklic, T., & Vidovic, I. (2002). Photosynthetic productivity of two winter wheat varieties (Triticum aestivum L.). *ROSTLINNÁ VÝROBA*, *48*(2), 80–86. Available from https://www.researchgate.net/publication/323106946.

Sala, F., & Boldea, M. (2011). On the optimization of the doses of chemical fertilizers for crops. *AIP Conference Proceedings*, *1389*(1), 1297. Available from https://doi.org/10.1063/1.3637856.

Sala, F., Rujescu, C., & Constantinescu, C. (2016). Causes and solutions for the remediation of the poor allocation of P and K to wheat crops in Romania. *AgroLife Scientific Journal*, *5*(1), 184–193.

Salvador, P., Gómez, D., Sanz, J., & Casanova, J. L. (2020). Estimation of potato yield using satellite data at a municipal level: A machine learning approach. *ISPRS International Journal of Geo-Information*, *9*(6), 343. Available from https://doi.org/10.3390/ijgi9060343.

Sayago, S., & Bocco, M. (2018). Crop yield estimation using satellite images: Comparison of linear and non-linear models. *AgriScientia*, *35*, 1–9. Available from https://doi.org/10.31047/1668.298x.v1.n35.20447.

Shewry, P. R. (2009). Wheat. *Journal of Experimental Botany*, *60*(6), 1537–1553. Available from https://doi.org/10.1093/JXB/ERP058.

Sims, D. A., & Gamon, J. A. (2002). Relationships between leaf pigment content and spectral reflectance across a wide range of species, leaf structures and developmental stages. *Remote Sensing of Environment*, *81*(2–3), 337–354. Available from https://doi.org/10.1016/S0034-4257(02)00010-X.

Somers, B., Asner, G. P., Tits, L., & Coppin, P. (2011). Endmember variability in spectral mixture analysis: A review. *Remote Sensing of Environment*, *115*(7), 1603–1616. Available from https://doi.org/10.1016/J.RSE.2011.03.003.

Sönmez, I., Üstündaȩ, B.B., Baȩiş, S., & Çetin, A. (2015). Agro-meteorological data Quality Control System design for Turkey's agricultural monitoring and information system (TARBIL) (pp. 276–279). In: *2015 4th international conference on agro-geoinformatics*, Agro-Geoinformatics 2015. Available from https://doi.org/10.1109/Agro-Geoinformatics.2015.7248118.

Steyn, W., Wand, S., Holcroft, D., & Jacobs, G. (2002). Anthocyanins in vegetative tissues: a proposed unified function in photoprotection. *New Phytologist*, *155*(3), 349–361.

Suzuki, E. (2019). World's population will continue to grow and will reach nearly 10 billion by 2050. version. 1. Available from https://blogs.worldbank.org/opendata/worlds-population-will-continue-grow-and-will-reach-nearly-10-billion-2050. Accessed 19.04.23.

Suzuki, T., Ueoka, Y., & Sato, H. (2009). Estimating structure of multivariate systems with genetic algorithms for nonlinear prediction. *Physical Review E – Statistical, Nonlinear, and Soft Matter Physics*, *80*(6). Available from https://doi.org/10.1103/PhysRevE.80.066208.

Thenkabail, P. S., Smith, R. B., & De Pauw, E. (2000). Hyperspectral vegetation indices and their relationships with agricultural crop characteristics. *Remote Sensing of Environment*, *71*(2), 158–182. Available from https://doi.org/10.1016/S0034-4257(99)00067-X.

Thenkabail, P. S., Smith, R. B., & De Pauw, E. (2002). Evaluation of narrowband and broadband vegetation indices for determining optimal hyperspectral wavebands for agricultural crop characterization. *Photogrammetric Engineering & Remote Sensing*, *68*

(6), 607–621. Available from https://www.gisresources.com/wp-content/uploads/2014/03/Evaluation-of-Narrowband-and-Broadband-Vegetation-Indices.pdf.

Ustin, S. L., Gitelson, A. A., Jacquemoud, S., Schaepman, M., Asner, G. P., Gamon, J. A., & Zarco-Tejada, P. (2009). Retrieval of foliar information about plant pigment systems from high resolution spectroscopy. *Remote Sensing of Environment*, *113*(SUPPL. 1), 67–77. Available from https://doi.org/10.1016/j.rse.2008.10.019.

Wang, Q., Shi, W., & Atkinson, P. M. (2014). Sub-pixel mapping of remote sensing images based on radial basis function interpolation. *ISPRS Journal of Photogrammetry and Remote Sensing*, *92*, 1–15. Available from https://doi.org/10.1016/j.isprsjprs.2014.02.012.

Wang, Y., Xu, X., Huang, L., Yang, G., Fan, L., Wei, P., Chen, G., Wang, Y., Xu, X., Huang, L., Yang, G., Fan, L., Wei, P., & Chen, G. (2019). An improved CASA model for estimating winter wheat yield from remote sensing images. *Remote Sensing*, *11*(9), 1088. Available from https://doi.org/10.3390/rs11091088.

WMO. (2010). Chapter 3 – Agricultural meteorological data, their presentation and statistical analysis. *Guide to agricultural meteorological practices WMO-No, 134*, 1–34. Available from http://www.wmo.int/pages/prog/wcp/agm/gamp/documents/WMO_No134_en.pdf.

Watt, J. (2013). *3D crop modelling. MSc Thesis* (p. 317) Department of Geography, University College London.

Wu, Q., Jin, Y., Bao, Y., Hai, Q., Yan, R., Chen, B., Zhang, H., Zhang, B., Li, Z., Li, X., & Xin, X. (2015). Comparison of two inversion methods for leaf area index using HJ-1 satellite data in a temperate meadow steppe. *International Journal of Remote Sensing*, *36*(19–20), 5192–5207. Available from https://doi.org/10.1080/01431161.2015.1040135.

Yamasaki, H., Sakihama, Y., & Lkehara, N. (1997). Flavonoid-peroxidase reaction as a detoxification mechanism of plant cells against H_20_2. *Plant Physiology*, *11*(2), 1405–1406. Available from http://www.plantphysiol.org.

Yue, J., Feng, H., Yang, G., Li, Z., Yue, J., Feng, H., Yang, G., & Li, Z. (2018). A comparison of regression techniques for estimation of above-ground winter wheat biomass using near-surface spectroscopy. *Remote Sensing*, *10*(2), 66. Available from https://doi.org/10.3390/rs10010066.

Zare, A., Gader, P., Bchir, O., & Frigui, H. (2013). Piecewise convex multiple-model endmember detection and spectral unmixing. *IEEE Transactions on Geoscience and Remote Sensing*, *51*(5), 2853–2862. Available from https://doi.org/10.1109/TGRS.2012.2219058.

Zhang, Y. (2011). *Chapter 7 Forest leaf chlorophyll study using hyperspectral remote sensing. Hyperspectral remote sensing of vegetation* (pp. 263–264). CRC Press. Available from https://doi.org/10.1201/b11222-17.

Zheng, C., Abd-elrahman, A., & Whitaker, V. (2021). Remote sensing and machine learning in crop phenotyping and management, with an emphasis on applications in strawberry farming. *Remote Sensing*, *13*(3), 1–29. Available from https://doi.org/10.3390/rs13030531.

Zhou, X., Huang, W., Zhang, J., Kong, W., Casa, R., & Huang, Y. (2018). *A novel combined spectral index for estimating the ratio of carotenoid to chlorophyll content to monitor crop physiological and phenological status*. Available from https://doi.org/10.1016/j.jag.2018.10.012.

CHAPTER

Land salinity mapping using Sentinel-1/2 remote sensing imagery: a case study in Da'an, Jilin Province

15

Qianqian Zhang[1] and Li Li[2]

[1]*Department of Earth System Science, Institute for Global Change Studies, Ministry of Education Ecological Field Station for East Asian Migratory Birds, Tsinghua University, Beijing, P.R. China*

[2]*College of Land Science and Technology, China Agricultural University, Beijing, P.R. China*

15.1 Introduction

Soil is a Natural-Environment-Society-Economic complex ecosystem. In the past five decades, the process of global soil salinization has advanced at an unprecedented rate (Schofield & Kirkby, 2003; Zhuang et al., 2021). According to the report of the Intergovernmental Science-Policy Platform on Biodiversity and Ecosystem Services (IPBES) in 2010, the global salinized land was 849 million hm^2, accounting for 5.7% of the global terrestrial land. Scientists recently warned that if no action is taken, the global salinized land will increase by 319 million hectares by 2050, and the degradation rate will reach 95%.

As the largest developing country in China, the salinized land area in 2014 was 2.62 million km^2, accounting for about 27% of total land area, and the situation of soil salinization is severe. The salinization areas in my country are mainly concentrated in the oases in the arid and semiarid climate regions in the north, the plain areas with poor runoff, and the coastal areas. Agricultural production is most significantly affected by soil salinization. The prerequisite for high and stable agricultural yield is to ensure irrigation. If the irrigation water in (semi) arid areas cannot be discharged in time, it will face serious soil salinization. Soil salinization can cause agricultural yield reduction, food shortage, environmental disasters and even threaten people's life, health, and safety. It is of great significance to study soil salinization in these areas. The soil and water in the soil salinization area contain a large number of salt ions, and the untreated water in the area is used in large quantities for irrigation, coupled with the backward drainage system, which leads to the large-scale expansion of salinization, and the salt ions enter the soil of agricultural land and lead to fertile fields. The salinization of the

Remote Sensing in Precision Agriculture. DOI: https://doi.org/10.1016/B978-0-323-91068-2.00007-2

soil will enter the soil of the construction land and cause the destruction of the residence, infrastructure, and social civilization of human life.

Two main types of satellite remote sensing data are used for soil salinization surveys: optical and radar (SAR). The two sensors work based on different fundamental physical principles and can provide rich surface information. Optical sensors passively measure electromagnetic radiation reflected from the Earth's surface and thus can map land cover based on spectral reflectance, making it very sensitive to cloud cover. Radar sensors actively emit wavelengths of electromagnetic radiation that can penetrate the clouds and receive signals back, so they are not affected by clouds. The signal returned by the radar sensor responds to the pattern and dielectric properties of the soil landscape at wavelengths comparable with the radar. Sentinel-1 SAR data (2014–) and Sentinel-2 (2015–) optical data are long-term, global coverage satellite optical remote sensing imagery and satellite microwave remote sensing imagery (radar) freely available to the public. Due to the serious inconsistency of Sentinel-1 SAR data and Sentinel-2 hyperspectral data, the two kinds of data cannot be directly used together, which limits the application potential of satellite remote sensing data. The inconsistency problem is mainly reflected in the following aspects:

1 The spatial resolution (the same pixel represents the size of the ground object) is different.
2 The electromagnetic spectrum detected by Sentinel-1 and Sentinel-2 is different for the same pixel at the same time.
3 The results of Sentinel-1 and Sentinel-2-based applications (e.g., flood monitoring, yield estimation) are different. Fig. 15.1 shows the difference between Sentinel-1 and Sentinel-2 false-color composite data of the study areas.

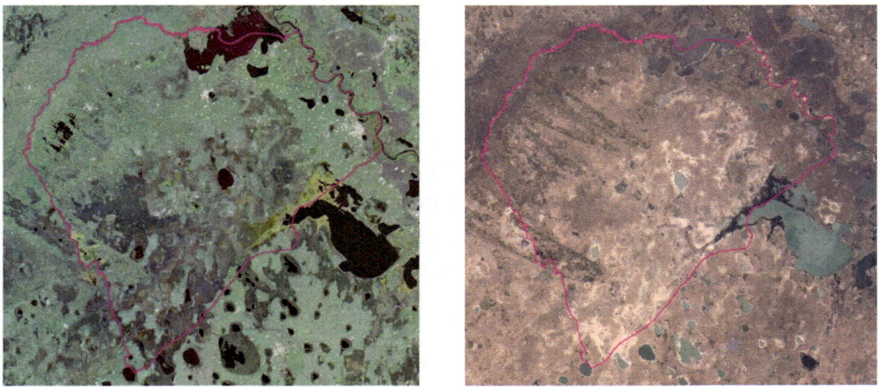

(A) R:VV G:VH B:VV-VH　　(B) R:B4 G:B3 B:B2

FIGURE 15.1

Sentinel-1(A) and Sentinel-2(B)of study area.

The discontinuity of Sentinel-1 and Sentinel-2 data stems from two aspects:

15.1.1 Sentinel-1 data are inconsistent

Sentinel-1 data are a dataset composed of a series of Sentinel-1 satellite (A/B) data. As the Sentinel-1 satellite itself does not have an onboard radiance calibration system, the data collected by different Sentinel-1 satellites and the data collected by the same Sentinel-1 satellite (such as Sentinel-1B) at different times may be caused by interference due to interference. The data acquisition methods in a wide format are inconsistent.

15.1.2 Inconsistency between Sentinel-1 and Sentinel-2 data

Compared with the inconsistency of Sentinel-1 data, the inconsistency between Sentinel-1 and Sentinel-2 data is more serious (Gritsevich et al., 2020). There are two reasons for the phenomenon:

15.1.3 Sentinel-1 and Sentinel-2 sensor performance differences

Sentinel-1 backscattered intensities are recorded as dimensionless intensity values (Digital Number values), while Sentinel-2 data are recorded as radiometrically corrected Surface Reflectance. In terms of satellite revisit period and coverage, the revisit periods of Sentinel-1 and Sentinel-2 study areas are about 12d and 10d, respectively, and the widths are 290 and 375 km, respectively.

15.1.4 Sentinel-1 and Sentinel-2 data are also defective

The self-discontinuity problem of Sentinel-1 due to systematic errors cannot be completely solved even after correction. Sentinel-1 data will appear as data distortion in the pixels of the image in the complex terrain area.

Remote sensing has been used for soil salinity mapping for many years. However, there is no generally accepted method to extract soil salinity parameters suitable for different environments from remote sensing data. At field and local scales, many studies have proposed translation models from remotely sensed variables to ground-level soil salinity levels. However, these models usually use different data sources in different parts of the world and do not have the same high precision and generality (Hoa et al., 2019; Taghadosi et al., 2019; Tripathi & Kamal Tiwari, 2020), so a unified data source is needed to establish a universal model.

Data fusion techniques to integrate information from two complementary sensor types are emerging as an efficient way to unify data sources (Ghamisi et al., 2019). If the salinized land is located in an area with high cloud cover, its distribution can be more reliably mapped by fusing radar data. However, there are many different data fusion techniques, and little information is available for salinization mapping. To fill this knowledge gap, this study compared three common data fusion methods and applied them to the soil salinization distribution map in Da'an City, western Jilin Province. The first data fusion method is raw pixel-oriented classification, the second method uses object-oriented classification, and the third method uses principal component extraction followed by pixel-oriented classification.

This chapter aims to establish a soil salinization extraction based on fusion between Sentinel-1 radar and Sentinel-2 optical sensor under the condition of surface vegetation cover. Specifically, it mainly solves the following problems:

1. Generate continuous optical and radar data features;
2. Establish a robust and efficient salinization classification model;
3. Combine Sentinel-1 and Sentinel-2 data. The resulting uniform soil salinization degree distribution.

15.2 Research data

The data used in this research were listed next:

15.2.1 Sentinel-1 data

In the research, Sentinel-1 preprocessing uses the same steps as the Sentinel-1 Toolbox (SNAP) process. The specific principles are explained in Chapters 2 and 3. The data are processed here as follows: apply orbital file, radiometric scaling, terrain correction, mosaicking, stacking, scatter filtering, and finally, resampling the data to a resolution of 10 m.

15.2.2 Sentinel-2 data

Sentinel-2 is a multispectral optical satellite launched by ESA in 2015 with a width of 290 km. The spectral imaging sensor it carries can cover 13 bands from visible light to near-infrared, among which the spatial resolution of the b1, b9, and b10 bands is too low (>30 m), and only 11 of them are used in this chapter, see Table 15.1 for details. Sentinel-2 in China has a revisit cycle of about 10 days. The unique vegetation red-edge band of Sentinel-2 is helpful for obtaining surface vegetation information. This chapter downloaded two Sentinel-2 optical products from the Copernicus data center around April 10, 2018. The obtained Sentinel-2 data are an L1C-level multispectral product (MSI) as the basic input data. Sentinel-2 products were preprocessed using the Sentinel-2 tool of SNAP desktop software. Atmospheric correction was done using Sen2Cor, outputting an L2A grade product. Sentinel-2 bands 5, 6, 7, 8a, 11, and 12 (all SWIR) have a native resolution of 20×20 m, which was resampled to 10×10 m using the bilinear interpolation method in the SNAP

Table 15.1 Feature set and formula.

Name	Symbol	Formula
Vegetation	NDVI	$(b8\text{-}b4)/(b8+b4)$
Radar feature	VV	γ_{vv}^{0}
	VH	γ_{vh}^{0}
	SPAN	$\mathrm{SPAN}=\sqrt{(\gamma_{vv}^{0^2}+\gamma_{vh}^{0^2})}$
	RVI	$\mathrm{RVI}=\gamma_{vh}^{0}/(\gamma_{vv}^{0}+\gamma_{vh}^{0})$
	NDI	$\mathrm{NDI}=\gamma_{vv}^{0}-\gamma_{vh}^{0}$
	SDI	$\mathrm{SDI}=(\gamma_{0_{vv}}^{2}-\gamma_{0_{vh}}^{2})/(\gamma_{0_{vv}}^{2}+\gamma_{0_{vh}}^{2})$
Red-edge	REII	$(b3+b6)/2$
	RENDS1	$(b5\text{-}b8a)/(b5+b8a)$
	RENDS2	$(b6\text{-}b8a)/(b6+b8a)$
	RENDS3	$(b7\text{-}b8a)/(b7+b8a)$
	RESI1	$(b3\text{-}b5)/(b3+b5)$
	RESI2	$(b3\times b6)/b2$
	RESI3	$(b5\times b8a)/b3$
	RESI4	$(b2+b7)^{1/2}$

tool, so that they match the resolution of the other frequency bands. Finally, Sentinel-1 and Sentinel-2 data paths are registered and trimmed and mosaicked according to the scope of the study area.

15.2.3 Land cover label

According to the distribution of land types in the research area, the most common land cover types in the area are selected, and the research areas are divided into slightly, moderately, highly, and extremely salinized land and water bodies in nonsaline-alkali land, construction land, and other seven categories. Reference data were collected by visually interpreting Google Earth imagery at approximately the same time period as Sentinel-2 (April 7–17), the Google Earth imagery used was SPOT imagery (from Digital Globe) at its spatial resolution about 1.5 m. The most common types of salinization degree in the study area were selected and other land types interpreted after investigation, including slight salinization, moderate salinization, high salinization, extreme salinization, water body and construction land, and others.

15.2.4 The workflow of this method

Fig. 15.2 shows the procedures of the research.

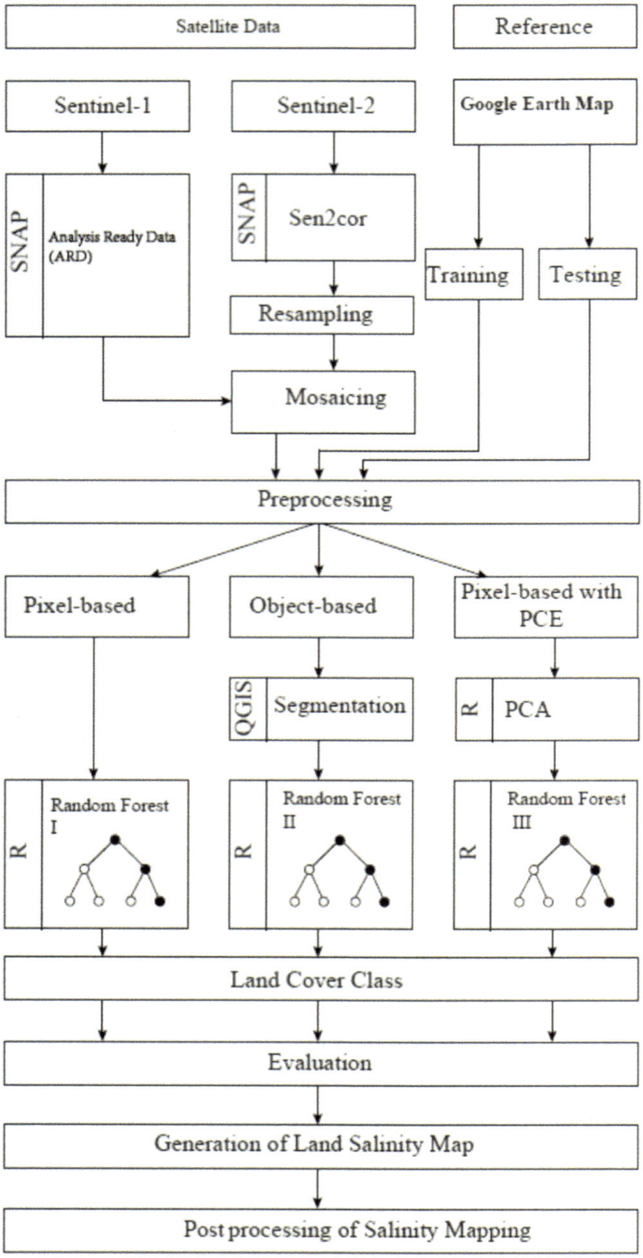

FIGURE 15.2

Soil salinity mapping using fused Sentinel-1/2.

15.3 Method

15.3.1 Feature set

Build a feature set data cube with optical vegetation index, red-edge index, and radar features. In soil salinization monitoring, vegetation is one of the key factors affecting the prediction accuracy. Although vegetation will obscure soil information, under high vegetation coverage, the vegetation index is more sensitive to changes in soil salinity. The radar characteristic variables are the back-term scattering coefficients of the VV polarization and the VH polarization. In remote sensing images, various salinity indices are used for monitoring and mapping of soil salinization due to the different blue, green, red, and near-infrared spectral characteristics of salinized soil. However, most of the indices currently constructed are concentrated in the visible light and near-infrared bands, and there are few studies in the short-wave near-infrared and red-edge spectral bands. The three red-edge spectra (b5–b7) unique to Sentinel-2 optical data are used to improve soil salinity monitoring and offer new opportunities. Eight red-edge salinity indices were calculated, and all the features are shown in Table 15.1.

15.3.2 Classification standard and sample making

1 Classification criteria

The samples unrelated to salinization are mainly based on the visual identification of field survey knowledge and high-resolution remote sensing images. The soil salinization classification is based on the sampling. In areas covered by vegetation, the visual interpretation method of salt-tolerant plants shall be adopted, and the classification standard can be referred to in Table 15.2.

2 Sample making in this study

The same reference dataset is used for all classifications, which consists of 210 training points and 70 testing points, and is equally divided among seven categories (Table 15.3), to ensure that the reference data are evenly distributed throughout the study area.

Table 15.2 Visual interpretation references for land salinity.

Level	Salt-tolerant plants
Slight	Foxtail, Purple dandelion
Moderate	Reed, Suaeda thorn
High	Light gray
Extreme	Bare soil, crusted, whitish

Table 15.3 Number of samples of various landcovers.

Item	No.							Sum
	Slight	Moderate	High	Extreme	Water	Construction land	Others	
Train	40	31	29	32	12	31	35	210
Test	14	10	12	13	6	7	8	70
Sum	44	41	41	45	18	38	43	280

Table 15.4 Key parameters of random forest.

Parameter	Value
n_estimator	25
max_depth	15
min_samples_leaf	1

15.3.3 Random forest–based classification

Based on the random forest method, this study uses three different modes to map land salinization: raw pixel-oriented classification, object-oriented classification, and pixel-oriented classification after principal component extraction. There are three key parameters of random forest model parameters, the maximum tree depth (max_depth), the number of features (n_estimator), and the minimum number of samples contained in the minimum leaf node (min_samples_leaf). The parameters of random forest classification are automatically obtained according to the Grid Search parameter test (See Table 15.4).

For each method, fused data were compared with individual optical and radar data sources. All three methods use random forest classification, which has shown excellent performance in land-use and cover change studies using various remote sensing datasets.

The object-oriented classification method considers the spectral, spatial, and other information of the image and suppresses the "salt and pepper phenomenon." This study uses open-source image segmentation tools in GRASS 7.0.5 to create objects and associated summary statistics. The GRASS image segmentation tool uses a region growing and merging technique, where all pixels are used as seeds, and the similarity between fragments and unmerged objects is used to determine which objects to merge. According to various attempts, the threshold for binning is 0.01, and the minimum target is five pixels. The features included in a random forest for object-oriented classification are the mean and standard deviation of the objects in each layer of the feature cube.

For pixel-oriented classification using principal component extraction, after principal component analysis, the extracted components are input into random forest classification as predictors. Finally, the classification using only optical data includes seven principal components, and the classification using only radar data, two principal components, and nine components for classification using data fusion.

To compare the performance, overall accuracy (proportion of correctly classified cases) and category-specific producer and user accuracy were calculated for saline land using the validation sample. Producer accuracy is calculated by dividing the number of correctly classified test points by the total number of test points for that category, and it represents the frequency with which the ground is correctly displayed on the classification map. The user's accuracy is calculated by dividing the number of correctly classified test points by the total number of test points classified into that category, and it expresses how often the category on the map actually appears on the ground.

15.4 Results

Data fusion improved overall accuracy by a statistically significant level compared with the case of pixel-oriented classification using only optical data (Table 15.5). Fusion does not significantly improve overall accuracy compared with using only optical data in the case of using Object-based classification and Pixel-based classification with Principal Component Extraction (PCE). In all three methods, fusion has a positive effect on some classes, while it has a mixed or even negative effect on others. Results using only radar data show lower overall accuracy, but the highest accuracy for object-based classification.

Object-based classification using data fusion had higher overall accuracy than the other two fusion methods ($P < .02$ in both cases). It also has the highest user and producer accuracy for classification of salinity land (Table 15.6).

Based on the random forest method and training data, soil salinization mapping was performed in the study area, and Fig. 15.3 was obtained.

The land cover map generated by data fusion using the classification method can be seen in Fig. 15.3. All methods detected the main various saline soils in the region. The resolution provided by remote sensing data fails to identify many roads as land subject to salinized imagery. After postprocessing, the road layer is

Table 15.5 Overall accuracy of different methods (%).

Method	Pixel-based	Object-based	Pixel-based with PCE
Fusion	91.3*	92.1	87.6
Optical (Sentinel-2)	86.3	91.5	86.9
SAR (Sentinel-1)	67.5	66.4	57.9

**Significant difference compared with optical at 95% confidence level (McNemar's Test $|z| > 1.96$)*

Table 15.6 Accuracy for three different methods of salinization classification (%).

Method	Accuracy%	
Pixel-based	Producer accuracy	88.7
	User accuracy	92.1
Object-based	Producer accuracy	92.3
	User accuracy	90.4
Pixel-based with PCE	Producer accuracy	89.5
	User accuracy	92.8

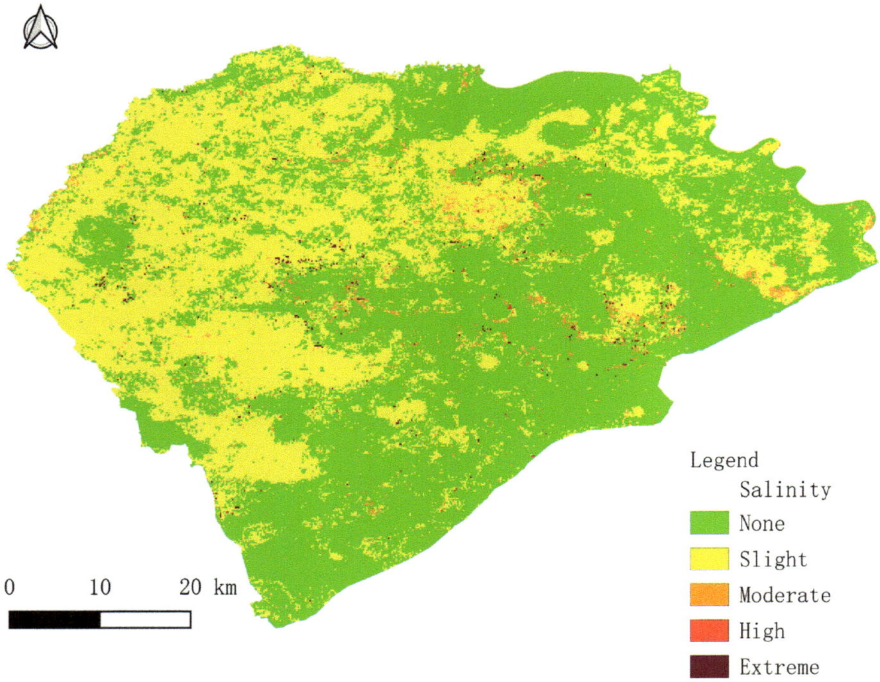

FIGURE 15.3

Sentinel-1/2-based soil salinity classification.

identified and masked, the classification effect is further improved, and the degree of distinction of each salinized area is also increased visually.

In a conclusion, by further considering the surface vegetation and making full use of the characteristics of optical data, the backscattering features of Sentinel-1 images are extracted and the multiple remote sensing indices and normalized vegetation indices of Sentinel-2 are fused. Input into the random forest model of automatic parameter adjustment for classification, compared with the single method using SAR, the effect of soil salinization classification and extraction is improved.

Reference

Ghamisi, P., Rasti, B., Yokoya, N., Wang, Q., Hofle, B., Bruzzone, L., ... Benediktsson, J. A. (2019). Multisource and multitemporal data fusion in remote sensing: A comprehensive review of the state of the art. *IEEE Geoscience and Remote Sensing Magazine*, *7*(1), 6–39.

Gritsevich, M., Nico, G., Conde, V., Mateus, P., & Catalao, J. (2020). On the computation of phase inconsistencies of Sentinel-1 interferograms over snow-covered areas, May *In EGU General Assembly Conference Abstracts*, 21946.

Hoa, P. V., Giang, N. V., Binh, N. A., Hai, L. V. H., Pham, T. D., Hasanlou, M., & Tien Bui, D. (2019). Soil salinity mapping using SAR sentinel-1 data and advanced machine learning algorithms: A case study at Ben Tre Province of the Mekong River Delta (Vietnam). *Remote Sensing*, *11*(2), 128.

Schofield, R. V., & Kirkby, M. J. (2003). Application of salinization indicators and initial development of potential global soil salinization scenario under climatic change. *Global Biogeochemical Cycles*, *17*(3).

Taghadosi, M. M., Hasanlou, M., & Eftekhari, K. (2019). Soil salinity mapping using dual-polarized SAR Sentinel-1 imagery. *International journal of remote sensing*, *40*(1), 237–252.

Tripathi, A., & Kamal Tiwari, R. (2020). *A simplified sub-Surface soil salinity estimation using synergy of Sentinel-1 SAR and Sentinel-2 multispectral satellite data, for early stages of wheat crop growth in Rupnagar, Punjab, India*. Land Degradation & Development. Available from <http://doi.org/10.22541/au.159863386.65555086>.

Zhuang, Q., Shao, Z., Huang, X., Zhang, Y., Wu, W., Feng, X., ... Altan, O. (2021). Evolution of soil salinization under the background of landscape patterns in the irrigated northern slopes of Tianshan Mountains. *Xinjiang, China. Catena*, *206*, 105561.

CHAPTER

Precision nutrient management for field and horticultural crops

16

Bhaswati Sarmah[1,2], Vishwa Jyoti Baruah[3], Marami Dutta[4], Mahima Begum[2,5] and Bipul Deka[6]

[1]*Department of Plant Breeding and Genetics, Assam Agricultural University, Jorhat, Assam, India*

[2]*AICRP on Potato, Assam Agricultural University, Jorhat, Assam, India*

[3]*Centre for Biotechnology and Bioinformatics, Faculty of Biological Sciences, Dibrugarh University, Dibrugarh, Assam, India*

[4]*Department of Soil Science, Assam Agricultural University, Jorhat, Assam, India*

[5]*Department of Agronomy, Assam Agricultural University, Jorhat, Assam, India*

[6]*AICRP on Irrigation Water Management, Assam Agricultural University, Jorhat, Assam, India*

16.1 Introduction

The motive of the Green Revolution has been to maximize the yield from available land and water resources without deteriorating the environment (Gustafsson et al., 2013). Several innovative technologies and approaches for crop production and management have been developed and evolved over the years to achieve this target. Precision agriculture is an identified strategy that merges the developed agricultural industry with the new technological know-how of the information age (Dige, 2020). As opposed to conventional agriculture, precision agriculture is an amalgamation of different crop management systems that involve managing whole fields as small areas within fields by complementing the exact kind and amount of inputs with the actual crop needs (Patil Shirish & Bhalerao, 2013). With the advent of new mechanical and artificial intelligence technology, the concept of precision agriculture realized its practical use with information as the key ingredient (Beloev et al., 2021). Economic and environmental aspects along with on-field management and proper infrastructure facility need to be considered if we want the precise management of agronomic practices to be feasible (Davis et al., 1998).

Precision nutrient management or site-specific nutrient management (SSNM) is a critical component of precision agriculture. It is an alternate strategy for fulfilling the nutrient demand of crops in a very precise way for the timely application of fertilizers (Johnston et al., 2009). It can help to overcome the problem associated with the blanket application of fertilizers, such as

Remote Sensing in Precision Agriculture. DOI: https://doi.org/10.1016/B978-0-323-91068-2.00013-8

wastage of nutrients, which leads to poor nutrient use efficiency, low profit, improved productivity, sustainability, as well as create more awareness toward environmental pollution to check climate change (Pampolino et al., 2012). As the timely application of agronomical inputs is the prime factor in determining the quality and quantity of produce, the principle of precision nutrient management relies on selecting the right product, accurate dose or rate, proper time of application, and place (Pathak et al., 2003). The best part of this is that the farmers need not perform these activities on their own, nor any supervision by experts is required, as there are modern versions of smarter robots and mobile autonomous systems, which have the decision-making ability based on the input parameters (Amer et al., 2015; Beloev et al., 2021). The word "Precision" refers to the ability of the technology to repeat its own measurement in the same location and time. In contrast, the accuracy of the technology discusses how well the sensor measurements can correlate to an actual field condition that is determined using the conventional measurement technique (Shafi et al., 2019).

16.2 Benefits of precision nutrient management

16.2.1 Precise nutrient uptake and utilization

The poor nutrient-use efficiency (NUE) of the major elements, including N, is the primary reason for poor crop production. Theoverall nutrient use efficiency of N, P, and K is lower than 50%, 15%, and 40%, respectively, and in the case of micronutrient, it is even lesser than 5% under normal soil-applied conditions. NUE in the case of cereal such as maize is only 30%–40% which is mainly due to more temporal as well as spatial variability of soil-applied *N* that restricts efficient use of N fertilizer (Freeman et al., 2007). Therefore, to enhance the NUE in cereals, several strategies such as leaf color chart (LCC) (Singh et al., 2006), GreenSeeker (GS), and SPAD meter have been practiced that helped in précised nutrient management (Prakasha et al., 2020). For example, applying nitrogenous fertilizers through SSNM resulted in 63% greater NUE in wheat compared with conventional practice (Sapkota et al., 2015). The precise application of nutrients reduced the loss of nutrients under SSNM that helps in achieving better nutrient use efficiency as compared with the blanket dose of fertilizers (Majumdar et al., 2012).

In the maize-wheat system, the use of precision nutrient management approaches based on nutrient expert (NE) system and GS, along with precise drilling of fertilizers in permanent beds, significantly increased the nutrient use efficiency, crop productivity, and profitability with a drastic reduction in global water potential (GWP) in comparison to farmers' fertilizer practice (broadcast) under conventional tillage (CT) system (Jat et al., 2021).

16.2.2 Judicious use of fertilizers

Application of nitrogenous fertilizer based on 90% sufficiency index in direct-seeded rice was able to save almost 50 kg/N ha without any grain yield reduction (Singh et al., 2006). Almost 9.5 and 30 kg/N could be saved through precision nutrient management with the help of GS and SPAD in Maize (Swinton & Lowenberg-DeBoer, 1998). The strong correlation between the NDVI with the total biomass accumulation and green leaf area (Bosquet et al., 2011) resulted in higher photosynthetic efficiency, and eventually, more crop growth and yield were observed in wheat. Singh et al. (2010) also reported a yield advantage with 20%–30% less N fertilizer use due to real-time nitrogen management (RTNM) based on SPAD Value in rice.

Chlorophyll meter (SPAD) and LCC are found to be very efficient noninvasive methods for regulation of N status in high nutrient feeder crops such as potatoes (Singh et al., 2017). Almost 10%–20% of fertilizers could be saved than the recommended dose of fertilizers by increasing the number of split applications of N with the help of Chlorophyll meter (SPAD) (Singh et al., 2019).

16.2.3 Maximization of yield per unit area

Precision nutrient management is the primary strategy for maximizing productivity under limited available resources as both excessive and suboptimal doses of chemical fertilizers negatively impact farmers' economies. Therefore, applying need-based nutrients to the crop through remote sensing enhances the farmer's profitability by maximizing crop productivity (Scharf et al., 2011). It significantly increased the grain yields of maize, rice, wheat, and other important crops over conventional fertilizers application (Saggar et al., 2013). In maize, the supply of nutrients with a NE with a targeted yield of 10 t/ha significantly increased the crop yield agronomic efficiency of N, P, and K with a more economic return (Joshi et al., 2018). Compared with farmers' fertilizer practice, this method based on nutrient management strategy increased maize, rice, and wheat crop yield by 10%–12%, 5%–8%, and 12%–15%, respectively. N management based on GS accounted for an increase of 8% grain yield in rice and 12% in wheat (O'Connor et al., 2019).

Compared with the blanket dose of fertilizer, split application of fertilizers based on GS and NE promoted the plant growth and yield in wheat (Kaur, 2017). The NE decision support system (DSS) also produced a similar grain yield from the 150% recommended dose of fertilizers, significantly higher than the recommended dose of fertilizers (Kumar & Singh, 2019). Varying results in terms of grain yield were also reported by Singh et al. (2020). According to their report, SSNM produced more yield than recommended dose of fertilizer (RDF) in a few locations, while yield produced in some other locations was found to be equivalent to RDF. In addition to increasing grain and stover yield and harvest index, the nutrients applied based on SSNM produced more plant height and leaf area index than the RDF in Maize (Joshi & Chandrashekar, 2017). The improvement in plant growth and other yield attributes leads to higher production. Apart from saving fertilizers up to 20–30 kg/ha in rice/wheat-based cropping system, the practicing of SSNM-based approaches also increased the grain yield up to 8%–12% (Parihar et al., 2020).

In RTNM, N is generally top-dressed (urea) after the actual SPAD value drops below the critical SPAD limit. The fertilizers application based on RTNM greatly curtailed the amount of fertilizers compared with a fixed rate of fertilizers and farmers' practice without affecting yield. Apart from increasing the soil available N, leaf N content, the top-dressing of 25 kg/N ha (SPAD value S36 in rice and S40 in wheat) significantly increased grain yield of both rice (5215 kg/ha) and wheat (4483 kg/ha) (Mainak et al., 2020). This indicated that both crops performed better under medium SPAD value as compared with lower SPAD value. GS is another RTNM tool that helps efficiently use nitrogen. In maize, the precision nutrient management practices such as the NE, GS, enhanced the plant growth parameters and increased the seed yield with a maximum economic return over the state recommendation (Shyam et al., 2021). The increase in plant growth parameters in Mmaize through SSNM was also reported by Majumdar et al. (2014).

16.2.4 Environmental sustainability

In the case of blanket dose of fertilizers, excessive use of fertilizer nutrients may lead to groundwater contamination due to leaching of NO_3 as well as environmental hazards (Sapkota et al., 2015), because of emission of nitrous oxide (N_2O) (Segarra et al., 2020; Vistoso et al., 2012). Apart from increasing the grain yield, balanced use of nutrients through NE-based SSNM helped save energy balance and soil moisture (Parihar et al., 2017).

By applying sufficient quantities of plant nutrients through SSNM, the excessive use of fertilizers in the environment can be minimized (Jat et al., 2013). In addition to this, SSNM based on the NE DSS reduced the greenhouse gas emission from the field under no-tillage (NT) as compared with CT (Sapkota et al., 2014). The more GWP under CT might be due to excessive use of machinery and fuel for repeated tillage operations that emitted a considerable amount of greenhouse gases.

16.2.5 Enhanced profit

Many researchers reported that the adoption of SSNM was economically profitable over farmers' fertilizer practice (Parihar et al., 2017). They reported a savings of Rs 2500–3200 ha^{-1} under SSNM compared with farmers' practice. However, the benefit of precision nutrient management varied depending upon factors such as field type, soil condition, and crop management (Majumdar et al., 2014). The SSNM performed better under conservation agriculture (permanent bed) than conventional agriculture and, consequently, more economic productivity and economic return in maize-wheat system (Jat & Bijay-Singh, 2014). Nutrient management based on NE and green added extra benefit of Rs 10,000 in Maize over state recommendation (Shyam et al., 2021). In terms of monetary benefit, Singh et al. (2020) also offered their remarks that nutrient management based on NE-SSNM could be an efficient approach for the maize, particularly in the North-Western Plain Zone (NWPZ), North-Eastern Plain Zone (NEPZ), and Peninsular Zone (PZ) of India. In earlier studies, several researchers reported the yield advantage with NE-SSNM (Parihar et al., 2017; Pooniya et al., 2015).

Similarly, a more economic return and higher B: C ratio could be obtained by increasing the number of split applications of N in potato-based on SPAD value (Singh et al., 2019). The reduction of volatilization/leaching losses of N through splitting resulted in higher NUE that resulted in more crop growth and, ultimately, more monetary benefit.

16.3 Tools and techniques in precision nutrient management

Agricultural production has increased dramatically in India due to the combined efforts of farmers, researchers, industry workers, and nutrient management practices. Conjectures made in the 20th century predicted that nutrient management in the 25th century would rely increasingly on computer models and satellite-based technologies such as precision nutrient management (Soil and Plant Analysis Council, 1994). Precision nutrient management is a practice in which, in a site-specific manner, nutrients are applied to plants to optimally match their inherent spatial and temporal needs for supplemental nutrients by using different tools such as remote sensing, global positioning system (GPS), geographic information system (GIS) systems, variable-rate technology (VRT), and yield monitoring. In agricultural crops, underfertilization may reduce yield, and overfertilization can harm the environment. With the adoption of this technology, it has become possible to manage soil nutrient variations over a stretch of the field with prescription fertilizer applications. It has been observed that the cultivable soils in India are mostly acidic in nature with varying pH in different locations. Employing methods such as remote sensing and combining data in a GIS can help detect nutrient stresses and site-specific applications of fertilizers and soil amendments. It also increases fertilizer use efficiency and reduces nutrient losses.

Precision nutrient management techniques have been developed over the last 25 years and were initially applied to field crops. Many small and portable tools such as the GS chlorophyll meter have been used for precision nitrogen management in rice, wheat, and maize (Martin et al., 2012). These are considered the most efficient and cheapest tools used in precision nutrient management. Similarly, LCC has demonstrated to be an efficient and quick tool for assessing N requirement in standing rice crops (Singh et al., 2002), while the Chlorophyll meter plays a significant role and is the most effective N management tool in potato as the chlorophyll content of leaves enormously relies upon the N status of the crop (Jongschaap & Booji, 2004). Remote sensing is another cost-effective technique broadly used for large areas that consider both spectral and thermal variation for quick determination of crop nitrogen status (Yousfi et al., 2019) and applying a different dose of fertilizers in the crop field (Olfs et al., 2005). The visible and near-infrared (NIR) spectral radiation that has been used in remote sensing provides information on leaf chlorophyll content. As the chlorophyll content of crop canopy is highly positively corelated to nitrogen content in leaves, therefore by determining the chlorophyll content through remote sensing, the nitrogen deficiency in plants can be identified early (Guerif et al., 2007).

Applying precision nutrient management tools and techniques to tree and vegetable crops started later, while fruit applications were not realized until the 1990s. The reason for this delay in technology transfer may be due to an absence of an automated way of recording yield and quality data (Ehsani & Karim, 2010). India's precision nutrient management system is in its infancy, yet there are numerous opportunities to adopt the technology to explore its potential, especially in high-value crops. The different technologies will help make fine-tuned decisions related to nutrient management and SSNM plans appropriate for each field.

16.3.1 Electrochemical and optical sensors

Electrochemical sensors are devised to directly or indirectly assess the spatial variability of soil fertility. Ion-selective membrane sensors are used to determine soil chemical properties, nutrients, and pH mapping (Yao et al., 2010). Electrical and electromagnetic sensor technology uses various measurement systems based on electrical circuits to determine the ability of soil media to conduct or accumulate electrical charge. Generally, soil's physical and chemical characteristics can affect circuit behavior and, thus, the measured electrical response. Adamchuk et al. (2004) opined that due to rapid response, low cost, and high durability, electrical and electromagnetic sensors had become the most attainable technique for soil mapping. The maps generated from electrical and electromagnetic sensing correlate well with soil properties such as texture, salinity, organic matter, and moisture content. The salt concentration of soil is commonly estimated via electrical resistivity or electrical conductivity (EC). Electromagnetic sensors to measure electrical resistivity and conductivity have been demonstrated on crop fields (Sudduth et al., 2003) (Fig. 16.1).

Optical sensing technology uses visible and NIR wavelength ranges to quantify soil properties rapidly. The principle of this approach is the interaction between incident light and soil surface properties, such that the reflected light varies as a function of soil physical and chemical properties (Mouazen et al., 2005). Several multispectral and hyperspectral sensors are available, which together form a wide array of optical sensors that can measure the spectral responses from

FIGURE 16.1

An electrochemical sensor (Sivakumar et al., 2022).

visible and NIR light reflected from plant canopies (Ma et al., 1996; Peñuelas et al., 1994). Determination of nitrogen, biomass variation, and leaf area index can be easily assessed by multispectral sensors such as Crop Circle and Crop Scan (Roberts et al., 2009). ASD FieldSpec, a type of hyperspectral sensor, has been used to obtain biochemical and biophysical information of the crop. Normalized-difference vegetation index (NDVI) is a spectral vegetation index that can provide the necessary information about the crop's photosynthetic efficiency and yield potential (Bronson et al., 2011; Raun et al., 2002).

The GS canopy sensor is an optical sensor that has been used in various crops such as rice, wheat, maize, barley, sugarcane, and cotton to measure leaf area index, green biomass, and photosynthetic efficiency (Aparicio et al., 2002; Heege et al., 2008; Portz et al., 2012; Raper et al., 2013; Singh et al., 2006; Söderström et al., 2010; Tremblay et al., 2009). A passive multispectral scanner, The Yara N-Sensor (Yara International ASA, Oslo, Norway), is effective in measuring the crop's chlorophyll content and biomass and can also determine the N status of the crop, thereby adjusting the N-fertilizer rates (Raper & Varco, 2015; Raper et al., 2013). Although this sensor effectively identifies soil physical, chemical, and mechanical properties (Adamchuk et al., 2004), it lacks the precision to determine the variation between cultivars of the crops (Fig. 16.2).

Optical nutrient-sensing techniques are nondestructive and are often more favored in comparison to electrochemical sensing (Chang et al., 2001; Rossel et al., 2006). The prominent limitation of electrochemical technology is that the values obtained may not be as accurate as a laboratory test, and a high sampling density may be required to increase the overall accuracy of the resulting soil nutrient or pH maps. On the other hand, optical soil sensors have a high potential for estimating soil organic matter content based on soil color (Adamchuk et al., 2004). In optical sensing of soil, the visual and NIR spectral reflectance can

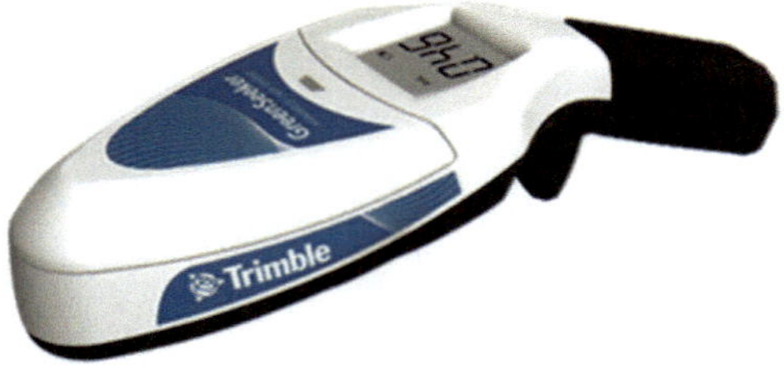

FIGURE 16.2

A GreenSeeker.

Adapted from Onwuchekwa-Henry et al. (2022).

potentially estimate texture, moisture, cation exchange capacity (CEC), and other soil parameters if proper data analysis techniques are applied.

16.3.2 Chlorophyll meters

Chlorophyll meters are reliable alternatives to traditional tissue analysis as plant N nutritional diagnostic tools. The most widely used chlorophyll meter is the handheld Minolta SPAD-502. The SPAD 502 chlorophyll meter (soil–plant Analysis Development) is a quick, nondestructive handheld device developed by Minolta, Osaka, Japan in the year 1989. It instantly provides an estimate of leaf nitrogen status as chlorophyll content (Boggs et al., 2003; FeiBo et al., 1998) by clamping the unplucked leafy tissue in the meter using two LEDs (light-emitting diodes) emitting red ($\ddot{e}$ = 650 nm) and infrared light ($\ddot{e}$ = 940 nm). The chlorophyll content of the leaf is presented in random units (0–99.9), and the meter readings are unitless, which need to be calibrated with chlorophyll or N content and leaf greenness. Another type of chlorophyll meter is Fields count CM 1000 chlorophyll meter (Spectrum Technologies, Inc. 2009). It uses point-and-shoot technology to estimate relative chlorophyll content instantly. It integrates a 1.27 cm diameter area at a distance of 30 cm from the plant leaf.

Fertiliser-N is applied whenever chlorophyll meter reading is less than the preset threshold value. The SPAD threshold value, which represents the limit below which a reduction in yield occurs, must be preestablished. Fertiliser-N applications are necessary below this threshold value to avoid yield loss. Peng et al. (1996) gave critical SPAD value that farmers could refer to in the rice field. The SPAD values of the index leaf are monitored at 7–10 days intervals starting from 15 days after transplanting till the initiation of flowering. In-season top dressing of 30 kg/N ha was recommended whenever SPAD value fell below the critical value of 35 for rice cultivar IR72 grown in the dry season in the Philippines. The threshold SPAD value of 35 correlated to 1.4 g/Nm^2 leaf area, a number that was found to be fairly stable for the high-yielding IR72 cultivar during the dry

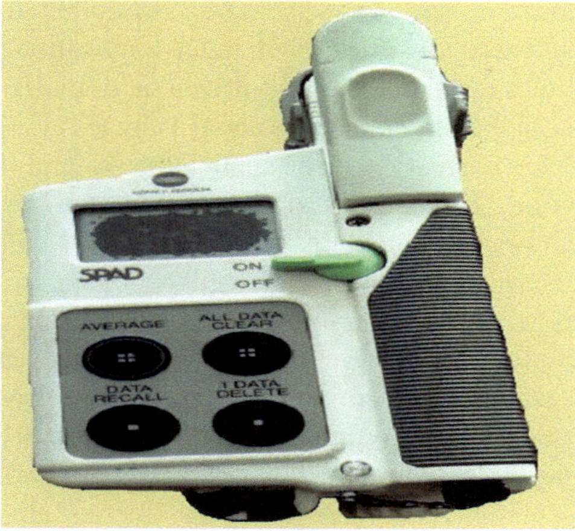

FIGURE 16.3

A chlorophyll meter.

growing season. The use of critical SPAD 35 reading resulted in similar yields with less fertilizer N and higher agronomic efficiency compared with fixed split-timing applications (Peng et al., 1996). The SPAD value of 35 was also found to be the appropriate threshold value for guiding need-based N management in transplanted rice in South India. Nevertheless, the SPAD meter threshold value of 35 is not universal and may vary in different rice-growing environments. For example, it was 32 during the wet season when there is continuous cloud cover for most of the growing season (Peng et al., 1996) (Fig. 16.3).

It was observed that the critical SPAD value of 37.5 is appropriate for guiding need-based N top dressing in rice in Pakistan (Hussain et al., 2003). In Bangladesh, significantly higher yields were obtained with 3–12% less fertilizer-N use in comparison to the blanket recommendation by using SPAD value 35 as the threshold SPAD value (Win et al., 2003). Maiti et al. (2004) found the SPAD value of 37 to be appropriate for achieving high yields with 27.5–45.5 kg/N ha less fertilizer as compared with a blanket dose of 150 kg/N ha. Huan et al. (2002) worked with 32 and 29 SPAD threshold values for need-based N management in direct wet-seeded rice. The SPAD 32 and 29 based N applications produced grain yield equivalent to the blanket fertilizer N recommendation of 120 kg/N ha but with 20 and 40 kg/N ha less than the blanket recommendation.

16.3.3 Variable-rate technology

An integral part of precision agriculture is VRT, where agricultural inputs are used for site-specific requirements in different parts of the field. If the field area is small, inputs can be applied manually. Variable-rate machineries use (1) correct positioning in the field with a differential GPS, (2) correct spatial variability information at the location, and (3) variable-rate controllers for automatic precise fertilizer application based on prederived input application maps or on-the-go sensing and calibrated fertilization. Map-driven and on-the-go sensor tractor-based VRT systems, including unmanned aerial vehicles (UAVs) such as drones, are effectively used for precise fertilizer application in large fields (Abbas et al., 2020; Doddamani et al., 2020). The crop sensing system can be operated at any time irrespective of weather condition, foggy or cloudy based on principles of active and passive sensing. It can be mounted on booms in most sprayers and spreaders and works with most variable rate controllers and delivery systems. It uses optical sensors for quantitative measurement of the crop variability and finally prepares a targeted prescription to manage the crop variability. This variable-rate application has many implications in addition to reducing the application of agrochemicals, including fertilizer and irrigation water, but major constraints with the technology are that VRT machineries are costly at the farm level, which should be tackled through adequate policy instruments (Fabiani et al., 2020).

In north-central Missouri, the profitability and water quality were evaluated with VRT for corn production on four claypan soil fields under average to better than average weather conditions for nitrogen and lime (Wang et al., 2003). Variable doses of N and lime were applied based on the topsoil depth and pH of the soil, respectively, and the VRT rates among the two modes of application were compared. The higher variation in topsoil depth and soil pH resulted in higher profitability and greater water quality benefits with VRT (Hedley & Yule, 2009). They found that the thickness of the topsoil is an indicator of soil productivity and crop yields for claypan soils. Similar findings reported that the varying N rate based on topsoil depth potentially increases N use efficiency and corn yield and reduces N losses (Gantzer & McCarty, 1987; Kitchen et al., 1998). Soil profile EC measurements provide an accurate and efficient mapping of topsoil depth in claypan fields (Kitchen et al., 1999). Soil EC mapping has also been used in other regions to help farmers assess soil quality and apply precision agricultural practices (Doerge, 2000). A distinctive aspect of this investigation was that it evaluated the economic and environmental benefits of VRT using topsoil depth data measured by soil EC. It also builds on earlier research that developed variable N application based on EC-measured topsoil depth for claypan soils (Kitchen et al., 1995, 1998, 1999).

Variable-rate application is the major target for precision nutrient management in fruit cultivation. Lopez-Granados et al. (2004) prepared site-specific fertilization maps based on leaf nutrient spatial variability in olive trees. They reported that consistent saving in N, K, and P fertilizers could be achieved if a differential fertilization program was based on spatial variability of leaf nutrient status of the trees. Fountas et al. (2011) manually applied to each olive tree P, K, and pH based on prescription maps from soil analysis. Soil analysis data and nutrient removal from the soil by the crop was used to create prescription maps for fertilizer application in apples (Aggelopoulou et al., 2010, 2011) and blueberry (Farooque et al., 2012).

Where variable soil exists under one irrigation system, the application of VRT to the irrigation system was useful to reduce drainage and nutrient leaching loss along with drainage (Hedley & Yule, 2009). Variable-rate modification for sprinkler systems was developed by (Bradbury et al., 2013) within 3 years of developing the irrigation prescription map method. It facilitates individual sprinkler control and is programmed to vary irrigation to different soil management zones under the same irrigation system (Hedley et al., 2009). The mapping method uses electromagnetic soil mapping to quantify soil variability based on soil texture and soil moisture (Hedley et al., 2004). These maps are then converted to irrigation prescription maps by characterizing soil water-holding properties within each management zone defined on the electromagnetic map. Irrigation varies with each management zone due to the difference in soil texture. Each textural class soil reaches the critical soil moisture deficit required for irrigation on different days. In this variable-rate irrigation, calculated saved water is modeled in different sites, showing 9%–26% water saving with equivalent energy savings. Modeled drainage and runoff were also reduced by 0%–55% during the irrigation period, with a subsequent reduction in risk of N leaching. The technology positively impacts crop water productivity, water and energy conservation, and the environment, yet it is still underutilized. Technological modification is commercially available, but further research is needed to develop a user-friendly decision-support system that can define management zone, sensing within-field variability in real time, and adaptively control site-specific variable-rate water applications.

16.3.4 Geographic information system and global positioning system

Increasing NUE and minimizing its adverse impacts on soil and the environment have been the focal points in the world for the past few decades. The NUE or FUE can be optimized by applying the fertilizer with the right source of nutrients, at the right rate, at the right time, at right place, and with the right tools. In such context, geoinformatics-aided SSNM, etc., are some of the most viable options for improving NUE and maintaining a sustainable food production system (Bronson et al., 2011; Lowenberg-DeBoer & Aghib, 1999).

Geospatial technologies such as GIS and GPS have an enormous potential for spatiotemporal analysis and monitoring in agriculture (Sesha Sai et al., 2010; Verma et al., 2020). Retrieval of soil properties, namely soil organic carbon, salinity, moisture, pH, nutrient status, etc., using spectral reflectance along with spectral indices from satellite data and its spatial variability using different spatial interpolation methods are used for site-specific input management for agricultural sustainability (Kaur et al., 2020).

A GIS is the systematic collection of computer hardware, software, geographic data, and personnel designed to efficiently capture, store, update, manipulate, analyze, and display all forms of geographically referenced information. Precision farming deals with spatial and temporal variability. The spatial analysis capability of GIS enables precision agriculture.

Computer GIS software, mobile GIS apps, and geo-web services had been developed for geospatial analysis. GIS software also incorporates a database and DSS that uses soil and plant nutrient attributes and location data for precision nutrient management. Multiple layers of information, namely remotely sensed data, crop yield, soil survey maps, thematic nutrient variability map, etc., could be stored, analyzed, retrieved at will, and displayed in both dataset form and map. Spatial variability mapping of soil nutrients using geostatistical techniques and management zone (MZ)-based soil test crop response are recently developed approaches for precision nutrient management (Moharana et al., 2019). Development in information and communication technologies (ICTs), remote sensing technologies, data fusion, namely k-means clustering algorithm (Franzen et al., 2016; Hartigan & Wong, 1979) and geostatistical interpolation techniques (kriging) have been delineated with good precision and reliability into MZ, making it an essentially practical procedure in commercial precision agriculture.

Spatial variability in the soil is important to identify the nutrient constraint zones concerning zones of production to explain the use of nutrients and use and improve the productivity factor. "Khasi" mandarin (Citrus reticulata Blanco) (geographical indicator) widely grown in Northeast India, Sikkim, and West Bengal was extensively investigated from 52 sites (grid points) of different geographical origins across 108 orchards (Srivastava & Malhotra, 2014). These grid sampling sites were georeferenced, and digitization was done. A huge database was created based on leaf nutrient composition (N, P, K, Ca, Mg, Fe, Mn, Cu, and Zn) and fruit yield. The produced data were analyzed through two diverse kinds of software-based DSS in two tiers of interpretation. The nutrients maps were generated and superimposed to identify different production zones showing high fruit yield and the lowest frequency of nutrient constraints. Leaf nutrient optima were computed in relation to targeted fruit yields using diagnosis and recommendation integrated system (DRIS) based analysis. Based on these indices, the spatial distribution of each nutrient constraint map was combined to obtain the three most productive citrus zones (Colaço & Molin, 2017). Different production zones with the extent of deficiency of nutrients and productivity in different states of NE India were identified. Identification of such production zones has huge potentiality for developing performance-based citrus industry under any geographical unit. Studies carried out in Khasi mandarin have revealed some distinct success in raising the productivity level of different production zones with the combined application of GIS and DRIS.

The spatial variability of some soil chemical properties soils under paddy cultivation in the Barat Laut Paddy Project area in Selangor was evaluated (Aishah et al., 2010). Geostatistical analyses were used to evaluate the spatial variability within-field using semivariograms and kriged maps. Kriged maps for each property were created using a geostatistical software package based on spatial dependence results, which recommended the necessity for a site-specific methodology for the nutrient management of paddy soils.

DSSs are interactive computer-based systems that help decision-makers use data and models to solve unstructured problems (Turban, 1995). Central tuber crop research institute has developed nutrient DSS software for SSNM in cassava, an important industrial crop. Cassava showed good yield with blanket fertilizer recommendation over vast areas in India, but in farmer's fields, yield cannot be increased beyond a level due to the high temporal and spatial variability of plant and soil properties. Thereafter, on-farm experiments were conducted in major cassava-growing areas of India, with the objective to validate the QUEFTS model (Janssen et al., 1990). Using ArcGIS and QGIS, prescription maps were developed for field-specific nutrient management of cassava. In this study, Soman et al. (2013) found that nutrient use efficiency parameters showed significant improvement in SSNM plots compared with the farmers' fertilizer practice.

Soil resource inventory helps create awareness of soil's potentialities and limitations for its effective use. With GIS and GPS, digital maps can be generated on the GIS platform. These digital maps are precise tools to relate the feature to any given geographical location as a strong visual impact. The concept of grid-based fertilizer application, through fertilizer recommendation, helps supply the nutrients as per the crop demand to improve crop growth and yield. The ready reckoner nutrient tables and nutrient recommendation maps were generated using interpolation technique for selected major field crops such as maize, rice, ragi, and potato (Raddy et al., 2021). These nutrient thematic maps can save a good amount of fertilizer for achieving a target yield by farmers.

Assessment of spatial variability of available plant nutrients in the soil is a viable option to identify and delineate critical nutrient deficiency zones. Large-scale mapping was recommended to identify farm-level nutrient deficiencies (Vasuki, 2010). The three-dimensional soil variability can be recorded using the interpolation technique (Cambardella & Karlen, 1999). Vasu et al. (2017) also mapped spatial variability of parameters by ordinary kriging in Thimmajipet near Hyderabad and part of Mahabubnagar district, Telangana, India. They found multinutrient deficiencies in most parts of the study area, and N was acutely deficient. Available nutrient status at the farm level was derived from spatial variability maps and identified the critical nutrient deficiency zones. Based on soil test results, nutrient management recommendations were issued to the farmers as well as policymakers for SSNM.

16.3.5 Remote sensing

Remote sensing systems used for precision agriculture refers to sensors that are mounted on satellites or aerial platforms such as aircraft and UAVs as well as Ground-based platforms (also called proximal remote sensing systems) (Santosh et al., 2014). Numerous recent satellites provide high-spatial- and temporal-resolution images; most satellite products are relatively coarse for many applications in precision agriculture. The appropriate spatiotemporal resolution required for precision agriculture is dependent on various management factors such as field size, efficiency, and adjustability of the farm equipment to apply varied rates of irrigation, fertilizer, and pesticides (Freeman et al., 2007). UAVs typically deliver advanced spatial resolution images as compared with satellites. Sensors mounted on satellites, aeroplanes, UAVs are passive sensors, and however, some satellites have active sensors. Many of the ground-based sensor platforms such as variable fertilizer rate application system, GS, and Crop circle have active proximity sensors. They can provide more accurate and reproducible NDVI and other vegetation indices used in crop nutrient status assessment. Thermal infrared and microwave sensors mounted on satellites are recently being used in different fields in agriculture (Babaeian et al., 2019; Sadeghi et al., 2017). However, the coarse spatial resolution of the microwave satellite sensors, especially passive sensors, limits their application in precision agriculture (Fang et al., 2018; Im et al., 2016).

The timely and appropriate application of nutrients is vital for optimum crop growth and yields, and preventing the losses of nutrients to the ground, and surface water minimizes damage to the environment. In general, a recommended dose of fertilizer and nutrients is applied during the crop cycle. However, this requirement varies spatially and during and among seasons due to variations in topography, weather, soil, hydrology, and management practices (Hendricks et al., 2019; Melkonian & ES, 2008). Therefore planning for precision farming using conventional tools such as chlorophyll meters can be very challenging (Marino & Alvino, 2015). Mapping of vegetation indices such as NDVI, SAVI (derived from remote sensing data) can help understand the spatial variability for crop nutrient status, which is important for real-time application of nutrient to the plant and also provide meaningful information on photosynthetic activity, chlorophyll content, and yield (Ali et al., 2017). These have also been helpful in developing site-specific management tools for assessing the soil's organic matter and phosphorus content (Blasch et al., 2015; Kalambukattu et al., 2018).

New technologies and methods are being adopted in modern agriculture to increase production and nutrient use efficiencies, reducing nutrient leaching and greenhouse gas emission. Such model-based technology development has shown new possibilities for precision nutrient management. Remote and mobile technologies allow it to obtain spatial information on fertilizer and provide valuable information to farmers to adjust any field management operations. Additional information on more precise dozes of nitrogen fertilizer may be obtained from simulation models. One simulation model, namely Daisy model, could illustrate the spatial and temporal variability in soil nitrate content in the autumn after spring barley (Rashid et al., 2021). This variation was obtained by running this model 57 times with 57 different sets of measured soil hydraulic properties. The variation in hydraulic properties had a marked influence on other model output—such as water balance, crop's dry matter and nitrogen yield, and nitrogen leaching. This result clearly shows the potential of this model to apply in precision nutrient management in which input of spatially distributed zones characterized by different sets of hydraulic properties governs nitrogen uptake and hence

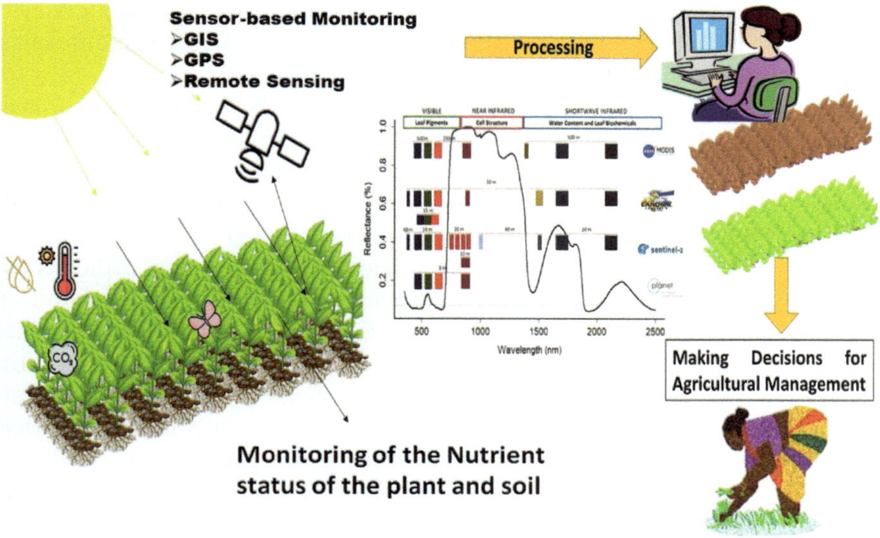

FIGURE 16.4

Geographic information system, global positioning system, and remote sensing tools for nutrient status monitoring and decision-making (adapted from Segarra et al., 2020).

varying need for nitrogen fertilizer input. The Daisy model was used to simulate potato (var. Folva) grown under various water and nitrogen regimes (Cammarano et al., 2020). Mapping of soil and crop parameters shows great potential for precision farming. Spatial data on soil hydraulic parameters and humus content is important to physically based SVAT and C/N turnover models. They concluded that this Daisy model was quite sensitive to parameterization of hydraulic properties as well as soil humus content. In a validation study, Daisy simulated well the dry matter yield and N uptake in potatoes.

Overall, precision agriculture based on remote sensing–based mapping of plant nutrient requirements can aid in increasing nutrient use efficiency and maintaining and increasing crop yield by reducing unwanted nutrient loss and lowering the threat to the environment (Castaldi et al., 2019; Cao et al., 2017) (Fig. 16.4).

16.3.6 Decision support systems

NE is an easy-to-use, interactive computer-based decision support tool that provides nutrient recommendations for individual farmers' individual fields with or without soil test data. This DSS is being validated and also promoted for enhanced yield, economic returns, and NUE while reducing environmental footprints (Jat et al., 2013; Majumdar et al., 2012; Sapkota et al., 2014; Satyanarayana et al., 2013). The software estimates the nutrient requirement based on attainable yield by combining information such as fertilizer/manure applied, crop residue retained in the previous crop, and crop growing conditions with expected N, P, and K responses in target fields to generate location-specific nutrient recommendations for crops (rice, wheat, and maize).

International Plant Nutrition Institute, along with CIMMYT and other partners, has developed the NE for wheat and hybrid maize for smallholder farmers of South Asia (http://software.ipni.net). These decision-support tools synthesized the on-farm research data into a simple delivery system that helps the farmers rapidly implement SSNM for their individual fields (Pampolino et al., 2012). This tool estimates the expected yield for a farmer's field based on growing conditions and calculates the cropping system's nutrient balance based on yield and fertilizer/manure applied for the previous crop grown. Such information can be combined with the expected N, P, and K response in the concerned field to have a location-specific nutrient recommendation for wheat (Bushong et al., 2016; Jat et al., 2013). It uses information provided by a farmer to suggest a meaningful yield goal for a given location and formulates a fertilizer management strategy required to attain the targeted yield. These types of software also calculate a simple profit analysis, comparing costs and benefits between the farmer's practice and the recommended alternative practice. Satyanarayana et al. (2013) evaluated NEs in conservation agriculture and CT maize during summer and winter. They found that NE-based fertilizer recommendations performed better than farmers' practice and state recommendations for maize. Bhende and Kumar (2014) recorded significantly higher wheat yields from NE-based nutrient management. In this study, the gross return over fertilizer cost was higher by about USD 300 per ha in NE than in farmers' fertilizer practice. NE and GS, when used combined, minimize the environmental footprint of wheat production under an NT system in northwestern India. Here, in terms of per tonne of wheat grain production and per dollar of net return, the estimated global warning potential was lower for NE-based strategies compared with other nutrient management strategies (Sapkota et al., 2014). In maize also, NE-based nutrient management could be an effective strategy for realizing the potential yield and enhancing net returns in the NWPZ, NEPZ, and PZ of India (Singh et al., 2019).

The International Rice Research Institute (IRRI), in collaboration with different partners, has developed another DSS called Crop Manager for Rice (released in the Philippines and Bangladesh and under evaluation in different states of India), Maize, and Rice-Wheat System (http://cropmanager.irri.org/home). It is a computer and mobile phone–based technology that enables small-scale rice, rice-wheat, and maize farmers with site and season-specific recommendations for fertilizer application. Recommendations are based on user-input information about farm location and management that can also be used by extension workers, crop advisors, and service providers to advise farmers specific to their growing condition. In addition to the nutrient application, crop managers can also provide general crop management based on local growing conditions.

The application of plant nutrients based on LCC also proved satisfactory in hybrid Maize in lateritic soils of India (Banerjee et al., 2014). The agronomic and recovery efficiency of nitrogen was also highest in the NE-treated plot, which is also significant among all other treatments. The physiological efficiency was highest in the NE-treated plot, which was statistically at par with LCC-based application. These results conform with the results obtained for rice (Das et al., 2009), maize and wheat (Kumar & Singh, 2019). NE-based practices recorded 58.80 q/ha wheat production in Hisar, Haryana. They concluded that NE-based practices produced higher productivity in comparison with the recommended dose of fertilizer.

16.3.7 Machine learning and Internet of Things

With the advent of the latest technologies, farming has been revolutionized over the past decade. The introduction of smart agriculture gadgets has helped farmers achieve maximum control on the crop production cycle with exceptional efficiency (Khanal et al., 2018). When most of the devices and items that we use in our everyday life are based on the Internet of Things (IoT), sensors based on IoT can create a new era in the precision agricultural scenario (Chun-Ting et al., 2020). IoT-based sensors can assist farmers in optimizing the harvesting time, checking the health status of the crop, checking the soil quality, measurement of moisture/water requirement, and allotting fertilizers and pesticides at precise time and location.

Nowadays, several applications are available on smartphones that employ IoT ideals that can aggregate data from diverse sources, remote sensors, and weather stations to be precise and provide up-to-date information on crop management operations such as sowing, weeding, watering, fertilizer application, and irrigation and analyze the data as well as provide suitable recommendations (Dagar et al., 2018).

Machine learning models are trained by the historical data of the agricultural field through which it gains experience and makes wise decisions for the data signals received from the IoT devices. The data collected from these IoT devices must be secured and ensure confidentiality for accurate prediction results. The number of nutrients and quantity of fertilizers needed by the plant can be determined based on the leaf color, soil quality, and pH (Sushanth & Sujatha, 2018). The application of machine learning in precision agriculture has reshaped the plan such as field-based crop suggestion, fertilizer recommendation, water supply prediction, harvest prediction, thereby controlling the water usage by assisting the agriculturalists or farmers for better yield in a smart way (Sivakumar et al., 2022).

16.4 Conclusion

Precision nutrient management is adopted to assimilate assorted information to create precise and efficient management decisions for comprehensive sustainable agricultural practices. This strategy is being adopted across the globe and in many parts of India to improve agricultural production and productivity in a brief span. Crop motoring and management become considerably effortless and accurate if the precision tools and techniques can be employed in the production system. These technologies have given a new ray of hope when the global resources of water and nutrients are deteriorating, and the food crisis is emerging. Although these techniques need to be checked and validated for most of the food crops available around the world, its utility shows a stable growth in crops that have been already employed. Adoptions of precision nutrient management techniques have proven to be economically profitable over conventional farmers' practice, saving up to Rs 2500–3200 ha^{-1}. The optimum dose of fertilizer application leads to controlled leaching of nutrients and fertilizers to the groundwater, checking the detrimental environmental hazards.

Acknowledgment

VJB and BS thankfully acknowledge the Department of Biotechnology, Government of India, for providing research grants (Grant No. BT/PR25099/NER/95/1014/2017). BS thanks Dr. Bidyut Chandan Deka, Vice Chancellor, AAU, Jorhat, India, for his constant encouragement and support.

References

Abbas, I., Liu, J., Faheem, M., Noor, R. S., Shaikh, S. A., Solangi, K. A., et al. (2020). Different sensor based intelligent spraying systems in agriculture. *Sensors and Actuators A: Physical*, *316*, 112–265. Available from https://doi.org/10.1016/j.sna.2020.112265.

Adamchuk, V. I., Hummel, J. W., Morgan, M. T., & Upadhyaya, S. K. (2004). On-the-go soil sensors for precision agriculture. *Computers and Electronics in Agriculture*, *44*(1), 71–91. Available from https://doi.org/10.1016/j.compag.2004.03.002.

Aggelopoulou, K., Bochtis, D., Fountas, F., Swain, K. C., Gemtos, T., & Nanos, G. (2011). Yield prediction in apples based on image processing. *Precision Agriculture*, *12*, 448–456. Available from https://doi.org/10.1007/s11119-010-9187-0.

Aggelopoulou, K. D., Wulfsohn, D., Fountas, S., Gemtos, T. A., Nanos, G. D., & Blackmore, S. (2010). Spatial variation in yield and quality in a small apple orchard. *Precision Agriculture*, *11*, 538–556. Available from http://doi.org/10.1007/s11119-009-9146-9.

Aishah, A. W., Zauyah, S., Anuar, A. R., & Fauziah, C. I. (2010). Spatial variability of selected chemical characteristics of paddy soils in Sawah Sempadan, Selangor, Malaysia. *Malaysian Journal of Soil Science*, *14*, 27–39.

Ali, M. M., Al-Ani, A., Eamus, D., & Tan, D. K. Y. (2017). Leaf nitrogen determination using non-destructive techniques—A review. *Journal of Plant Nutrition*, *40*, 928–953. Available from https://doi.org/10.1080/01904167.2016.1143954.

Amer, G., Mudassir, S.M. M., & Malik, M.A. (2015, May). Design and operation of Wi-Fi agribot integrated system. In: *2015 international conference on industrial instrumentation and control* (ICIC) (pp. 207–212). IEEE. Available from https://doi.org/10.1109/IIC.2015.7150739.

Aparicio, N., Villegas, D., Araus, J. L., Casadesus, J., & Royo, C. (2002). Relationship between growth traits and spectral vegetation indices in durum wheat. *Crop Science*, *42*(5), 1547–1555. Available from https://doi.org/10.2135/cropsci2002.1547.

Babaeian, E., Sidike, P., Newcomb, M. S., Maimaitijiang, M., White, S. A., Demieville, J., Ward, R. W., Sadeghi, M., LeBauer, D. S., & Jones, S. B. (2019). A new optical remote sensing technique for high resolution mapping of soil moisture. *Frontiers in Big Data*, *2*, 37. Available from https://doi.org/10.3389/fdata.2019.00037.

Beloev, I., Kinaneva, D., Georgiev, G., Hristov, G., & Zahariev, P. (2021). Artificial intelligence-driven autonomous robot for precision agriculture. *Acta Technologica Agriculturae*, *24*(1), 48–54. Available from https://doi.org/10.2478/ata-2021-0008.

Bhende, S. N., & Kumar, A. (2014). Nutrient expert®-based fertilizer recommendation improved wheat yield and farm profitability in the Mewat. Better Crops-South. *Asia*, *8* (1), 21–23.

Blasch, G., Spengler, D., Hohmann, C., Neumann, C., Itzerott, S., & Kaufmann, H. (2015). Multitemporal soil pattern analysis with multispectral remote sensing data at the field scale. *Computers and Electronics in Agriculture*, *113*, 1–13. Available from https://doi.org/10.1016/j.compag.2015.01.012.

Boggs, J. L., Tsegaye, T. D., Coleman, T. L., Reddy, K. C., & Fahsi, A. (2003). Relationship between hyperspectral reflectance, soil nitrate-nitrogen, cotton leaf chlorophyll, and cotton yield: A step toward precision agriculture. *Journal of Sustainable Agriculture*, *22*(3), 5–16. Available from https://doi.org/10.1300/J064v22n03_03.

Bosquet, L. C., Molero, G., Stellacct, A. M., Bort, J., Nogues, S., & Araus, J. L. (2011). NDVI as a potential tool for predicting biomass, plant nitrogen content and growth in wheat genotypes subjected to different water and nitrogen conditions. *Cereal Research Communications*, *39*(1), 147–159. Available from https://doi.org/10.1556/CRC.39.2011.1.15.

Bradbury, S., Mackenzie, C., Mackenzie, J., & Hedley, C. B. (Eds.), (2013). *Precision irrigation as a tool to reduce nutrient leaching and runoff. Accurate and efficient use of nutrients on farms*. New Zealand: Fertilizer and Lime Research Centre, Palmerston North.

Bronson, K. F., Scharf, P. C., & Kitchen, N. R. (2011). In David E. Clay, & John F. Shanahan (Eds.), *Use of GIS-based site-specific nitrogen management for improving energy efficiency* (USDA-ARS/UNL Faculty. 677. Published in GIS Applications in Agriculture, Volume II: Nutrient Management for Energy Efficiency, pp. 360–384). Boca Raton: CRC Press. Available from https://digitalcommons.unl.edu/usdaarsfacpub/677.

Bushong, J. T., Mullock, J. L., Miller, E. C., Raun, W. R., & Arnall, D. B. (2016). Evaluation of mid-season sensor based nitrogen fertilizer recommendations for winter wheat using different estimates of yield potential. *Precision Agriculture*, *17*, 470–487. Available from https://doi.org/10.1007/s11119-016-9431-3.

Cambardella, C. A., & Karlen,D., L. (1999). Spatial analysis of soil fertility parameters. *Precision Agriculture*, *1*, 5–14. Available from https://doi.org/10.1023/A:1009925919134.

Cammarano, D., Holland, J., & Ronga, D. (2020). Spatial and temporal variability of spring barley yield and quality quantified by crop simulation model. *Agronomy*, *10*(3), 393. Available from https://doi.org/10.3390/agronomy10030393.

Cao, Q., Miao, Y., Li, F., Gao, X., Liu, B., Lu, D., & Chen, X. (2017). Developing a new crop circle active canopy sensor based precision nitrogen management strategy for winter wheat in North China Plain. *Precision Agriculture*, *18*, 2–18. Available from https://doi.org/10.1007/s11119-016-9456-7.

Castaldi, F., Hueni, A., Chabrillat, S., Ward, K., Buttafuoco, G., Bomans, B., Vreys, K., & Brell, M. (2019). Van Wesemael, B. Evaluating the capability of the Sentinel 2 data for soil organic carbon prediction in croplands. *ISPRS Journal of Photogrammetry*, *147*, 267–282. Available from https://doi.org/10.1016/j.isprsjprs.2018.11.026.

Chang, C. W., Laird, D. A., Mausbach, M. J., & Hurburgh, C. R. (2001). Near-infrared reflectance spectroscopy-principal components regression analyses of soil properties. *Soil Science Society of America Journal*, *65*(2), 480–490. Available from https://doi.org/10.2136/sssaj2001.652480x.

Chun-Ting, P., Meng-Ju, L., Nen-Fu, H., Jhong-Ting, L., & Jia-Jung, S. (2020, January). Agriculture blockchain service platform for farm-to-fork traceability with IoT sensors. In: *2020 international conference on information networking* (ICOIN) (pp. 158–163). IEEE. Available from https://doi.org/10.1109/ICOIN48656.2020.9016535.

Colaço, A. F., & Molin, J. P. (2017). Variable rate fertilization in citrus: A long term study. *Precision Agriculture*, *18*(2), 169–191. Available from https://doi.org/10.1007/s11119-016-9454-9.

Dagar, R., Som, S., & Khatri, S.K. (2018, July). Smart farming-IoT in agriculture. In: *2018 international conference on inventive research in computing applications* (ICIRCA) (pp. 1052–1056). IEEE. Available from https://doi.org/10.1109/ICIRCA.2018.8597264.

Das, D. K., Maiti, D., & Pathak, H. (2009). Site-specific nutrient management in rice in eastern India using a modelling approach. *Nutrient Cycling in Agroecosystems*, *83*, 85–94. Available from https://doi.org/10.1007/s10705-008-9202-2.

Davis, G., Casady, W., & Massey, R. (1998). In U. Extension (Ed.), *Precision agriculture: An introduction* (pp. 1–7).

Dige, K. T. (2020). Precision agriculture in India: Opportunities and challenges. International Journal of Research in Engineering. *Science and Management*, *3*(8), 395–397.

Doddamani, A., Kouser, S., & Ramya, V. (2020). Role of drones in modern agricultural applications. *Current Journal of Applied Science and Technology*, *39*(48), 216–224. Available from https://doi.org/10.9734/cjast/2020/v39i4831224.

Doerge, T. (2000). *EC mapping helps farmers know soil*. Vol. 18(2), 7. Lafayette, Indiana: CTIC Partners, Conservation Technology Information Center.

Ehsani, R., & Karim, D. (2010). Yield monitors for specialty crops. In S. Upadhyaya, K. Giles, S. Haneklaus, & E. Schnug (Eds.), *Advanced engineering systems for specialty crops: A review of precision agriculture for water, chemical, and nutrient*, *VTI Agriculture and Forestry Research* 59, 309.2009, pp. 31–43. Braunschweig, Germany: Johann Heinrich von Thünen-Institut.

Fabiani, S., Vanino, S., Napoli, R., Zajicek, A., Duffkova, R., Evangelou, E., et al. (2020). Assessment of the economic and environmental sustainability of Variable Rate Technology (VRT) application in different wheat intensive European agricultural areas. a water energy food nexus approach. *Environmental Science & Policy*, *114*, 366–376. Available from https://doi.org/10.1016/j.envsci.2020.08.019.

Fang, B., Lakshmi, V., Bindlish, R., & Jackson, T. J. (2018). AMSR2 soil moisture downscaling using temperature and vegetation data. *Remote Sensing*, *10*, 1575. Available from https://doi.org/10.3390/rs10101575.

Farooque, A. A., Zaman, Q. U., & Schumann, A. W. (2012). Delineating management zones for site-specific fertilization in wild blueberry fields. *Applied Engineering in Agriculture*, *28*, 57–70. Available from http://doi.org/10.13031/2013.41286.

FeiBo, W., Lianghuan, W., & Fuhua, X. (1998).). Chlorophyll meter to predict nitrogen side-dress requirements for short-season cotton (Gossypium hirsutum L.). *Field Crops Research*, *56*(3), 309–314. Available from https://doi.org/10.1016/S0378-4290(97)00108-1.

Fountas, S., Aggelopoulou, K., Bouloulis, C., Nanos, G. D., Wulfsohn, D., Gemtos, T. A., Paraskevopoulos, A., & Galanis, M. (2011). Site-specific management in an olive tree plantation. *Precision Agriculture*, *12*, 179–195. Available from http://doi.org/10.1007/s11119-010-9167-4.

Franzen, D., Kitchen, N., Holland, K., Schepers, J., & Raun, W. (2016). Algorithms for in-season nutrient management in cereals. *Agronomy Journal*, *108*, 1775–1781. Available from https://doi.org/10.2134/agronj2016.01.0041.

Freeman, K. W., Arnall, D. B., Mullen, R. W., Garima, K. G., Martin, K. L., Teal, R. K., & Raun, W. R. (2007). By- plant prediction of corn forage biomass and nitrogen uptake at various growth stages using remote sensing and plant height measures. *Agronomy Journal*, *99*, 530–536. Available from https://doi.org/10.2134/agronj2006.0135.

Gantzer, C. J., & McCarty, T. R. (1987). Predicting corn yields on a claypan soil using a soil productivity index. *Transactions of the American Society of Agricultural Engineers (ASAE)*, *30*, 1347–1352. Available from https://doi.org/10.13031/2013.30569.

Guerif, M., Houlès, V., Baret, F. (2007). Remote sensing and detection of nitrogen status in crops. Application to precise nitrogen fertilization. In: *4th international symposium on intelligent information technology in agriculture*, Beijing, 26–29.

Gustafsson, J., Cederberg, C., Sonesson, U., & Emanuelsson, A. (2013). *The methodology of the FAO study: Global food losses and food waste-extent, causes and prevention*. FAO, 2011.

Hartigan, J., & Wong, M. (1979). A K-means clustering algorithm. *Journal of the Royal Statistical Society*, *28*, 100–108. Available from https://doi.org/10.2307/2346830.

Hedley, C. B., & Yule, I. J. (2009). Soil water status mapping and two variable-rate irrigation scenarios. *Precision Agriculture*, *10*, 342–355. Available from https://doi.org/10.1007/s11119-009-9119-z.

Hedley, C. B., Yule, I. J., Eastwood, C. R., Shepherd, T. G., & Arnold, G. (2004). Rapid identification of soil textural and management zones using electromagnetic induction sensing of soils. *Australian Journal of Soil Research*, *42*, 389–400. Available from https://doi.org/10.1071/SR03149.

Hedley, C. B., Yule, I. J., Tuohy, M. P., & Vogeler, I. (2009). Key performance indicators for simulated variable-rate irrigation of variable soils in humid regions. *Trans ASABE*, *52*, 1575–1584. Available from https://doi.org/10.13031/2013.29146.

Heege, H. J., Reusch, S., & Thiessen, E. (2008). Prospects and results for optical systems for site-specific on-the-go control of nitrogen-top-dressing in Germany. *Precision Agriculture*, *9*(3), 115–131. Available from https://doi.org/10.1007/s11119-008-9055-3.

Hendricks, G. S., Shukla, S., Roka, F. M., Sishodia, R. P., Obreza, T. A., Hochmuth, G. J., & Colee, J. (2019). *Economic and environmental consequences of overfertilization under extreme weather conditions. Journal of Soil and Water Conservation*, *74*, 160–171. Available from https://doi.org/10.2489/jswc.74.2.160.

Huan, T. T. N., Tan, P. S., & Hiraoka, H. (2002). Optimum application rates of nitrogen fertilizer for high-yielding rice cultivars based on growth diagnosis in the Mekong delta. *JIRCAS Working Report*, *26*, 3–9.

Hussain, F., Zia, M.S., Akhtar, M.E., & Yasin, M. (2003). Nitrogen management and use efficiency with chlorophyll meter and leafcolour chart. *Pakistan Journal of Soil Science* (Pakistan), *22*, 1−10.

Im, J., Park, S., Rhee, J., Baik, J., & Choi, M. (2016). Downscaling of AMSR-E soil moisture with MODIS products using machine learning approaches. *Environmental Earth Sciences*, *75*, 1−19. Available from https://doi.org/10.1007/s12665-016-5917-6.

Janssen, B. H., Guiking, F. C. T., & Van der Eijk, D. (1990). *Geoderma*, *46*, 299−318. Available from https://doi.org/10.1016/0016-7061(90)90021-Z.

Jat, M. L., & Bijay-Singh, G. B. (2014). Nutrient management and use efficiency in wheat systems of South Asia. *Advances in Agronomy*, *125*, 171−259. Available from https://doi.org/10.1016/B978-0-12-800137-0.00005-4.

Jat, M. L., Satyanarayana, T., Majumdar, K., Parihar, C. M., Jat, S. L., Tetarwal, J. P., & Jat, R. K. (2013). Fertilizer best management practices for maize systems. *Indian Journal of Fertilisers*, *9*(4), 80−94.

Jat, R. K., Bijarniya, D., Kakraliya, S. K., Sapkota, T. B., Kakraliya, M., & Jat, M. L. (2021). Precision nutrient rates and placement in conservation maize-wheat system: Effects on crop productivity, profitability, nutrient-use efficiency, and environmental footprints. *Agronomy*, *11*, 2320. Available from https://doi.org/10.3390/agronomy11112320.

Johnston, A. M., Khurana, H. S., Majumdar, K., & Satyanarayana, T. (2009). Site specific nutrient management- concept, current research and future challenges in Indian agriculture. *Journal of the Indian Society of Soil Science*, *57*, 1−10.

Jongschaap, R. E. E., & Booji, R. (2004). Spectral measurements at different spatial scales in potato; Relating leaf, plant and canopy nitrogen status, Intern. *Journal of Applied Earth Observation and Geoinformation*, *5*(3), 205−218. Available from https://doi.org/10.1016/j.jag.2004.03.002.

Joshi, N., & Chandrashekar, C. P. (2017).): Precision nutrient management in Maise (*Zea mays* L.) under northern transition zone of Karnataka. *Journal of Farm Sciences*, *30*(3), 343−348.

Joshi, N., Chandrashekara, C. P., & Potdar, M. P. (2018). Assessment of precision nutrient management techniques in Maise and their effect on yield, nutrient use efficiency, and economics. *International Journal of Applied and Pure Science and Agriculture*, *4*(1), 13−20. Available from https://doi.org/10.22623/IJAPSA.2018.4003.YAL3Z.

Kalambukattu, J. G., Kumar, S., & Raj, R. A. (2018). Digital soil mapping in a Himalayan watershed using remote sensing and terrain parameters employing artificial neural network model. *Environmental Earth Sciences*, *77*, 203. Available from https://doi.org/10.1007/s12665-018-7367-9.

Kaur, H., Kaur, A., Singh, B., & Bhatt, R. (2020). Application of geospatial technology in assessment of spatial variability in soil properties: A review. *Current Journal of Applied Science and Technology*, *39*(39), 57−71. Available from https://doi.org/10.9734/cjast/2020/v39i3931104.

Kaur, J. (2017). *Precision nutrient management in wheat (*Triticum aestivum *L.) using nutrient expert and GreenSeeker. Department of Agronomy*. Ludhiana: Punjab Agricultural University.

Khanal, S., Fulton, J., Klopfenstein, A., Douridas, N., & Shearer, S. (2018). Integration of high resolution remotely sensed data and machine learning techniques for spatial prediction of soil properties and corn yield. *Computers and Electronics in Agriculture*, *153*, 213−225. Available from https://doi.org/10.1016/j.compag.2018.07.016.

Kitchen, N. R., Hughes, D. F., Donald, W. W., & Alberts, E. E. (1998). Agrichemical movement in the root-zone of claypan soils: Ridge- and mulch-tillage systems compared. *Soil & Tillage Research*, *48*, 179–193. Available from https://doi.org/10.1016/S0167-1987(98)00144-5.

Kitchen, N. R., Sudduth, K. A., & Drummond, S. T. (1999). Soil electrical conductivity as a crop productivity measure for claypan soils. *Journal of Production Agriculture*, *12*, 607–671. Available from https://doi.org/10.2134/jpa1999.0607.

Kitchen, N.R., Sudduth, K.A., Hughes, D.F., Drummond, S.T., & Birrell, S.J. (1995). On-the-go changes in fertilizer rates to agree with claypan soil productivity. In: *Proceedings of clean water-clean environment 2lst century conference*, Vol. 2, p. 103–106. St. Joseph, MI, USA: American Society of Agricultural Engineers.

Kumar, M., & Singh, B. (2019). Precision nutrient management in wheat using nutrient expert. *International Journal of Current Microbiology and Applied Sciences*, *8*(2), 2265–2269. Available from https://doi.org/10.20546/ijcmas.2019.802.263.

Lopez-Granados, F., Jurado-Exposito, M., Alammo, S., & GarciaTorres, L. (2004). Leaf nutrient spatial variability and site-specific fertilization maps within olive (*Olea europaea* L.) orchards. *European Journal of Agronomy*, *21*, 209–222. Available from http://doi.org/10.1016/j.eja.2003.08.005.

Lowenberg-DeBoer, J., & Aghib, A. (1999). Average returns and risk characteristics of site specific P and K management: Eastern corn belt on-farm trial results. *Journal of Production Agriculture*, *12*(2). Available from https://doi.org/10.2134/jpa1999.0276.

Ma, B. L., Morrison, M. J., & Dwyer, L. M. (1996). Canopy light reflectance and field greenness to assess nitrogen fertilization and yield of maize. *Agronomy Journal*, *88*(6), 915–920. Available from https://doi.org/10.2134/agronj1996.00021962003600060011x.

Mainak, G., Dillip, K. S., Madan, K. J., Virendra, K. T., & Bohra, A. (2020). Optimizing chlorophyll meter (SPAD) reading to allow efficient nitrogen use in rice and wheat under rice-wheat cropping system in eastern India. *Plant Production Science*. Available from https://doi.org/10.1080/1343943X.2020.1717970.

Maiti, D., Das, D. K., Karak, T., & Banerjee, M. (2004). Management of nitrogen through the use of leaf color chart (LCC) and soil plant analysis development (SPAD) or chlorophyll meter in rice under irrigated ecosystem. *Science World Journal*, *4*, 838–846. Available from https://doi.org/10.1100/tsw.2004.137.

Majumdar, K., Jat, M. L., Pampolino, M., Kumar, A., Vishal, V., Gupta, N., Singh, V., Satyanarayana, T., Dwivedi, B. S., Singh, V. K., Kumar, D., Kamboj, B. R., Sidhu, H. S., Meena, M. C., & Johnston, A. (2014). Economics of potassium fertilizer application in rice, wheat and maize grown in the Indo-Gangetic Plains. *Indian Journal of Fertilisers*, *8*(5), 44–53.

Majumdar, K., Jat, M. L., & Shahi, V. B. (2012). Effect of spatial and temporal variability in cropping seasons and tillage practices on maize yield responses in eastern India. *Better Crop.-South Asia*, *6*, 4–6.

Marino, S., & Alvino, A. (2015). Hyperspectral vegetation indices for predicting onion (Allium cepa L.) yield spatial variability. *Computers and Electronics in Agriculture*, *116*, 109–117. Available from https://doi.org/10.1016/j.compag.2015.06.014.

Martin, K., Raun, W., & Solie, J. (2012). By-plant prediction of corn grain yield using optical sensor readings and measured plant height. *Journal of Plant Nutrition*, *35*(9), 1429–1439. Available from http://doi.org/10.1080/01904167.2012.684133.

Melkonian, J.J., ES, H.M. V. (2008, July). Adapt-N: Adaptive nitrogen management for Maise using high resolution climate data and model simulations. In Proceedings of the 9th international conference on precision agriculture, Denver, CO, USA, 20–23.

Moharana, P. C., Jena, R. K., Pradhan, U. K., Nogiya, M., Tailor, B. L., Singh, R. S., et al. (2019). Geostatistical and fuzzy clustering approach for delineation of site-specific management zones and yield-limiting factors in irrigated hot arid environment of India. *Precision Agriculture*, *21*(2), 426–448. Available from https://doi.org/10.1007/s11119-019-09671-9.

Mouazen, A. M., De Baerdemaeker, J., & Ramon, H. (2005). Towards development of on-line soil moisture content sensor using a fibre-type NIR spectrophotometer. *Soil and Tillage Research*, *80*(1–2), 171–183. Available from https://doi.org/10.1016/j.still.2004.03.022.

O'Connor, A. C., Gallaher, M. P., Clark-Sutton, K. B., Lapidus, D., Oliver, Z., Scott, T. J., Wood, D. W., & Brown, E. G. (2019). Economic benefits of the Global Positioning System (GPS). *RTI International*.

Olfs, H. W., Blankenau, K., Brentrup, F., Jasper, J., Link, A., & Lammel, J. (2005). Soil and plant-based nitrogen fertilizer recommendations in arable forming. *Journal of Plant Nutrition and Soil Science*, *168*(4), 414–431. Available from http://doi.org/10.1002/jpln.200520526.

Onwuchekwa-Henry, C. B., Ogtrop, F. V., Roche, R., & Tan, D. K. (2022). Model for predicting rice yield from reflectance index and weather variables in lowland rice fields. *Agriculture*, *12*(2), 130. Available from https://doi.org/10.3390/agriculture12020130.

Pampolino, M. F., Witt, C., Pasuquin, J. M., Johnston, A., & Fisher, M. J. (2012). Development approach and evaluation of the nutrient expert softwarefor nutrient management in cereal crops. *Computers and Electronics in Agriculture*, *88*, 103–110. Available from https://doi.org/10.1016/j.compag.2012.07.007.

Parihar, C. M., Jat, S. L., Singh, A. K., Ghosh, A., Rathore, N. S., Kumar, B., Pradhan, S., Majumdar, K., Satyanarayana, T., Jat, M. L., Saharawat, Y. S., Kuri, B. R., & Saveipune, D. (2017). Effects of precision conservation agriculture in a maize-wheat-mungbean rotation on crop yield, water-use and radiation conversion under a semiarid agro-ecosystem. *Agricultural Water Management*, *192*, 306–319. Available from https://doi.org/10.1016/j.agwat.2017.07.021.

Pathak, H., Singh, Y., & Singh, B. (2003). Site-specific nutrient management for improving the productivity of rice-wheat cropping system. In Y. Singh, et al. (Eds.), *Nutrient management for sustainable rice-wheat cropping system. New Delhi (India): National Agricultural Technology Project, Indian Council for Agricultural Research and Ludhiana (India)* (pp. 79–98). Punjab Agricultural University.

Patil Shirish, S., & Bhalerao, S. A. (2013). Precision farming: The most scientific and modern approach to sustainable agriculture. *International Research Journal of Science and Engineering*, *1*(2).

Peng, S., Garcia, F. V., Laza, R. C., Sanico, A. L., Visperas, R. M., & Cassman, K. G. (1996). Increased N-use efficiency using a chlorophyll meter on high-yielding irrigated rice. *Field Crops Research*, *47*(2–3), 243–252. Available from https://doi.org/10.1016/0378-4290(96)00018-4.

Peñuelas, J., Gamon, J. A., Fredeen, A. L., Merino, J., & Field, C. B. (1994). Reflectance indices associated with physiological changes in nitrogen-and water-limited sunflower leaves. *Remote Sensing of Environment*, *48*(2), 135–146. Available from https://doi.org/10.1016/0034-4257(94)90136-8.

Pooniya, V., Jat, S. L., Choudhary, A. K., Singh, A. K., Parihar, C. M., Bana, R. S., Swarnalakshmi, K., & Rana, K. S. (2015). Nutrient expert' assisted site-specific-nutrient-management: An alternative precisionfertilization technology for maize-wheat cropping system in South-Asian Indo-Gangetic Plains. *Indian Journal of Agricultural Sciences*, *85*(8), 996−1002.

Portz, G., Molin, J. P., & Jasper, J. (2012). Active crop sensor to detect variability of nitrogen supply and biomass on sugarcane fields. *Precision Agriculture*, *13*(1), 33−44. Available from https://doi.org/10.1007/s11119-011-9243-4.

Prakasha, G., Mudalagiriyappa., Somashekar, K. S., & Goudra, Shivanand (2020). A novel approach for increasing productivity under precision nitrogen management in Maise (*Zea mays* L.) through crop sensors. *Journal of Pharmacognosy and Phytochemistry*, *9* (5), 97−103, 2020.

Raddy, G., Lalitha, B. S., & Jayadeva, H. M. (2021). Spatial fertilizer recommendation mapping based on soil test crop response equations for important crops using GIS and GPS. *Communications in Soil Science and Plant Analysis*, *52*(1), 58−75. Available from https://doi.org/10.1080/00103624.2020.1849259.

Raper, T. B., & Varco, J. J. (2015). Canopy-scale wavelength and vegetative index sensitivities to cotton growth parameters and nitrogen status. *Precision Agriculture*, *16*(1), 62−76. Available from https://doi.org/10.1007/s11119-014-9383-4.

Raper, T. B., Varco, J. J., & Hubbard, K. J. (2013). Canopy-based normalized difference vegetation index sensors for monitoring cotton nitrogen status. *Agronomy Journal*, *105* (5), 1345−1354.

Rashid, M. A., Bruun, S., Styczen, M. E., Ørum, J. E., Borgen, S. K., Thomsen, I. K., & Jensen, L. S. (2021). Scenario analysis using the Daisy model to assess and mitigate nitrate leaching from complex agro-environmental settings in Denmark. *Science of the Total Environment*, 151518. Available from https://doi.org/10.1016/j.scitotenv.2021.151518.

Raun, W. R., Solie, J. B., Johnson, G. V., Stone, M. L., Mullen, R. W., Freeman, K. W., & Lukina, E. V. (2002). Improving nitrogen use efficiency in cereal grain production with optical sensing and variable rate application. *Agronomy Journal*, *94*(4), 815−820. Available from https://doi.org/10.2134/agronj2002.8150.

Roberts, D. F., Adamchuk, V. I., Shanahan, J. F., Ferguson, R. B., & Schepers, J. S. (2009). Optimization of crop canopy sensor placement for measuring nitrogen status in corn. *Agronomy Journal*, *101*(1), 140−149. Available from https://doi.org/10.2134/agronj2008.0072x.

Rossel, R. V., Walvoort, D. J. J., McBratney, A. B., Janik, L. J., & Skjemstad, J. O. (2006). Visible, near infrared, mid infrared or combined diffuse reflectance spectroscopy for simultaneous assessment of various soil properties. *Geoderma*, *131*(1−2), 59−75. Available from https://doi.org/10.1016/j.geoderma.2005.03.007.

Sadeghi, M., Babaeian, E., Tuller, M., & Jones, S. B. (2017). The optical trapezoid model: A novel approach to remote sensing of soil moisture applied to Sentinel-2 and Landsat-8 observations. *Remote Sensing of Environment*, *198*, 52−68. Available from https://doi.org/10.1016/j.rse.2017.05.041.

Saggar, S., Jha, N., Deslippe, J., Bolan, N. S., Luo, J., & Giltrap, D. L. (2013). Denitrification and N2O:N2 production in temperate grasslands: Processes, measurements, modeling and mitigating negative impacts. *Science of the Total Environment*, *465*, 173−195. Available from https://doi.org/10.1016/j.scitotenv.2012.11.050.

Santosh, K. M., Sundaresan, J., Roggem, R., Déri, A., & Singh, R. P. (2014). Geospatial technologies and climate change; Springer international publishing: Dordrecht. *The Netherlands*. Available from https://doi.org/10.1007/978-3-319-01689-4.

Sapkota, T. B., Majumdar, K., Jat, M. L., Kumar, A., Bishnoi, D. K., McDonald, A. J., & Pampolino, M. (2014). Precision nutrient management in conservation agriculture basedwheat production of Northwest India: Profitability, nutrient useefficiency and environmental footprint. *Field Crops Research*, *155*, 233–244. Available from https://doi.org/10.1016/j.fcr.2013.09.001.

Sapkota, T., Jat, M., Shankar, V., Singh, L., Rai, M., Grewal, M., & Stirling, C. (2015). Tillage residue and nitrogen management effects on methane and nitrous oxide emission from rice-wheat system of Indian Northwest Indo-Gangetic Plains. *Journal of Integrative Environmental Sciences*, *12*(1), 31–46. Available from https://doi.org/10.1080/1943815X.2015.1110181.

Satyanarayana, T., Majumdar, K., Pampolino, M., Johnston, A. M., Jat, M. L., Kuchanur, P., & Patil, S. G. (2013). Nutrient ExpertTM: A tool to optimize nutrient use and improve productivity of maize. *Better Crops-South Asia*, *97*(1), 21–24.

Scharf, P. C., Shannon, D. K., Palm, H. L., Sudduth, K. A., Drummond, S. T., Kitchen, N. R., Mueller, L. J., Hubbard, V. C., & Oliveira, L. F. (2011). Sensor-based nitrogen applications out-performed producer-chosen rates for corn in on-farm demonstrations. *Agronomy Journal*, *103*, 1684–1691. Available from https://doi.org/10.2134/agronj2011.0164.

Segarra, J., Buchaillot, M. L., Araus, J. L., & Kefauver, S. C. (2020). Remote sensing for precision agriculture: Sentinel-2 improved features and applications. *Agronomy*, *10*(5), 641. Available from https://doi.org/10.3390/agronomy10050641.

Sesha Sai, M. V. R., Ramana, K. V., & Hebbar, R. (2010). Agriculture. In P. S. Roy, R. S. Dwivedi, & D. Vijayan (Eds.), *Remote sensing applications* (46, pp. 1–20). Hyderabad: NRSC/ISRO.

Shafi, U., Mumtaz, R., García-Nieto, J., Hassan, S. A., Zaidi, S. A. R., & Iqbal, N. (2019). Precision agriculture techniques and practices: From considerations to applications. *Sensors*, *19*(17), 3796. Available from https://doi.org/10.3390/s19173796.

Shyam, C. S. L., Rathore, S. S., Shekhawat, K., & Singh, R. K. (2021). Precision nutrient management in Maise (*Zea mays*) for higher productivity and profitability. *Indian Journal of Agricultural Sciences*, *91*(6), 933–935.

Singh, A. K., Jat, S. L., Parihar, C. M., Kumar, M., Singh, C. S., Hallikeri, S. S., Sreelatha, D., Manjulatha, G., & Maha, D. M. (2020). Precision nutrient management for enhanced yield and profitability of Maize (*Zea mays*). *Indian Journal of Agricultural Sciences*, *90*(5), 952–960.

Singh, B., Gupta, R. K., Singh, Y., Gupta, S. K., Singh, J., Bains, J. S., & Vashishta, M. (2006). Need-based nitrogen management using leaf color chart in wet direct-seeded rice in north-western India. *Journalof New Seeds (New York, N.Y.)*, *8*(1), 35–47. Available from https://doi.org/10.1300/J153v08n01_03.

Singh, B., Singh, Y., Ladha, J. K., Bronson, K. F., Balasubramanian, V., Singh, J., & Khind, C. S. (2002). Chlorophyll meter-and leaf colour chart-based nitrogen management for rice and wheat in Northwestern India. *Agronomy Journal*, *94*(4), 821–829. Available from https://doi.org/10.2134/agronj2002.8210.

Singh, S. P., Kumar, M., Dua, V. K., Sharma, S. K., Sadawarti, M. J., & Roy, S. (2019). Leaf chlorophyll meter—A non-destructive method for scheduling nitrogen in potato crop. *Potato Journal*, *46*(1), 73–80.

Singh, U., Praharaj, C. S., Kumar, N., Deo, M. M., & Kumar, L. (2017). *Enhancing nutrient use efficiency through next generation fertilizers in field crops. Training Manual* (p. 229) Kanpur, Uttar Pradesh, India: ICAR-Indian Institute of Pulses Research, 208 024.

Singh, V., Singh, B., Singh, Y., Thind, H. S., & Gupta, R. K. (2010). Need based nitrogen management using the chlorophyll meter and leaf colour chart in rice and wheat in south Asia: A review. *Nutrient Cycling in Agroecosystems*, *88*, 361–380. Available from https://doi.org/10.1007/s10705-010-9363-7.

Sivakumar, R., Prabadevi, B., Velvizhi, G., Muthuraja, S., Kathiravan, S., Biswajita, M., & Madhumathi, A. (2022). Internet of things and machine learning applications for smart precision agriculture. *IoT Applications Computing*, 135. Available from https://doi.org/10.5772/intechopen.97679.

Söderström, M., Börjesson, T., Pettersson, C. G., Nissen, K., & Hagner, O. (2010). Prediction of protein content in malting barley using proximal and remote sensing. *Precision Agriculture*, *11*(6), 587–599. Available from https://doi.org/10.1007/s11119-010-9181-6.

Soman, S., Byju, G., & Bharathan, R. (2013). GIS based decision support system for precision farming of cassava in India. *Acta Biologica India*, *2*(2), 394–399.

Srivastava, A. K., & Malhotra, S. K. (2014). Nutrient management in fruit crops. Issues and Strategies. *Indian Journal of Fertilisers*, *10*(12), 72–88.

Sudduth, K. A., Kitchen, N. R., Bollero, G. A., Bullock, D. G., & Wiebold, W. J. (2003). Comparison of electromagnetic induction and direct sensing of soil electrical conductivity. *Agronomy Journal*, *95*(3), 472–482. Available from https://doi.org/10.2134/agronj2003.4720.

Sushanth, G. & Sujatha, S. (2018). IoT based smart agriculture system. In: *International conference on wireless communications, signal processing and networking (WiSPNET)*, IEEE Xplore Digital Library. https://doi.org/10.1109/WiSPNET.2018.8538702.

Swinton, S. M., & Lowenberg-DeBoer, J. (1998). Evaluating the profitability of site-specific farming. *Journal of Production Agriculture*, *11*(4), 439–446. Available from https://doi.org/10.2134/jpa1998.0439.

Tremblay, N., Wang, Z., Ma, B. L., Belec, C., & Vigneault, P. (2009). A comparison of crop data measured by two commercial sensors for variable-rate nitrogen application. *Precision Agriculture*, *10*(2), 145–161. Available from https://doi.org/10.1007/s11119-008-9080-2.

Turban, E. (1995). *Decision support and expert system*. USA: Prentice Hall.

Vasu, D., Singh, S. K., Sahu, N., Tiwary, P., Chandran, P., Duraisami, V. P., Ramamurthy, V., Lalitha, M., & Kalaiselvi, B. (2017). Assessment of spatial variability of soil properties using geospatial techniques for farm level nutrient management. *Soil and Tillage Research*, *169*, 25–34. Available from https://doi.org/10.1016/j.still.2017.01.006.

Vasuki, N. (2010). Micronutrient management for enhancing crop production-Future strategy and requirement. *Journal of the Indian Society of Soil Science*, *58*(1), 32–36.

Verma, P., Chauhan, A., & Ladon, T. (2020). Sitespecific nutrient management: A review. *Journal of Pharmacognosy and Phytochemistry Sp*, *9*(5), 233–236, 29.

Vistoso, E., Alfaro, M., Saggar, S., & Salazar, F. (2012). Effect of nitrogen inhibitors on nitrous oxide emissions and pasture growth after an autumn application in volcanic soil. *Chilean Journal of Agricultural Research*, *72*, 133–139. Available from https://doi.org/10.4067/S0718-58392012000100021.

Wang, D., Prato, T., Qiu, Z., Kitchen, N. R., & Sudduth, K. A. (2003). Economic and environmental evaluation of variable rate nitrogen and lime application for claypan soil fields. *Precision Agriculture*, *4*, 35–52. Available from https://doi.org/10.1023/A:1021858921307.

Win, K.K. (2003). Plot-specific N fertilizer management for improved N-use efficiency in rice-based systems of Bangladesh (Vol. 12). In C. VerlagYao, X., Du, W., Feng, S., & Zou, J. (2010). *Image-based plant nutrient status analysis: An overview*. Proc. of IEEE international conference on intelligent computing and intelligent systems (ICIS). Xiamen, Taiwan: IEEE. https://doi.org/10.1109/ICICISYS.2010.5658601. PMid:20309941.

Yousfi, S., Peira, J., Horra, G., & Ablan, P. (2019). *Remote sensing: Useful approach for crop nitrogen management and sustainable agriculture, sustainable crop production* (pp. 1–13). IntechOpen. Available from https://doi.org/10.5772/intechopen.89422.

SECTION 5

Precision agriculture practices for crop protection

CHAPTER

17 Smart farming to support agricultural crop damage assessment: interweaving Earth Observation and IoT data

Anastasia Dagla[1], Panagiota Louka[1,2], Yorgos Efstathiou[1], Nikos Kalatzis[1], Vassilis Protonotarios[1] and Argyros Argyridis[1]

[1]*NEUROPUBLIC SA, Piraeus, Greece*

[2]*Department of Natural Resources & Agricultural Engineering, Agricultural University of Athens, Athens, Greece*

17.1 Introduction

Agricultural sector is an important provider for humankind, as it covers needs for food and fiber. There is a constantly growing demand for food and agricultural products, which is projected to be even more intense during the next decades. This is attributed mainly to population growth, limited availability of arable land, and human-induced climate change, which is accompanied by a higher frequency and intensity of extreme weather events (Lieder & Schröter-Schlaack, 2021).

Population growth is expected to reach almost 10 billion by 2050, raising global food security issues (FAO, 2017). More precisely, current estimates of the Food and Agriculture Organization of the United Nations (FAO) show that feeding a world population of 10 billion people in 2050 would require increasing overall food production by some 70% until 2050. Developing countries are mostly influenced as the agricultural production would need to almost double (Alexandratos & Bruinsma, 2012; Desa, 2015).

At the same time, innovative technologies, especially in the field of Earth Observation, Remote Sensing, and Artificial Intelligence, are already put together to meet global food demand, by continuously monitoring crop growth and hence, enhance crop productivity. However, FAO projections prove that crop yields are not following the same rising trend, but instead they are kept in critically low levels (FAO, 2017).

To meet the growing demand for agricultural products, there is an urgent need to minimize crop losses, which may significantly affect agricultural productivity.

Remote Sensing in Precision Agriculture. DOI: https://doi.org/10.1016/B978-0-323-91068-2.00024-2

Crop losses occur due to either biotic or abiotic factors. The main abiotic factors concern available water and nutrients as well as temperature. The biotic factors are linked to infestation from pests, diseases, and weeds (Chawla et al., 2020).

The changing climate conditions induce more intense crop losses from all the aforementioned factors and bring about even higher tension on the agricultural sector. More frequent, intense, and difficult-to-predict extreme weather events, linked to abiotic factors, take place, which usually exceed stress tolerance of crops that remain exposed to severe weather conditions. Also, biotic factors have more intense negative effects due to climate change—this may refer to the spreading of more resistant pests that cause disastrous diseases to crops, especially during great heat waves and long draught periods (FAO, 2017).

Extreme weather events that result in significant cumulative losses in agricultural production worldwide during the last decades include floods, extreme temperatures, and windstroms. For instance, extreme rainfall may cause flooding, which is considered one of the major causes of crop damage, and it is estimated that losses due to frost may aggravate because of climate change. According to FAO, during the years 2003–13, a total of 93,319 ha and 1.6 million tons of crops were damaged or destroyed by floods worldwide, which account for 57% of crop damage by all natural disasters (FAO, 2015). Furthermore, extreme temperature events, either concerning frost or heat, are a significant challenge for agricultural production as they are predicted to increase under future climate scenarios (Zheng et al., 2012). Last, but not least, extreme wind may lead to crop lodging, which results in the reduction of the crop height and crop yield. Crop lodging has resulted in losses of £47–120 million for rapeseed crop and $80 million for winter wheat crop in the United Kingdom per year (Berry et al., 1998; Kendall et al., 2017).

Taking into consideration the current increase in extreme weather event frequency and the high spatial variability of crop damages, the critical need for immediate and accurate crop loss assessment after extreme events is evident (Harrison et al., 2016; Rahman et al., 2021). The main benefit from urgent assessment mechanisms is the automation of side processes and hence, the faster processing of compensations. Also, it is linked to the determination of the prices of agricultural products, agricultural policymaking, and insurance appraisal (Rahman & Di, 2020). Apart from this, there are also valuable long-term benefits that result from the construction of reliable crop damage time series datasets. It becomes possible to suggest more precise treatments mitigating the damage and to model more reliable crop yields.

Nowadays, with the evolution and adoption of smart systems in agriculture, integrating state-of-the-art methodologies, tools, and practices, such as the Internet of Things (IoT), Big Data analytics, Artificial Intelligence (Neural Networks, Machine Learning, etc.), Cloud Computing, Remote Sensing, geographic information systems (GIS), and other contemporary technologies, leads to a significant transformation, which is being characterized as the fourth revolution in agricultural production (Walter et al., 2017; Zhai et al., 2020).

The incorporation of information and data-based technologies for collection, analysis, and application in agriculture is referred to as smart farming (SF) and is being described as a panacea in terms of efficiency and sustainability for the agriculture of the future (Alfred et al., 2021; Jouanjean, 2019; Lajoie et al., 2020; Zaballos et al., 2019). All the aforementioned tools are promising and are expected to optimize agricultural operations aimed to enhance production and reduce inputs and yield losses (Delgado et al., 2019; Elijah et al., 2018; Jha et al., 2019).

The use of Remote Sensing data and technologies in smart farming has increased rapidly during the past few decades. There is unprecedented availability of high-resolution (spatial, spectral, and temporal) satellite images that are exploited in applications including crop monitoring, irrigation management, nutrient application, disease and pest management, and yield prediction (Sishodia et al., 2020).

An important application in which Remote Sensing data are exploited involves the estimation of crop yield and assessment of crop losses. Most studies use for these applications, as plethora of research studies indicated, the strong relationship between crop yield and vegetation indices (Anderson et al., 1993; Shanahan et al., 2001; Wiegand & Richardson, 1990; Wylie et al., 1991). In most cases, the vegetation index is aggregated over a period of time and then related to the corresponding crop yield instead of using instantaneous (satellite overpass) observations (Chawla et al., 2020). In this approach, the variation of the crops' spectral reflectance during the growing season is taken into account and the "noise" due to soil, clouds, or other factors is reduced, and the total effect of photosynthesis is included as well (Benedetti & Rossini, 1993; Labus et al., 2002; Rudorff & Batista, 1990).

Although the use of Remote Sensing in smart farming has great potential, these are significant restrictions that must be addressed to reach its potential. The majority of satellite data are freely available, but high-level technical expertise is often required for the processing of the data. Also, for specific applications such as the estimation of crop losses, higher temporal resolution is necessary. Furthermore, cloudy days and variable or inconsistent irradiance or sunlight may render many satellite images unsuitable for use (Chawla et al., 2020).

Taking into consideration the potential and restrictions of using Remote Sensing data for estimating crop losses and smart farming in general, the research presented herein was designed. To ensure that farming is a profitable activity and boosts crop yields and productivity, an application has been developed that estimates and visualizes near-real time climate-related crop damages by suggesting a rather rational and scientific approach. This chapter presents a newly developed crop damage-assessment service integrating in situ IoT-acquired field-level data and public-domain EO data to support the decision-making of farmers, paying agencies, and agro-insurance companies. Its major scientific contribution is the suggested automated, data-driven, and interactive user approach for remotely managing crop damages and avoiding on-site visits.

The so-called ResAgri (Resilient and Sustainable Agriculture) application is designed and developed by NEUROPUBLIC (NP) (https://www.neuropublic.gr/) in the framework of the European project e-shape Horizon 2020. NEUROPUBLIC is an ICT SME offering a wide portfolio of Information Services particularly focusing on technologies for the small farms. ResAgri is part of NP's package of Smart Farming digital services, and it targets mainly agri-experts, appraisers, insurance companies, and paying agencies, who need to collaborate with farmers to quantify the reported crop damages and arrange compensation payments. Similar services are currently developed to support the agri-food sector, but most of them lack reliable results on parcel level of detail, due to the resolution of the available datasets. ResAgri exploits the field data acquired from NP's dense IoT network of weather stations, gaiatrons (Adamides et al., 2020; Kalatzis et al., 2019) that allow us to detect damages caused even by local-based and small-scale weather events.

ResAgri focuses on evaluating noncotton, tomato, and grapevines crop parcel cultivations for the growing season of 2020 and 2021, in parcels located in three regions of Greece with increased agricultural activity and with significant financial impact (Rhodope, Thessaly and Crete). There are three available functions presented in a webGIS environment: (1) crop monitoring, (2) damage assessment, and (3) high-level monitoring. Crop monitoring is designed for advanced users who desire direct access to the daily observations of every crop. Observations include parcels' field measurements and the corresponding NDVI values, as well as the available satellite images at any selected time moment. The user is able to track crop growth and compare it with other crops or with different growing seasons. Damage assessment function is developed to inform the user, both visually (damage maps) and quantitatively (damage indicator), regarding the crop damages for five main disastrous weather events (heat, rainfall, hail, frost, and windstorm). To get an overview of the damaged crops per prefecture of Greece and generalize the influence of damaged crops to a regional level of detail, the function of high-level monitoring function was developed. The platform of ResAgri is constantly updated with new data field measurements, and satellite images and is designed to give results per growing season (2020, 2021 etc.). Hence, apart from its direct use, we construct reliable time series of damaged crops per growing season, weather event, location, and crop type, which we aim to use as a powerful dataset to further improve and automate the farming practices in Greece.

The overall structure of this book chapter takes the form of distinct sections, including this introductory section. The rest is organized as follows.

17.2 Use cases and data used

The chosen study areas are located in the regions of (1) Thessaly, (2) Rhodope, and (3) Crete in Greece. Cotton, tomato, and grapevine crops of 2020 and 2021 are processed. The use cases, the user and technical requirements of the developed service were extracted through a number of iterative interviews involving the following:

- Farmers, agronomists, and advisors.
- An insurance company with a dominant role in agricultural domain.

An overview is presented in Fig. 17.1.

The study areas are known as main agricultural plains of Greece. Thessaly and Rhodope are major cotton-growers, both in national level and within the EU borders. Thessaly is based in mainland Greece. It has a rather flat landscape and mild weather conditions. Floods and extensive rainfalls are the most common and dangerous weather events that could potentially cause crop damages (Ghionis et al., 2014). Rhodope is on the northern part of Greece. Its terrain is mountainous, and it experiences more frequent and intense hail and windstorm events (Kazakis et al., 2015). Crete is located in southern Greece and is constantly hit by long-term droughts and severe heats (Koutroulis et al., 2013).

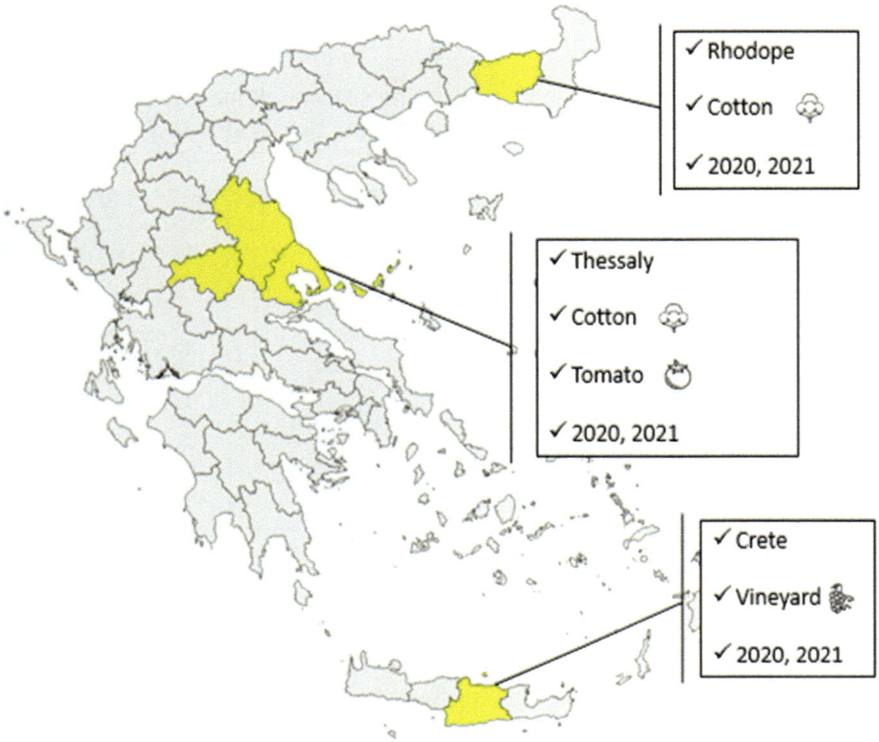

FIGURE 17.1

Overview of the chosen study areas, crop types, and study years.

Some typical characteristics of the selected crops (cotton, tomato, and grapevines) are presented in Table 17.1 and the respective typical NDVI curves in Fig. 17.2. The growing season of crops was identified based on data provided from our experienced agronomist colleagues, regarding the selected crop types. Additionally, they were verified by the modeled NDVI curve that we designed to investigate crop spectral behavior per region. Consequently, cotton crops have a growing season of almost 200 days, from April to November and almost the same applies for tomato crops—150 days, from March to September. Grapevines have a less stable and shorter growing season starting from September to May, lasting approximately 120 days. The optimal growing temperatures vary from 20 to almost 35°C, and therefore, the selected crop types are considered summer crops.

To assess precisely the crop damages, it is necessary to consider crop's special characteristics, as well as the existing framework for crop damage compensations. For this, Hellenic Organization of Agricultural Insurances (HOAI) is responsible, which is a state organization and the main insurance carrier of plant production and livestock capital in Greece (http://www.elga.gr/). According to HOAI standards, the extreme weather events that take place out of the crop's growing season or during the seeding process are not compensated as it is assumed that they are not influencing crop's growth, and hence, they were not considered in our analysis.

Table 17.1 Typical crop properties; growing season, climatic requirements, optimal temperature, and potential most destructive weather event.

Crop characteristics	Cotton	Tomato	Vineyard
Growing season	200 days, April – November	100-150 days, March – September	120 days, May – September
Climatic requirements	✓ Warm and humid climate ✓ High rainfall (not intense) ✓ Long warm summer	✓ Warm and cool climate ✓ Low-medium rainfall	✓ Hot and dry climate ✓ Long warm summer ✓ Rainy winter
Optimal temperature	21 - 37 °C	18 - 26 °C	25 – 32 °C
Most harmful weather event	Frost	Frost	Hail

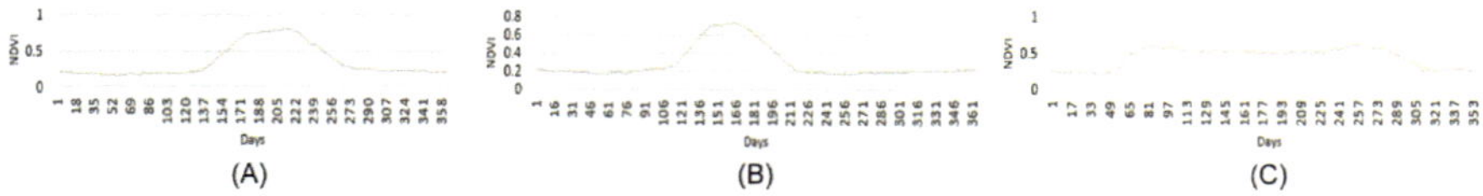

FIGURE 17.2

NDVI curves of the selected crop types: (A) cotton, (B) tomato, and (C) vineyard. *NDVI*, Normalized difference vegetation index.

The input data that were used for the analysis presented in ResAgri platform consist of the following:

1. remotely sensed Earth Observation (EO) data,
2. in situ field data collected from IoT weather stations, and
3. a set of crop parcel polygons.

The main data sources of the input EO and IoT-sensed data are the Copernicus Sentinel-2 Earth Observation (EO) mission that consistently acquires optical satellite imagery and is operated by the European Space Agency (ESA) (https://sentinel.esa.int/web/sentinel/missions/sentinel-2) and the "gaiasense field" network of "gaiatron" weather stations. The weather stations are located close to the boundaries of every crop parcel, and they transmit hourly data from the field to the giasense cloud service (Adamides et al., 2020; Kalatzis et al., 2019). The dataset of crop polygons changes for every growing season and is based on our developed crop classification algorithm.

The data pipeline is updated automatically with present satellite images and field measurements, when they become available. This means that ResAgri application has the potential to cover all areas of Greece with near-real-time services. The input data properties are furthered described as follows (Table 17.2):

1. *EO data*: Optical multispectral Sentinel-2 satellite images with a spatial resolution of 10 m and temporal resolution of approximately 5 days. Specific tiles of the study areas; Thessaly, Rhodope, and Crete were selected and prepared (Harrison et al., 2016).
2. *IoT-acquired data*: Hourly meteorological measurements that include: air temperature (°C), precipitation (mm), wind speed (km/h), and relative air humidity (%), as well as the locations (*x*, *y*) of the weather stations (Adamides et al., 2020; Kalatzis et al., 2019).
3. *Polygons of crops*: Crop parcel polygons of cotton, tomato, and grapevines for 2020 and 2021. The geometry of the selected parcels resulted from an in-house developed crop classification algorithm.

Table 17.2 Input Earth Observation and Internet of Things-acquired data and their properties.

Input data	Data provider	Observations	Dimensional analysis	Spatial resolution	Temporal resolution
GaiaTron weather stations	NEUROPUBLIC	• Air temperature • Precipitation • Wind speed • Humidity	0D	-	per 1 hour
Sentinel-2 satellite images	esa	Multispectral optical images	2D	10m	per 4 days

17.3 Methodology and implementation

Firstly, the input data were preprocessed with respect to the following steps:

1. Weather data noise removal
2. Satellite image preparation
3. Crop parcel selection
4. Modeling NDVI distribution.

During the main data processing procedure, the major tasks were subdivided into smaller individual tasks. These are:

1. weather event detection,
2. satellite image selection,
3. damage map computation and
4. damage quantification per crop parcel, by introducing a damage indicator.

17.3.1 Data preprocessing

17.3.1.1 Weather data noise removal

The field observations, obtained from the IoT weather stations, were preprocessed by performing distance-based noise removal and anomaly detection (Zemicheal & Dietterich, 2019). After this, the daily values of the field observations were extracted from the hourly data to compute crop parcel statistics. To be more precise, for formulating the daily data, the mean values per day were extracted for the surface temperature, wind speed, and humidity. For the daily precipitation values, the hourly observations per day were summed up.

17.3.1.2 Satellite image preparation

The Sentinel-2 satellite images were preprocessed using the "Earth Insight Engine," an in-house developed data pipeline infrastructure, dedicated to prepare the Sentinel-2 satellite images after they become available from ESA (Diagram 17.1).

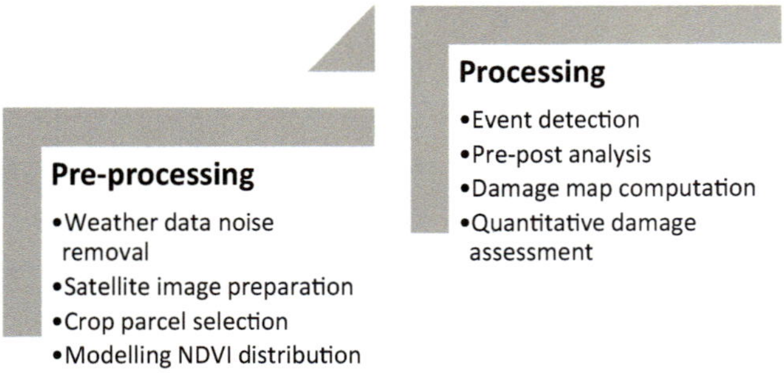

DIAGRAM 17.1

Technology pipeline.

It includes the necessary preprocessing steps of the satellite image processing. Apart from downloading and unzipping the satellite images, they are atmospherically corrected, and the clouds are masked by applying ESA's Sen2Cor dedicated algorithm (https://step.esa.int/main/snap-supported-plugins/sen2cor/). Additionally, Vegetation Indices (VIs) are computed, and the respective images are archived. VIs images were used as secondary data, during the main processing. The VIs computed are:

1. Normalized difference vegetation index (NDVI) and
2. Color infrared (CIR) images.

17.3.1.3 Crop parcel selection

A certain number of crop parcels are included in ResAgri application, and the selection was done condition-based. For this purpose, a dedicated model was designed and implemented. It takes as input data (1) the locations of the weather stations, (2) the crop parcels resulting from the crop classification algorithm, and (3) the prefectures of the study areas. Crop parcels within the borders of the prefectures and with an 8 km distance from the weather stations were selected. This was done to match every parcel with a neighboring weather station and its corresponding field observations. After this, conditions regarding parcel size and geometry were defined, taking into account the spatial resolution of the satellite images that were used. More specifically, crop parcels with an area greater than 0.3 ha were chosen. An overview of the parcel selection model follows in Diagram 17.2.

17.3.1.4 Modeling normalized difference vegetation index distribution

To define whether an extreme weather event influences crop growth, the crop type, the study area, and its microclimate should be considered. Therefore, it was necessary to model past NDVI time series. This was done to investigate the behavior of a typical NDVI distribution for each crop type per study area and helped us to individualize the analysis under the special characteristics of each crop type based on its growing region.

Crop parcels of 2018, 2019, and 2020 were used to formulate the NDVI time series. For each region and crop type, the NDVI values were computed per day that an appropriate satellite image was available. After this, the NDVI values were interpolated, to obtain an NDVI value for every day of the year. Finally, by combining the yearly NDVI distributions, a typical NDVI distribution, which shows the expected crop spectral behavior, is derived. To be more precise, the distribution represents the average behavior of every crop type in its growing region and was therefore used for determining the damage indicator, during the main data processing. The distributions of the damaged parcels were compared with the modeled distribution, to ascertain the magnitude of the crop damages and also validate the damaged crops that resulted from the pre–post analysis that

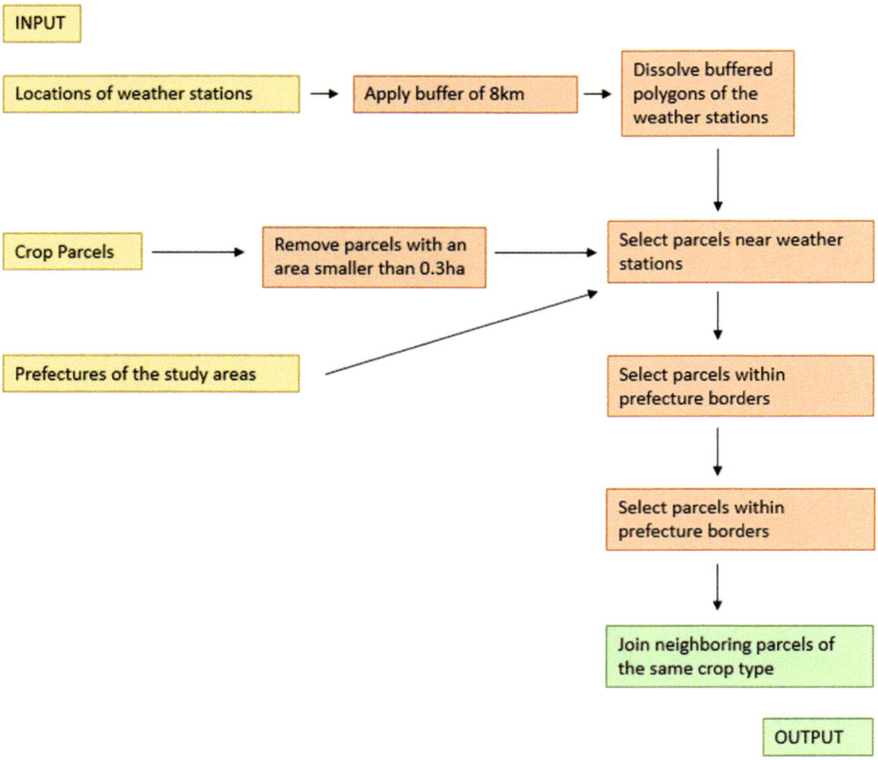

DIAGRAM 17.2

Parcel selection model.

was performed. An example of the modeled NDVI distribution for cotton and tomato crops in Thessaly follows in Fig. 17.3.

17.3.2 Processing

17.3.2.1 Event detection

The field data, surface temperature, precipitation, wind speed, and humidity, acquired from the weather stations are used to build a near-real-time alarm system that detects and reports the extreme weather events. Determining which weather conditions are considered hazardous for the crops and which crop damages are eligible for compensation, the HOAI standards were adopted. This means that certain threshold values were applied to the field observations to detect the potential disastrous weather events.

According to HOAI, the extreme weather events for crop growth are (i) heat, (ii) frost/hail, (iii) windstorm, and (iv) flood /excessive rainfall. More specifically,

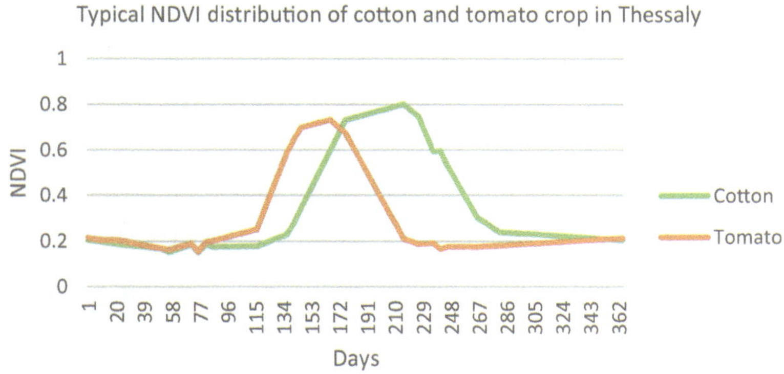

FIGURE 17.3

Example of modeled NDVI distribution for cotton and tomato crops in Thessaly. *NDVI*, Normalized difference vegetation index.

the weather events take place when the following conditions are satisfied:

1. *Heat* when air temperature is above 40°C
2. *Frost* when air temperature is below 0°C
3. *Windstorm* when the wind speed exceeds 62 km/h.
4. *Flood* when the precipitation values exceed 15 mm.

These values are considered threshold values. The designed event detection system generates alarms when the field observations exceed a threshold value. After this, the weather events are stored and categorized based on their type, duration, and location. For each weather station that reports an event, an Area of Interest (AoI) is created. This is the influenced by the weather event area, and therefore, the crop damages within this area should be confirmed.

17.3.2.2 Satellite image selection for pre–post analysis

To assess the climate-related crop damages, a pre–post analysis is held, based on the date of the reported weather event. For this purpose, suitable satellite images were selected to support the pre–post analysis. The selection depends on the duration of the weather event (start/end date) as well as on the influenced by the weather event area, the AoI.

For the pre–post analysis, two NDVI images were chosen that focus on the AoI: one before the starting date of event (preimage) and the other after the ending date of the event (postimage). The post-NDVI image was selected within a time window of 10–15 days after the ending date of the event, to guarantee that the detected damage actually influences the crop performance. Occasionally it needs a certain period of time for the damaged crop to appear in the NDVI image, considering the intensity of the weather event and the respective crop damage.

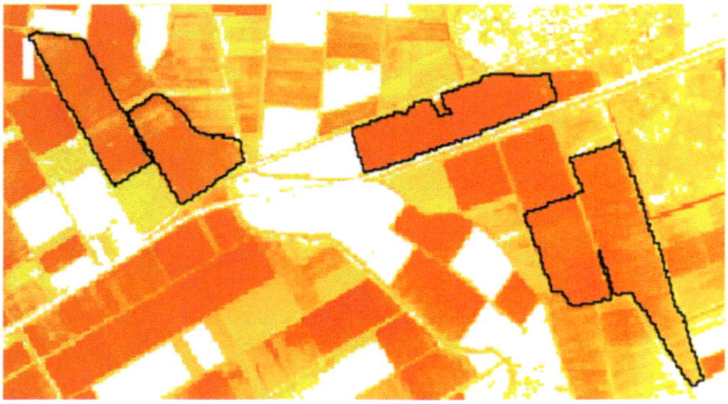

FIGURE 17.4

Extent of an example damage map.

17.3.2.3 Damage map computation

By comparing the pre–post-NDVI images, thematic maps were created focusing on the damaged crop parcels after every reported extreme weather event. A pixel-based classification algorithm was implemented to generate the damaged maps. The difference between the pre- and post- NDVI values of every pixel was estimated. Crop damage is associated with drops of NDVI values during the cultivation period of the plant. Hence, only the negative NDVI differences were taken into account, to classify the occurred damages. Fig. 17.4 shows an example damage map, which was created after a severe flood of 2020, which destroyed plenty of cotton and tomato parcels in the region of Thessaly.

Crops damage grade varies from yellow to orange and red color for the mild, moderate, and severe damages, respectively. The user is able to obtain pixel-level information and hence, to identify certain locations inside the parcel's boundaries that are mostly influenced by the extreme weather events.

17.3.2.4 Quantitative damage assessment

To quantify the occurred damage of every crop parcel, a damage indicator is defined and estimated. For its computation, the results of the pre–post analysis and the typical NDVI distributions were used. The resulted damage is based not only on the differences of the NDVI values before and after a disastrous weather event (dNDVIpp) but also on the difference between the post-NDVI value and the NDVI value of the typical NDVI distribution (dNDVItc) that resulted from the modeling of the past NDVI time series. Weights were given to dNDVIpp and dNDVItc that were decided after fine-tuning the model.

The damage indicator was computed based on the following formula:

$DI = [(dNDVIpp\ 0.4) + (dNDVItc \cdot 0.6)]\ 100\%$

where

DI: percent damage indicator

dNDVIpp: The mean value of the NDVI differences that resulted from the pre–post analysis per crop parcel.
dNDVItc: The mean value of the NDVI differences that resulted from the comparison with the typical NDVI distribution per crop parcel.

17.4 Results

The results of the suggested method are available for the end users (agro-experts, insurance companies, farmers, etc.) to assess them in a dedicated web service that is here presented, the ResAgri platform. Additionally, some validated extreme weather events and the corresponding crop damages are discussed.

17.4.1 ResAgri dashboard

To communicate the applied method in a user-friendly way, a special web service is developed, which consists of three functions:

1. crop parcel statistics (Crop Monitoring),
2. crop damages per parcel (Damage Assessment), and
3. crop damages per region (High Level Monitoring).

ResAgri is a webGIS environment, and it is available at https://damage.resagri.eu/en#. The UI of ResAgri is based on OpenLayers library, which uses georeferenced map images in the form of Web Mapping Service (WMS), which are served from MapServer open-source platform.

ResAgri functions are categorized in the following information layers Fig. 17.5:

- *Crop monitoring*: A feature designed for the advanced users who need to directly access the primary data at a parcel level of detail. The users have the opportunity to locate and monitor the crop performance at any available timestamp. More specifically, crop monitoring includes information related to

FIGURE 17.5

The UI of ResAgri.

the parcel's NDVI values and field weather measurements (surface temperature, wind speed, precipitation, and humidity). Furthermore, Red-Green-Blue satellite images and the VI images can be selected for the desired timestamp.

- *Damage assessment*: This function presents the damage maps and the quantitative damage analysis of every parcel that was influenced by an extreme weather event (heat, frost, windstorm, and flood). The user can visually evaluate the spatial extend of the damage, the type and date of the weather event, as well as how much a parcel was damaged by assessing its damage indicator. The damage indicator refers to parcel level of detail. However, thematic damage maps were generated pixel-based, and hence, the user is able to decide upon specific regions inside parcel boundaries.
- *High-level monitoring*: Through this service the user can get informed about the overall picture of the damaged parcels in a regional level of detail. For every prefecture of Greece, the crop damages of every event are generalized. The number of the estimated claims and the size of the affected area are presented per prefecture.

17.4.2 Extreme weather events

The event detection alarm system reported 44 weather events for the region of Thessaly during the growing seasons of 2020 and 2021, 15 for the region of Rhodope and 16 for Crete region. Considering the type of the occurred events, it has been observed that Thessaly is mostly influenced for both growing seasons by frost and extensive rainfalls, Rhodope almost only by extensive rainfalls, and Crete both by frost and severe heat waves. By measuring the duration of the event, the number of the weather stations that reported it and combining this information with the generated damaged maps, we were able to determine the disastrous events and finally, present them to ResAgri. The great number of the reported events shows that the method is efficient and that it can detect even the mild and small-scale events.

The weather events that we obtained were validated with (1) appraisers from the project-involved insurance company, (2) local agro-focused newspapers, and (3) HAOI standards. The event selection was also based on HAOI standards regarding the compensation payments after climate-related crop damages occurred.

Consequently, the events that caused crop damages are:

1. A severe large-scale rainfall and hail took place during the growing season of 2020 (September 18, 2020) in Thessaly, that was named "Ianos." It was reported from more than 10 weather stations, and it lasted 6 hours. Many crops got flooded, and it is considered as one of the most disastrous events in the area.
2. A smaller-scale rainfall and hail took place on July 3, 2020 in a Kranonas, a small subregion of Thessaly. This event lasted 2 days, and it was reported by

three neighboring stations. Although, it influenced a very small area compared with the previous one, plenty of crops got damaged.

3. Finally, a severe heat occurred on May 15, 2020 in Crete island that destroyed many grapevine crops. Although, this event lasted only 2 hours, it caused great crop damages.

17.4.3 Crop damages

The damage map that was created for the event that occurred in Thessaly on September 18, 2021 (Fig. 17.6), indicates event's intensity and the severity of the damages that influenced a rather large area.

To be more precise, almost half of the crops located in Thessaly were damaged and some of them were completely destroyed. In ResAgri application, only the parcels with an estimated damage more than 20% are included because based on HOAI standards only these crops are eligible for compensation. Fig. 17.7 shows the available information for every damaged crop pacel.

From the event that occurred in Kranonas, Thessaly, on July 4, 2020 (Fig. 17.8), we obtain that the AoI was very small compared with the first event. However, all the crops inside the AoI were reported with a damage greater than 20%, and hence, all of them are eligible for compensation.

The heat event in Crete (Fig. 17.9) is the most diverse in terms of crop damage spatial variability. The AoI is the largest one compared with the other two events, and not all neighboring crops are damaged in a similar way. Hence, the spatial pattern of the crop damages from this heat event differs significantly from the pattern of the flood-related crop damages.

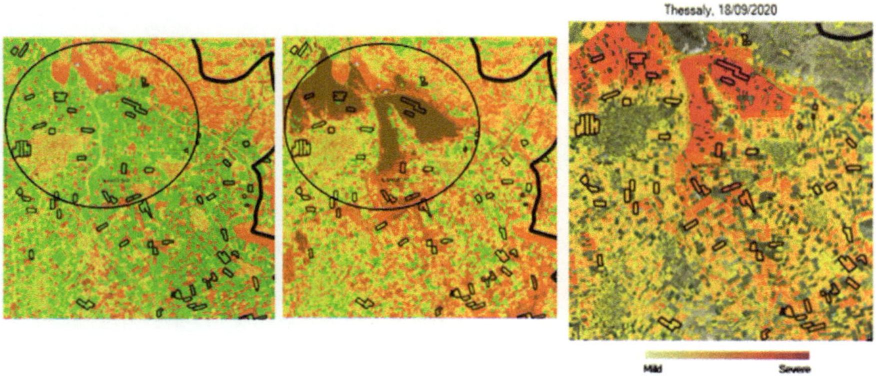

FIGURE 17.6

From left to right: pre-NDVI image, post-NDVI image, damage map of Thessaly, 18/09/2020. *NDVI*, Normalized difference vegetation index.

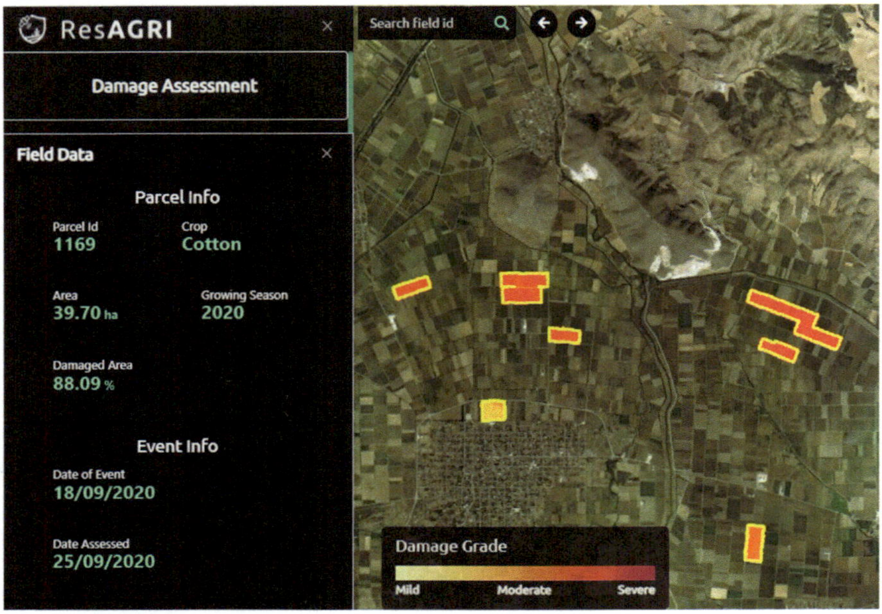

FIGURE 17.7

Crop parcel damage statistics.

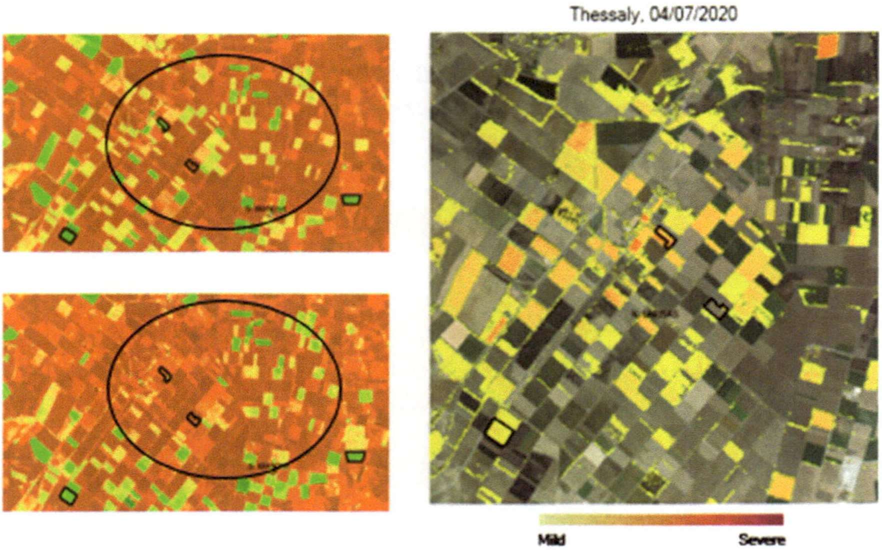

FIGURE 17.8

From upper left: pre-NDVI image, post-NDVI image, damage map of Thessaly, Kranonas, 04/07/2020. *NDVI*, Normalized difference vegetation index.

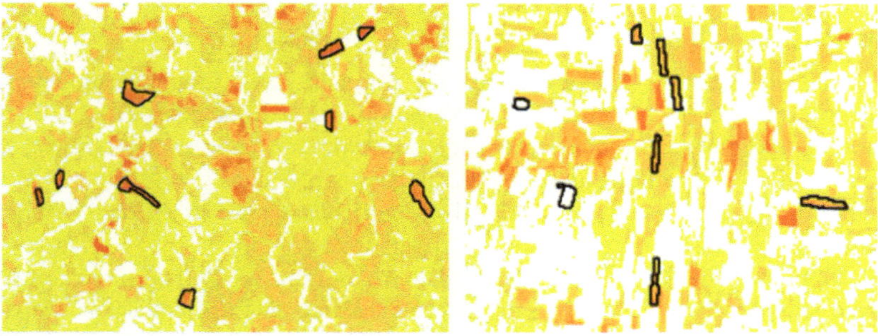

FIGURE 17.9

Example extents of Crete damage map, May 15, 2020.

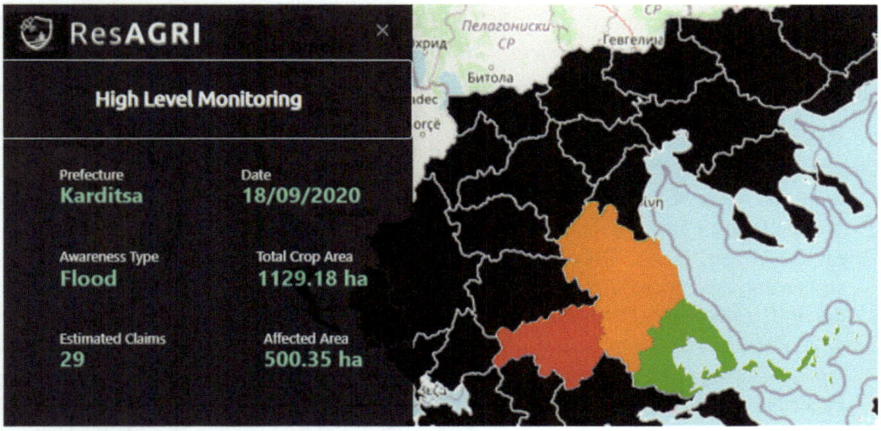

FIGURE 17.10

Crop damage assessment in Thessaly, September 18, 2020, at a regional level of detail.

The crop damages have generalized in a prefecture level for every event. An example of the generalization of the first event can be seen in Fig. 17.10, in which we visually define the mostly damaged regions and give statistics regarding the occurred crop damages.

Acknowledgment

The e-shape project has received funding from the European Union's Horizon 2020 Research and Innovation Programme under Grant Agreement 820852.

References

Adamides, G., Kalatzis, N., Stylianou, A., Marianos, N., Chatzipapadopoulos, F., Giannakopoulou, M., Papadavid, G., Vassiliou, V., & Neocleous, D. (2020). Smart farming techniques for climate change adaptation in cyprus. *Atmosphere*, *11*(6), 557. Available from https://doi.org/10.3390/atmos11060557.

Alexandratos, N., & Bruinsma, J. (2012). *World agriculture towards 2030/2050: The 2012 revision.*

Alfred, R., Obit, J. H., Chin, C. P. Y., Haviluddin, H., & Lim, Y. (2021). Towards paddy rice smart farming: A review on big data, machine learning, and rice production tasks. *IEEE Access*, *9*, 50358–50380. Available from https://doi.org/10.1109/ACCESS.2021.3069449.

Anderson, G. L., Hanson, J. D., & Haas, R. H. (1993). Evaluating landsat thematic mapper derived vegetation indices for estimating above-ground biomass on semiarid rangelands. *Remote Sensing of Environment*, *45*(2), 165–175. Available from https://doi.org/10.1016/0034-4257(93)90040-5.

Benedetti, R., & Rossini, P. (1993). On the use of NDVI profiles as a tool for agricultural statistics: The case study of wheat yield estimate and forecast in Emilia Romagna. *Remote Sensing of Environment*, *45*(3), 311–326. Available from https://doi.org/10.1016/0034-4257(93)90113-C.

Berry, P., Spink, J., Griffin, J., Sylvester-Bradley, R., Baker, C., Clare, R., & Scott, R. (1998). *Research to understand, predict and control factors affecting lodging in wheat, 131.*

Chawla, I., Karthikeyan, L., & Mishra, A. K. (2020). A review of remote sensing applications for water security: Quantity, quality, and extremes. *Journal of Hydrology*, 585. Available from https://doi.org/10.1016/j.jhydrol.2020.124826.

Delgado, J., Short, N. M., Roberts, D. P., & Vandenberg, B. (2019). Big data analysis for sustainable agriculture. *Frontiers in Sustainable Food Systems*, *3*. Available from https://doi.org/10.3389/fsufs.2019.00054.

Desa, U. (2015). *World population prospects: The 2015 revision, key findings and advance tables*. Department of Economic and Social Affairs Population Division. Working Paper No. ESA/P/WP.241.

Elijah, O., Rahman, T. A., Orikumhi, I., Leow, C. Y., & Hindia, M. N. (2018). An overview of Internet of Things (IoT) and data analytics in agriculture: Benefits and challenges. *IEEE Internet of Things Journal*, *5*(5), 3758–3773. Available from https://doi.org/10.1109/JIOT.2018.2844296.

FAO. (2015). *The impact of disasters on agriculture and food security*. Food and Agriculture Organization of the United Nations.

FAO. (2017). The future of food and agriculture–Trends and challenges. Annual Report.

Ghionis, G., Sifnioti, D., Lazogiannis, K., Poulos, S. E., Kotsopoulos, S., & Nastos, P. T. (2014). *Climatic conditions of the delta of the PiniosRiver*. COMECAP 2014 e-book proceedings, Herakleion, Greece.

Harrison, M. T., Cullen, B. R., & Rawnsley, R. P. (2016). Modelling the sensitivity of agricultural systems to climate change and extreme climatic events. *Agricultural Systems*, *148*, 135–148. Available from https://doi.org/10.1016/j.agsy.2016.07.006.

Jha, K., Doshi, A., Patel, P., & Shah, M. (2019). A comprehensive review on automation in agriculture using artificial intelligence. *Artificial Intelligence in Agriculture*, *2*, 1–12. Available from https://doi.org/10.1016/j.aiia.2019.05.004.

Jouanjean, M. A. (2019). *Digital opportunities for trade in the agriculture and food sectors.*

Kalatzis, N., Marianos, N., & Chatzipapadopoulos, F. (2019). IoT and data interoperability in agriculture: A case study on the gaiasenseTM smart farming solution. *In Global IoT Summit, GIoTS 2019 - Proceedings. Institute of Electrical and Electronics Engineers Inc*. Available from https://doi.org/10.1109/GIOTS.2019.8766423.

Kazakis, N., Kougias, I., & Patsialis, T. (2015). Assessment of flood hazard areas at a regional scale using an index-based approach and analytical hierarchy process: Application in Rhodope–Evros region, Greece. *Science of The Total Environment*, *538*, 555–563. Available from https://doi.org/10.1016/j.scitotenv.2015.08.055.

Kendall, S. L., Holmes, H., White, C. A., Clarke, S. M., & Berry, P. M. (2017). Quantifying lodging-induced yield losses in oilseed rape. *Field Crops Research*, *211*, 106–113. Available from https://doi.org/10.1016/j.fcr.2017.06.013.

Koutroulis, A. G., Tsanis, I. K., Daliakopoulos, I. N., & Jacob, D. (2013). Impact of climate change on water resources status: A case study for Crete Island, Greece. *Journal of Hydrology*, *479*, 146–158. Available from https://doi.org/10.1016/j.jhydrol.2012.11.055.

Labus, M. P., Nielsen, G. A., Lawrence, R. L., Engel, R., & Long, D. S. (2002). Wheat yield estimates using multi-temporal NDVI satellite imagery. *International Journal of Remote Sensing*, *23*(20), 4169–4180. Available from https://doi.org/10.1080/01431160110107653.

Lajoie, -O. 'Malley, Bronson, A., Burg, K., & Klerkx, L. (2020). The future (s) of digital agriculture and sustainable food systems: An analysis of high-level policy documents. *Ecosystem Services*, *45*, 101183.

Lieder, S., & Schröter-Schlaack, C. (2021). Smart farming technologies in arable farming: Towards a holistic assessment of opportunities and risks. *In Sustainability*, *13*(12). Available from https://doi.org/10.3390/su13126783.

Rahman, Md. Shahinoor, & Di, L. (2020). A systematic review on case studies of remote-sensing-based flood crop loss assessment. *Agriculture*, *10*(4), 131. Available from https://doi.org/10.3390/agriculture10040131.

Rahman, M. S., Di, L., Yu, E., Lin, L., & Yu, Z. (2021). Remote sensing based rapid assessment of flood crop damage using novel disaster vegetation damage index (DVDI. *International Journal of Disaster Risk Science*, *12*(1), 90–110. Available from https://doi.org/10.1007/s13753-020-00305-7.

Rudorff, B. F. T., & Batista, G. T. (1990). Spectral response of wheat and its relationship to agronomic variables in the tropical region. *Remote Sensing of Environment*, *31*(1), 53–63. Available from https://doi.org/10.1016/0034-4257(90)90076-X.

Shanahan, J. F., Schepers, J. S., Francis, D. D., Varvel, G. E., Wilhelm, W. W., Tringe, J. M., Schlemmer, M. R., & Major, D. J. (2001). Use of remote-sensing imagery to estimate corn grain yield. *Agronomy Journal*, *93*(3), 583–589. Available from https://doi.org/10.2134/agronj2001.933583x.

Sishodia, R. P., Ray, R. L., & Singh, S. K. (2020). Applications of remote sensing in precision agriculture: A review. *Remote Sensing*, *12*(19), 1–31. Available from https://doi.org/10.3390/rs12193136.

Walter, A., Finger, R., Huber, R., & Buchmann, N. (2017). Opinion: Smart farming is key to developing sustainable agriculture. *Proceedings of the National Academy of Sciences*, *114*(24), 6148–6150. Available from https://doi.org/10.1073/pnas.1707462114.

Wiegand, C. L., & Richardson, A. J. (1990). Use of spectral vegetation indices to infer leaf area, evapotranspiration and yield: I. rationale. *Agronomy Journal*, *82*(3), 623–629. Available from https://doi.org/10.2134/agronj1990.00021962008200030037x.

Wylie, B. K., Harrington, J. A., Prince, S. D., & Denda, I. (1991). Satellite and ground-based pasture production assessment in Niger: 1986–1988. *International Journal of Remote Sensing*, *12*(6), 1281–1300. Available from https://doi.org/10.1080/01431169108929726.

Zaballos, A.G., Rodriguez, E.I., & Adamowicz, A. (2019). The impact of digital infrastructure on the sustainable development goals: A study for selected Latin American and Caribbean countries. In *OECD. Digital Opportunities for Better Agricultural Policies.*

Zhai, Z., Martínez, J. F., Beltran, V., & Martínez, N. L. (2020). *Decision support systems for agriculture 4.0: Survey and challenges. Computers and Electronics in Agriculture, 170*, 105256.

Zemicheal, T., & Dietterich, T. G. (2019). Anomaly detection in the presence of missing values for weather data quality control. *COMPASS 2019 - Proceedings of the 2019 conference on computing and sustainable societies*, 65–73. Available from https://doi.org/10.1145/3314344.3332490, Association for Computing Machinery, Inc.

Zheng, B., Chenu, K., Fernanda Dreccer, M., & Chapman, S. C. (2012). Breeding for the future: What are the potential impacts of future frost and heat events on sowing and flowering time requirements for Australian bread wheat (Triticum aestivium) varieties. *Global Change Biology*, *18*(9), 2899–2914. Available from https://doi.org/10.1111/j.1365-2486.2012.02724.x.

CHAPTER

18 Detection of grapevine yellows using multispectral imaging

Uroš Žibrat and Matej Knapič

Plant Protection Department, Agricultural Institute of Slovenia, Ljubljana, Slovenia

18.1 Introduction

Protection of vines from diseases and pests is one of the most important measures in grape production. Viticulture is considered to be one of the most profitable agricultural sectors, with its main aim being the production of large quantities of high-quality grapes. Grapevines are perennial plants, with a productive lifespan of more than 30 years, in the absence of diseases and pests. The implementation of protection measures can have a significant impact on the economics of production. Optimal production quantity and quality can be achieved by timely and precise application of disease and pest management practices.

As with most crops, many diseases can infect grapevines (Table 18.1). Most diseases are caused by fungi, followed by viruses and phytoplasmas. The most common fungal diseases, such as downy and powdery mildew, are considered to

Table 18.1 Economically important grapevine diseases.

Fungi	Viruses and phytoplasmas	Bacteria
Downy mildew *Plasmophara viticola*	Grapevine *Flavescence dorée*	Bacterial necrosis *Xylophilus ampelinus*
Powdery mildew *Uncinula necator*	Grapevine Bois Noir *Candidatus Phytoplasma solani*	Crown gall *Agrobacterium tumefaciens*
Gray rot *Botrytis cinerea*	Grapevine leafroll-associated Virus GLRav	
Black rot of grapevine *Diaporthe neoviticola*	Grapevine Fanleaf Virus GFLV	
Red fire disease of grapevine *Pseudopeziza tracheiphila*	Grapevine Point gris Virus GPGV	
Esca		
Black rot of grapes *Guignardia bidwellii*		

Remote Sensing in Precision Agriculture. DOI: https://doi.org/10.1016/B978-0-323-91068-2.00001-1

be manageable by using fungicides. On the other hand, the achievement of production targets is increasingly limited by viral and phytoplasma-associated diseases, which cannot be adequately managed using plant protection products (Wilcox et al., 2015).

Grapevine *F. dorée* is a phytoplasma-associated disease that occurs in many grape-growing areas worldwide and is considered one of the most important grapevine diseases in Europe and is therefore subject to quarantine restrictions (Angelini et al., 2018; Fontaine et al., 2016). In Slovenia, it was first found in 2005 and has by 2009 spread to all important vine-growing regions in the country (Mehle et al., 2011) (Fig. 18.1). In the absence of control measures, its rapid spread over large areas is enabled by comparatively high abundances of its vector, the American grapevine leafhopper *Scaphoideus titanus* Ball (Hemiptera: Cicadelidae), which spreads the disease by sucking on plants and transmitting the phytoplasma through its saliva. This leafhopper feeds and reproduces almost exclusively on grapevines and is the only known vector of grapevine yellows (Chuche & Thiéry, 2014). The only effective suppression actions are removal of infected vines and regular extermination of *S. titanus* in all vineyards. In case these actions were abandoned, grapevine yellows could destroy an entire vineyard within 2 years after infection (Prezelj et al., 2013).

In-field detection of grapevine yellows is currently performed with two common techniques: (1) visual inspections and (2) molecular analysis. Visual inspections are the most widely used technique, as they are comparatively easy, don't require any special equipment, and are often performed by growers themselves. However, their accuracy is subject to an individual's experience and knowledge. Furthermore, an expert should be available for long-term monitoring, which is both impractical and expensive in larger fields. Visible signs of infection, especially in early stages, can be overlooked or mistaken for other stressors, both abiotic (e.g., severe nutrition problems, such as magnesium deficiency) and biotic (e.g., GLRav). Moreover, symptoms also depend on grapevine variety, general physiological status, and environmental conditions. Another characteristic of grapevine yellows is that plants can become temporarily asymptomatic, which further confounds visual inspections. Accurate knowledge is therefore essential; field assessments have to be performed by experienced experts. Identification is further complicated by the presence of Bois noir phytoplasma, with almost identical visual symptoms. Confirmatory, molecular tests have to be carried out to ensure accurate detection (Alemu, 2015; Margaria et al., 2009). While molecular tests are more accurate, they are more time-consuming and expensive and as such not feasible to be used on all plants in even a single vineyard. Because grapevine *F. dorée* can pervade entire vineyards and grape-growing areas comparatively rapidly, early and accurate detection is essential.

The current Slovenian action plan is focused on already infected areas and their immediate surroundings, in the so-called safety zones (UVHVVR, 2019). This type of action plan neglects all other areas and thus easily overlooks new foci. In addition, the number of experts is limited, and already there are not enough to cover all newly established foci. This problem is further enhanced by most field work being performed in high season, when the availability of inspectors and other experts is lowest. A remote sensing early detection system could assist the experts and contribute to effective management of grapevine yellows foci, including limiting their spread. Field work, including sampling for molecular tests, could hence be centered on those areas, allowing for more efficient control by experts.

18.2 Remote sensing approaches for grapevine yellows detection

An ideal remote sensing plant health assessment system should (1) enable noninvasive and accurate detection of pests and diseases, (2) allow for early detection, (3) be coupled with accurate GPS data and thus gather data over any area, and (4) detect and distinguish between several diseases. Various remote sensing methods have been developed to detect different diseases in selected crops, but the majority of them identify only one disease, or a limited number, in one crop. Only recently more research is dedicated toward developing classification models for detecting different diseases in crops.

Standard RGB cameras are the cheapest and most readily and widely available remote sensing systems and are therefore of great interest for disease detection. Even though these sensors provide spectral information in only three bands (red, green, and blue), appropriate data preprocessing can still enable comparatively accurate plant health assessment. For example, Camargo and Smith (2009) used texture-related features in combination with support vector machines to detect bacterial angular (*Xanthomonas campestris*) and ascochyta blight (*Ascochyta gossypii*) in cotton. Another approach, by calculating enhanced color information, such as LAB (L—light, AB—color-component dimensions), YCBCR (Y—luma component, CB—blue difference, CR—red difference chroma components), and HSV (H—hue, S—saturation, V—volume), was used by Bock et al. (2010) to detect Citrus canker (*Xanthomonas axonopodis*) in grapefruit. One of the most extensive studies of remote sensing for plant disease detection was performed by Neumann et al. (2014). They used various extracted features, such as gray levels, texture, dispersion, and shape in pattern recognition to detect five diseases in sugar beet.

Multispectral sensors are probably the most popular for plant health assessment. They are similar in size to standard RGB cameras and can also be easily mounted on unmanned aerial vehicles (UAVs). However, they provide additional spectral information, either with additional bands in the visible part of the spectrum or by adding bands in the near infrared (700–1000 nm, NIR) spectral region. Plants reflect approximately half of incoming light in the NIR region; chlorophyll is almost transparent in wavelengths above 700 nm. Plant cells function as elementary corner reflectors, thus changes in plant tissue structure can be measured. For example, plants in stress will have a lower reflectance in the NIR region. Vegetation indices use this effect. These indices are simple spectral transformations of more than two bands and are designed to allow comparisons of photosynthetic activity and canopy structure (Huete et al., 2000). Development of these indices began more than half a decade ago. The most well-known and often used index, the normalized difference vegetation index (NDVI), was developed by Kriegler et al. in 1969. Even though it's linked to photosynthetic capacity and energy absorption in the plant canopy, it has been used for a variety of purposes, including satellite imagery in agriculture (Blaes et al., 2016) and for vegetation recover after fires (Lanorte et al., 2014). The efficiency of multispectral imaging to detect foliar symptoms has been studied for a variety of diseases and crops, for example, panicle blast (*Magnaporthe grisea*) in rice (Kobayashi et al., 2001), citrus greasy spot disease (Du et al., 2008), late blight in tomato (Zhang et al., 2005), soybean cyst nematodes (*Heterodera glycines*) (Nutter et al., 2002), light leaf spot (*Pyrenopeziza brassicae*) in winter oilseed rape (Veys et al., 2018), and greenbug (*Schizapis graminum*) (Yang et al., 2009), aphids (*Diuraphis noxia*), and streak mosaic virus (Mirik et al., 2011) in wheat.

Remote sensing for disease detection in vineyards has seen increased use in recent years, with several studies focusing on grapevine yellows. Gennaro et al. (2016) and Junges et al. (2018) investigated esca infections at the crop and leaf levels, respectively. Infections with leafroll-associated virus were investigated by MacDonald et al. (2016). Albetis et al. (2017) developed methods for detecting *F. dorée* using multispectral imaging and in a later study distinguished between symptomatic and asymptomatic plants (Albetis et al., 2018). Using hyperspectral imaging, Al-Saddik et al. (2017, 2019) developed spectral indices and determined the most important wavelengths for detecting grapevine yellows. Cruz et al. (2019) combined color photographs of clipped leaves and convolutional neural networks to detect grapevine yellows infected plants.

All of these methods share the same drawback, that is, they are applicable over limited areas. Multirotor UAVs are limited by battery capacity and weight limits, with flight times up to 30 minutes. This is adequate for individual or limited groups of vineyards. Over large areas, such as entire regions or countries, satellite multispectral data are more suitable. The analysis of satellite imagery for agricultural purposes has been used since the 1970, when Landsat data became available. Contemporary high-performance commercial satellites, such as WorldView-2, provide high-quality imagery of large areas throughout the world (Dixon & McCann, 1997). By combining high spatial and spectral resolution, accurate assessment of plant health over large areas is feasible. Accurate plant health information is crucial for precision agriculture and targeted phytosanitary controls, facilitating targeted management of quarantine diseases. The potential of WorlView-2 satellite imaging for identifying grapevine yellows foci in a selected vineyard in Slovenia's coastal region was explored.

18.3 Methods

18.3.1 WorldView-2 image acquisition and preprocessing

WorldView-2 8-band images (Table 18.2) were obtained on August 11, 2011 from an area of 217 km^2 in the Primorje (grapevine yellows foci Viližan) vine-growing region of Slovenia. The images were radiometrically corrected, georeferenced, and orthorectified. Ground control points were determined on official digital orthophoto imagery of the same area, with a ground resolution of 0.25 m (DF025, http://www.e-prostor.gov.si). Atmospheric and topographic corrections were applied using the ACTOR3 method, and reflectance values were calculated. Because the spatial resolution of multispectral imaging with this particular satellite is at minimum 1.84 m, the images were pan-sharpened (i.e., the multispectral spatial resolution has been increased) to a spatial resolution of 0.46 m, using the on-board panchromatic sensor. Preprocessed images were supplied by company Kaliopa d.o.o. from Slovenia.

18.3.2 Plant health assessment

One vineyard with one of the most common red varieties in Slovenia (Refošk) from grapevine yellows foci Viližan was selected for this study. In this vineyard, three trellis lines with a good mix of healthy and infected plants were selected (Fig. 18.1). In these lines, each plant's health status was assessed the same week, and satellite images were recorded. The assessment was performed by official phytosanitary examiners, trained by the Administration for food safety, veterinary sector, and plant protection of Slovenia, in accordance with official guidelines for assessment of grapevine yellows infection severity. Coordinates of the first and last column of each Trellis line were determined, and the distance (in m) from the first column to the trunk of each tree was measured. Each plant was assigned to one of five classes: N—no visible signs, L—slightly infected (rare and low expressed signs), M—moderately infected (some visible signs), S—severely

Table 18.2 WorldView-2 spectral band edges and center wavelengths.

Band	Center wavelength (nm)	Lower band edge (nm)	Upper band edge (nm)
Coastal	427	401	453
Blue	478	447	508
Green	546	511	581
Yellow	608	588	627
Red	659	629	689
Rededge	724	704	744
NIR1	831	772	890
NIR2	908	862	954

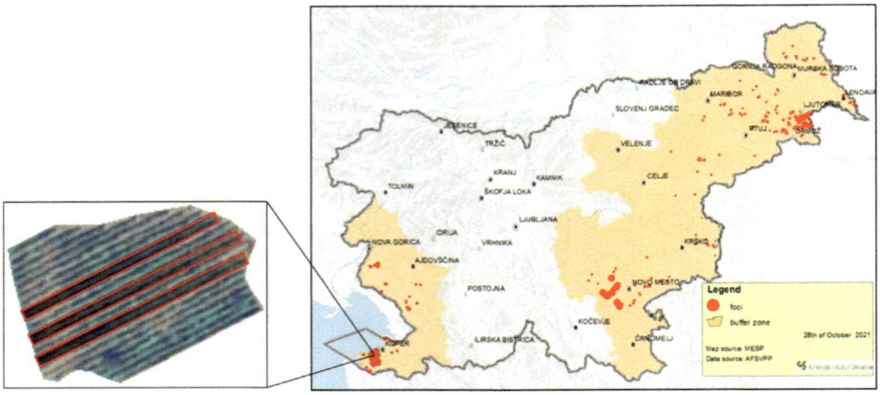

FIGURE 18.1

Map of grapevine yellows foci and buffer zones in Slovenia at the end of growing season 2021. On the left-hand side is an RGB image of the selected vineyard. The red rectangles denote the selected trellis lines.

Agricultural Institute of Slovenia.

infected (almost the entire plant shows visible signs), X—dead (Fig. 18.2). Pixels denoting dead plants were removed prior to analysis. In total, 243 plants from three lines (77, 83, and 83 plants, respectively) were included for further analyses. Of these, 68 were without signs of infection, and 175 showed different levels of symptom expression (112 with slight, 37 with moderate, and 26 with severely expressed symptoms). Individual plants were represented with one to six pixels (mean four pixels/plant). Grapevine yellows infections were confirmed in a random subsample of 10 plants (five without and five with symptoms), using molecular analyses. DNA was extracted based on the binding of DNA to magnetic beads as described by Mehle et al. (2013a). To discriminate between *F. dorée* and "*Candidatus* Phytoplasma solani," validated real-time PCR tests were used (Mehle et al., 2013b).

18.3.3 Spectral data preprocessing

Using the GPS data of the first column of each line and the measured distance for each plant, a shapefile was generated using ArcGIS, with individual polygons denoting the location of each plant. Because plants grow along the Trellis system, the polygon was shaped as a rectangle, along the line. The separation between individual pairs of plants was determined halfway between these pairs. These polygons were then used to extract per-pixel spectral data from the multispectral image. Ground truth data of plant health were used to label spectral data on a per-plant basis. Each plant was represented by three to five pixels, to a total of 986 pixels.

FIGURE 18.2

Grapevine infected with grapevine yellows. Almost the entire plant shows symptoms of infection and was therefore classified as severely infected, even though some leaves appear to be without symptoms.

Agricultural Institute of Slovenia.

From the original eight spectral bands, 39 spectral indices were calculated (Table 18.3). Together with the eight bands, 47 spectral features were used. As individual spectral indices were developed on different sensors, they were adapted to the spectral properties of WorldView-2. All indices have been used by researchers for agricultural remote sensing applications and are linked to plant physiological status. All features were then standardized to zero mean and unit variance.

Sparse partial least squares discriminant analysis (sPLS-DA) was used for feature extraction. PLS is a dimensionality reduction method, which transforms the original feature space of highly collinear variables to a set of uncorrelated features, with maximum correlation with target variables. The sparse variant of PLS discriminant analysis performs both variable selection and classification. The number of features and latent variables (LV) to retain was determined using a grid search and 10-times repeated fivefold cross-validation, for each latent variable. sPLS classification performance (measured as balanced error rate) was evaluated using three types of prediction: (1) maximum, (2) centroids, and (3) mahalanobis distance. The final number of LV was determined by calculating classification error rates for each number of LV. The importance of each index for each LV was determined using variable in projection analysis (VIP). Selected LV were used for classifications. sPLS-DA was performed twice; first all four

Table 18.3 Indices used in sparse partial least squares discriminant analysis, ordered alphabetically from their abbreviations.

Nr.	Index label	Full name	Equation
1	CLG	Green-band chlorophyll index	$\frac{rededge}{green} - 1$
2	DVI	Difference vegetation index	$2.4 * \frac{nir1}{red}$
3	EVI	Enhanced vegetation index	$2.5 * \frac{nir1 - red}{nir1 + 6*red - 7.5*blue + 1}$
4	EVI2	Enhanced vegetation index 2	$2.5 * \frac{nir1 - red}{nir1 + 2.4*red + 1}$
5	GEMI	Global environment monitoring index	$\frac{2*(nir1^2 - red^2) + 1.5*nir1 + 0.5*red}{nir1 + red + 0.5} * \left(1 - 0.25 * \frac{2*(nir1^2 - red^2) + 1.5*nir1 + 0.5*red}{nir1 + red + 0.5}\right) - \frac{red - 0.125}{1 - red}$
6	GNDVI	Green NDVI	$\frac{nir1 - green}{nir1 + green}$
7	MCARI	Modified chlorophyll absorption in reflectance index	$\frac{rededge - red}{rededge - green} * \frac{rededge}{red}$
8	MNDWI	Modified NDWI	$\frac{green - nir2}{green + nir2}$
9	MSAVI	Modified SAVI	$\frac{2*nir1 + 1 - \sqrt{2*nir1 + 1^2 - 8*(nir1 - red)}}{2}$
10	MSAVI2	Modified SAVI 2	$nir1 + 0.5 - (0.5*\sqrt{2*nir1 + 1^2 - 8*(nir1 - 2*red)})$
11	TCARI	Transformed chlorophyll absorption ratio	$3*\left((rededge - red) - 0.2*(rededge - green)*\left(\frac{rededge}{red}\right)\right)$
12	GOSAVI	Ground optimized SAVI	$\frac{nir1 - green}{nir1 + green + 0.16}$
13	ATSAVI	Adjusted transformed SAVI	$1.22 * \frac{nir1 - 1.22*red + 0.03}{1.22*nir1 + red + 1.22*0.03 + 0.08*(1 + 1.22^2)}$
14	PSSR	Simple ratio nir/red	$\frac{nir1}{red}$
15	ARVI	Atmospherically resistant vegetation index	$\frac{nir1 - red - 1*(red - blue)}{nir1 + red + 1*(red - blue)}$
16	NBRI	Normalized burn ratio index	$\frac{nir1 - nir2}{nir1 + nir2}$
17	NDVI	Normalized difference vegetation index	$\frac{nir1 - red}{nir1 + red}$
18	CTVI	Corrected TVI	$\frac{NDVI + 0.5}{\sqrt{abs(NDVI + 0.5)}}$
19	NDVIC	NDVI C	$\frac{(nir1 - red)}{(nir1 + red)} * \left(1 - \frac{nir2 - min(nir2)}{max(nir2) - min(nir2)}\right)$

20	NDWI	Normalized difference water index	$\frac{green - nir1}{green + nir1}$
21	NRVI	Normalized RVI	$\frac{red/nir1 - 1}{red/nir1 + 1}$
22	RVI	Simple ratio red/nir	$\frac{red}{nir1}$
23	SAVI	Soil-adjusted vegetation index	$nir1 - red^{*}\frac{(1 + 0.5)}{nir1 + red + 0.5}$
24	SLAVI	Specific leaf area vegetation index	$\frac{nir1}{red + nir2}$
25	SR	Simple ratio nir2/red	$\frac{nir2}{red}$
26	TTVI	Thiam's transformed vegetation index	$\sqrt{abs\left(\frac{nir1 - red}{nir1 + red} + 0.5\right)}$
27	TVI	Transformed vegetation index	$\sqrt{\frac{nir1 - red}{nir1 + red} + 0.5}$
28	VARI	Visible atmospherically resistant index	$\frac{green - red}{green + red - blue}$
29	ExG	Excess green	$2^{*}green - (red + blue)$
30	ExR	Excess red	$1.4^{*}red - green$
31	ExB	Excess blue	$1.4^{*}blue - green$
32	ExGR	Excess green-blue	$ExG - ExR$
33	NGRDI	Normalize difference green/red index	$\frac{green - red}{green + red}$
34	MGRVI	Modified NGRDI	$\frac{green^2 - red^2}{green^2 + red^2}$
35	WI	Water stress index	$\frac{green - blue}{red - green}$
36	IKAW	Kawashima index	$\frac{red - blue}{red + blue}$
37	GLI	Green leaf index	$\frac{2^{*}green - red - blue}{2^{*}green + red + blue}$
38	RGBVI	Red green blue vegetation index	$\frac{green^2 - blue^{*}red}{green^2 + blue^{*}Red}$
39	VEG		$\frac{green}{red^{0.333}{}^{*}blue^{0.333}}$

Blue, red, green, rededge, nir1, and nir2 are spectral bands of satellite WorldView-2. Band characteristics are provided in Table 18.2. Further details about indices are available on the Index database (http://www.indexdatabase.de).

plant health classes were used, and the second time infected plants were pooled into new class I—infected.

18.3.4 Classification

Each of the two datasets (original and after dimensionality reduction using sPLS) was divided into train and test sets, using a 70/30 split. Classifications were performed using support vector machines (SVM), using the Gaussian radial basis function kernel. This algorithm searches for hyperplanes (decision boundaries) in feature space with the maximum distance between two data points belonging to two classes. Hyperparameters (cost function C and gamma) were tuned using a grid search. Model training on the train set was performed using five-times repeated fivefold cross-validation. A final validation was performed on the test set of each dataset. This procedure was repeated twice, first as a four-class problem (classes N, L, M, and S), and secondly as a binary classification (N and I—infected).

All data preprocessing, analysis, and visualization were performed in R (R Core Team, 2018), using packages MixOmics (Rohart et al., 2017), caret (Kuhn, 2021), kernlab (Karatzoglou et al., 2004), doParallel (Microsoft Corporation & Weston, 2022), and ggplot2 (Wickham, 2016).

18.4 Results and discussion

Selected spectral indices were calculated for each pixel, and by using sparse PLS discriminant analysis, 23 LV were extracted, explaining more than 99% of variation in the data (Fig. 18.3). The best achieved balanced classification error rate was 0.37 for binary classification and 0.53 with for classes; in both cases, maximum distance achieved the best result.

The sPLS-DA tuning procedure identified relevant variables for each latent variable, and their importance is visualized using VIP analysis. With four classes, relevant bands and indices were coastal, rededge, nir2, CLG, NBRI, NDVIC, SLAVI, VARI, ExR, ExB, NGRDI, MGRVI, GLA, and RGBVI (Fig. 18.4). On the other hand, with two classes the only regularly relevant band was nir2. Among indices, similar patterns were observed in LVs 2–8 (Fig. 18.5). In almost all LVs' relevant indices were CLG, EVI2, GEMI, GNDVI, MSAVI, MSAVI2, GOSAVI, ATSAVI, NDWI, and SAVI. Several of these are linked with the red and blue parts of the spectrum, as well as the near-infrared part. This outcome was expected, as a red grape variety (Refošk) that undergoes a color change to red due to grapevine yellows, was used in this study. Similar conclusions in regard to identifying plant health and stress status were reached by Jones & Vaughan (2010). Diseases influence spectral reflectances of plants in a specific way, resulting in different spectral signatures. The shoots of susceptible

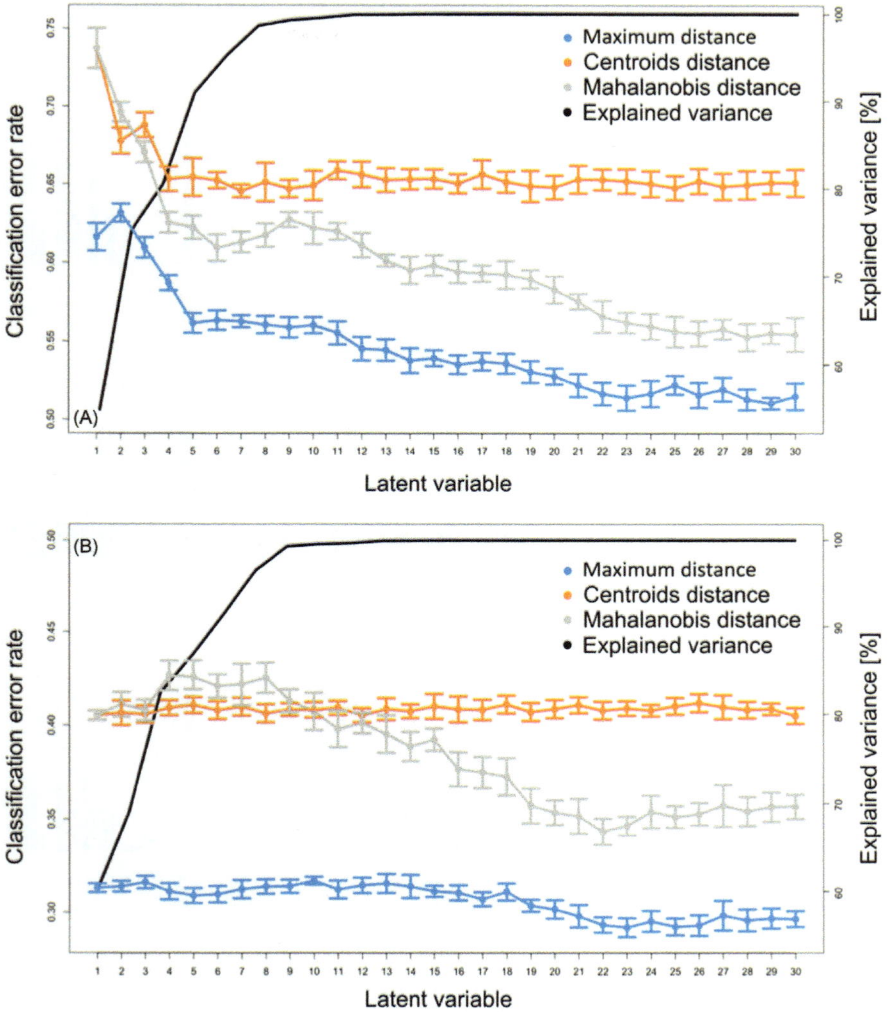

FIGURE 18.3

Results of sparse partial least squares discriminant analysis tuning. Balanced error rates were evaluated using three types of prediction: maximum, centroids, and Mahalanobis distance. For these measures, 95% confidence bars are also given. The black line denotes cumulative explained variance for all LVs. (A). sPLS-DA with four classes. (B) sPLS-DA with two classes. In both cases, 23 LVs were selected as optimal.

grapevines, when infected early, fail to lignify, are thin, rubbery, and hang pendulously. The leaves show color aberrations and downward-rolled margins. Leaves of infected red vines turn red in latter stages of infection, while leaves of white vines turn yellow. Grapevine yellows in later stages also affect leaf structure,

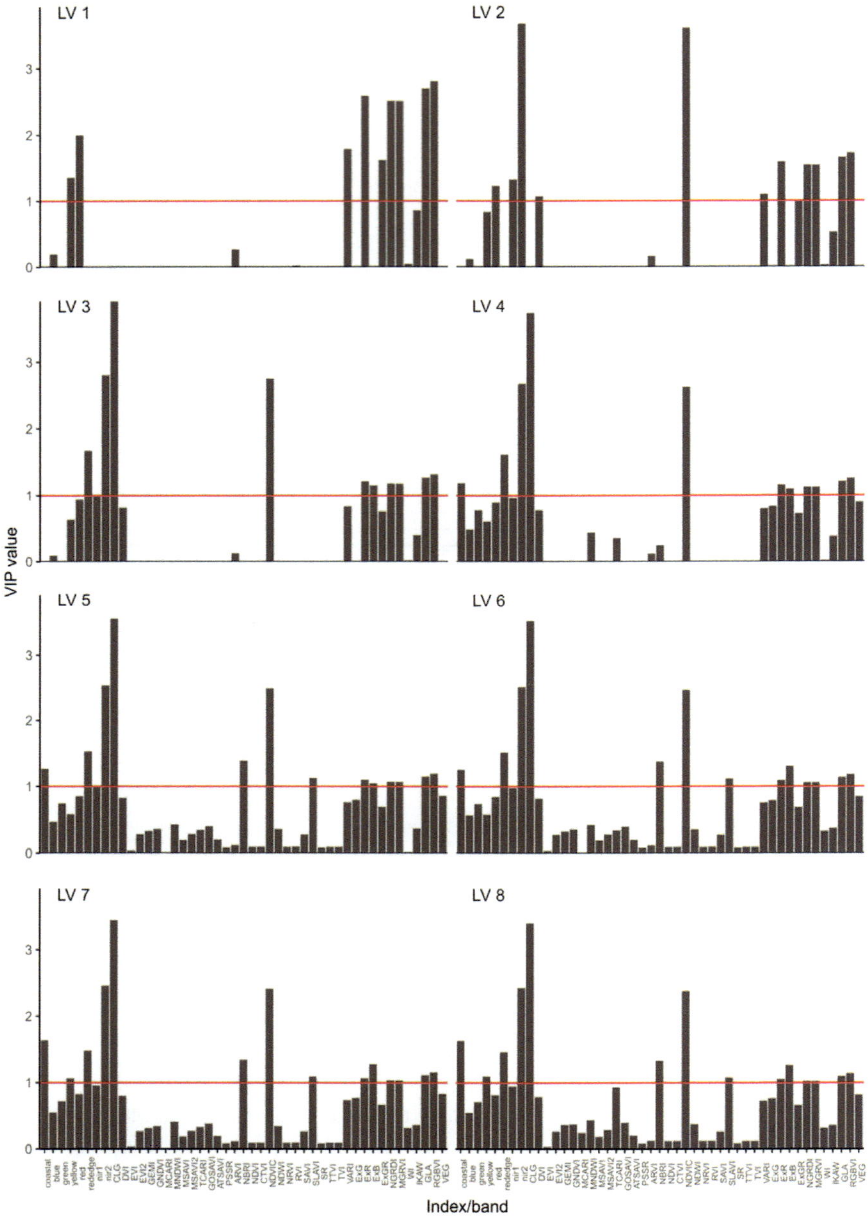

FIGURE 18.4

Variable importance in projection for the first eight latent variables of sparse partial least squares discriminant analysis with four classes. The red line signifies the boundary of importance, indices and/or bands with VIP values above 1 are considered relevant. Only the first eight LVs are presented, as they already explained more than 98% of variation.

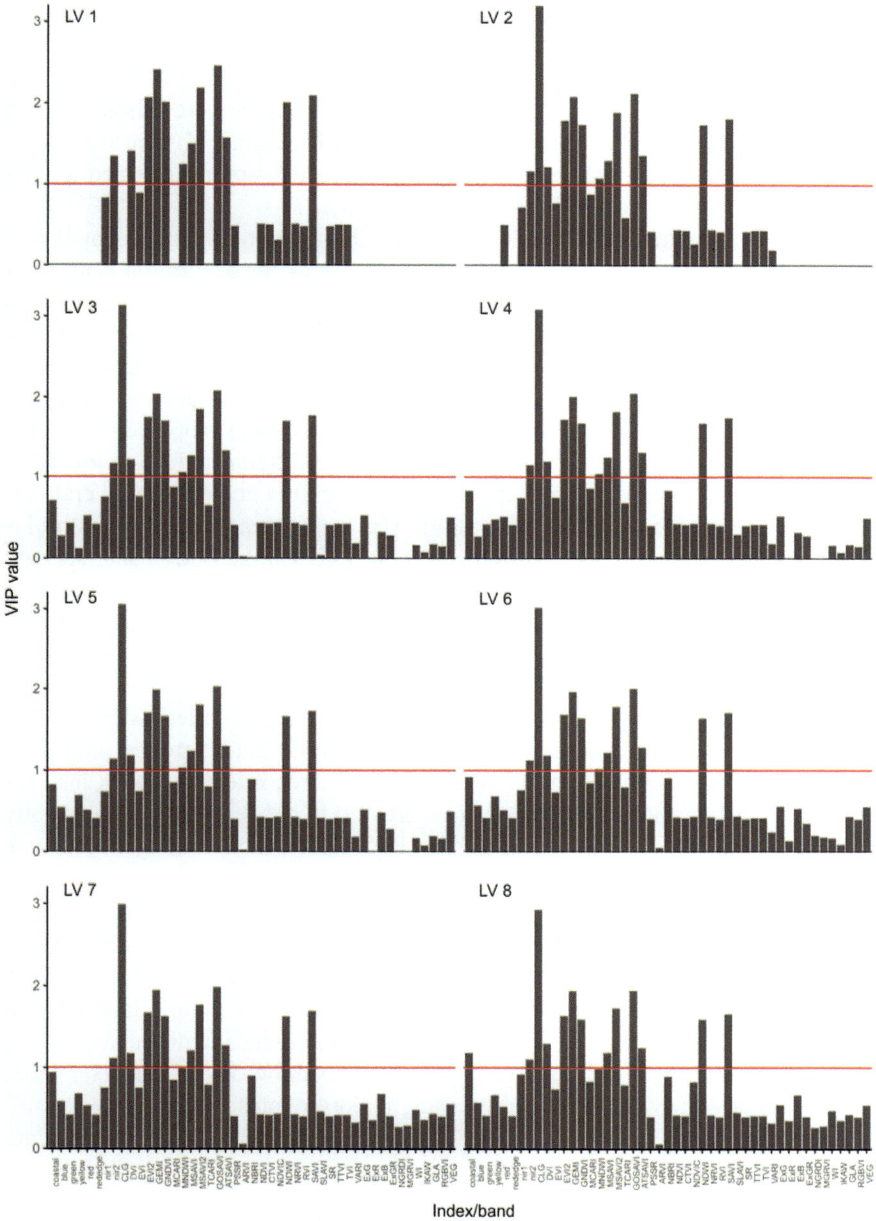

FIGURE 18.5

Variable importance in projection for the first eight latent variables of sparse partial least squares discriminant analysis with two classes. The red line signifies the boundary of importance, indices and/or bands with VIP values above 1 are considered relevant. Only the first eight LVs are presented, as they already explained more than 98% of variation.

which changes the reflectance in the near-infrared spectral region. The more signs of infection are developed, the better they can be identified, both with on-site and remote sensing methods. Generally, the same patterns in identification of symptom expression levels, where very few pixels from group N were misclassified as S (and vice versa), and misclassifications between N, L, and M were more common, were observed. Most misclassifications occurred among infected groups, which was further showcased in binary classifications.

Vegetation spectra of the Worldview-2 satellite are not of typical shape, compared with bottom-of-atmosphere reflectance; they lack the absorption peak of the blue band. This peak is suppressed by atmospheric Rayleigh scattering (Tarantino et al., 2012), leading to converging reflectance profiles at shorter wavelengths. Furthermore, at longer wavelengths (e.g., near-infrared bands), atmospheric water vapor absorption becomes dominant. Considering the importance of short and long wavelengths for plant health assessment, it becomes apparent that these effects will lead to increased difficulty in identifying different infection levels.

Per-pixel classification using support vector machines achieved an overall success of 75.3% on the test set (Table 18.4). The lowest sensitivity was with class M (moderately expressed signs of infection), at 0.32. On the other hand, specificity with this class was 0.987, and accuracy was 0.65. Highest accuracies were observed in classes with opposite levels of symptom expression, N and S (0.799 and 0.811, respectively). A majority vote procedure increased overall success to 80%. Patterns remained the same, with class M being the least reliably classified and S most accurate. No plants from group N were misclassified as S, and vice versa. Most misclassifications occurred among the infected groups, and some occurred between groups N and L. By pooling all three infected groups into new group "I," classification accuracies were increased to above 84%, with a sensitivity of 0.946 and specificity of 0.61 (Table 18.5). Even though images were pansharpened, ground resolution was comparatively rough. At 0.5 m pixel size, there was still a not inconsiderable amount of spectral mixing. On a trellis system, neighboring plants can interleave their branches, which makes obtaining spectrally clean data impossible, even with better resolutions (e.g., from a UAV). In such cases, classification accuracy will decrease and lead to misclassifications. This presents a problem when infected plants would be identified as healthy, which would lead to poor disease management. This drawback of per-pixel classification can be somewhat mitigated by using majority votes.

By implementing a simple majority vote system, overall classification accuracies in both sets (four and two classes) were increased to above 80% (Tables 18.4 and 18.5). Importantly, specificity increased to above 0.88 for all classes in per-pixel classification, that is, the number of false-positives decreased. Sensitivity was lowest for group M, where several plants were misclassified as either L or S. A similar pattern was observed in binary classification, where overall accuracy was 87.8%, sensitivity was 0.91, and specificity was 0.84. A majority voting system increased classification accuracies and true positive and negative rates, mitigating the effect of mixed pixels in images with comparatively course resolutions.

Table 18.4 Support vector machines classification results with four classes.

				Per-pixel classification			Majority vote classification		
	Calibration			Validation			Validation		
Class	Accuracy	Sensitivity	Specificity	Accuracy	Sensitivity	Specificity	Accuracy	Sensitivity	Specificity
N	0.984	1	0.968	0.799	0.728	0.871	0.832	0.778	0.887
L	0.936	1	0.872	0.747	0.826	0.668	0.797	0.705	0.89
M	0.712	0.424	1	0.654	0.32	0.988	0.663	0.444	0.881
S	0.994	0.988	1	0.812	0.639	0.984	0.92	0.9	0.94

The train set was used for model calibration, and test set for validation. Majority voting was implemented only for the test set classification.

Table 18.5 Support vector machines classification results with two classes.

Class				Per-pixel classification			Majority vote classification		
	Calibration			Validation			Validation		
	Accuracy	Sensitivity	Specificity	Accuracy	Sensitivity	Specificity	Accuracy	Sensitivity	Specificity
N	0.997	0.991	1	0.841	0.946	0.61	0.878	0.912	0.844
I	0.997	1	0.991	0.841	0.61	0.946	0.878	0.844	0.912

The train set was used for model calibration, and test set for validation. Majority voting was implemented only for the test set classification.

Classification accuracy could be further increased by using sensors with better spatial and spectral resolution. While UAVs with hyperspectral systems would provide the best data, they are limited by their batteries and applicable only on individual vineyards. This limitation is somewhat mitigated by airborne hyperspectral imaging, which can provide the best cost−effect ratio, when several vineyards need to be assessed. In these systems, operating time is limited by fuel and data storage space, yet more than 10 vineyards can still be imaged in a few hours. Multispectral satellite imaging would be most applicable over entire regions, and in use cases where speed of data acquisition and access to results is more important than classification accuracy. Such a case could be governmental control of quarantine diseases.

18.5 Conclusions

Multispectral satellite imaging has proved to be suitable for the identification of grapevine yellows infected grapevines. This approach has the potential to predict infections over large areas, thus enabling more focused field work, making monitoring grapevines on a national level more time- and cost-effective.

Feature extraction and selection using sparse PLS discriminant analysis combined with support vector machines enabled comparatively good classification accuracies. Instead of calculating mean reflectance for each plant, per-pixel classifications on plants and implemention of a simple majority voting system are recommended. As spatial resolution presents the biggest obstacle and limits the use of satellite imaging for crop health assessment, majority voting of per-pixel classifications can mitigate the effect of mixed edge pixels and increase identification accuracy.

While the results of this research are promising, it was performed on a very limited dataset from only one vineyard. More studies are needed to develop a truly viable satellite remote sensing system for grapevine yellows detection. These studies should include more grapevine varieties of both red and white grapes and vineyards from different regions and positions. Furthermore, reliable ground truth data are critical, but can be expensive to obtain. Absolutely accurate detection of grapevines yellows can be performed only using molecular methods, which is time-consuming and costly and therefore, not viable in entire vineyards. This, in turn, showcases the importance of remote sensing for plant health assessment.

References

Albetis, J., Duthoit, S., Guttler, F., Jacquin, A., Goulard, M., Poilvé, H., Féret, J. B., & Dedieu, G. (2017). Detection of *Flavescence dorée* grapevine disease using unmanned aerial vehicle (UAV) multispectral imagery. *Remote Sensing*, *9*, 308.

Albetis, J., Jacquin, A., Goulard, M., Poilvé, H., Rousseau, J., Clenet, H., Dedieu, G., & Duthoit, S. (2018). On the potentiality of UAV multispectral imagery to detect *Flavescence dorée* and grapevine trunk diseases. *Remote Sensing*, *11*, 23. Available from https://doi.org/10.3390/rs11010023.

Alemu, K. (2015). Detection of diseases, identification and diversity of viruses: A review. *Journal of Biology, Agriculture and Healthcare*, *5*(1), 204–213.

Angelini, E., Constable, F., Duduk, B., Fiore, N., Quaglino, F., & Bertaccini, A. (2018). Grapevine phytoplasmas. In G. Rao, A. Bertaccini, N. Fiore, & L. Liefting (Eds.), *Phytoplasmas: Plant pathogenic bacteria-I* (pp. 123–151). Singapore: Springer.

Al-Saddik, H., Simon, J. C., & Cointault, F. (2017). Development of spectral disease indices for 'Flavescence Dorée' grapevine disease indentification. *Journal of Sensors*, *17*, 1–25.

Al-Saddik, H., Simon, J. C., & Cointault, F. (2019). Assessment of the optimal spectral bands for designing a sensor for vineyard disease detection: The case of '*Flavescence dorée*'. *Precision Agriculture*, *20*, 398–422. Available from https://doi.org/10.1007/s11119-018-9594-1.

Blaes, X., Chomé, G., Lambert, M.-J., Traoré, P., Schut, A., & Defourny, P. (2016). Quantifying fertilizer application response variability with VHR satellite NDVI time series in a rainfed small holder cropping system of Mali. *Remote Sensing*, *8*, 531.

Bock, C. H., Poole, G. H., Parker, P. E., & Gottwald, T. R. (2010). Plant disease severity estimated visually, by digital photography and image analysis, and by hyperspectral imaging. *Critical Reviews in Plant Science*, *29*, 59–107.

Camargo, A., & Smith, J. S. (2009). Image pattern classification for the identification of disease causing agents in plants. *Computers and Electronics in Agriculture*, *66*(2), 121–125.

Chuche, J., & Thiéry, D. (2014). Biology and ecology of the Flavescence dorée vector Scaphoideus titanus: A review. *Agronomy and Sustainable Development*, *34*, 381–403.

Cruz, A., Ampatzidis, Y., Pierro, R., Materazzi, A., Panattoni, A., De Bellis, L., & Luvisi, A. (2019). Detection of grapevine yellows symptoms in *Vitis vinifera* L. with artificial intelligence. *Computers and Electronics in Agriculture*, *157*, 63–76. Available from https://doi.org/10.1016/j.compag.2018.12.028.

Dixon, J., & McCann, M. (Eds.), (1997). *Precision agriculture in the 21st century* (1st edn, p. 168). Washington: National Academy Press.

Du, Q., Chang, N. B., Yang, C., & Srilakshmi, K. R. (2008). Combination of multispectral remote sensing, variable rate technology and environmental modelling for citrus pest management. *Journal of Environmental Management*, *86*, 25–26.

Fontaine, F., Gramaje, D., Armengol, J., Smart, R., Nagy, Z. A., Borgo, M., Rego, C., & Corio-Costet, M. F. (2016). *Grapevine trunk diseases: A review*. Paris, France: Cahiers de recherche, OIV Publications.

Gennaro, S. F. D., Battiston, E., Marco, S. D., Facini, O., Matese, A., Nocentini, M., Palliotti, A., & Mugnai, L. (2016). Unmanned aerial vehicle (UAV)-based remote sensing to monitor grapevine leaf stripe disease within a vineyard affected by esca complex. *Phytopathologia Mediterranea*, *55*, 262–275.

Huete, A., Didan, K., Miura, T., Rodriquez, E. P., Gao, X., & Ferreria, L. G. (2000). Overview of the radiometric and biophysical performance of the MODIS vegetation indices. *Remote Sensing of the Environment*, *83*(5), 195–213. Available from https://doi.org/10.1006/anbo.2000.1262.

Jones, Hamlyn G., & Vaughan, Robin A. (2010). *Remote sensing of vegetation: Principles, techniques, and applications* (pp. 1–353). New York: Oxford University Press.

Junges, A., Ducati, J., Scalvi Lampugnani, C., & Almanca, M. A. (2018). Detection of grapevine leaf stripe disease symptoms by hyperspectral sensor. *Phytopathologia Mediterranea*, *57*(3), 399–406. Available from https://doi.org/10.14601/Phytopathol_Mediterr-22862.

Karatzoglou, A., Smola, A., Hornik, K., & Zeileis, A. (2004). kernlab - An S4 package for kernel methods in R. *Journal of Statistical Software*, *11*(9), 1–20. Available from http://www.jstatsoft.org/v11/i09/.

Kobayashi, T., Kanda, E., Kitanda, K., Ishiguro, K., & Torigoe, Y. (2001). Detection of rice panicle blast with multispectral radiometer and the potential of using airborne multispectral scanners. *Phytopathology*, *91*, 316–323.

Kriegler, F.J., Malila, W.A., Nalepka, R.F., Richardson W. (1969). Preprocessing transformations and their effects on multispectral recognition. In: *Proceedings of the sixth international symposium on remote sensing of environment*, pp. 97–131.

Kuhn, M. (2021). *caret: Classification and regression training*. R package version 6.0–90. https://CRAN.R-project.org/package = caret.

Lanorte, A., Lasaponara, R., Lovallo, M., & Telesca, L. (2014). Fisher–Shannon information plane analysis of SPOT/VEGETATION normalized difference vegetation index (NDVI) time series to characterize vegetation recovery after fire disturbance. *International Journal of Applied Earth Observation and Geoinformatics*, *26*, 441–446.

MacDonald, S. L., Staid, M., & Cooper, M. L. (2016). Remote hyperspectral imaging of grapevine leafroll-associated virus 3 in cabernet sauvignon vineyards. *Computers and Electronics in Agriculture*, *130*, 109–117.

Margaria, P., Turina, M., & Palmano, S. (2009). Detection of *Flavescence dorée* and *Bois noir* phytoplasmas, Grapevine leafroll associated virus-1 and -3 and Grapevine virus A from the same crude extract by reverse transcription-RealTime Taqman assays. *Plant Pathology*, *58*(2), 838–845.

Mehle, N., Nikolić, P., Rupar, M., Boben, J., Ravnikar, M., & Dermastia, M. (2013a). Automated DNA extraction for large numbers of plant samples. In M. Dickinson, & J. Hodgetts (Eds.), *Phytoplasma: Methods in molecular biology (methods and protocols)* (vol 938, p. 193). Totowa, NJ: Humana Press, 145. Available from https://doi.org/10.1007/978-1-62703-089-2_12.

Mehle, N., Prezelj, N., Hren, M., Boben, J., Gruden, K., Ravnikar, M., & Dermastia, M. (2013b). A real-time PCR detection system for the *Bois Noir* and *Flavescence Dorée* phytoplasmas and quantification of the target DNA. In M. Dickinson, & J. Hodgetts (Eds.), *Phytoplasma. Methods in molecular biology (methods and protocols)* (vol 938, pp. 253–268). Totowa, NJ: Humana Press. Available from https://doi.org/10.1007/978-1-62703-089-2_12.

Mehle, N., Ravnikar, M., Seljak, G., Knapič, V., & Dermastia, M. (2011). The most widespread phytoplasmas, vectors and measures for disease control in Slovenia. *Phytopathogenic Mollicutes*, *1*(2), 65–76.

Microsoft Corporation & Weston, S. (2022). *doParallel: Foreach parallel adaptor for the 'parallel' package*. R package version 1.0.17. https://CRAN.R-project.org/package = doParallel.

Mirik, M., Jones, D. C., Price, J. A., Workneh, F., Ansley, R. J., & Rush, C. M. (2011). Satellite remote sensing of wheat infected by wheat streak mosaic virus. *Plant Disease*, *95*, 4–12.

Neumann, M., Hallau, L., Klatt, B., Kersting, K., Bauckhage, C. (2014). Erosion band features for cell phone image based plant disease classification. In: *Proceeding of the 22nd international conference on patternrecognition (ICPR)*, Stockholm, Sweden, pp. 3315–3320.

Nutter, F. W., Tylka, G. L., Guan, J., Moreira, A. J. D., Marett, C. C., Rosburg, T. R., Basart, J. P., & Chong, C. S. (2002). Use of remote sensing to detect soybean cyst nematode-induced plant stress. *Journal of Nematology*, *34*(3), 222–231.

Prezelj, N., Nikolić, P., Gruden, K., Ravnikar, M., & Dermastia, M. (2013). Spatiotemporal distribution of flavescence doree phytoplasma in grapevine. *Plant pathology*, *62*(4), 760–766.

R Core Team. (2018). *R: A language and environment for statistical computing*. R Foundation for Statistical computing, Vienna, Austria. https://www.R-project.org/.

Rohart, F., Gautier, B., Singh, A., & Le Cao, K.-A. (2017). mixOmics: An R package for 'omics feature selection and multiple data integration. *PLoS Computational Biology*, *13* (11)e1005752.

Tarantino, C., Adamo, M., Pasquariello, G., Lovergine, F., Blonda, P., & Tomaselli, V. (2012). Chapter 6: 8-band data processing of the Worldview-2 satellite in a wide area of applications. In R. B. Rustamov, & S. E. Salahova (Eds.), *Earth observation*. Intech, open acces at. Available from http://www.intechopen.com/books/earth-observation.

UVHVVR. (2019). *Flavescence dorée* action plan. Ministry of Agriculture, Forestry and Food, Republic of Slovenia. https://www.gov.si/assets/organi-v-sestavi/UVHVVR/Zdravje-rastlin/Skodljivi-organizmi-rastlin/Zlata-trsna-rumenica/2019_Nacrt_ukrepov_zlata_trsna_rumenica_splet.pdf.

Veys, C., Fokion, C., AlSuwadi, A., Hibbert, J., Hansen, M., Bernotas, G., Smith, M., Yin, H., Rolfe, S., & Grieve, B. (2018). Multispectral imaging for presymptomatic analysis of light leaf spot in oilseed rape. *Plant Methods*, *14*, 4. Available from https://doi.org/10.1186/s13007-019-0389-9.

Wickham, H. (2016). *ggplot2: Elegant graphics for data analysis*. New York: Springer-Verlag, 2016.

Wilcox, W. F., Gubler, W. D., & Uyemoto, J. K. (Eds.), (2015). *Compendium of grape diseases, disorders, and pests* (2nd edn, p. 232). St. Paul, Minnesota: The American Phytopathological Society. Available from http://doi.org/10.1094/9780890544815.

Yang, Z., Rao, M. N., Elliott, N. C., Kindler, S. D., & Popham, T. W. (2009). Differentiating stress induced by green bugs and Russian wheat aphids in wheat using remote sensing. *Computers and Electronica in Agriculture*, *67*, 64–70.

Zhang, M., Qin, Z., & Liu, X. (2005). Remote sensed spectral imagery to detect late blight in field tomatoes. *Precision Agriculture*, *6*, 489–508.

SECTION 6

Advanced modeling practices in precision agriculture

CHAPTER

Geostatistical modeling—a tool for predictive soil mapping

19

Aatira Hilal[1], Shabir Ahmed Bangroo[1,2], Nayar Afaq Kirmani[1], Javaid Ahmed Wani[1], Asim Biswas[2], Mohammad Iqbal Bhat[1], Khushboo Farooq[1], Owais Bashir[1] and Tajamul Islam Shah[1]

[1]*Division of Soil Science, Sher-e-Kashmir University of Agricultural Sciences & Technology, Shalimar, Jammu & Kashmir, India*

[2]*School of Environmental Sciences, University of Guelph, Guelph, ON, Canada*

19.1 Introduction

Geostatistical methodology and the basic concept of regionalized variables have been commonly applied in soil research and digital soil mapping (DSM) since the early 1980s, increasing the use of statistical analysis to assess huge quantities of soil data collected throughout the world. Soil scientists have made significant evaluations, reconsiderations, and improvements on practical and theoretical elements such as nonstationarity and the evaluation of spatial uncertainty as the soil is an appropriate medium for practical geostatistics. Geostatistics is a rapidly evolving discipline of applied spatial statistics that comprises a diverse set of modeling techniques and methodologies for machine learning (ML), processing, and representation, allowing for better decision-making based on spatially distributed data. Geostatistics has many applications, ranging from old prospecting for mineral ores to current uses in finance, medicine, and epidemiology (Meshalkina, 2007). Because of the relevance of spatial variability on land use, soil heterogeneity has been examined on a global basis (Food and Agricultural Organisation (FAO), 1974) as well as on a microscale (Variations in the chemical and structural composition of soil minerals) (Sawhney, 1977). As soil-forming factors and processes interact and are spatially correlated over many distinct spatial and temporal scales, the source of soil variability is scale-dependent. Soil spatial variability is examined using geostatistical methods in soil science because soil varies considerably from place to place (Nielsen et al., 1973), and also its spatial variability can be quantified using variograms. Geostatistics, a widely used statistical tool, has been determined by examining the spatial pattern in soil parameter variability (Carvalho et al., 2002; Vieira, 2000; Vieira et al., 2002). Geostatistics is a collection of statistical models that combine the spatial information of soil measurements in data acquisition, enabling spatiotemporal data description and modeling,

Remote Sensing in Precision Agriculture. DOI: https://doi.org/10.1016/B978-0-323-91068-2.00011-4

prediction at unsampled points, and evaluation of the uncertainties associated with these estimates. Understanding the spatial variability of soil is essential for environmental modeling (Burrough, 1983; Corwin et al., 2006) and estimation (Trangmar et al., 1986), precision farming (Goderya, 1998), soil assessment processes (Heuvelink & Pebesma, 1999; McBratney et al., 2000), and natural resource management. An appropriate understanding of soil variability as a function of time and space is essential in developing rational, scientific, and quantitative models of soil and ecosystem processes (Foussereau et al., 1993; Wilding et al., 1994).

Geostatistics is a subfield of applied statistics concerned with spatiotemporal datasets. Initially proposed to predict ore grade probability distributions for mining activities (Krige, 1951), it is now used in various disciplines including mineral exploration, hydrogeology, weather forecasting, earth sciences, geomorphology, geometallurgy, forest management, pollution monitoring, industrial ecology, soil science, and agricultural sectors. Geostatistics provides two key tools: semivariograms to describe the relationship between all pairs of observational points and kriging models to predict values at unsampled locations. In geostatistics, spatiotemporal data evaluation refers to a group of highly specialized procedures that provide a systematic outcome for the assessment of spatially (and temporally) referenced measurement methods such as image processing methods used to analyze remote sensing data; point pattern analysis is often used to analyze discrete point/line objects; geostatistics is used to analyze continuous spatial information, and geomorphometry is a field of science that specializes in the analysis of quantitative topography analysis (Fig. 19.1). As a result, spatiotemporal data analysis (STDA) is a hybrid of two major branches: geoinformation science and spatiotemporal statistics or STDA = GIS + spatial statistics.

With the advances in data computing and information technologies, there is an increasing trend in using advanced approaches to construct soil inference systems,

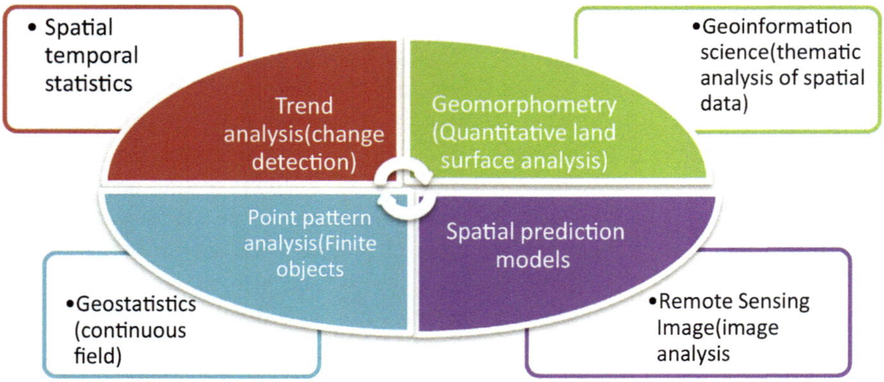

FIGURE 19.1

Spatiotemporal data analysis as a collection of different research fields.

spatial soil data management, digital soil maps, and other applications on a variety of platforms. The complexity of interpreting large datasets has resulted in the emergence of new statistical approaches as well as advances in a global domain such as ML and data mining (Hastie et al., 2009). Furthermore, the emerging potential of technologies such as geographic information systems (GIS), remote and proximal sensors, GPS, and data sources such as digital elevation models (DEM) is providing new opportunities, particularly in soil research. One of the most rapidly expanding areas of geostatistics is stochastic modeling. Within the framework of this method, multiple realizations (maps) of the examined phenomena are generated. Each one corresponds to the original data and exhibits a distinct spatial distribution pattern of the phenomena under consideration.

Geostatistical methods (kriging) are often regarded as a fundamental spatial interpolation tool in land resource assessments (Bhunia et al., 2016; Hengl et al., 2003). Based on the spatial size of the study area, the distance between sampling sites, and the spatial pattern of modeling semivariograms, these approaches evaluate spatial distribution and variability. These have been frequently used to assess the spatial variation in soil variables including physical, chemical, and biological characteristics and to determine spatial correlation coefficients in soils (Fromm et al., 1993; Liu et al., 2014; Vieira et al., 2007; Wigginton et al., 2000; Zheng et al., 2009). Geostatistical methods use a set of algorithms that incorporate various types of a priori knowledge into the prediction procedure, which are used in many places, including GIS and the R statistical environment (e.g., Danilov et al., 2018; Olea, 2012; Zhu et al., 2012). Geostatistical techniques involve the use of statistical approaches based on random function (random variable) theory, to explain the complexities associated with geospatial quantification and prediction. Before geostatistics, several simpler interpolation methods/algorithms were well known, such as inverse distance weighting (IDW), bilinear interpolation, and nearest-neighbor interpolation (Isaaks & Srivastava, 1989). The geostatistical approach facilitates the quantification of spatial correlations between sample values, which may subsequently be used to interpolate values at an unsampled location. These advancements are explained by the paradigm of regionalized variables (Matheron, 1963). Geostatistics, which is based on regional variables, has proven to be one of the best approaches for assessing the spatial distribution and variability of soil characteristics (Webster & Oliver, 2007).

Geostatistics is a framework for the spatial characterization of soil variables and a prediction method that incorporates the spatial continuous behavior of soil variables into the prediction model and has gained popularity in soil science over the last two decades (Delbari, 2007; Gaston et al., 2001; Goovaerts, 1998; Trangmar et al., 1986). Ordinary kriging (OK), often known as the best linear unbiased estimator (BLUE), is the most widely used geostatistical interpolation method (Isaaks & Srivastava, 1989). OK has a wide range of applications in soil science (Burgess & Webster, 1980). When integrated with OK, conventional classification techniques for mapping soil parameters exhibit greater accuracy (Voltz et al., 1997).

19.2 Digital soil mapping using geostatistics (spatial statistics)

In the last decade, there has been great progress in the development and use of quantitative techniques and soil mapping (Lagacherie et al., 1995). Systematic soil sampling and mapping processes were developed in the 1990s, based on the investigation of the correlation between soil parameters and soil formation elements (environmental variables). A soil map is a visual illustration of the spatial variability of soil properties (Yaalon, 1989). The first soil map was developed for agricultural uses in the mid-18th century. The concept of DSM has been used from the beginning of time; for example, the pioneer of GIS, Roger Tomlinson (Tomlinson, 1978), classified digitized polygonal feature maps as digital soil maps. Inherent soil variation is a fundamental impediment to obtaining reliable digital soil models (Burrough et al., 1994). To account for soil variation, two broad but independent techniques have been proposed: discrete (polygon-based) modeling of soil variation and continuous (pixel-based) modeling of soil variation (Heuvelink & Webster, 2001). The former classifies soil into more and less homogeneous subgroups, whereas the latter depicts the soil–landscape interaction as a continuum. There are various drawbacks to conventional soil categorization, which adopts a polygon-based soil map unit model. The conventional soil classification technique, according to Hartemink et al. (2010), generates dynamic, inflexible maps that need additional processes to correlate with grid-based digital soil data. Polygon-based models usually fail to express uncertainty (Grunwald, 2006), although assessing uncertainty to quantify map unit composition has recently gained popularity.

Furthermore, conventional soil mapping methods, which rely primarily on ground-based surveys, may not provide sufficient soil information, lack precision in spatial boundaries, and interpret data that do not reflect the current state of soil conditions and are inconsistent with grid-based geospatial data in many land surface processes (McBratney et al., 2003). Moreover, conventional methodologies for modeling soil spatial variations are usually prohibitively expensive or impractical, particularly at regional, national, or global dimensions, and particularly in areas with inadequate spatial soil data, such as dry and semiarid regions (Forkuor et al., 2017). Therefore, quantitative objective modeling of soil properties or classes replaced qualitative subjective modeling as the gradual development for soil scientists (Goovaerts, 2001).

These advancements result in the regionalized variable theory (RVT) formulating a unifying model of soil spatial variation using the following equation (after Burrough, 1986)

$$Z(x) = \mu(x) + \varepsilon'(x) + \varepsilon''(x) \tag{19.1}$$

where x is the location of any random variable; $Z(x)$ is the random variable Z at location x; $\mu(x)$ is the deterministic structural component (trend or drift); $\varepsilon'(x)$ is the stochastic component, spatially dependent residual from $\mu(x)$ [the regionalized variable] but locally differing both in a lateral and vertical direction; $\varepsilon''(x)$ represents a nonspatially correlated component, noise, unexplained variability.

As a consequence of different elements, including spatial resolution, various soil attributes or processes, and information about spatial location and time (Lin et al., 2005), spatial variability in soil provides a range of variability extending from micro to megascopic scale (Wright & Wilson, 1979). As indicated by the deterministic variable ($\mu(x)$ in Eq. (19.1)), these components can combine to generate a trend at various scales. However, the mechanisms that create soil variation are often unknown, and soil variability is difficult to be described analytically in many spatial and temporal domains (Heuvelink & Webster, 2001). Soil attribute values acquired from surveying collected at a nearby spatial location are often identical or spatially connected (Oliver, 1987). In Eq. (19.1), the spatially dependent random component $\varepsilon'(x)$ is based on this assumption. Under the stationarity assumption, semivariograms have been used to describe the stochastic structural component as a function of the distance between two neighboring points. Nonspatially correlated component of the variable, the noise component of variation is the unexplained variability ($\varepsilon''(x)$ in Eq. (19.1)) that exists in the model with a mean of zero and variance of σ^2 (Webster, 2000).

To evaluate the association between soil and its forming elements, several studies have used the soil factorial concept, an empirical-deterministic model of soil formation created by V.V. Dokuchaev (Glinka & Marbut, 1927) and modified by Jenny (1941). On the other hand, RVT (Matheron, 1971) enabled scientists to predict the values of a wide range of soil parameters at unknown places (Webster, 1994). Since the 1960s, several statistical methods, mostly geostatistical tools, have been widely integrated under the new discipline of soil science known as "pedometrics." Webster (1994) defines pedometrics as "the use of probability and statistical approaches to soil science" or "the application of mathematics and statistics to investigate the distribution and origin of the soil." Since its inception in the early 1800s, pedometric mapping, also known as predictive mapping, that is, the spatial prediction of soil variables at unknown locations using statistical inference, has gained importance. The advancement of geostatistics, which allowed researchers to more accurately interpolate spatial patterns of soil attributes, enhanced the relevance of such maps (Webster, 1994). The framework of DSM is based on a foundation of soil science, quantitative approaches (statistics and geostatistics), geographic information science, and cartography. Since the 1990s, pedometrics has been systematically studied under the discipline of deterministic and stochastic variability of soil attributes and classes.

DSM is a technique for creating and integrating spatial soil information systems by combining field and laboratory observational methods with spatial and nonspatial soil inference systems (Lagacherie & McBratney, 2006). DSM is a powerful process for obtaining geospatial soil databases at a given resolution by establishing quantitative relationships (The International Working Group on Digital Soil Mapping (WG-DSM)) between auxiliary data collected from terrain features (e.g., digital elevation model, DEM) and satellite data with measurements made in the field and laboratory (Costa et al., 2018; McBratney et al., 2003).

In September 2004, Montpellier (France) conducted the first global DSM workshop. It offered a chance to merge high-level expertise and technology relevant to DSM, including soil surveys, soil database management, GIS, pedometric methods, data mining methods, and remote sensing data. DSM is a raster data model that consists of two-dimensional cells (pixels) arranged in a grid, each of which has been assigned a unique spatial position and contains soil information. The uncertainty of the prediction can be quantified using DSM, which also displays the spatial distribution of soil attributes or classes. Additionally, this approach may be used to construct initial soil survey maps, update or alter previous soil surveys, provide specific soil interpretations, and assess risk (Carré et al., 2007). In spatial modeling, the DSM approach is commonly used to investigate the spatial variation of soil attributes in farmland, arable areas, and other agricultural resources (Goldman et al., 2020; Poppiel et al., 2019; Taghizadeh-Mehrjardi et al., 2019; Zeraatpisheh et al., 2017). Various studies have been made to determine the soil characteristics, including organic matter (Angelopoulou et al., 2019; Emadi et al., 2020; Malik et al., 2020; Tajik et al., 2020), texture (Bousbih et al., 2019; Liu et al., 2020), pH (Pahlavan-Rad & Akbarimoghaddam, 2018), surface soil moisture (Khellouk et al., 2021; Tripathi & Tiwari, 2020), soil weathering process (Goydaragh et al., 2021), and electrical conductivity (Yao et al., 2016) that have been successfully predicted. DSM extends far beyond the digitization of already-existing soil maps to continually map soil properties. Additionally, other terms such as pedometric mapping, environmental correlation, predictive soil mapping, soil-landscape modeling, computer-assisted soil cartography, and spatial extrapolation using models have been used or proposed (Gessler et al., 1995; Kempen et al., 2010; Legros, 2006; Scull et al., 2003).

Generally, DSM may be broadly divided into three categories: (1) the pedotransfer function approach, (2) the geostatistical technique, and (3) the state-factor (clorpt) approach. In soil mapping, a mathematical function known as a "pedotransfer function" is used to predict one soil feature by another, making it simpler to evaluate various soil attributes (Bouma & van Lanen, 1987). Geostatistical methods, such as those developed by Krige (1951) and Matheron, use measured and processed data to interpolate soil-attribute maps, whereas clorpt or scorpan methods, which are derived from the work of Dokuchaev (1883) and Jenny (1941), use regression (Minasny et al., 2004), classification (Behrens & Scholten, 2006), or fuzzy-set theory methods (Zhu, 2000) to create soil-attribute or soil-class maps. Although SCORPAN and geostatistical approaches (GSA) are the core objectives of DSM, practically all methodological combinations are valid as empirical quantitative soil-spatial prediction models (McBratney et al., 2003). According to McBratney et al. (2000), the clorpt method is a deterministic model that has to be augmented with geostatistics to account for a random component of the spatial variance of the model's residuals. It is known as a hybrid model because it offers strong spatial prediction tools, especially for catchment and regional studies (Fig. 19.2).

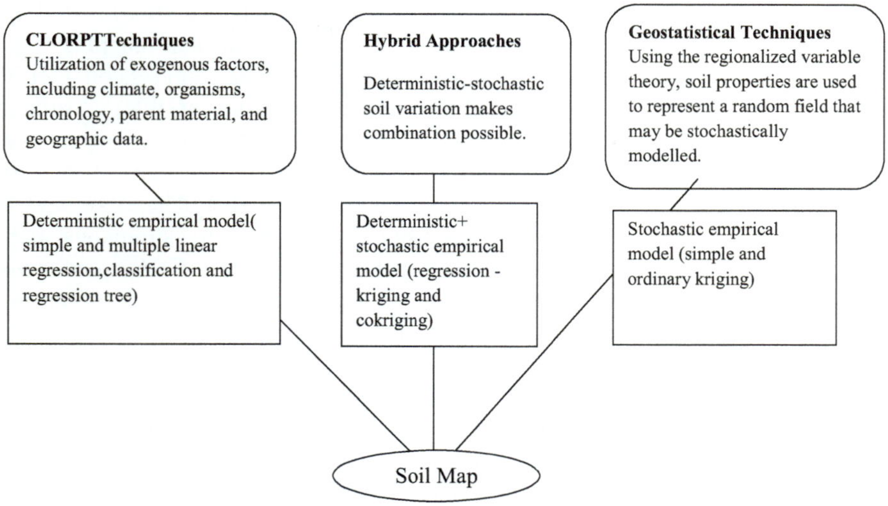

FIGURE 19.2

The different types of pedometric techniques, their hybrids, and styles of spatial prediction.

19.3 Brief overview of approaches to soil spatial prediction

One of the earliest documented applications of geostatistics in soil research was the salinity experiment in southwest Iran (Hajrasuliha et al., 1980). Recent technological advancements have provided significant potential for improving the way soil maps are created (McKenzie et al., 2000). Applying GIS to estimate spatial soil variability for more accessible mapped environmental factors is essential for the future of soil surveying. To enhance soil survey activities, the Geospatial Research Unit (GRU) of the National Soil Survey Center (NSSC) has highlighted DSM as an important field of study. The GRU has provided funding for many DSM research initiatives. Numerous techniques, such as statistical classification (hierarchical and fuzzy), spatial and temporal interpolation (geostatistics), statistical analysis (visualization, regression, and classification), uncertainty analysis (accuracy evaluation), and the inclusion of auxiliary data, are used to generate prediction maps of soil classes and soil attributes. The development and expansion of geostatistical techniques such as regression kriging (Hengl et al., 2004), nonstationarity in the variance and mean (Lark, 2009; Wadoux et al., 2018), and sampling design optimization (Brus, 2014) have been significantly aided by soil science pedometricians and spatial statisticians. From Linear Models to geostatistical techniques such as OK, simple kriging (SK), universal kriging (UK), and hybrid models (Regression Kriging), to various ML algorithms such as random forest and support vector regression (SVR), innovation has occurred in the field of predictive algorithms. Apart from GSAs, several sophisticated mapping algorithms and multivariate methods, such as Multiple Linear Regression (MLR) (Alvarez et al., 2011), geographically weighted regression (GWR), Artificial Neural Network (ANN) (Malone et al., 2009), K-Nearest Neighbor, support vector machine (SVM) (Khalil et al., 2005), are currently used in many studies for predicting different soil properties (Lamichhane et al., 2019). There are several methods for generalizing soil attributes from discrete data to continuous maps (i.e., DSM) (McBratney et al., 2003; Scull et al., 2003). GSA, predictive-statistical approaches (PSA- machine learning methods), and hybrid approaches (HA) are the three types of spatialization techniques.

Spatial autocorrelation is taken into account in GSAs such as OK and GWR (Brunsdon et al., 1996; Webster & Oliver, 2001). OK is one of the most commonly used methods for spatial interpolation as it is a kriging-based approach that is based on the assumption of an unknown mean and is one of the most often used methods of spatial interpolation (Pająk et al., 2017). However, this method ignores the effect of environmental factors (Mishra et al., 2009) and is unable to attain the requisite precision owing to the stationarity assumption (Eldeiry & Garcia, 2010). GWR is based on interactions between several environmental factors within a local domain. It can incorporate the spatial location of samples into a regression using a locally weighted least square approach (Fotheringham et al., 2003). Harris et al. (2010) compared the spatial prediction performance of GWR and OK techniques using simulated datasets (Harris et al., 2010). The research found that GWR gave additional information on spatial processes, although residuals remained spatially dependent. Kriging techniques are frequently used to predict spatial features having high spatial autocorrelation (more similarity in values), such as climatic conditions, natural soil qualities, and geological aspects (Cambardella et al., 1994).

Predictive-statistical techniques (PSA) employ the statistical correlation between soil qualities and a set of observed environmental factors over a range of observational locations. There are many more predictive strategies (Grunwald, 2009), such as those based on partitioning or splitting of data (i.e., decision tree approaches) (Lagacherie & Holmes, 1997) or ML algorithms (Liess et al., 2016), which have a higher probability of interpreting for the intricate interactions between soil-forming processes and soil attributes that have generally been ignored are used to derive these models because of their simplicity and computational efficiency. ML algorithms can handle nonlinearity and multicollinearity with limited soil observational and auxiliary environmental data, and they can also overcome overfitting (Drake et al., 2006; Gautam et al., 2011). These are defined by SVMs and ANNs, which are extensively used in DSM (Krishna et al., 2019; Nguyen et al., 2017). By projecting sample data into a new hyperspace using kernel functions, SVMs may construct an optimal hyperplane. The hyperplane separates classes and creates more flexibility in the classification process. It may also incorporate data (SVMs for regression, SVR) using a mathematical function that reduces empirical risk and complexity when describing nonlinear patterns (Gunn, 1998). ANN models human learning mechanisms. Nonlinear and interconnected nodes in ANN build and strengthen the links between the input and output data (Haykin, 1998).

There has been a substantial increase in HA, which combine both predictive and geostatistical methodologies in recent years. Geographically weighted regression kriging (GWRK) and artificial neural network kriging (ANNK) are two HAs. To improve better predictions and minimize errors, GWRK and ANNK can include spatial autocorrelation of observed variables (Kumar et al., 2012). Both GWRK and ANNK are capable of accounting for spatial parametric nonstationarity and also an association between soil variables and other environmental factors. Furthermore, residuals can be predicted using SK and contributed as a spatially stochastic variable to the calculated trend. Guo et al. (2017) found that GWRK performs better partial least square regression kriging for predicting soil organic matter using visible and near-infrared spectra.

Moreover, pedometric examines the application of the fuzzy system to quantitative soil classification and mapping (McBratney & Odeh, 1997). Using quantitative grouping of terrain attributes or rules based on expert knowledge, the fuzzy theory may be used to generate rules for delineating the landscape. Zhu et al. (1996) used fuzzy membership to infer the presence of appropriate soil series from topographic data. These fuzzy criteria are then used to map the distribution of soil series in a given area and showed that the resulting series was both logical and spatially coherent.

A significant number of reviews with multiple objectives on DSM have been published, which summarizes different modeling techniques (Table 19.1). McBratney et al. (2000) were among the first to provide a brief description of studies in DSM. McBratney et al. (2003) then evaluated methods for merging environmental GIS data into soil mapping. Grunwald (2009) assessed the efficacy of DSM and modeling methodologies for attaining specific objectives at the local, national, and global scales, and Grunwald et al. (2011) reviewed research on DSM at the continental scale. Regression kriging, a specific type of hybrid approach, was used in a review of the literature given by Keskin and Grunwald (2018) to quantify the variables influencing the effectiveness of the proposed approach. This article concisely examines geostatistical techniques for quantifying spatial dependency and various methods of interpolation under different types of spatial variation, as well as the potential future of geostatistical application to spatial research in DSM.

19.3.1 Geostatistics and related methods

The theory of regionalized variables, which serves as the basis of geostatistical techniques (Matheron, 1965), provides the opportunity to learn about spatial variability in soil characteristics as the realization of a random component recognized by a stochastic process. Spatial information may be used in kriging algorithms to

Table 19.1 Recent literature on predictive soil mapping, summarizing the modeling approach, dependent variables, and environmental factors included in the model.

Modeling method	Criteria of the research	Dependent and environmental variables	References
Random forest	To predict the SOC concentration under different land use and climate change (China)	Terrain factors, vegetation index, Vis-NIR spectrometer, and climate factors as predictors	Zhou et al. (2019)
Deterministic (IDW), geostatistical spherical, and exponential kriging (OK), and EBK and Machine Learning (RF, SVM)	To compare deterministic, geostatistical, and machine learning models for predicting SOC (India)	SOC, ppt., annual temp., terrain attributes (slope, El.), erosion	Sahu and Ghosh (2021)
RF, OK, GLM	To predict organic carbon stock in farmland topsoil in East China	Total nitrogen content, slp., WTD CEC, Lat., Longitude, pH, text., NDVI	Deng et al. (2018)
OK	To measure different soil properties and their spatial variability (India)	[EC, pH, K, P, and OC] and climate (ppt., annual temp.), parent material, topographic attributes	Shit et al. (2016)
OK and cokriging	To evaluate prediction map of soil CEC spatial variability with the PCA (Iran)	Texture, OC, EC, CEC, Ph	Seyedmohammadi and Matinfar (2018)
Regression-kriging	Digital mapping of soil phosphorus (Iran)	topographic indicators (terrain attributes-DEM; pH, texture, OC, available P, PL	Sarmadian et al. (2014)
SK	To map different soil properties) and predict their spatial variability using the geostatistical technique (Nigeria)	pH, SOC, Sand, Clay, Phosphorus, Nitrogen, and CEC	Kingsley et al. (2019)

(Continued)

Table 19.1 Recent literature on predictive soil mapping, summarizing the modeling approach, dependent variables, and environmental factors included in the model. *Continued*

Modeling method	Criteria of the research	Dependent and environmental variables	References
Regression kriging	Estimation of the SOC distribution by a hybrid interpolation method (Central India)	Predictor variables such as BI, greenness index (GI), WI, NDVI, VTCI, DEM, slp., compound topographic index (CTI), and SOC	Mondal et al. (2017)
OK universal kriging (UK)/kriging with external drift (KED), RF, and regression kriging (RK) approaches	To prepare digital maps of sand contents in arid western India	Auxiliary and environmental variables like terrain attributes (slp, El., hillshade, TWI and bioclimatic variables) (annual mean temp, annual ppt)	Santra et al. (2017)
Regression kriging	To quantify SOC and total nitrogen using predictor variables (India)	SOC, TSN, and Elevation, slope, NDVI, CTI, SPI, STI	Bangroo et al. (2020)
Ordinary kriging technique	Quantifying spatial variability of soil properties in apple orchards using geospatial techniques (India)	pH, EC, OC, Av. N, P, K, Ca, Mgandelv., annual temp., annual rainfall	Bangroo et al. (2021)
OK, GWR, SVR, ANN, GWRK, ANNK	Mapping Topsoil Organic Carbon Content using Geostatistical, Machine Learning, and Hybrid Approaches (China)	SOC, BD, av. Temp., av. ppt, slope, aspect, NDVI, TWI, MSAVI, SPI, RLD, SOA	Chen et al. (2019)
Multinomial logistic regression and random forest	Comparing the efficiency of digital and conventional soil mapping to predict soil types in a semiarid region in Iran	Topographic attribute (EI, TWI, PICu, NDVI, SAVI, RVI, slope, and soil depth)	Zeraatpisheh et al. (2017)
ANN	Digital soil mapping in a Himalayan watershed using remote sensing and terrain parameters	SOC, av. N, P and terrain parameters TWI, SPI, Slope	Kalambukattu et al. (2018)

(*Continued*)

Table 19.1 Recent literature on predictive soil mapping, summarizing the modeling approach, dependent variables, and environmental factors included in the model. *Continued*

Modeling method	Criteria of the research	Dependent and environmental variables	References
SK	Digital Mapping of Soil Profile Properties to predict soil fertility for Precision Agriculture (Mexico)	CEC, pH, Ca, Mg, Na, K and altitude, S, and surface drainage	López-Castañeda et al. (2022)
OK, LOK, COK, and IK	To predict the spatial distribution of soil texture fraction (Austria)	Mean annual temp, ppt, soil textures (sand, silt, clay)	Delbari et al. (2011)
Geostatistical model (kriging)	To examine spatial variability in soil properties affecting salinity and alkalinity (Turkey)	pH, EC, ESP, CEC, K_s, and surface elevation	Cemek et al. (2007)
GLM, PCA, RF, ML with OK, IDW	Mapping soil properties with soil-environmental covariates using geostatistics and multivariate statistics	El, slope, As, PlCur, TWI, and pH, OC, av. P, ex. Acidity	John et al. (2021)
MLR, RF, SVM, RFRK	To evaluate the DSM approach for predicting soil fertility	pH, EC, OC, av. P, K, Fe, Mn, Cu, Zn, B, and an. Av. Ppt, max. temp, min, temp, aspect, slope, El	Dharumarajan et al. (2022)

ANN, *Artificial neural networks;* ANNK, *artificial neural networks kriging;* BD, *bulk density;* BI, *brightness index;* Ca, *calcium;* CEC, *cation exchange capacity;* COK, *cokriging;* CTI, *compound topographic index;* Cu, *copper;* DEM, *digital elevation model;* El., *elevation;* ESP, *exchangeable sodium;* Fe, *Iron;* GLM, *general linear model;* GWR, *geographically weighted regression;* GWRK, *geographically weighted regression kriging;* IDW, *inverse distance weightage;* IK, *indicator kriging;* K, *potassium;* K_s, *hydraulic conductivity;* lat., *latitude;* LOK, *lognormal ordinary kriging;* Mg, *magnesium;* Mn, *manganese;* MSAVI, *modified soil adjusted vegetation index;* Na, *sodium;* NDVI, *Normalized difference vegetation index;* OK, *ordinary kriging;* P, *phosphorus;* PCA, *principal component analysis;* PL, *plan curvature;* Plcur, *plane of curvature;* ppt, *precipitation;* RF, *random forest;* RFRK, *random forest regression kriging;* RLD, *relief of land surface;* RVI, *ratio vegetation index;* SAVI, *soil adjusted vegetation index;* SK, *simple kriging;* SOA, *slope of aspect;* SOC, *soil organic carbon;* SPI, *stream power index;* STI, *sediment transport index;* SVM, *Support vector machine;* temp., *temperature;* tex., *soil texture;* TSN, *total soil nitrogen;* TWI, *topographic wetness index;* VTCI, *vegetation temperature condition index;* WI, *wetness index;* WTD, *water table depth.*

estimate the output of characteristics at unsampled locations. A variogram is a statistical function that measures the structure, magnitude, and intensity of spatial patterns and is an important tool in geostatistics. Kriging is a type of weighted local averaging in which the average weights are used to compute variogram, as a measure of spatial dependency. In predictive soil mapping research, GSAs are used to spatially predict soil property values from field-collected data at unsampled locations. In geostatistics, the random variable theory (say any soil property) $Z(x)$ that is both random and spatially autocorrelated indicates that the estimated value of a soil property Z at any point x within a sample region is

$$Z(x) = \mu + \varepsilon(x) \tag{19.2}$$

where μ represents the mean, and ε represents a random, spatially independent dispersion value about the mean. Deviations from the population mean are assumed to have a normal distribution with a mean of zero and a variance of σ^2. Many soil parameters have skewed probability distributions that must be changed to the normal distribution (e.g., natural log) before statistical analysis. Dispersion statistics, such as coefficient of variance, standard deviation, and standard error, are used to demonstrate the accuracy mean of the estimator as mean values are used to estimate attributes at unsampled locations within sampling units. These statistical data have been used widely to show the variety of soil characteristics within sample sites. The geostatistical analytic technique is primarily based on two operational steps: (1) semivariogram calculus and modeling and (2) kriging interpolation technique.

19.3.2 Theory of regionalized variables

The fundamental concept of geostatistics consists of regionalized variables, random functions, and stationarity. Recent advances in statistical modeling lead to a proper assessment of the spatial dependency of soil parameters during interpolation. These advancements are based on the theory of regionalized variables, which considers both the structured and random components of spatially distributed variables to provide quantitative tools for their description and best, unbiased assessment.

19.3.2.1 Regionalized variables and random functions

A random or stochastic variable is a variable whose values randomly change according to some probability distribution law. A regionalized variable $z(x)$ is a random component that has distinct values z based on its sample location x within a region (Journel & Huijbregts, 1978). A regionalized variable $z(x)$ may therefore be considered as a distinct realization of a random variable Z at a given point x within the region. When all $z(x)$ values are examined at all locations, the regionalized variable $z(x)$ comprises an infinite set of random variables $z(x)$ for all locations within the area. Because it correlates a random variable Z with any point x, it is known as a random function.

19.3.2.2 Semivariogram modeling

A semivariogram is a geostatistical tool that reflects the degree of spatial correlation of a random component with spatial dimensions. Spatial patterns are frequently characterized in the geostatistical literature in terms of dissimilarity (rather than similarity) across datasets as a function of distance. The experimental semivariogram $\gamma(h)$ quantifies the average difference between data points separated by a vector h, as half the average squared difference between the components of every data pair. The semivariogram is calculated as follows:

$$\gamma(h) = \frac{1}{2}\sum_{i=1}^{N(h)} [(Z(x_i) - Z(x_i + h))]^2 \tag{19.3}$$

where $\gamma(h)$ represents experimental semivariance, h is the lag distance, Z is any soil property parameter, $N(h)$ is the number of pairs of locations separated by a lag distance h; $Z(x_i)$, and $Z(x_i + h)$ are values of Z at different locations x_i and $x_i + h$ (Eq. 19.3). The spatial continuity between data points is measured by the structural semivariance function. The semivariogram (Fig. 19.3) is obtained by plotting semivariance against the lag distance or separation distance between data points. The nugget, sill, and range are the three elements that characterize the structure of the semivariogram (Fig. 19.3). These spatial semivariogram structures aid in the identification of suitable data evaluation techniques and data interpretation (Lambert et al., 2003), improved evaluation of simulation and uncertainty analysis (Papritz & Dubois, 1999), and designing experiments (Fagroud & Van Meirvenne, 2002), and monitoring networks (Prakash & Singh, 2000). Predicting the spatial variability of attribute values is simplified with the modeling of observed semivariance values (Goovaerts, 1998). The spatial extent of variables is

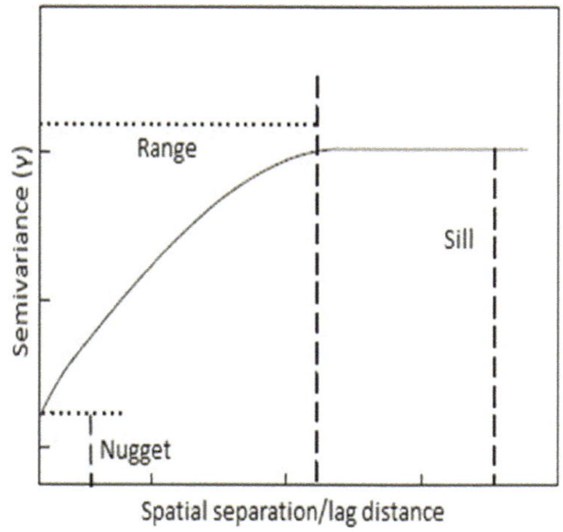

FIGURE 19.3

A typical example of semivariogram.

critical in differentiating random error in semivariance, as well as interpolation and mapping analysis such as kriging (Nielsen & Wendroth, 2003).

In Table 19.2, a represents the range, C_0 represents the nugget semivariance, and $C_0 + C_1$ is the sill or total semivariance. The range is the lag distance up to which there is a spatial autocorrelation beyond range, the values between variables do not influence nearby values, that is, the distance at which values of one variable become spatially independent of another and still represent the maximum value of semivariogram (variance of a dataset) used to estimate the range. Nugget represents the shorter distance or spatial variation/measurement error. The ratio of the nugget (C_0) to the sill ($C_0 + C_1$) illustrates the relevance of the random variable in the overall spatial variability of the data and provides a systematic assessment of spatial dependency for each soil variable. The semivariogram slope will be near zero if the semivariogram function is $\gamma(h) = C$; the soil variable is regarded as nonspatially correlated (pure nugget).

The selection of theoretical frameworks and their fitting technique is essential for better prediction of unknown areas (McBratney & Webster, 1986). According to Table 19.2, the spherical, exponential model, the Gaussian model, and the linear plateau model are the most widely used semivariogram models (Burgess & Webster, 1980) (Fig. 19.4). The maximum likelihood (Cressie, 1991) and least square regression (Trangmar et al., 1986) approach, including weighted least square methods, improve model parameter values by lowering prediction variation from experimental semivariances. As a result, among the fitting methods, the weighted nonlinear least square method is the most robust and efficient (Cressie, 1985) following the principle of minimum variance estimate. The goal is to choose the best fit for the model and then use the model to make kriging predictions. The reason that the sum of the differences between the model-predicted value and experimental values should be small shows that the model algorithm outperforms better. As a result, choosing a model is important for better prediction (Burnhan & Anderson, 2002). Semivariogram structures have to be "positive-definite" to have nonnegative explanatory variables, which limits their capacity to fit into the modeling data.

Table 19.2 Commonly used models for semivariogram fitting (Huang et al., 2012).

Name of the model	Equation
Linear model	$\gamma(h) = C_0 + C_1 h$
Spherical model	$\lambda(h) = \left\{C_0 + C_1\left[1.5\left(\frac{h}{a}\right) - 0.5\left(\frac{h}{a}\right)^3\right]\right\}, 0 \leq h \leq a$
Exponential model	$\gamma(h) = C_0 + C_1\left(1 - e^{\frac{h}{a}}\right)$
Gaussian model	$\gamma(h) = C_0 + C_1\left(1 - e^{\left(\frac{-h}{a}\right)^2}\right)$

Source: *Huang, Z., Wang, H., & Zhang, R. (2012). An improved kriging interpolation technique based on SVM and its recovery experiment in oceanic missing data.* American Journal of Computational Mathematics, 2, *56–60.*

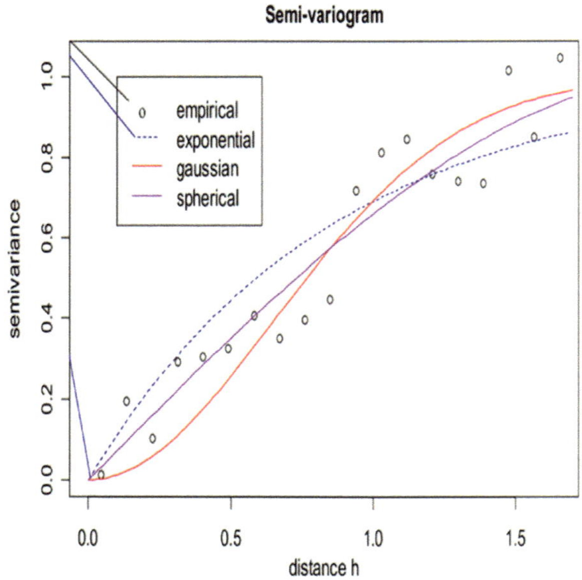

FIGURE 19.4

Basic variogram models showing different components.

19.3.3 Kriging interpolation technique

Kriging is a geostatistical interpolation technique commonly used to estimate the value of a random variable (for example, soil variables as a function of spatial area) at an unknown location based on known observations of its value at nearby locations. Kriging consists of multiple steps. It starts with an explanatory statistical analysis of the data, then proceeds on to variogram modeling, surface construction, and variance surface exploration. The basic statistical assumption behind kriging is one of stationarity, which indicates that statistical features (such as mean and variance) are independent of exact spatial locations, thus the mean and variance of a variable at one site are similar to the mean and variance of the variable at another position. Kriging uses autocorrelation, or statistical correlations between the observed points, to predict values in the spatial field, in contrast to other deterministic models such as IDW and Spline. Kriging may provide prediction surfaces with some level of uncertainty. The two fundamental conditions for kriging to yield the best linear unbiased prediction are stationarity (constant mean and variance) and isotropy (uniformity in all directions); however, these assumptions might vary depending on the kind and method of kriging used. The basic equation for the average weighted sum of the data is as follows (Eq. 19.4):

$$\hat{Z}_{(S_0)} = \sum_{i=1}^{N} \lambda_i Z_{(s_i)} \tag{19.4}$$

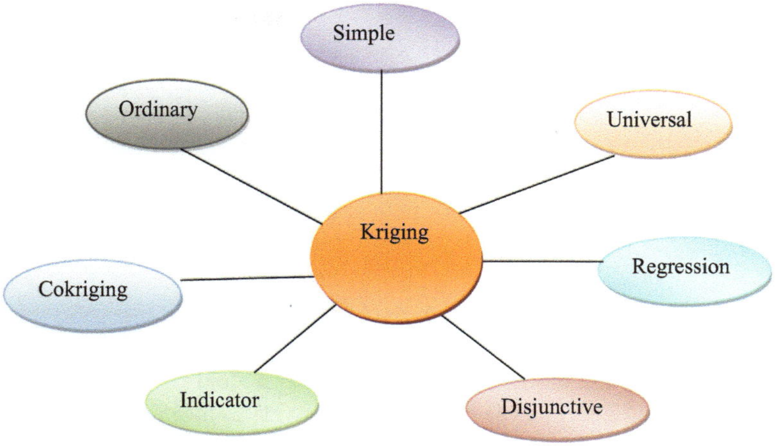

FIGURE 19.5

Different types of kriging methods.

where $Z_{(si)}$ represents the measured value at the ith location; λ_i is the unknown weight; S_o is the prediction location; N is the number of measured values. The kriging weights are determined by the spatial arrangement of the known points as well as the distance between the measured and the predicted point. Fig. 19.5 depicts several kriging approaches, the most commonly used are OK, UK, cokriging, and the regression kriging interpolation method.

19.3.3.1 Kriging types

19.3.3.1.1 Ordinary kriging

The most widely used form of kriging is OK, which assumes a constant mean value but specifically emphasis on the spatial location of the sample points within the nearest neighborhood for estimating unknown values. As the error means is zero, it is a linear unbiased estimator. OK filters the local mean from the linear estimator by allowing the kriging weights equal to 1. The OK approach models the association between value variation and the separation of input paired locations. In general, OK may provide smooth maps with low standard errors. The OK is commonly preferred over SK as it does not need knowledge of mean stationarity over the entire area.

$$Z_{OK}(\mu) = \sum_{\alpha=1}^{n(u)} \lambda_{\alpha}^{OK}(u) Z(u_{\alpha}) \text{ with} \sum_{\alpha=1}^{n(u)} \lambda_{\alpha}^{OK} = 1 \tag{19.5}$$

where $\lambda_u^{OK}(u)$ is the weight assigned to the known variable at location (u), $n(u)$ is the number of measured values used in the estimation of the neighborhood u.

19.3.3.1.2 Universal kriging

UK is a nonstationary form of OK in which the mean fluctuates deterministically at various locations (local trend or drift), but only the variance remains constant. This second-order stationarity (weak stationarity) is although important assumption when dealing with environmental factors. In the United Kingdom, the first trend is estimated using geographic coordinates, and then the variance is what remained (the residuals) as a random field is added to the trend to generate a final output.

$$Z(\vec{x}) = \sum_{k=0}^{K} \beta_k f_k(\vec{x}) + \varepsilon(\vec{x}) \tag{19.6}$$

where f_x represents the global function of position $\vec{x}$, and β_k is the coefficient of the base function. The $\varepsilon(\vec{x})$ is the spatially correlated error, which is modeled with a variogram considering the residuals.

19.3.3.1.3 Cokriging

Cokriging (CK) is a variant of OK in which auxiliary variables (known as covariates that are often related to the variable of interest) are used to enhance the accuracy of the interpolation of the variable of interest. Cokriging, unlike regression and UK, does not need the availability of secondary information at all prediction sites. The covariable might be evaluated at the same locations as the target (co-located samples), at distinct locations, or both. Cokriging is most commonly used when the covariable is less expensive to measure than the target variable.

$$Z_{\mathrm{CK}}(u) = \sum_{\alpha_1=1}^{n(u)} \lambda_{\alpha_1}^{\mathrm{CK}}(u) Z_1(u_{\alpha_1}) - \sum_{\alpha_2=1}^{n_2(u)} \lambda_{\alpha_2}^{\mathrm{CK}}(u)\left[Z_2(u_{\alpha_2}) - m_2 + m_1\right] \tag{19.7}$$

where m_1 and m_2 are the means λ_{α_1} and λ_{α_2} and the weights of primary and secondary variables.

19.3.3.1.4 Regression kriging

The term RK was introduced by Odeh et al. (1992, 1995). Regression kriging (RK) is the mathematical extension of UK or kriging with external drift, wherein auxiliary variables are used directly to generate kriging weights. Regression kriging combines a simple regression model with basic regression residual kriging. Following the computation and modeling of the experimental residual variogram, SK is applied to the residuals to offer a spatial prediction of the residual.

$$Z^*{}_{RK}(u) = m^*{}_{RK}(u) + \sum_{\alpha=1}^{n(u)} \lambda_{\alpha}^{RK}(u) R(u_{\alpha}) \tag{19.8}$$

where $m^*{}_{RK}(u_{\dot{\alpha}})$ is the regression estimate for location u, $R(u_{\alpha})$ are the residuals $[R(u_{\alpha}) - m(u_{\alpha})]$ of the observed location $n(u)$.

The RK technique predicts soil properties by coupling a regression of the target soil characteristics on environmental variables with Kriging of the estimated error (Hengl, 2007). Regression analysis may be used in this hybrid technique to evaluate the parameter trend between the target and environmental factors, while stochastic and spatially autocorrelated residuals are interpolated using a Kriging function (Yigini et al., 2018).

19.3.4 Digital soil mapping paradigms (The SCORPAN model)

The most prominent soil science paradigm is $S = f(\mathrm{cl}, o, r, p, t)$, which was initially developed at the end of the 19th century by Dokuchaev (1967) and was later modified by Jenny (1941). According to the equation, the soil is the product of its forming elements, such as climate, organism, parent material, relief, and time and variation in these soil-forming components could be used to distinguish between different soil types. The equation has been used in soil surveys to better understand soil–landscape relationships at a given location. Many investigations sought to statistically quantify Jenny's conceptual equation after her work. Jenny's concept is often used in conventional soil mapping techniques. Jenny's approach has limited applicability in DSM due to its limitations in quantitative data representation and spatial data interpolation. As a result, McBratney et al. (2003) created the SCORPAN soil formation, which uses spatial-based interpolation and quantitative expressions to represent soil and related environmental variables for their wider applicability in DSM. In DSM, the SCORPAN model incorporates five environmental variables: soil characteristics (s), climate (c), organisms (o), relief (r), and parent material (p). The SCORPAN soil development model is represented as follows:

$$Sc, a = f(s, c, o, r, p, a, n) + \varepsilon \tag{19.9}$$

where S_c and S_a represent soil classes or soil attributes as a function of soil, climate, organisms, relief, parent material, age, and spatial position, "ε" is spatially correlated residuals. The following are the key developments of the SCORPAN model for their application in DSM: (1) comprehension of environmental variables that are dependent on one another and are described as environmental covariates, and integration of soil as an environmental predictor, (2) spatially distributed pattern of the model, and (3) numerical structure of the functional relationships. Soil, in the form of point observation, existing soil maps, or remotely sensed data can be used as input data in the SCORPAN model. Environmental covariates are raster data that are spatially distributed and are examined by a geographic information system (GIS). The SCORPAN model quantifies the nature of the relationship between spatially explicit digital environmental factors and soil classes or attributes that must be inferred in a spatial context. It also provides the ability to estimate error or uncertainty in the spatial prediction of soil classifications or properties.

SCORPAN is a theoretical framework of a spatial soil inference system (SSINFERS). The regression-kriging method, which illustrates the entire process of data through the system to estimate the unknown soil properties, is a common spatial prediction technique used by Hengl (2007) and may be used to apply the SCORPAN model. These models assume a stochastic relationship between a large number of predictors and desired soil qualities, although they may also be used to enhance deterministic soil genesis models.

19.3.4.1 Digital soil mapping models

DSM models are a group of inferences that uses a subset of soil data and auxiliary spatial variables to estimate and predict secondary soil attributes to describe the changes in soil-forming elements throughout the mapped zone. Since the early 1990s, substantial research has been conducted on these models, and comprehensive evaluations of different models are also available (McBratney et al., 2003). These models are successfully used to generate digital soil information for regions, nations, and continents (Henderson et al., 2005).

19.3.4.2 Different types of digital soil mapping models

Various types of DSM models are adopted, depending on the nature of the inference. It is therefore classified into two types: (1) SCORPAN models or spatial inference models, which generate soil classification maps and mapping of individual soil properties from soil observations and auxiliary spatial variables; and (2) soilscape inference models, which generate additional properties from predefined outputs. The following sections provide a quick overview of these two types:

McBratney et al. (2003) provided a comprehensive list of these models (scorpan functions). They suggest a generic formulation for these models as follows: $S = f(Q) + e$, where S represents the soil class or soil attribute, Q represents the scorpan predictor variables, and e is the prediction error. The common approach for developing these functions is to collect m field observations of S at known locations $[x, y]$ and connect them with some type of function to a set of pedologically significant predictor variables Q, which will typically be a set of variables or data layers of size M at locations $[x, y]$ from the auxiliary database with the $[x, y]$ (X, Y). $(x, y) \subset (X, Y)$. A digital map is produced after the model has been fitted at the m observation sites and the predictions have been applied to the M points or cells in the raster layer. The method's effectiveness is predicated on the assumptions that, ideally, $m > M$ and that S is far more challenging and expensive to evaluate than Q.

In the framework of the proposed spatial soil inference system (SSINFERS), the SCORPAN functions described above might be generated in training regions where a certain data configuration is available and then adapted to specific areas as soon as they match the required data configuration. Early examples of the implementation of SCORPAN functions in the context of the reference area approach have previously been proposed (Lagacherie et al., 1995). This has to be developed within a more comprehensive context.

19.3.4.2.1 Soilscape inference models

This section provides an overview solely of the pedotransfer functions, or soilscape inference models, which are commonly used in DSM. Reviews on the development and implementation of pedotransfer functions, especially for predicting soil hydraulic parameters, were published by Wosten et al. (2001) and McBratney et al. (2000). Based on the quantity of relevant evidence, Wosten (1997) distinguished two categories of PTFs: class and continuous PTFs. In class PTFs, specific soil properties are estimated based on the class of the soil sample (textural, horizon, etc.). In continuous PTFs, particular soil properties are predicted based on one or more observable factors. McBratney et al. (2000) devised a more detailed categorization that takes into account the crisp/fuzzy structure of the inputs and outputs.

19.4 Conclusion

Following the initial research by Burgess & Webster (1980), geostatistics has become a key tool in soil science during the last 30 years. An essential tool of geostatistics in soil science is the structural analysis of soil variation and its application for local prediction. The semivariogram provides a numerical tool for relating the underlying pattern of variation in specific attributes to the spatial effect of soil-forming variables and processes. Geostatistics may be used in soil survey operations to analyze reconnaissance data to determine future sample size and configurations both within and across terrain units, as well as for the structural studies of soil variation to aid in understanding soil genesis. The resources of the geostatistics-based sampling procedure in functional soil investigation must be field-tested and compared with traditional sampling approaches in diverse terrain types. Interpolation of interpreted data particularly observed and estimated attributes, which may vary independently of mapping unit boundaries, can be used to improve general-purpose information provided in standard soil maps using kriging. Technological advances such as GPS and GIS, as well as the advancement of geostatistics and other statistical algorithms, have enhanced the processing of large databases and provided additional tools for spatial modeling and prediction, resulting in more accurate mapping of soil characteristics, while developments in proximal and remote sensing have enabled for the quick and relatively affordable acquisition of enormous quantities of georeferenced information that can be connected to a model wide variety of soil properties.

References

Alvarez, R., Steinbach, H. S., & Bono, A. (2011). An artificial neural network approach for predicting soil carbon budget in agroecosystems. *Soil Science Society of America Journal*, *75*(3), 965–975.

Angelopoulou, T., Tziolas, N., Balafoutis, A., Zalidis, G., & Bochtis, D. (2019). Remote sensing techniques for soil organic carbon estimation: A review. *Remote Sensing*, *11*(6), 676.

Bangroo, S. A., Najar, G. R., Achin, E., & Truong, P. N. (2020). Application of predictor variables in spatial quantification of soil organic carbon and total nitrogen using regression kriging in the North Kashmir forest Himalayas. *Catena*, *193*, 104632.

Bangroo, S. A., Sofi, J. A., Bhat, M. I., Mir, S. A., Mubarak, T., & Bashir, O. (2021). Quantifying spatial variability of soil properties in apple orchards of Kashmir, India, using geospatial techniques. *Arabian Journal of Geosciences*, *14*(19), 1−10.

Behrens, T., & Scholten, T. (2006). A comparison of data-mining techniques in predictive soil mapping. *Developments in Soil Science*, *31*, 353−617.

Bhunia, G. S., Shit, P. K., & Maiti, R. (2016). Comparison of GIS-based interpolation methods for the spatial distribution of soil organic carbon (SOC). *Journal of the Saudi Society of Agricultural Sciences*. Available from https://doi.org/10.1016/j.jssas.2016.02.001.

Bouma, J., & van Lanen, H. A. J. (1987). Transfer functions and threshold values: From soil characteristics to land qualities. In: *Proceedings of the international workshop on quantified land evaluation procedures*. Washington, DC, 27 April−2 May 1986, pp. 106−110.

Bousbih, S., Zribi, M., Pelletier, C., Gorrab, A., Lili-Chabaane, Z., Baghdadi, N., & Mougenot, B. (2019). Soil texture estimation using radar and optical data from Sentinel-1 and Sentinel-2. *Remote Sensing*, *11*(13), 1520.

Brunsdon, C., Fotheringham, A. S., & Charlton, M. E. (1996). Geographically weighted regression: A method for exploring spatial nonstationarity. *Geographical Analysis*, *28* (4), 281−298.

Brus, D. J. (2014). Statistical sampling approaches for soil monitoring. *European Journal of Soil Science*, *65*(6), 779−791.

Burgess, T. M., & Webster, R. (1980). Optimal interpolation and isarithmic mapping of soil properties. I. The semi-variogram and punctual kriging. *Journal of Soil Science*, *31*, 315−331.

Burnhan, K. P., & Anderson, D. R. (2002). *Model selection and multimodel inference*. New York: Springer.

Burrough, P. A. (1983). Multiscale sources of spatial variation in soil. I. The application of fractal concepts to nested levels of soil variation. *Journal of Soil Science*, *34*(3), 577−597.

Burrough, P. A. (1986). *Principles of geographical information systems for land resource assessment*. Oxford: Oxford University Press.

Burrough, P. A., Bouma, J., & Yates, S. R. (1994). The state of the art in pedometrics. *Geoderma*, *62*(1−3), 311−326.

Cambardella, C. A., Moorman, T. B., Novak, J. M., Parkin, T. B., Karlen, D. L., Turco, R. F., & Konopka, A. E. (1994). Field-scale variability of soil properties in central Iowa soils. *Soil science society of America Journal*, *58*(5), 1501−1511.

Carré, F., McBratney, A. B., Mayr, T., & Montanarella, L. (2007). Digital soil assessments: Beyond DSM. *Geoderma*, *142*(1−2), 69−79.

Carvalho, J. R. P. D., Silveira, P. M. D., & Vieira, S. R. (2002). Geostatistics to determine spatial variability of soil chemical properties using different preparation systems. *Pesquisa Agropecuária Brasileira*, *37*, 1151−1159.

Cemek, B., GÜLer, M., KiliÇ, K., Demir, Y., & Arslan, H. (2007). Assessment of spatial variability in some soil properties as related to soil salinity and alkalinity in Bafra plain in northern Turkey. *Environmental Monitoring and Assessment*, *124*(1), 223–234.

Chen, L., Ren, C., Li, L., Wang, Y., Zhang, B., Wang, Z., & Li, L. (2019). A comparative assessment of geostatistical, machine learning, and hybrid approaches for mapping topsoil organic carbon content. *ISPRS International Journal of Geo-Information*, *8*(4), 174.

Corwin, D. L., Hopmans, J., & de Rooij, G. H. (2006). From field-to landscape-scale vadose zone processes: Scale issues, modeling, and monitoring. *Vadose Zone Journal*, *5*(1), 129–139.

Costa, E. M., Samuel-Rosa, A., & Anjos, L. H. C. D. (2018). Digital elevation model quality on digital soil mapping prediction accuracy. *Ciência e Agrotecnologia*, *42*, 608–622.

Cressie, N. (1985). Fitting variogram models by weighted least squares. *Journal of the International Association for Mathematical Geology*, *17*(5), 563–586.

Cressie, N. A. C. (1991). *Statistics for spatial data*. John Wiley and Sons, Inc.

Danilov, A., Pivovarova, I., & Krotova, S. (2018). Geostatistical analysis methods for estimation of environmental data homogeneity. *The Scientific World Journal*, 2018.

Delbari, M. (2007). *Estimation and Stochastic Simulation of Soil Properties for Case Studies in Lower Austria and Sistan Plain, Southeast of Iran* (Ph.D. thesis).University of Natural Resources and Applied Life Sciences, Vienna.

Delbari, M., Afrasiab, P., & Loiskandl, W. (2011). Geostatistical analysis of soil texture fractions on the field scale. *Soil and Water Research*, *6*(4), 173–189.

Deng, X., Chen, X., Ma, W., Ren, Z., Zhang, M., Grieneisen, M. L., & Lv, X. (2018). Baseline map of organic carbon stock in farmland topsoil in East China. *Agriculture, Ecosystems & Environment*, *254*, 213–223.

Dharumarajan, S., Lalitha, M., Niranjana, K. V., & Hegde, R. (2022). Evaluation of digital soil mapping approach for predicting soil fertility parameters—A case study from Karnataka Plateau, India. *Arabian Journal of Geosciences*, *15*(5), 1–21.

Dokuchaev, V.V. (1967). *Russian chernozem* (Vol. 1). Israel Program for Scientific Translations.

Dokuchaev, V.V. (1883). *Russian Chernozems (Russkii Chernozems)*. Israel Program for Scientific Translations, Jerusalem, transl. from Russian by N. Kaner (1967), available from U.S. Dept. of Commerce, Springfield, VA.

Drake, J. M., Randin, C., & Guisan, A. (2006). Modeling ecological niches with support vector machines. *Journal of Applied Ecology*, *43*(3), 424–432.

Eldeiry, A. A., & Garcia, L. A. (2010). Comparison of ordinary kriging, regression kriging, and cokriging techniques to estimate soil salinity using LANDSAT images. *Journal of Irrigation and Drainage Engineering*, *136*(6), 355–364.

Emadi, M., Taghizadeh-Mehrjardi, R., Cherati, A., Danesh, M., Mosavi, A., & Scholten, T. (2020). Predicting and mapping of soil organic carbon using machine learning algorithms in Northern Iran. *Remote Sensing*, *12*(14), 2234.

Fagroud, M., & Van Meirvenne, M. (2002). Accounting for soil spatial autocorrelation in the design of experimental trials. *Soil Science Society of America Journal*, *66*(4), 1134–1142.

Food and Agricultural Organisation (FAO). (1974). *Soil map of the world* (Vol. I). Paris: Legend UNESCO.

Forkuor, G., Hounkpatin, O. K., Welp, G., & Thiel, M. (2017). High-resolution mapping of soil properties using remote sensing variables in south-western Burkina Faso: A comparison of machine learning and multiple linear regression models. *PLoS One*, *12*(1), e0170478.

Fotheringham, A. S., Brunsdon, C., & Charlton, M. (2003). *Geographically weighted regression: The analysis of spatially varying relationships*. John Wiley & Sons.

Foussereau, X., Hornsby, A. G., & Brown, R. B. (1993). Accounting for variability within map units when linking a pesticide fate model to soil survey. *Geoderma*, *60*(1–4), 257–276.

Fromm, H., Winter, K., Filser, J., Hantschel, R., & Beese, F. (1993). The influence of soil type and cultivation system on the spatial distributions of the soil fauna and microorganisms and their interactions. *Geoderma*, *60*(1–4), 109–118.

Gaston, L. A., Locke, M. A., Zablotowicz, R. M., & Reddy, K. N. (2001). Spatial variability of soil properties and weed populations in the Mississippi Delta. *Soil Science Society of America Journal*, *65*(2), 449–459.

Gautam, R., Panigrahi, S., Franzen, D., & Sims, A. (2011). Residual soil nitrate prediction from imagery and non-imagery information using neural network technique. *Biosystems Engineering*, *110*(1), 20–28.

Gessler, P. E., Moore, I. D., McKenzie, N. J., & Ryan, P. J. (1995). Soil-landscape modelling and spatial prediction of soil attributes. *International Journal of Geographical Information Systems*, *9*(4), 421–432.

Glinka, K. D., & Marbut, C. F. (1927). *Great soil groups of the world and their development*. Edwards Bros.

Goderya, F. S. (1998). Field scale variations in soil properties for spatially variable control: A review. *Journal of Soil Contamination*, *7*(2), 243–264.

Goldman, M. A., Needelman, B. A., Rabenhorst, M. C., Lang, M. W., McCarty, G. W., & King, P. (2020). Digital soil mapping in a low-relief landscape to support wetland restoration decisions. *Geoderma*, *373*, 114420.

Goovaerts, P. (1998). Geostatistical tools for characterizing the spatial variability of microbiological and Physico-chemical soil properties. *Biology and Fertility of Soils*, *27*(4), 315–334.

Goovaerts, P. (2001). Geostatistical modelling of uncertainty in soil science. *Geoderma*, *103*(1–2), 3–26.

Goydaragh, M. G., Taghizadeh-Mehrjardi, R., Golchin, A., Jafarzadeh, A. A., & Lado, M. (2021). Predicting weathering indices in soils using FTIR spectra and random forest models. *Catena*, *204*, 105437.

Grunwald, S. (2006). What do we really know about the space–time continuum of soil landscapes? In S. Grunwald (Ed.), *Environmental soil-landscape modeling: Geographic information technologies and pedometrics* (pp. 3–36). New York, NY: CRC Press.

Grunwald, S. (2009). Multi-criteria characterization of recent digital soil mapping and modeling approaches. *Geoderma*, *152*(3–4), 195–207.

Grunwald, S., Thompson, J. A., & Boettinger, J. L. (2011). Digital soil mapping and modeling at continental scales: Finding solutions for global issues. *Soil Science Society of America Journal*, *75*(4), 1201–1213.

Gunn, S. R. (1998). Support vector machines for classification and regression. *ISIS Technical Report*, *14*(1), 5–16.

Guo, L., Zhao, C., Zhang, H., Chen, Y., Linderman, M., Zhang, Q., & Liu, Y. (2017). Comparisons of spatial and non-spatial models for predicting soil carbon content based on visible and near-infrared spectral technology. *Geoderma*, *285*, 280–292.

Hajrasuliha, S., Baniabbassi, N., Metthey, J., & Nielsen, D. R. (1980). Spatial variability of soil sampling for salinity studies in southwest Iran. *Irrigation Science*, *1*(4), 197–208.

Harris, P., Fotheringham, A. S., Crespo, R., & Charlton, M. (2010). The use of geographically weighted regression for spatial prediction: An evaluation of models using simulated data sets. *Mathematical Geosciences*, *42*(6), 657–680.

Hartemink, A. E., Hempel, J., Lagacherie, P., McBratney, A., McKenzie, N., MacMillan, R. A., … Zhang, G. L. (2010). *GlobalSoilMap.net – a new digital soil map of the world. Digital soilmapping* (pp. 423–428). Dordrecht: Springer.

Hastie, T., Tibshirani, R., Friedman, J. H., & Friedman, J. H. (2009). *The elements of statistical learning: Data mining, inference, and prediction* (Vol. 2, pp. 1–758). New York: Springer.

Haykin, S. (1998). *Neural networks: A comprehensive foundation subsequent edition*. Upper Saddle River, NJ, United States: Prentice Hall PTR.

Henderson, B. L., Bui, E. N., Moran, C. J., & Simon, D. A. P. (2005). Australia-wide predictions of soil properties using decision trees. *Geoderma*, *124*(3–4), 383–398.

Hengl, T. (2007). *A practical guide to geostatistical mapping of environmental variables. EUR 22904 EN* (JRC38153). Luxembourg: Office for Official Publications of the European Communities.

Hengl, T., Heuvelink, G. B., & Stein, A. (2004). A generic framework for spatial prediction of soil variables based on regression-kriging. *Geoderma*, *120*(1–2), 75–93.

Hengl, T., Rossiter, D. G., & Stein, A. (2003). Soil sampling strategies for spatial prediction by correlation with auxiliary maps. *Soil Research*, *41*(8), 1403–1422.

Heuvelink, G. B., & Pebesma, E. J. (1999). Spatial aggregation and soil process modelling. *Geoderma*, *89*(1–2), 47–65.

Heuvelink, G. B. M., & Webster, R. (2001). Modelling soil variation: Past, present, and future. *Geoderma*, *100*(3–4), 269–301.

Huang, Z., Wang, H., & Zhang, R. (2012). An improved kriging interpolation technique based on SVM and its recovery experiment in oceanic missing data. *American Journal of Computational Mathematics*, *2*, 56–60.

Isaaks, E. H., & Srivastava, R. M. (1989). *An introduction to applied geostatistics* (p. 561) New York: Oxford University Press.

Jenny, H. (1941). *Factors of soil formation*. New York: McGraw Hill.

John, K., Afu, S. M., Isong, I. A., Aki, E. E., Kebonye, N. M., Ayito, E. O., … Penížek, V. (2021). Mapping soil properties with soil-environmental covariates using geostatistics and multivariate statistics. *International Journal of Environmental Science and Technology*, *18*(11), 3327–3342.

Journel, A. G., & Huijbregts, C. H. (1978). *Mining geostatistics*. New York: Academic Press.

Kalambukattu, J. G., Kumar, S., & Arya Raj, R. (2018). Digital soil mapping in a Himalayan watershed using remote sensing and terrain parameters employing artificial neural network model. *Environmental Earth Sciences*, *77*(5), 1–14.

Kempen, B., Heuvelink, G. B. M., Brus, D. J., & Stoorvogel, J. J. (2010). Pedometric mapping of soil organic matter using a soil map with quantified uncertainty. *European Journal of Soil Science*, *61*(3), 333–347.

Keskin, H., & Grunwald, S. (2018). Regression kriging as a workhorse in the digital soil mapper's toolbox. *Geoderma*, *326*, 22–41.

Khalil, A., Gill, M.K., & McKee, M. (2005, July). New applications for information fusion and soil moisture forecasting. In: *2005 7th international conference on information fusion* (Vol. 2, pp. 7). IEEE.

Khellouk, R., Barakat, A., Jazouli, A. E., Boudhar, A., Lionboui, H., Rais, J., & Benabdelouahab, T. (2021). An integrated methodology for surface soil moisture estimating using remote sensing data approach. *Geocarto International*, *36*(13), 1443–1458.

Kingsley, J., Lawani, S. O., Esther, A. O., Ndiye, K. M., Sunday, O. J., & Penížek, V. (2019). Predictive mapping of soil properties for precision agriculture using geographic information system (GIS) based geostatistics models. *Modern Applied Science*, *13*(10).

Krige, D. G. (1951). A statistical approach to some basic mine valuation problems on the Witwatersrand. *Journal of the Southern African Institute of Mining and Metallurgy*, *52* (6), 119–139.

Krishna, G., Sahoo, R. N., Singh, P., Bajpai, V., Patra, H., Kumar, S., ... Sahoo, P. M. (2019). Comparison of various modeling approaches for water-deficit stress monitoring in rice crop through hyperspectral remote sensing. *Agricultural Water Management*, *213*, 231–244.

Kumar, S., Lal, R., & Liu, D. (2012). A geographically weighted regression kriging approach for mapping soil organic carbon stock. *Geoderma*, *189*, 627–634.

Lagacherie, P., & Holmes, S. (1997). Addressing geographical data errors in a classification tree for soil unit prediction. *International Journal of Geographical Information Science*, *11*(2), 183–198.

Lagacherie, P., Legros, J. P., & Burfough, P. A. (1995). A soil survey procedure using the knowledge of soil patterns established on a previously mapped reference area. *Geoderma*, *65*(3–4), 283–301.

Lagacherie, P., & McBratney, A. B. (2006). Spatial soil information systems and spatial soil inference systems: Perspectives for digital soil mapping. *Developments in Soil Science*, *31*, 3–22.

Lambert, D. M., Lowenberg-DeBoer, J., & Bongiovanni, R. (2003). *Spatial regression models for yield monitor data: A case study from Argentina. 2003 Annual meeting, July 27–30, Montreal, Canada*. American Agricultural Economics Association (New Name 2008: Agricultural and Applied Economics Association).

Lamichhane, S., Kumar, L., & Wilson, B. (2019). Digital soil mapping algorithms and covariates for soil organic carbon mapping and their implications: A review. *Geoderma*, *352*, 395–413.

Lark, R. M. (2009). Kriging a soil variable with a simple nonstationary variance model. *Journal of Agricultural, Biological, and Environmental Statistics*, *14*(3), 301–321.

Legros, J. P. (2006). *Mapping of the soil*. Science Publishers.

Liess, M., Schmidt, J., & Glaser, B. (2016). Improving the spatial prediction of soil organic carbon stocks in a complex tropical mountain landscape by methodological specifications in machine learning approaches. *PLoS One*, *11*(4), e0153673.

Lin, H., Wheeler, D., Bell, J., & Wilding, L. (2005). Assessment of soil spatial variability at multiple scales. *Ecological Modelling*, *182*(3–4), 271–290.

Liu, L., Wang, H., Dai, W., Lei, X., Yang, X., & Li, X. (2014). Spatial variability of soil organic carbon in the forestlands of northeast China. *Journal of Forestry Research*, *25*(4), 867–876.

Liu, F., Zhang, G. L., Song, X., Li, D., Zhao, Y., Yang, J., & Yang, F. (2020). High-resolution and three-dimensional mapping of soil texture of China. *Geoderma*, *361*, 114061.

López-Castañeda, A., Zavala-Cruz, J., Palma-López, D. J., Rincón-Ramírez, J. A., & Bautista, F. (2022). Digital mapping of soil profile properties for precision agriculture in developing countries. *Agronomy*, *12*(2), 353.

Malik, A. A., Martiny, J. B., Brodie, E. L., Martiny, A. C., Treseder, K. K., & Allison, S. D. (2020). Defining trait-based microbial strategies with consequences for soil carbon cycling under climate change. *The ISME Journal*, *14*(1), 1–9.

Malone, B. P., McBratney, A. B., Minasny, B., & Laslett, G. M. (2009). Mapping continuous depth functions of soil carbon storage and available water capacity. *Geoderma*, *154* (1–2), 138–152.

Matheron, G. (1963). Principles of geostatistics. *Economic geology*, *58*(8), 1246–1266.

Matheron, G. (1965). *Les variables régionaliséesetleur estimation: une application de la théorie de fonctionsaléatoires aux sciences de la nature* (Vol. 4597). Masson et CIE.

Matheron, G. (1971). *The theory of regionalised variables and its applications* (5, p. 212). Les Cahiers du Centre de Morphologie Mathématique.

McBratney, A. B., Bishop, T. F. A., & Teliatnikov, I. S. (2000). Two soil profile reconstruction techniques. *Geoderma*, *97*(3–4), 209–221.

McBratney, A. B., & Odeh, I. O. (1997). Application of fuzzy sets in soil science: Fuzzy logic, fuzzy measurements, and fuzzy decisions. *Geoderma*, *77*(2–4), 85–113.

McBratney, A. B., Odeh, I. O., Bishop, T. F., Dunbar, M. S., & Shatar, T. M. (2000). An overview of pedometric techniques for use in soil survey. *Geoderma*, *97*(3–4), 293–327.

McBratney, A. B., Santos, M. M., & Minasny, B. (2003). On digital soil mapping. *Geoderma*, *117*(1–2), 3–52.

McBratney, A. B., & Webster, R. (1986). Choosing functions for semi-variograms of soil properties and fitting them to sampling estimates. *Journal of Soil Science*, *37* (4), 617–639.

McKenzie, N. J., Gessler, P. E., Ryan, P. J., & O'connell, D. A. (2000). The role of terrain analysis in soil mapping. *Terrain Analysis: Principles and Applications*, *10*.

Meshalkina, Y. L. (2007). A brief review of geostatistical methods applied in modern soil science. *Moscow University Soil Science Bulletin*, *62*(2), 93–95.

Minasny, B., Hopmans, J. W., Harter, T., Eching, S. O., Tuli, A., & Denton, M. A. (2004). Neural networks prediction of soil hydraulic functions for alluvial soils using multistep outflow data. *Soil Science Society of America Journal*, *68*(2), 417–429.

Mishra, U., Lal, R., Slater, B., Calhoun, F., Liu, D., & Van Meirvenne, M. (2009). Predicting soil organic carbon stock using profile depth distribution functions and ordinary kriging. *Soil Science Society of America Journal*, *73*(2), 614–621.

Mondal, A., Khare, D., Kundu, S., Mondal, S., Mukherjee, S., & Mukhopadhyay, A. (2017). Spatial soil organic carbon (SOC) prediction by regression kriging using remote sensing data. *The Egyptian Journal of Remote Sensing and Space Science*, *20*(1), 61–70.

Nguyen, P. M., Haghverdi, A., De Pue, J., Botula, Y. D., Le, K. V., Waegeman, W., & Cornelis, W. M. (2017). Comparison of statistical regression and data-mining techniques in estimating soil water retention of tropical delta soils. *Biosystems Engineering*, *153*, 12–27.

Nielsen, D. R., Biggar, J. W., & Erh, K. T. (1973). Spatial variability of field-measured soil-water properties. *Hilgardia*, 214–259.

Nielsen, D.R., & Wendroth, O. (2003). *Spatial and temporal statistics: Sampling field soils and their vegetation*. Catena Verlag.

Odeh, I. O., Chittleborough, D. J., & McBratney, A. B. (1992). Fuzzy-c-means and kriging for mapping soil as a continuous system. *Soil Science Society of America Journal*, *56* (6), 1848–1854.

Odeh, I. O., McBratney, A. B., & Chittleborough, D. J. (1995). Further results on prediction of soil properties from terrain attributes: Heterotopic cokriging and regression-kriging. *Geoderma*, *67*(3–4), 215–226.

Olea, R. A. (2012). *Geostatistics for engineers and earth scientists.* Springer Science & Business Media.

Oliver, M. A. (1987). Geostatistics and its application to soil science. *Soil Use and Management*, *3*(1), 8–20.

Pahlavan-Rad, M. R., & Akbarimoghaddam, A. (2018). Spatial variability of soil texture fractions and pH in a flood plain (case study from eastern Iran). *Catena*, *160*, 275–281.

Pająk, M., Halecki, W., & Gąsiorek, M. (2017). Accumulative response of Scots pine (Pinussylvestris L.) and silver birch (*Betula pendula* Roth) to heavy metals enhanced by Pb-Zn ore mining and processing plants: Explicitly spatial considerations of ordinary kriging based on a GIS approach. *Chemosphere*, *168*, 851–859.

Papritz, A., & Dubois, J. R. (1999). *Mapping heavy metals in soil by (non-) linear kriging: An empirical validation. geoENV II—geostatistics for environmental applications* (pp. 429–440). Dordrecht: Springer.

Poppiel, R. R., Lacerda, M. P., Safanelli, J. L., Rizzo, R., Oliveira., Novais, J. J., & Demattê, J. A. (2019). Mapping at 30 m resolution of soil attributes at multiple depths in midwest Brazil. *Remote Sensing*, *11*(24), 2905.

Prakash, M. R., & Singh, V. S. (2000). Network design for groundwater monitoring—a case study. *Environmental Geology*, *39*(6), 628–632.

Sahu, B., & Ghosh, A. K. (2021). Deterministic and geostatistical models for predicting soil organic carbon in a 60 ha farm on Inceptisol in Varanasi, India. *Geoderma Regional*, *26*, e00413.

Santra, P., Kumar, M., & Panwar, N. (2017). Digital soil mapping of sand content in arid western India through geostatistical approaches. *Geoderma Regional*, *9*, 56–72.

Sarmadian, F., Keshavarzi, A., Rooien, A., Iqbal, M., Zahedi, G., & Javadikia, H. (2014). Digital mapping of soil phosphorus using multivariate geostatistics and topographic information. *Australian Journal of Crop Science*, *8*(8), 1216–1223.

Sawhney, B. L. (1977). Interstratification in layer silicates. In JB Dixon, & SB Weed (Eds.), *Minerals in soil environments* (pp. 405–434). Madison, Wisconsin: Soil Science Society of America.

Scull, P., Franklin, J., Chadwick, O. A., & McArthur, D. (2003). Predictive soil mapping: A review. *Progress in Physical Geography*, *27*(2), 171–197.

Seyedmohammadi, J., & Matinfar, H. R. (2018). Statistical and geostatistical techniques for geospatial modeling of soil cation exchange capacity. *Communications in Soil Science and Plant Analysis*, *49*(18), 2301–2314.

Shit, P. K., Bhunia, G. S., & Maiti, R. (2016). Spatial analysis of soil properties using GIS based geostatistics models. *Modeling Earth Systems and Environment*, *2*(2), 1–6.

Taghizadeh-Mehrjardi, R., Minasny, B., Toomanian, N., Zeraatpisheh, M., Amirian-Chakan, A., & Triantafilis, J. (2019). Digital mapping of soil classes using ensemble of models in Isfahan region, Iran. *Soil Systems*, *3*(2), 37.

Tajik, S., Ayoubi, S., & Zeraatpisheh, M. (2020). Digital mapping of soil organic carbon using ensemble learning model in Mollisols of Hyrcanian forests, northern Iran. *Geoderma Regional*, *20*, e00256.

Tomlinson, R. F. (1978). Design considerations for digital soil map systems. In: *11th congress of soil science*. ISSS, Edmonton, Canada.

Trangmar, B. B., Yost, R. S., & Uehara, G. (1986). Application of geostatistics to spatial studies of soil properties. *Advances in Agronomy*, *38*, 45–94.

Tripathi, A., & Tiwari, R. K. (2020). Synergetic utilization of sentinel-1 SAR and sentinel-2 optical remote sensing data for surface soil moisture estimation for Rupnagar, Punjab, India. *Geocarto International*, *37*(8), 2215–2236.

Vieira, S.R. (2000). Geostatistics in studies of soil spatial variability. Tópicosemciência do solo. Viçosa: SociedadeBrasileira de Ciência do Solo, 1, 1–54.

Vieira, S. R., Millete, J., Topp, G. C., & Reynolds, W. D. (2002). Handbook for geostatistical analysis of variability in soil and climate data. *Tópicosemciência do solo*, *2*, 1–45.

Vieira, V. A. D. S., Mello, C. R. D., & Lima, J. M. D. (2007). Spatial variability of soil physical attributes in a small watershed. *Ciência e Agrotecnologia*, *31*, 1477–1485.

Voltz, M., Lagacherie, P., & Louchart, X. (1997). Predicting soil properties over a region using sample information from a mapped reference area. *European Journal of Soil Science*, *48*(1), 19–30.

Wadoux, A. M. C., Brus, D. J., & Heuvelink, G. B. (2018). Accounting for non-stationary variance in geostatistical mapping of soil properties. *Geoderma*, *324*, 138–147.

Webster, R. (1994). The development of pedometrics. *Geoderma*, *62*(1–3), 1–15.

Webster, R. (2000). Is soil variation random? *Geoderma*, *97*(3–4), 149–163.

Webster, R., & Oliver, M. A. (2001). *Geostatistics for experimental scientists*. Chichester, UK: John Wiley and Sons.

Webster, R., & Oliver, M. A. (2007). *Geostatistics for environmental scientists*. John Wiley & Sons.

Wigginton, J. D., Lockaby, B. G., & Trettin, C. C. (2000). Soil organic matter formation and sequestration across a forested floodplain chronosequence. *Ecological Engineering*, *15*, S141–S155.

Wilding, L. P., Bouma, J., & Goss, D. W. (1994). Impact of spatial variability on interpretive modeling. *Quantitative Modeling of Soil Forming Processes*, *39*, 61–75.

Wosten, J. H. M. (1997). Pedotransfer functions to evaluate soil quality. In *Developments in soil science*. (Vol. 25, pp. 221–245). Elsevier.

Wosten, J. H. M., Pachepsky, Y. A., & Rawls, W. J. (2001). Pedotransfer functions: bridging the gap between available basic soil data and missing soil hydraulic characteristics. *Journal of Hydrology*, *251*(3–4), 123–150.

Wright, R. L., & Wilson, S. R. (1979). On the analysis of soil variability, with an example from Spain. *Geoderma*, *22*(4), 297–313.

Yaalon, D. H. (1989). *The earliest soil maps and their logic*. Bulletin of the International Society of Soil Science.—International Society of Soil Science, Wageningen, pp. 24.

Yao, R., Yang, J., Wu, D., Xie, W., Gao, P., & Jin, W. (2016). Digital mapping of soil salinity and crop yield across a coastal agricultural landscape using repeated electromagnetic induction (EMI) surveys. *PLoS One*, *11*(5), e0153377.

Yigini, Y., Olmedo, G. F., Reiter, S., Baritz, R., Viatkin, K., & Vargas, R. (2018). *Soil organic carbon mapping: Cookbook* (p. 220) Rome: FAO.

Zeraatpisheh, M., Ayoubi, S., Jafari, A., & Finke, P. (2017). Comparing the efficiency of digital and conventional soil mapping to predict soil types in a semi-arid region in Iran. *Geomorphology*, *285*, 186–204.

Zheng, Z., Zhang, F., Ma, F., Chai, X., Zhu, Z., Shi, J., & Zhang, S. (2009). Spatiotemporal changes in soil salinity in a drip-irrigated field. *Geoderma*, *149*(3–4), 243–248.

Zhou, Y., Hartemink, A. E., Shi, Z., Liang, Z., & Lu, Y. (2019). Land use and climate change effects on soil organic carbon in North and Northeast China. *Science of the Total Environment*, *647*, 1230–1238.

Zhu, A. X. (2000). Mapping soil-landscape as spatial continua: The neural network approach. *Water Resources Research*, *36*(3), 663–677.

Zhu, A. X., Band, L. E., Dutton, B., & Nimlos, T. J. (1996). Automated soil inference under fuzzy logic. *Ecological Modelling*, *90*(2), 123–145.

Zhu, X., Liu, D., & Chen, J. (2012). A new geostatistical approach for filling gaps in Landsat ETM + SLC-off images. *Remote Sensing of Environment*, *124*, 49–60.

CHAPTER 20

A novel approach for training nonparametric statistical models to retrieve rapeseed fresh above-ground biomass using in situ and Sentinel-2 data

Dessislava Ganeva

Space Research and Technology Institute of the Bulgarian Academy of Sciences, Sofia, Bulgaria

20.1 Introduction

Nonparametric regression models for biophysical variables retrieval from remote sensing (RS) data are trained either with in situ or simulated data (Verrelst et al., 2019). When trained with in situ data, they are usually synchronized with the acquisition of the RS data. Because it is expensive and time-consuming to acquire ground data, the in situ measurements are often scares (Camps-Valls et al., 2006), and it is always a concern how much data are enough for training robust models. Design and training of regression models often result in a complex, time-consuming task with many parameters to be tuned. Common approach in improving performance algorithm is to diagnostic the bias and variance (Ng, 2011). If the model has a high bias (Geurts, 2010), it does not define well neither the training data nor the test data (underfit). When the model has a high variance, it fits very close to the training data, but does not define the test data (overfit). When the model has a high variance, additional training data can be collected and/or the number of features can be reduced.

New satellite missions have high temporal resolution. For example, when combining Sentinel-2A and Sentinel-2B, their revisit time is of 5 days or less when the studied area is in two adjacent Sentinel-2 swaths. Under certain conditions between revisit acquisitions, the variables to be retrieved do not change. In this context, the study evaluates the possibility to create additional training data by using the same in situ measurements with corresponding RS data from different dates close to the ground measurements.

Remote Sensing in Precision Agriculture. DOI: https://doi.org/10.1016/B978-0-323-91068-2.00025-4

The objective of this research is to compare the performance of several nonparametric regression methods for fresh above-ground biomass (AGBf) retrieval from Sentinel-2 when in situ data are scarce. We tested one linear method, regularized least squares linear regression (RLR), and five machine leaning methods, support vector regression (SVR), random forests regression (RFR), gaussian processes regression (GPR), kernel ridge regression (KRR), and neural networks (NN). We compared two cases when the model is trained with (1) one RS data coincident to one ground data, Scenario1 (S1) and (2) several RS data corresponding to one ground data, Scenario2 (S2).

20.2 Dataset

20.2.1 In situ data collection

We measured AGBf from winter rapeseed crop (*Brassica napus*) on four test fields in North-East Bulgaria (43.4928, 27.7255) for two growing seasons (2017/2018, 2018/2019). The measurements were from elementary sampling units (ESUs) of 1 × 1 m in size. All the AGBf within each ESU was cut and measured on the field. The measurements of AGBf were tested for outliers with Grubbs' test (Grubbs & Beck, 1972). A total of 46 ESU of vegetation measurements were included in the study with mean, min, and max values respectively of 1197, 146, and 2761 g/m^2.

20.2.2 Sentinel-2 data

Clear-sky Top-Of-Canopy Sentinel-2A/MSI reflectance data (L2A product) covering the study area were downloaded from European Space Agency (2022), combining Sentinel-2A and Sentinel-2B (tile 35TNJ, Relative Orbit R007 & R050). All spectral bands at 10 and 20 m spatial resolution were resampled to 10 m before training the models for AGBf retrieval (Table 20.1).

Table 20.1 In situ measurements and corresponding satellite images for fresh above-ground biomass retrieval.

In situ measurement dates	Number of ESUs	Satellite images dates
23.11.2017, 24.11.2017	13	12.11.2017, 14.11.2017, 29.11.2017*, 9.12.2017
01.4.2018	15	03.04.2018*
10.04.2019, 11.04.2019	18	06.04.2019*

*The dates marked with * are for S1. All the dates are for S2.*

Table 20.2 Visual monitoring of oilseed crop development before winter 2017.

November 11, 2017	November 23, 2017	December 13, 2017
	ESU P1S2	

	ESU P1S3	

	ESU P2S2	

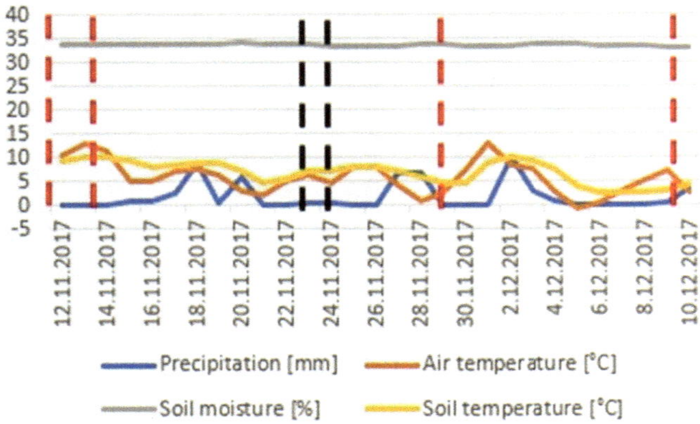

FIGURE 20.1

Meteorological conditions before winter 2017. The vertical dashed lines in red are the RS dates, and the black are the ground measurements dates.

20.3 Methodology

20.3.1 Verification of assumptions

A condition of quasi unchanged AGBf between November 11 and December 13, 2017 is a requirement to test S2. To verify this condition, we (1) visually monitored the oilseed crop development, Table 20.2, and (2) monitored the meteorological conditions on the test fields, Fig. 20.1. During this period, the rapeseed crop was in BBCH14-BBCH17 development stage with mean height of 11–13 cm depending on the ESU.

20.3.2 Nonparametric statistical models' development for fresh above-ground biomass retrieval and comparison

One linear (regularized least RLR), three kernel-based (GPR, KRR, and SVR) one decision tree (RFR), and one deep learning (NN) methods were trained for AGBf retrieval by the tool SimpleR (Camps-Valls et al., 2012; Lázaro-Gredilla et al., 2013). The datasets are composed of in situ measurements of vegetation and bare soil samples. The dataset for S1 contains 52 samples and for S2 100 samples. Each dataset is divided into training, 70%, and test, 30%. Each method is trained 100 times, resulting in 100 models, for each scenario, by randomly splitting the samples into training and test. For parameters optimization, different approaches are used depending on the method. For RLR, KRR, RFR, and NN, a random split of 2/3 for training and 1/3 for validation is applied. For the rest of the methods, the whole training set is used to determine the optimized parameters. For each model, we calculate the learning curve of the root mean-square error

(RMSE) for training and testing samples over a training set of input−output pairs increasing from 10 to the maximum number of samples, by groups of 10. We consider the learning curves of RMSE to estimate if a learning algorithm is suffering from high bias or high variance. To compare the models, we calculate the following goodness-of-fit metrics of the 100 realizations: mean error (ME), RMSE, mean absolute error (MAE), and the correlation coefficient (r) between the measured variable and the output offered by the models.

We did a one-way analysis of variance (ANOVA) followed by Tukey's HSD test (Midway et al., 2020; Miller, 1981) to compare the means of the six models with both scenarios in R (v 4.1.0) (R Core Team, 2021).

20.4 Results

20.4.1 Goodness-of-fit metrics

For all tested methods, the goodness-of-fit metrics improve from S1 to S2, Table 20.3, which indicates that the additional samples in S2 have added for that models' robustness. The best performing method for AGBf retrieval from Senstine-2 images is GPR and S2.

20.4.2 ANOVA and Tukey's HSD test

ANOVA results, Table 20.4, show that mean ME, mean RMSE, and mean MAE are statistically different with increasing F value from S1 to S2.

Table 20.3 ME, RMSE, MAE, and r between the measured and the estimated fresh above-ground biomass of models/scenario in the test set.

Method	Scenario	ME	RMSE	MAE	*r*
GPR	1	−62.59	441.05	337.36	0.81
	2	−2.88	371.30	243.32	0.84
KRR	1	−32.21	475.63	373.76	0.80
	2	6.16	391.62	275.09	0.84
NN	1	−41.03	531.19	403.91	0.76
	2	−2.52	416.84	285.61	0.81
RFR	1	−33.93	491.14	373.35	0.75
	2	2.45	419.65	272.20	0.80
RLR	1	−13.60	457.80	367.51	0.81
	2	7.69	396.60	305.77	0.82
SVR	1	−36.33	541.21	428.87	0.74
	2	94.92	501.88	340.26	0.76

For S1, n = *52. For S2,* n = *100.* P < *.001.*

Table 20.4 ANOVA F value at $P<.001$.

Scenario	ME	RMSE	MAE
1	15.21	23.58	22.08
2	67.83	72.75	78.12

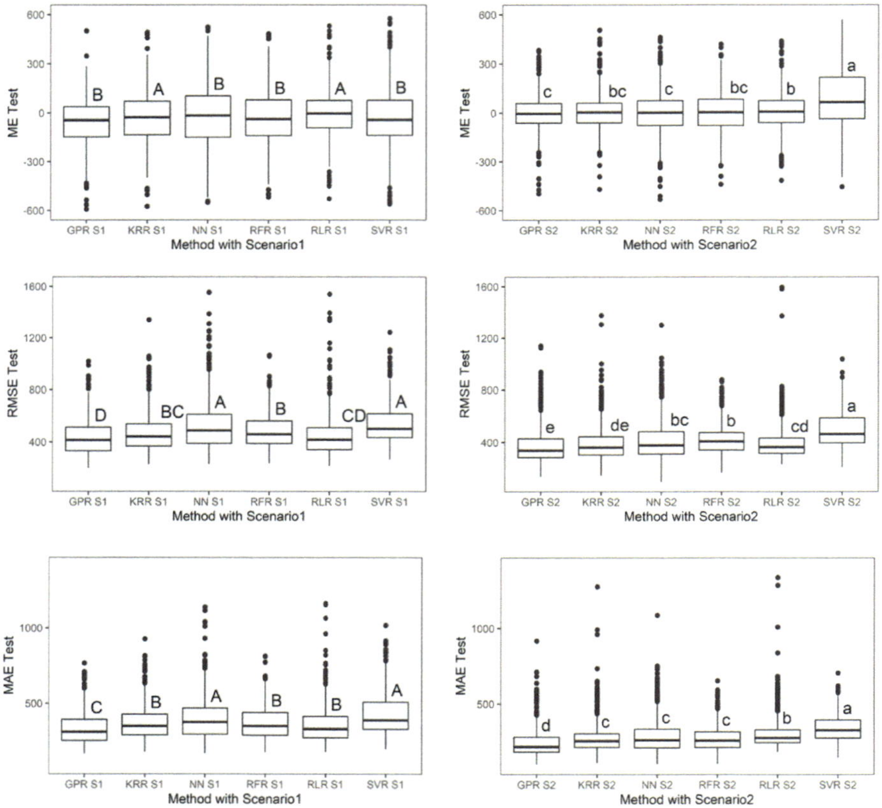

FIGURE 20.2

Mean ME, RMSE, and MAE Test for S1 (left) and S2 (right). The letters indicate the grouping of Tukey multiple comparisons of means at 95% confidence level.

Mean ME, RMSE, and MAE for all methods are less spread in S2 than in S1, Fig. 20.2. The SVR method performs the worse in both scenarios. GPR levels out with the other methods for ME between S1 and S2. However, GPR is significantly different from the other methods for RMSE and MAE for both scenarios, with decreasing values in S2 compared with S1.

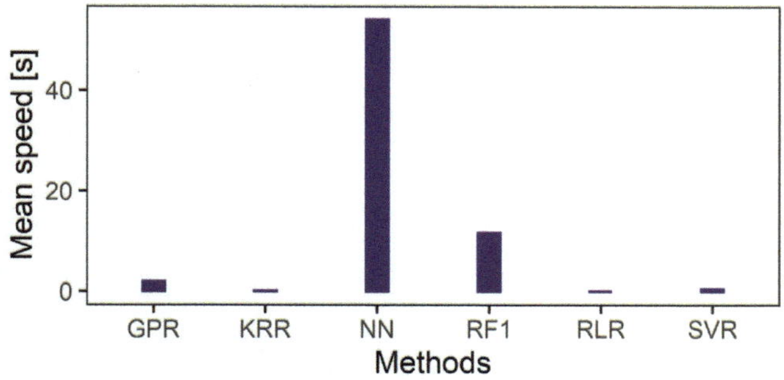

FIGURE 20.3

Mean speed in *seconds* for training the models.

20.4.3 Speed of training methods

The different tested methods have very different training speed, Fig. 20.3. The best-performing method, GPR in S2, is almost 50 times quicker than the slowest, NN.

20.4.4 Learning curve

From the learning curve results, Fig. 20.4, is also visible that SVR, and in lesser measure RLR, is underfit in both scenarios. For GPR, KRR, and in lesser measure RFR, the difference between RMSE Test and RMSE Train is maintained between S1 and S2 compared to S1, with −3.12%, 4.47%, and 7.78%, respectively. However, RMSE Test decreases between S1 and S2 and provides very similar good results for GPR and KRR in S2. Even if KRR and RFR have also similar results in S1, KRR outperforms RFR in S2.

20.5 Discussion

From the presented results, we can conclude that RLR and SVR have high bias with S1 and the additional samples with S2 did not improve their performance. One possible way to improve SVR performance is to select cost function that does not implicitly assumes a specific density model for the errors, independently of their statistical nature (Camps-Valls et al., 2006), because RS data contain uncertainty and different kinds of noise. Another possible way to improve SVR and RLR is in fine-tuning the hyper parameters of the methods.

The best performing methods are GPR and KRR, Table 20.3, in accordance to other studies (Caicedo et al., 2014; Verrelst et al., 2012) that retrieve biophysical variables from Sentinel-2. The additional samples from S1 to S2 allowed both methods to lower the test metrics: ME, RMSE, and MAE and increase the r. For GPR, it also permitted to statistically differentiate the mean test RMSE from all other methods.

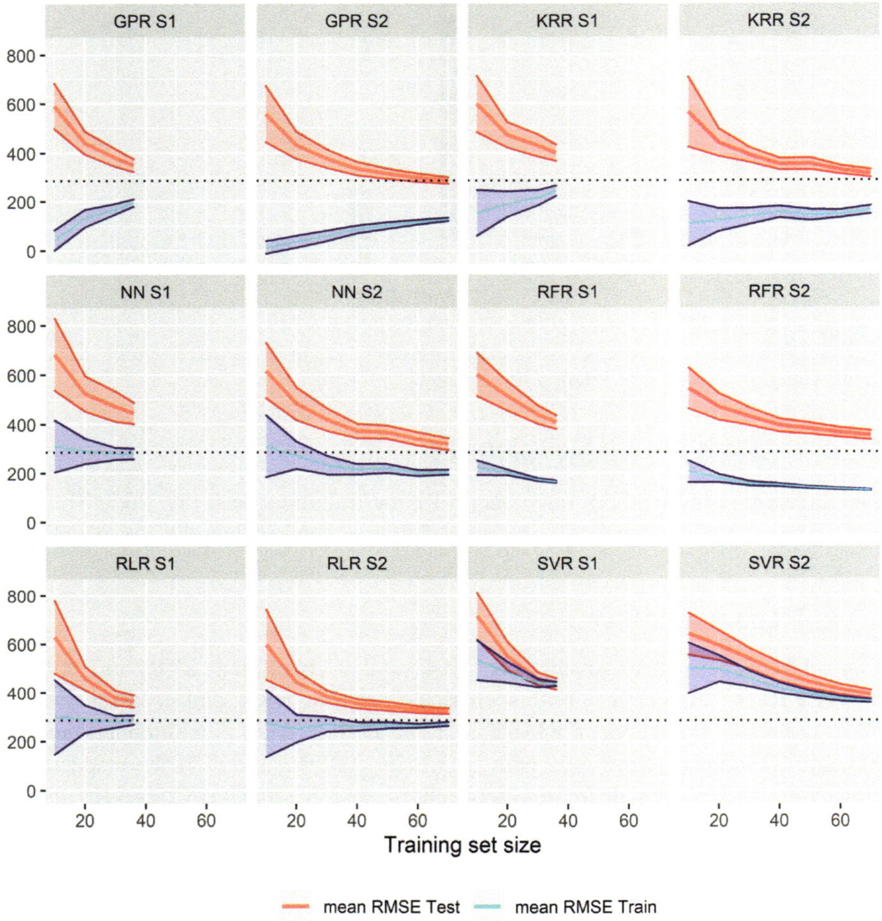

FIGURE 20.4

Learning curve for tested methods for S1 and S2 with ±95% CI. The dotted line indicates the min value of the mean RMSE Test for S2.

When studying the effect of the training set size on the model performance, (Kamir et al., 2020) found that GPR was the best-performing model and that the model performance improved when increasing the training set from 25 to 125 samples, which support our results.

In some studies (Camps-Valls et al., 2006), support vector machines are considered as promising alternative to NNs, because NN performance can be degraded when working with low-sized datasets. In our case, NN performed better than SVR. NN was the slowest method during training, in accordance to other studies (Camps-Valls et al., 2012; Verrelst et al., 2012).

20.6 Conclusion

Nonparametric regression methods construct the relationship between the output, AGBf, and the input, Sentinel-2 spectral bands, by a model derived from the information from the data. In situ AGBf measurements are costly and time-consuming. Therefore, under certain conditions, we propose as training/testing samples the pair AGBf/Sentinel-2 data from one in situ measurement and several Sentinel-2 dates. The condition is high temporal resolution for the RS data at the same time a period of nondevelopment of the plant. In this case, the best-performing model for AGBf retrieval, GPR S2, improved the measured goodness-of-fit metrics (ME, RMSE, MAE, *r*). The proposed method constitutes effective tool that could lead to improvements in other biophysical variables retrieval.

Acknowledgments

D. Ganeva was partially supported by Bulgarian Ministry of Education and Science under the National Research Programme "Young scientists and postdoctoral students" (ПМС 577/17.08.18г.). Furthermore, this research was supported by the Action CA17134 SENSECO (Optical synergies for spatiotemporal sensing of scalable ecophysiological traits) funded by COST (European Cooperation in Science and Technology, http://www.cost.eu, accessed on December 30, 2021).

References

Caicedo, J. P. R., Verrelst, J., Munoz-Mari, J., Moreno, J., & Camps-Valls, G. (2014). Toward a semiautomatic machine learning retrieval of biophysical parameters. *IEEE Journal of Selected Topics in Applied Earth Observations and Remote Sensing*, *7*(4), 1249–1259. Available from https://doi.org/10.1109/JSTARS.2014.2298752.

Camps-Valls, G., Bruzzone, L., Rojo-Rojo, J. L., & Melgani, F. (2006). Robust support vector regression for biophysical variable estimation from remotely sensed images. *IEEE Geoscience and Remote Sensing Letters*, *3*(3), 339–343. Available from https://doi.org/10.1109/LGRS.2006.871748.

Camps-Valls, G., Munoz-Mari, J., Gomez-Chova, L., Guanter, L., & Calbet, X. (2012). Nonlinear statistical retrieval of atmospheric profiles from MetOp-IASI and MTG-IRS infrared sounding data. *IEEE Transactions on Geoscience and Remote Sensing*, *50*(5), 1759–1769. Available from https://doi.org/10.1109/TGRS.2011.2168963.

European Space Agency. The Copernicus Open Access Hub. 2022. https://scihub.copernicus.eu/dhus/#/home [Accessed 20 January 2022].

Geurts, P. (2010). Bias vs variance decomposition for regression and classification. In O. Maimon, & L. Rokach (Eds.), *Data mining and knowledge discovery handbook* (pp. 733–746). US: Springer. Available from https://doi.org/10.1007/978-0-387-09823-4_37.

Grubbs, F. E., & Beck, G. (1972). Extension of sample sizes and percentage points for significance tests of outlying observations. *Technometrics*, *14*(4), 847–854.

Kamir, E., Waldner, F., & Hochman, Z. (2020). Estimating wheat yields in Australia using climate records, satellite image time series and machine learning methods. *ISPRS Journal of Photogrammetry and Remote Sensing*, *160*, 124–135. Available from https://doi.org/10.1016/j.isprsjprs.2019.11.008.

Lázaro-Gredilla, M., Titsias, M. K., Verrelst, J., & Camps-Valls, G. (2013). Retrieval of biophysical parameters with heteroscedastic Gaussian processes. *IEEE Geoscience and Remote Sensing Letters*, *11*(4), 838–842. Available from https://doi.org/10.1109/LGRS.2013.2279695.

Midway, S., Robertson, M., Flinn, S., & Kaller, M. (2020). Comparing multiple comparisons: Practical guidance for choosing the best multiple comparisons test. *PeerJ*, *8*, e10387. Available from https://doi.org/10.7717/peerj.10387.

Miller, R. G. (1981). *Simultaneous statistical inference*. New York: Springer. Available from https://doi.org/10.1007/978-1-4613-8122-8.

Ng, A. *Advice for applying machine learning*. 2011. https://see.stanford.edu/materials/aimlcs229/ML-advice.pdf. [Accessed 20 January 2022].

R Core Team. R: A Language and Environment for Statistical Computing. version. Retrieved January 20, 2022. 2021. https://www.r-project.org/.

Verrelst, J., Malenovský, Z., Van der Tol, C., Camps-Valls, G., Gastellu-Etchegorry, J.-P., Lewis, P., North, P., & Moreno, J. (2019). Quantifying vegetation biophysical variables from imaging spectroscopy data: A review on retrieval methods. *Surveys in Geophysics*, *40*(3), 589–629. Available from https://doi.org/10.1007/s10712-018-9478-y.

Verrelst, J., Muñoz, J., Alonso, L., Delegido, J., Rivera, J. P., Camps-Valls, G., & Moreno, J. (2012). Machine learning regression algorithms for biophysical parameter retrieval: Opportunities for Sentinel-2 and -3. *Remote Sensing of Environment*, *118*, 127–139. Available from https://doi.org/10.1016/j.rse.2011.11.002.

CHAPTER

Applications of drones in precision agriculture: future of smart and sustainable farming 21

Ritvik Tyagi[1] and Prem Chandra Pandey[2]

[1]School of Management, School of Management and Entrepreneurship (SME), Shiv Nadar Institution of Eminence (Deemed to be University), Greater Noida, Uttar Pradesh, India

[2]Department of Life Sciences, School of Natural Sciences, Shiv Nadar Institution of Eminence (Deemed to be University), Greater Noida, Uttar Pradesh, India

21.1 Introduction

The agricultural practices have always been a hectic work, which involves human involvement and hard work. Each year we come across a number of farmers who complain that they are unable to get a reasonable price for their crops. The drone technology can help achieve precise agriculture. Currently this technology can be termed to be in a developing phase, but once developed, it can easily provide information ranging from the water-holding capacity of soil to the presence of weed in a particular location.

The drone technology is packed with a number of advantages as it can help, such as:

- Random crop health monitoring—big farms can use the drone technology to monitor crop's health at any point in time.
- Irrigation monitoring—drones can help in monitoring the level of irrigational water required in any particular area of the farm.
- Weed identification and removal—drones can easily detect weeds in a farm and can remove them by itself.
- Fertility rate of different parts of the field—drones can monitor a particular nutrient requirement in each part of the field (Veroustraete, 2015).

It can be assessed that drones are the future of farming and have massive potential to improve agricultural practices with precision agriculture (PA) and SMART farming, contribute to Sustainable Land Management (SLM), thus avoiding land degradation. With increasing demands in food production and consumption, there is an all-time high supply and demand along with increased commodity prices, the agricultural practices are at a junction, with more yield

Remote Sensing in Precision Agriculture. DOI: https://doi.org/10.1016/B978-0-323-91068-2.00003-5

stress and less wastage of input resources. Therefore, there is an urgent need than ever before for this sector across the world to improve resource management in terms of higher yields and production with efficient inputs, and less wastage of input resources. Now consumers are looking at organic aspects for increased crop yields, such as use of organic manures during their growth stages, and the origin of those yields. Population increases exponentially and land conversion or land use changes from one to another cover are also acting as limiting factors for the more agricultural area (Pandey & Pandey, 2023).

- **Condition-1:** *Arable land remains the same with exponential increase in the population*: If we assume our farming area/land will remain the *same by 2050*, and population is growing at exponential rate. The increasing population that needs to be feed would increase by 15%, as it's expected that by 2050, the human population will reach 10 billion. This increased population would demand more food for which we need to start planning our resources in such a way that it provides more food production. In this condition, more pressure would be on the land, so farmers will use more and more water for irrigation, which will impact the soil nutrients and conditions.
- **Condition-2:** *Conversion of arable land to non-arable land with exponential increase in the population*: Urban expansion is another major problem, which will impact agriculture. The agricultural land that has been cultivated for years will be converted to other land use, maybe residential, industrial, or commercial purposes or even suddenly get converted to urban planned towns. This will not only reduce the land for farming/production but also decrease the productivity of the whole surrounding area.
- **In both conditions**, there will be high demands of foods, vegetables and cash crops in the market across the globe. Further, this will increase the price of the commodities, which will impact the lower-income group as well as third world countries, and thus hunger will cause the death of millions as predicted. As of this, the low-income group people will not be able to afford expensive foods. Thus, Sustainable Development Goals (SDGs) of the United Nations with 17 Goals, which include Zero HUNGER (SDG 01), NO POVERTY (SDG 02), Sustainable Agriculture and along with climate actions (SDG 13) will not be realized. Other SDGs are directly and indirectly linked to agricultural practices, so posing grand challenges towards their achivement (Pandey & Pandey, 2023).

In addition to population increase and land conversion factors, climate changes are playing an important role in the agriculture sectors, and it continues creating new complexity layers for agricultural sectors in getting food security (Gebbers & Adamchuk, 2010) to all by 2030 as per IPCC report. SDGs as announced by United Nations have 17 goals in their basket, including (1st) No poverty, (2nd) Zero hunger, (3rd) Good health and well-being, (12th) Responsible production and consumption, and (13th) Climate actions that are directly or indirectly linked to agricultural practices, in one or the other way. To meet food security, there is need and requirement to increase crop yields with adequate nutritional amount to fulfill the needs and requirements of increasing population, only then zero hunger could be achieved with balanced nutritional diet to avoid nutrition deficiency among children. Climate actions can be achieved in agricultural practices by adopting Sustainable Land Management (SLM) to minimize or lessen land degradation and soil pollution which facilitate enhanced and improved crop yields per hectare compared to traditional practices. There are several environmental problems related to land degradation and soil contamination, a significant consequence of anthropogenic activities, outrageous climatic conditions resulting in pollution, heavy metal mixing with soil, effluent discharges, along with excessive use of pesticides and chemicals. Thus, SLM could be a way to achieve sustainable soil/land management while achieving climate actions.

Therefore, IPCC published a milestone report affirming the role of SLM in reversing the impact of climate change on land degradation and contamination. This report asserts that SLM could be cost-effective along with short-term and long-term benefits (IPCC). Given this, it is critical that farming should not only adopt advanced practices but also work to alleviate the impacts of environmental and anthropogenic activities and yet secure their economy against any financial loss. IPCC advised that SLM could be the key to mitigating climate change impact to avoid land degradation or due to natural consequences and human and agricultural activities. SLM is mainly to prevent soil and land degradation and maintain the quality of land and soil. Therefore, integrating drones in smart farming/PA could provide practical solutions in SLM, short-term and long-term duration, while assuring financial protection against loss. Remote sensing imaging technologies (including space-borne, airborne, terrestrial) help in the fine scale and temporal mapping of crop yields and crops in the agricultural field. Drone images deliver very high spatial resolution up to centimeter and are more efficient that can be deployed at the desired time for field and crop monitoring. Drone-acquired images provide more intense information regarding field and crops, with GIS simultaneously to help in irrigation, fertilizer prescriptions, and crop disease identification to facilitate more yields. The output generated from drone-acquired images allows farmers to save more on the expenses in the input resources, as it provides information by optimizing the optimal amount at the appropriate place and at the right time, reducing the significant cost to agricultural practices.

21.2 Precision farming

PA is a farming management concept based on observing, measuring, and responding to the inter- and intra-field variability in crops (Singh et al., 2020). PA aims to develop a smart system and big database to obtain the best outcome with minimum resources. PA can be considered a big wave in the modern agriculture revolution. With proper implementation of this technology, the expected food production is expected to increase by 1.5 times. The role of PA or Precision Farming practices is to avoid the wastage of input resources with increased crop yields using advanced technologies in agricultural sectors (refer to Fig. 21.1). PA is mainly a management concept based on observation, measurement, and response to inter- and intra-field variability in crops. PA helps farmers to use advanced technologies in the practices, right from sowing to harvesting, and to make better informed decisions for land management, soil management, crop managements based on the data types collected from the fields using remote sensing (RS), Global Positioning System (GPS) along with Geographical Information System (GIS) for taking decision (Maes & Steppe, 2019; Pandey et al., 2021). The use of geospatial technologies (RS, GPS, and GIS) forms an integral part of the PA practices, which allow fine spatial mapping and temporal monitoring of the different data types (soil condition, crop condition and parameter along with yield mapping) within agricultural fields. In PA, mostly drones equipped with GPS are used to collect field data (soil, extent, crops conditions, yield mapping) along with real time data transmitted through smart sensors installed in the field. There are unmanned aerial vehicles (UAVs)/unmanned aerial systems (UASs), which play a significant role in the majority of PA practices across the globe, mainly in European and American countries. PA is helping agribusiness experts lead the way toward SLM, while additionally ensuring

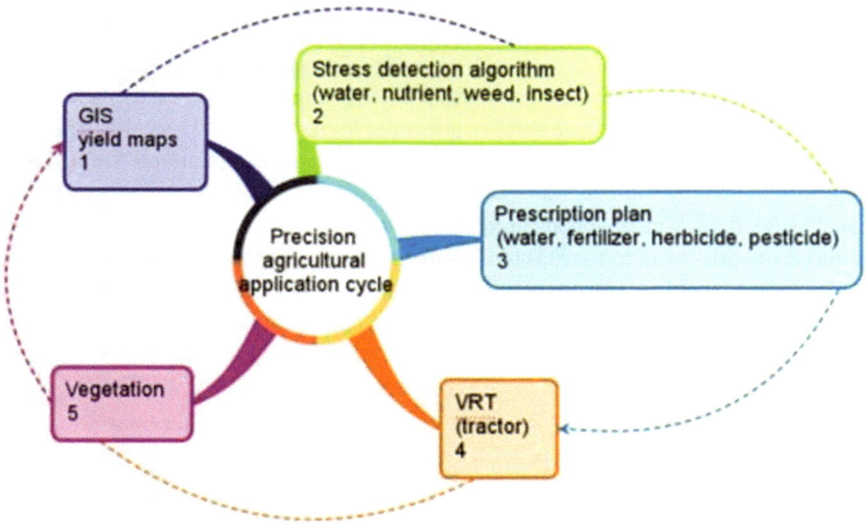

FIGURE 21.1

Cycle of precision agriculture. *GIS*, Geographic information system; *VRT*, variable-rate technology.

Adapted from Abdullahi, H. S., & Sheriff, R. E. (2017). Case study to investigate the adoption of precision agriculture in Nigeria using simple analysis to determine variability on a maize plantation. Journal of Agricultural Economics and Rural Development, 3*(3), 279–292.*

and expanding net profitability. The advanced decision-making by the farmers in their practices has resulted in significant increase in the global market estimated to reach around 50 billion by 2050, all possible by using PA in the agricultural practices. As it can be inferred from Fig. 21.1, there are five stages of PA cycle (Abdullahi and Sheriff, 2017): First stage is data collection and processing using GIS, the second stage is the assessment of stressed conditions related to soil conditions (water stress, soil nutrients) and related to crops (disease, nutrient deficiencies, insects attack, pest attacks, and so on), the third stage is the prescription plans based on the requirement of the field, the fourth stage is VRT implementation in the field, and fifth stage is the vegetation/crop assessment using advanced data collected using geospatial technologies, and again vegetation monitoring for their growth and developments and yield estimation. This cycle is repeated and helps in the continuous monitoring of the entire fields. PA practices, which can help farmers make better informed decisions, have evolved significantly over recent years, with the global market now estimated to reach $43.4 billion by 2025. Snow (2016) reported that PA and public safety will make up more than 90% of the market growth for UAVs (Black, 2016).

The main objective of PA is to provide advanced technologies, field information to farmers in advance, thus helping in intense and efficient agricultural practices as compared with traditional farming practices. The field information at the fingertips of farmers will help them to assess the soil conditions (soil moisture, soil nutrients, soil types), crop conditions (growth, crop stress of any kind such as to identify crop diseases before they become widespread in entire field) for more increased yield without crop damage due to those conditions. This information will help farmers to adjust the irrigational water to manage the water amount, adjust prescriptions of rightful amount of fertilizers, adjust and apply right amount of the pesticide/medicines to the only affected regions instead of the entire agricultural fields. Therefore, with these digital data and readily available information about the entire field condition, farmers can make prompt decisions based on economic and environmental factors such as input prescribed to the fields for irrigation, prescription of fertilizer, disease treatment by optimizing the inputs and applying them only the right amounts at the appropriate time, reducing significant input costs, and increasing profitability. The changing scenario of Indian agricultural paradigm with big data, smart farming, IoT is critically discussed in Shankarnarayan and Ramakrishna (2020).

Thus, PA applications help farmers to

1. Build a record system for the field,
2. Improve decision-making,
3. Foster greater traceability,
4. Enhance the marketing of farm products, and
5. Increase quality of farm products, with advanced technology. Therefore, high-tech PA can be highly beneficial for farmers in developing countries such as India (Rajvanshi, 2013).

21.3 Drones in precision agriculture

Drones are nowadays commonly used for different research activities as well as stylish photography, most widely used in inaccessible regions. The way it has revolutionized the research and collecting field samples at undulated terrains and hilly regions has pronounced the importance of drones. Recently, industries are integrating the smart technology, different types of sensors mounted on drones, better flight time, tracking of the flights with better control over the its path, having popularized the drones-acquired images for several research activities (Puri et al., 2017).

Drone-equipped agricultural fields can monitor thousands of acres of land in a single day and to just by keeping an eye on the screen and flying drones over the field (Daponte et al., 2019). Today's advanced drones have the ability to capture detailed crop and livestock data. These data not only help the farmers but also help other people involved in this business such as the seed companies and other farming equipment because the companies would know exactly on what parameters their product defect in a particular area. This would help in manufacturing customized farming equipment and seeds. Drones can help to monitor soil conditions and also different growth stages of the crops (crop condition, such as health) (Srivastava et al., 2019) right from vegetative stages (seeding, transplanting, tillering, stem elongation), reproductive stages (panicle initiation, booting stage, heading/flowering) and ripening stages (milk stage, dough stage, and maturity) (Thilakarathna & Raizada, 2018), and detailed information is depicted in Fig. 21.2. This is achieved through spectral signature and variation in spectral traits from hyperspectral sensors and thermal sensors. It is a different perspective to collect field observations for crop assessment and yield predictions of crops, which vary from crop to crop. Canopy hyperspectral measurements were conducted at different crop growth stages as T1-seeding stage, T2-panicle stage, T3-after booting stage, T4-milk stage, and T5-maturity (to estimate yields). Moreover, there are several limitations, it is quite difficult to predict the crop yields after the milk stage when the rice plant's leaf colour turns yellow, as flowering stage is considered the maximum/peak greenness stage of the crops, afterward which leaf color changes and phase when plants produce their flower sets.

Actually, digital data generated through remotely sensed images are used for several tasks, and it reduces the complexity of work by reducing entire field to specific regions as per needs and requirements using optimal route and optimized path (Srivastava et al., 2020). Prescribed maps are used to generate soil condition maps for nutrients, soil moisture, crop health and stress, weeds, pesticides, and accordingly VRA techniques and optimal route are implemented for site-specific applications of required input at right place and at appropriate time, such as planting, spraying, and harvesting. Because of the advantages and cost-effectiveness, the increasing use of drones in farming is growing rapidly, to effectively deal with the SLM approach. A few limitations of space-borne images could be

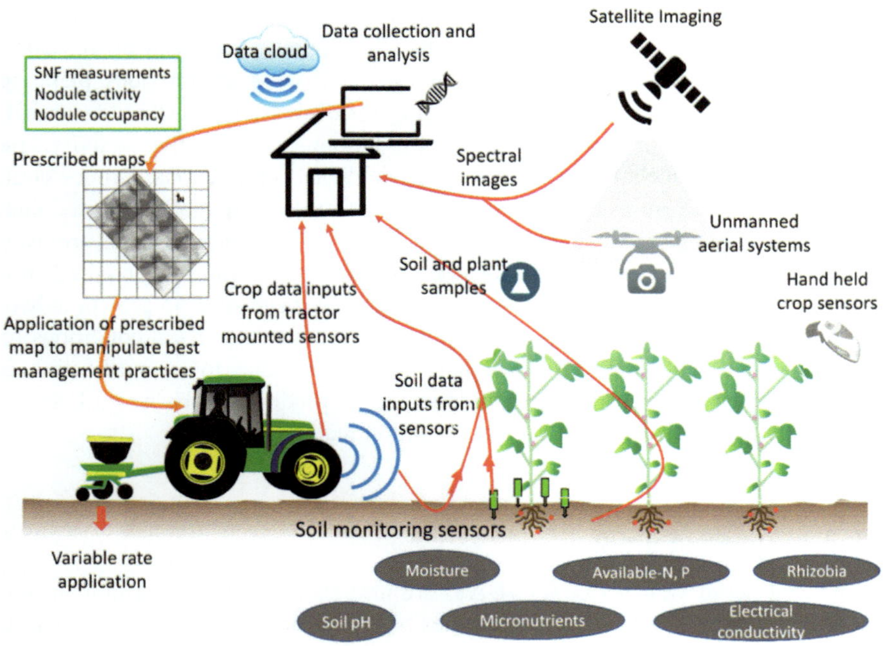

FIGURE 21.2

A conceptual model showing how precision agricultural tools can be integrated together at the field level, using on-the-go variable management, from spectral information to information transfer, and several applications in the field.

Adapted from Thilakarathna, M. S., & Raizada, M. N. (2018). Challenges in using precision agriculture to optimize symbiotic nitrogen fixation in legumes: Progress, limitations, and future improvements needed in diagnostic testing. Agronomy, 8*(5), 78.*

overcome by implementing drone technology in SLM, such as mountain-based regions where inaccessible area will be easily covered with drones as alternative techniques. It also helps in careful observation of large fields taking into account slope and elevation to generate best possible seeding prescription maps.

Drone technology provides information and helps monitor agriculture for soil conditions (soil health, soil nutrients, soil composition), crop conditions (crop health, disease, infestation, crop stress, crop yields), and weather information. Information received from drone images helps to develop and degenerate prescription maps for implementation of input resources and reduce wastage in the field and precise plan for irrigation systems for achieving improved crop yields and productivity. These are spatial maps generated using advanced remotely sensed images. At the same time, drones can efficiently perform temporal assessment compared with other remotely sensed images, which cannot provide data for each moment required during agricultural practices. With the help of drones, it is also possible to determine the values of plant factors, soil management, slope in hilly terrain, slope length, and soil erosion in land management using drone technology for soil conservation as one of the SLM goals. Rachman et al. (2019) reviewed the usefulness of drone technology in PA, which provides planning, operational support, soil and crop monitoring for evaluations. Further, it can help assess large fields for replanting decisions, thinning and pruning activity, and improving crop models for estimating crop yields and productions. Drones could prove a great platform as they can provide very fine-scale spatial data and collect temporal data at any time as needed and required to fulfil the particular provision at the right and accurate time. Using fine-resolution temporal data acquired from a drone could be beneficial in the analysis and assessment of crop growth as well as understanding proper growth of different crop types to study soil conditions for nutrient uptake by plants to avoid soil degradation due to overutilization of resources in the field. Agriculture drone market is predicted to be worth US$32.4 with irrigation mapping, crop and soil condition mapping.

21.3.1 Types of drones

Unmanned aerial vehicles (UAVs/UAS) are aircraft types without a human pilot aboard. There are different types of drones based on their usage and size (small to large) design and structure (Patel et al., 2018) and according to usage like photography, mapping, surveillance. The flight time and speed vary according to usage and payload capacity as per the user's needs and requirements (Circuit, 2021; Greenwood et al., 2019). UAVs are categorized on the above basis and the platform on which they are mounted to acquire the images. Based on the platform, they are divided into four major categories as mentioned as follows:

1. Multi-rotor drones
2. Fixed-wings drones
3. Single-rotor drones
4. Fixed-wings hybrid VTOL

Due to the aforementioned design and structure, drones' stability differs, and their capacity to take off with payload capacity also differs, which may vary from 2 to 15 kg, depending upon the user's requirements. More capacity payload take-off is required for spraying activities in agricultural practices for more efficient work on large farms (Vergouw et al., 2016). Few drones, such as single-rotor drones offer better stability and are more powerful with high-capacity payload flight time. Fixed-wing drones do not provide much stability; therefore it is not best for aerial photography as it is unable to withstand particular positions. Multi-rotor drones are widely used for mapping, surveillance, and photogrammetry, due to their stability and ability to stay in specific positions for longer duration. It may have four to six-rotors as per requirements. Drones are used to acquire different types of images with mounted sensors based on the user need and requirements, such as multispectral data, thermal data, and hyperspectral data for assessment of soil/crop conditions (Pandey et al., 2020, 2021). Detailed information about drone types is reviewed by (Velusamy et al., 2022).

The process of drone-based PA is to automate the whole working with the best solution for maximum output. Drones have features of being highly mobile; hence, it can manage the farms. Drones work on sensors and monitor with high-quality built-in cameras; there are different types of drone mapping, which are mentioned as follows:

- **Thermal sensors**—These sensors are capable of acquiring the long-wavelength radiation emitted by the earth surface and estimate the temperature information. The images acquired using Thermal Sensors working in 8–14 μm wavelength ranges are mounted on drones, are used for monitoring the temperature of the soil, the crop conditions, as well as they provide farmers information about their fields' irrigation requirements. Hence, farmers are able to schedule irrigation according to the need of their crops. This type of drone not only helps in increasing farm productivity but also helps in reducing water wastage. Even other techniques may be applied such as density slicing using color schemes to differentiate different hotspots in the agricultural fields. Moreover, color-coded mosaic images representing different range of pixel values provide information about the relative thermal radiometric values.
- **Multispectral sensors**—These sensors are capable of acquiring the optical data with few spectral bands with very high spatial resolution. These type of sensors mounted on drones are used for monitoring crop health. These sensors have few channels/bands such as Blue, Green, Red, and red edge-NIR to detect the major changes in features. For example, MICA SENSE red edge sensors are most widely used and popular in this group. They help in increasing the field productivity by gathering data and helping farmers to manage the amount of fertilizations required. These drones work on the basis of detection of wavelength because of which it can segregate objects in the captured image.
- **Hyperspectral sensors**—These types of seniors are capable of acquiring data with very high spectral resolution. These sensors are mounted on drones and commonly used in agricultural research-related work (Singh et al., 2020). These are very high-powered drones and have the ability to monitor the crops with such a high precision and accuracy that it can easily protect against yield loss by detecting and identifying diseases in field at a very initial stage. These sensors have more than 200 spectral channels ranging from 400 to 2500 nm with high spectral resolution about 5–10 nm for detailed information about the feature and can detect subtle changes in the features.
- **RGB camera**: RGB cameras are mounted to acquire the true images of the earth's surface condition, to check the positioning and validate with ground samples. These are high-resolution images up to 10 cm using RGB films. These RGB camera films are mainly used to cross-check features or identify in field for

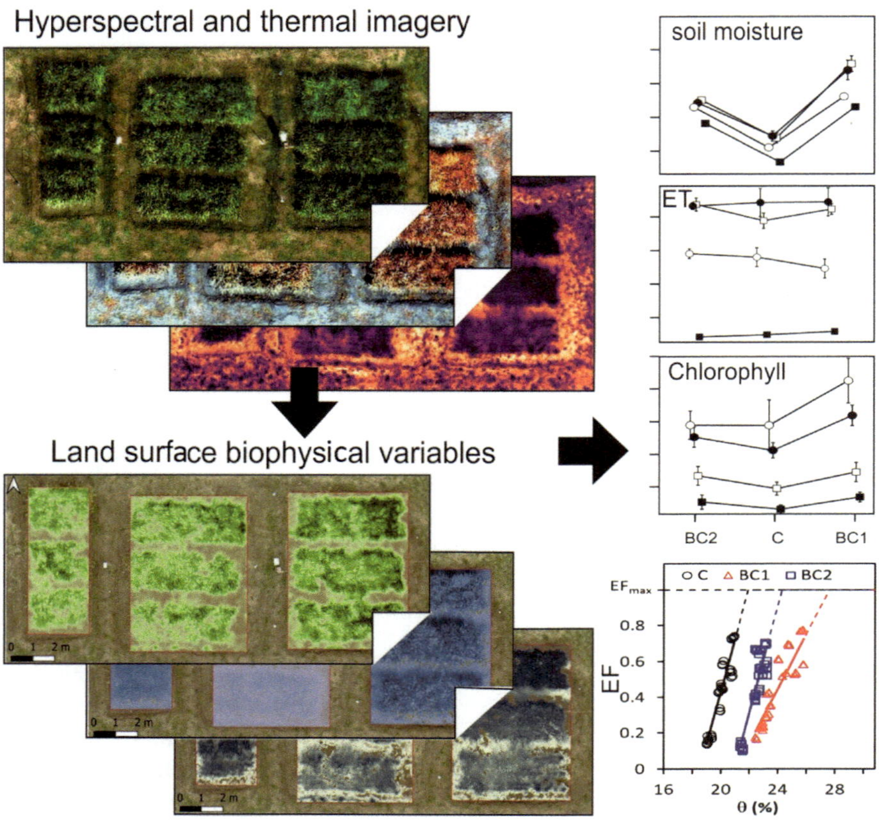

FIGURE 21.3

Hyperspectral and thermal images acquired from drone are being used to generate land surface biophysical variables, NDVI, temperature, soil water contents, soil matric potential, chlorophyll, evaporation rate, productivity.

Adapted from Jin, H., Köppl, C. J., Fischer, B., Rojas-Conejo, J., Johnson, M. S., Morillas, L., Lyon, S. W., Durán-Quesada, A. M., Suárez-Serrano, A., & Manzoni, S. (2021). Drone-based hyperspectral and thermal imagery for quantifying upland rice productivity and water use efficiency after biochar application. Remote Sensing, 13*(10), 1866.*

validations. Sometimes, drone-acquired images embedded with RGB sensors are used for extrapolating digital terrain model and digital surface models used to generate the height models related to the surveyed area.

These aforementioned sensors are capable of generating land surface biophysical variables, NDVI, temperature, soil water contents, soil matric potential, chlorophyll, evaporation rate, productivity of the crops in the field as seen in Fig. 21.3 (Jin et al., 2021), and thus very accurate outcome is expected, which is essential in PA.

21.4 Applications of sensors mounted on unmanned aerial vehicles

This section deals with the applications of the remotely sensed images acquired through UAVs/drones for the agricultural research. These include RGB images, thermal images, hyperspectral images, and LiDAR data.

21.4.1 Thermal applications

Thermal images acquired from drones are used to assess the surface temperature of the field for implications of other inputs in the field accordingly (refer to Fig. 21.4 for more information). As per thermal imagery analysis, bare soil around the field plots will have higher surface temperature than the cropped area in the field, this will provide an idea of how much area is covered with the crop and any gaps inside the whole field plots and also provides spatially continuous information concerning the water status of crops. Fig. 21.4 shows higher temperature and lower temperature according to the color of the pixels; here the yellow color indicates hot pixels and the dark color cool pixels. Even with the thermal images, we can generate CWSI (Crop Water Stress Index), which is used to detect the water stress conditions of a plant, because, at the foliar level, stomata closure

FIGURE 21.4

Land surface temperature (LST, 2.25 cm resolution) at 12:30 on November 16, 2018 estimated from the FLIR tau2 thermal camera and the Cubert-derived land surface emissivity. The yellow color indicates hot pixels and the dark color cool pixels.

Adapted from Jin, H., Köppl, C. J., Fischer, B., Rojas-Conejo, J., Johnson, M. S., Morillas, L., Lyon, S. W., Durán-Quesada, A. M., Suárez-Serrano, A., & Manzoni, S. (2021). Drone-based hyperspectral and thermal imagery for quantifying upland rice productivity and water use efficiency after biochar application. Remote Sensing, 13*(10), 1866.*

reduces transpiration and evaporative cooling, which results in an increased temperature of the crop leaf. This increased temperature can be detected by thermal imagery better than any other images and also to assess the regions having dry and wet soil, which helps in the better management of irrigation practices accordingly.

21.4.2 Hyperspectral applications

Hyperspectral images acquired from drones are used to assess the spectral differences among species, assess weeds among the crops, and generate the spectral profile of the crops (wheat, rice, barley, sorghum, and so on). The crop's spectral profile helps us understand any stress or diseased condition using techniques such as vegetation indices and red edge effect. Fig. 21.5 shows the spectral profile of crops and bare soil, and any other feature/object present in the field is also

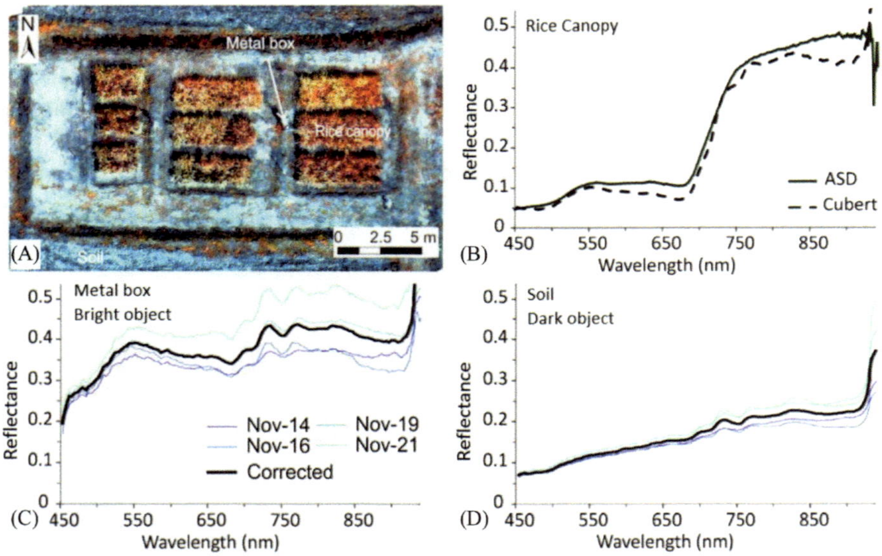

FIGURE 21.5

False-color composite image from Cubert hyperspectral camera band 104, 29, and 12. (A) Red color denotes vegetation and blue color bare soil. (B) Comparison of rice canopy reflectance measured by Analytical Spectral Devices, Inc. (ASD) spectroradiometer and Cubert camera after the empirical line correction using (C) reflectance of dark soil object, and (D) reflectance of bright metal box object (black line) in 4 days of measurement (colored lines).

Adapted from Jin, H., Köppl, C. J., Fischer, B., Rojas-Conejo, J., Johnson, M. S., Morillas, L., Lyon, S. W., Durán-Quesada, A. M., Suárez-Serrano, A., & Manzoni, S. (2021). Drone-based hyperspectral and thermal imagery for quantifying upland rice productivity and water use efficiency after biochar application. Remote Sensing, 13*(10), 1866.*

identified correctly. Methods with spectral vegetation indices based on hyperspectral or multispectral reflectance essentially estimate the Green Area Index or the chlorophyll content of the crop canopy surface, not accounting the other nongreen parts of the crops or vegetation. This gap can be fulfilled by the implementation of LiDAR data for PA (Bates et al., 2021).

21.4.3 LiDAR applications

In PA, LiDAR-derived products are used to generate the 3D structure or model of the fields. Crop height estimation is the most common application of LiDAR data for the PA. Therefore, developing 3D models of agricultural areas will facilitate the assessment of natural resources in and around the region. Even with the 3D model, farmers will have an idea about the exact farm terrain and will identify the water catchment area and be able to locate the flow of soil erosion. LiDAR sensors are being used to estimate the parameters for smaller areas and can be useful to serve as references for any remotely sensed images for assessment of similar parameters in larger agricultural fields (Zhang et al., 2021), as this has been demonstrated using the test in grasslands. Mainly LiDAR sensors are being used to estimate the plant area index (PAI) based on canopy density, modelled by generating the canopy heights of crops as illustrated in Fig. 21.6 (Bates et al., 2021; Nuijten et al., 2019). In conclusion, LiDAR sensors can estimate the large agricultural fields' crop canopy height, fractional vegetation coverage, and above-ground crop biomass (Bendig et al., 2015; Schut et al., 2018) with accurate and improved results.

21.5 Agricultural applications performed using drones-acquired images

21.5.1 Farm drone setup and surveying

Drones are used to monitor and observe the whole farm land all together and help the framers to focus on the particular area of interest of their field where crops need more attention. Precision agricultural farms that would work solely on the drone-based agriculture practices would be set up by the following expertise:

- Farm-based programming—Each area has its own specific crops, which are grown by following certain growing patterns, a drone must be programmed in such a way that it is able to function according to the area and the type of crop that is being cultivated.
- Self-learning programing—Humans have the ability to change their decisions according to the situation; in same way drones should be able to analyze themselves and implement artificial intelligence to perform with more accuracy.

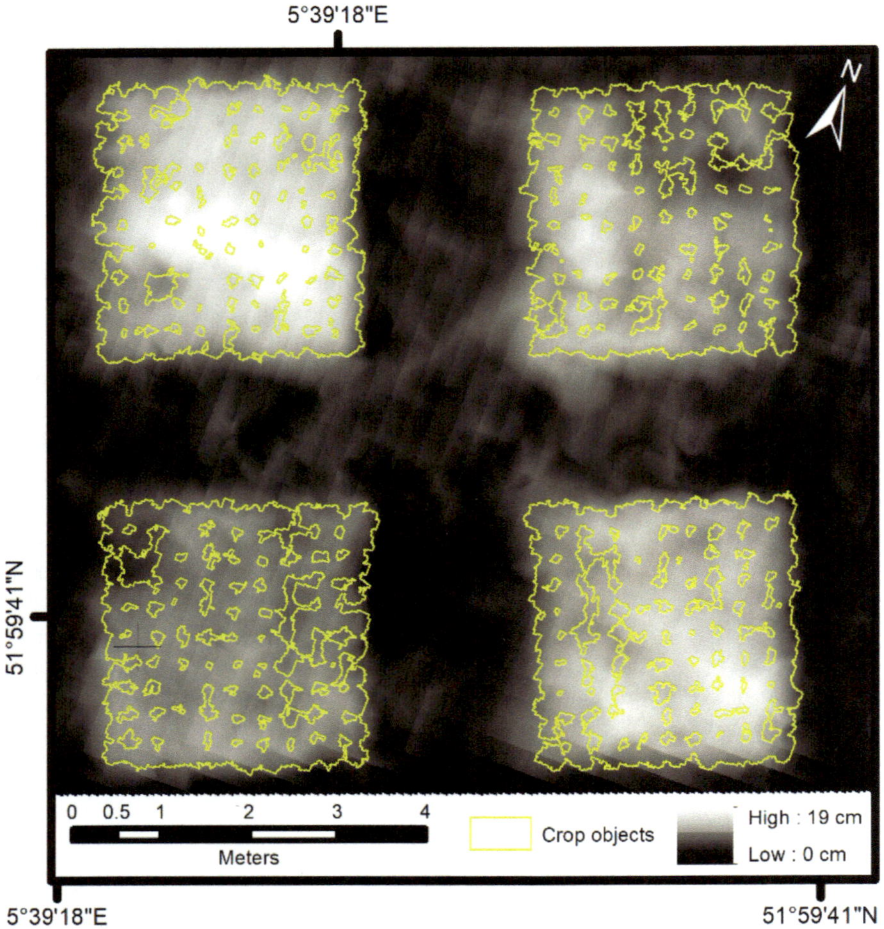

FIGURE 21.6

Map showing the crop height model for four experimental plots including multiresolution image segmentation-based vector objects representing main crop area. Highest height values occur at the center of spatial objects while lowest height values occur at the edges of objects.

Adapted from Nuijten, R. J., Kooistra, L., & De Deyn, G. B. (2019). Using unmanned aerial systems (UAS) and object-based image analysis (OBIA) for measuring plant-soil feedback effects on crop productivity. Drones, 3*(3), 54.*

- Modifications—As mentioned in this chapter, drones are of different types; each one of it has its own specifications. Hence, there would be requirement of all the different equipment to obtain maximum yield from the farm.

21.5.2 Crop monitoring, plant disease detection, and management

Crop monitoring and management are one of agriculture's most challenging tasks because it requires much time and patience. With drone-based working fields, this situation can be changed as drones can easily map the crops and fields, which helps in better estimate of production. A good-quality drone system can detect germs and diseases in advance, sometimes, before it is visible to the human eye, the drone detects issues in monitoring and alerts about its potential to spread among the nearby crops. Thus, scouting the crop fields has become an esssential aspect of PA. Crop scouting helps in field monitoring for the presence of any stress to the crops, that is, pests, weeds (Pérez-Ortiz et al., 2016), disease, or any other potential hindrances to the crop growth and developments such as trellis breakage, irrigation leaks. In a recent publication, Yang et al. (2020) briefly discussed about autonomous UAV scouting for rice lodging assessment. This information provides advanced and early treatment to plants. Currently the new infrared camera can provide the Normalized Difference Vegetation Index (NDVI), it enabling the analysis of solar radiation's intensity and absorption rate. By the use of this technology, we not only monitor the plants but also we can monitor particular parts of the plant (Mahajan & Bundel, 2016). Drones are being used in crop monitoring such as, crop health assessment, crop disease, and crop water stress. Even sometimes, security drones can be deployed to monitor fencing and perimeters of more valuable crops. Srivastava et al. (2019) developed GUI (Graphical User Interface)-based applications to analyze the drone-acquired images for crop health status. Other published articles also recognized the importance of health assessment using NDVI, the cheaper way of using drones (Mahajan & Bundel, 2016). Thermal sensors are being implemented by researchers for the early detection of soil-borne disease caused by the fungus *Verticillium dahliae* in crops, and this was remarkably demonstrated by Calderón et al. (2013) using thermal sensors for plant pathology. The concept behind the detection process is described in detail by Calderón et al. (2013). The authors used combinations of sensors (RGB, multispectral, and thermal) to estimate the crop yields based on the performance of features analyzed, such as crop height, vegetation indices, canopy cover, and temperature (Feng et al., 2020). Crop monitoring and management also include several other applications through drones, may be particularly crops and plant counting, field cover, and crop type classification mapping (Bah et al., 2018; Kitano et al., 2019; Nowakowski et al., 2021).

21.5.3 Soil analysis and practices using drones

There are a several applications of drone technology which can be beneficial in PA. The drone technology can be used to assess the soil conditions, thus leading to accurate calculation of the potential yield from the field. Drones can be connected with mapping screens by which we can determine the soil conditions in the 3D mapping of the field with precise soil color. This helps in monitoring the

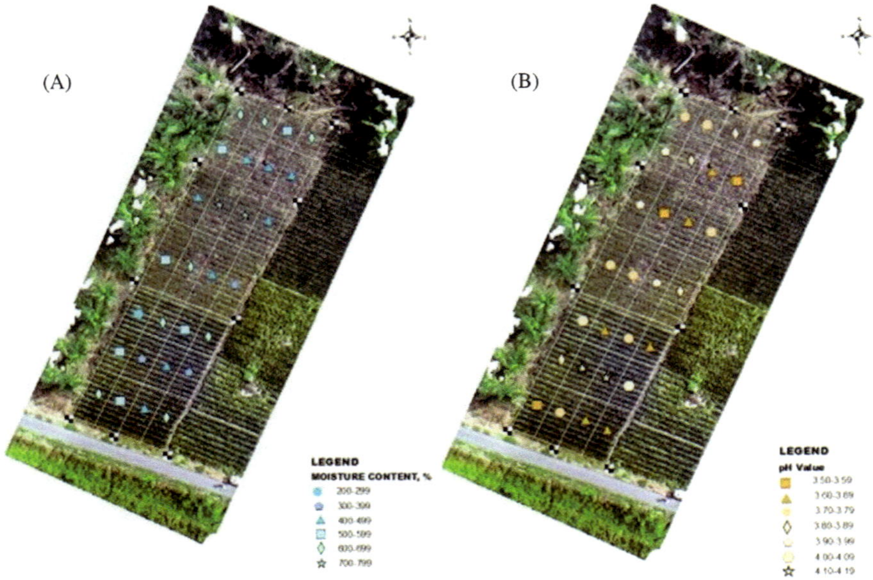

FIGURE 21.7

Physical properties map: (A) Moisture content (B) pH.

Adapted from Mustaffa, A., Mukhtar, A., Rasib, A., Suhandri, H., & Bukari, S. (2020). Mapping of peat soil physical properties by using drone-based multispectral vegetation imagery. In: IOP conference series: Earth and environmental science *(Vol. 498, Issue 1, pp. 012021). IOP Publishing.*

soil quality, soil moisture content, and water flow in the field. Even with the thermal images, we can generate CWSI, which is used to detect the water stress conditions of a plant, because, at the foliar level, stomata closure reduces transpiration and evaporative cooling, which results in an increased temperature of the crop leaf. This increased temperature can be detected better by thermal imagery than other images. Moreover, multiresolution image segmentation and template matching algorithms were used in an integrated workflow to detect individual crops with 99.8% accuracy and delineate crop-covered area with 85.4% accuracy (Nuijten et al., 2019). Soil conditions are being examined by implementing multispectral imagery, mainly assessing the soil physical properties such as soil's pH and soil moisture contents (refer to Fig. 21.7) to measure crop health index (Mustaffa et al., 2020).

21.5.4 Crop water stress and water-use efficiency

Drones provide an image of the whole land area in a few minutes, which usually takes a very long time to analyze and monitor the field without drones. Drone systems help in phenotyping and provide well-managed data for research applications. The periodic monitoring offers professional research by estimating the best possible production, which can gain maximum profits from that particular land. Drones are more efficient in assessing water-use efficiency (Jin et al., 2021) and provide information about crop water stress in the agricultural field.

21.5.5 Route planning and optimization using drones

Drones can help farmers plan and monitor their crops in such a way that with minimal efforts of farmers, and completely a drone-based field, the productivity would be maximum. With more and more shifting to this technology, the setup cost is expected to go down. Drones help farmers optimize the use of input such as seeds, fertilizers, irrigation, and pesticides effectively and protect crops from pests, reduce the monitoring time, and lead to an overall cost reduction with high-yielding crop fields. Optimal routes and paths are possible with the help of traveler's salesman problem (TSP) methods (Srivastava et al., 2020) and very effective in minimizing the input resources in the field, bagging more profits for farmers (Srivastava et al., 2020) as there are no more wastage of fertilizers and pesticides; these can be just applied where needed or in regions that are stressed or diseased (refer to Fig. 21.8). An article was recently published on applications of drone for optimization of route for resource input on the agricultural farm (Srivastava et al., 2020). It implemented drones with TSP to find the optimal path and route for input of the resources, such as fertilizers, pesticides, and insecticides in the farm. This will be easy to prevent resource wastage and avoid repetition of input resources at any region in the agricultural farm. Drone acquired images can be used to monitor the crops and soil conditions for fertility, thus allowing researchers to apply manures/fertilizers in the affected regions using TSP optimal path and route, to accurately apply and reduce wastage of the input resources. Even this helps farmers to plan accordingly the applications of irrigation water and input resources in the farm.

21.5.6 Prescription maps: crop dusting and spraying

Farmers have to spray fertilizers, pesticides, and other periodic requirements of their fields by attaching a sprayer to the tractor and then moving it across their field; this is very time-consuming work and requires farmers to be very patient and accurate in evenly spraying across the field. Different parameters such as soil sensors data, topography data, soil survey, and soil sampling data are incorporated in the drone-acquired images (thermal, hyperspectral, multispectral) to create a zonal map, which can be used to generate the prescription maps in the agricultural fields. Thus, accurate prescription maps may be generated per user needs and requirements as shown in Fig. 21.9. Drone technology will not only help in save time and resources but will also help boost productivity. The additional benefit of using a drone is that it has data about the requirements of every single crop; hence it can respond more appropriately.

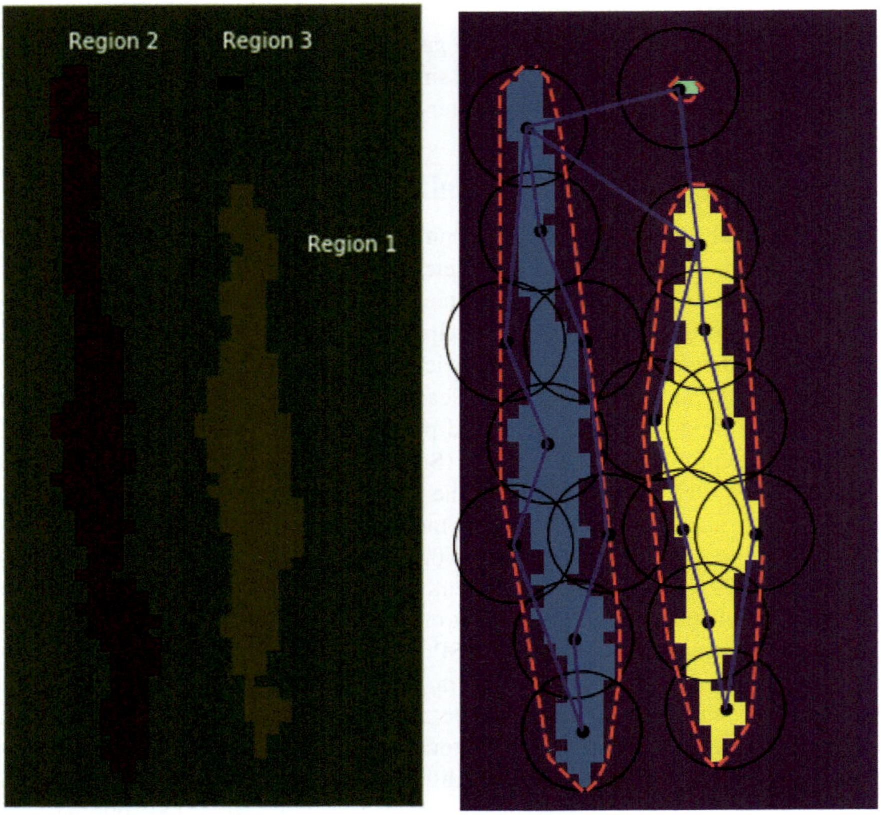

FIGURE 21.8

(A) Stressed regions in test site with different areas marked as regions I, II, and III (based on the area covered) and (B) path covered for all the stressed regions and their respective spray points.

Adapted from Srivastava, K., Pandey, P. C., & Sharma, J. K. (2020). An approach for route optimization in applications of precision agriculture using UAVs. Drones, *4(3), 58.*

21.5.7 Crop productivity and crop insurance

Nowadays, drone technology is more and more often used in the estimation of crop productivity (discussed in earlier sections) using thermal and hyperspectral imagery acquired from drones (Jin et al., 2021) and accordingly may be applied for crop insurance, with agriculture claims management is one of the key applications: effective and quick measures to track the performance of crops and especially in smaller parts of the field. The drone system data provide a change alert that tracks crop yield conditions. Drones are also increasingly used in the agricultural insurance and assessment sector, including in insurance claims forensics. In India, the Pradhan Mantri Fasal Beema Yojana (PMFBY) makes it mandatory to

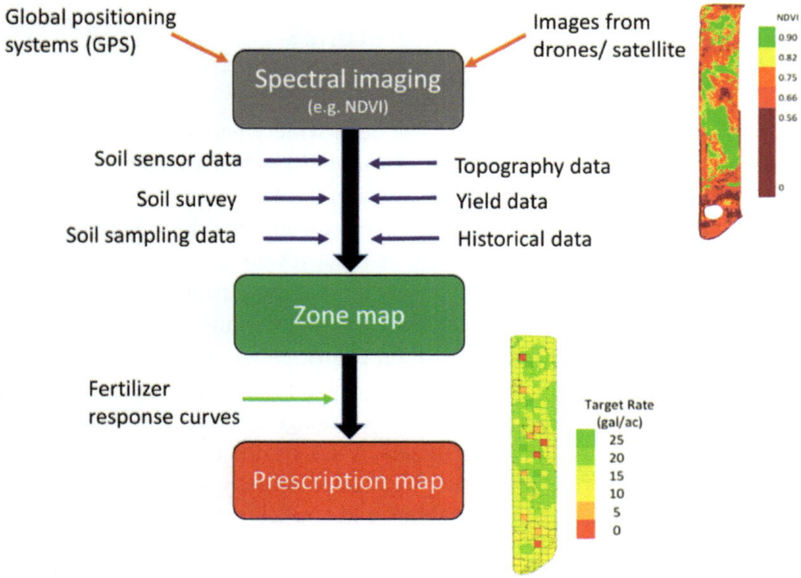

FIGURE 21.9

A conceptual model showing how prescription maps can be generated for field application of symbiotic nitrogen fixation-optimizing inputs. GPS-enabled spectral images (e.g., NDVI maps) can be generated by different software programs, by using in-season aerial imagery captured from satellite/unmanned aerial systems. Then, spectral images can be used to generate management zones (zone creation) within a given field, by integrating yield data, soil data (e.g., data from sensors, surveys, soil sample analysis), topography data, and other available historical data (e.g., cropping history). Finally, zonal maps can be used to generate the prescription maps by considering the fertilizer response curves (e.g., N-response curve) of different crops.

Adapted from Thilakarathna, M. S., & Raizada, M. N. (2018). Challenges in using precision agriculture to optimize symbiotic nitrogen fixation in legumes: Progress, limitations, and future improvements needed in diagnostic testing. Agronomy, 8*(5), 78.*

use exponential technology, including remote sensing and drone imaging to detect fraudulent claims and discrepancies. As drone imagery is instrumental in giving an accurate estimate of loss, so farmers can claim crop insurance by capturing drone feeds as evidence. Companies such as Skymet are using drones to provide agriculture survey services to insurance companies and the state governments of Maharashtra, Gujarat, Rajasthan, and Madhya Pradesh in the Republic of India. (Joshi et al., 2020). Drones can be easily used in monitoring the crops for insurance; without proper field images, it is challenging to calculate the insurance claim, but with this, one can easily estimate their losses and claim for insurance. This is possible because of constant monitoring and data archive for soil and crop status at different time periods.

21.6 Other applications using drones

Some more application of drones in PA are as follows:

- **Variable-rate application (VRA)** has been widely heralded as a means of applying crop inputs in a nonuniform manner based on varying needs throughout a field with advantages including higher average yields, lower farm input costs, and environmental benefits from applying fewer inputs. It helps in the applications of materials or input resources for specific sites. It is also called as precision fertilizer program planning through VRA (Abdullahi & Sheriff, 2017). It is used for fertilizers, pesticides and herbicides, soil amendments, seeding in the agricultural farms, either in dry or wet applications. The use of drone technology helps in monitoring the fertilizer requirements, hence leading to minimum damage to food chain, which usually gets affected by overusage of fertilizers. VRA can be through real-time-based sensors to provide prescription maps for automatically and electronically controlled site-specific field operations and applications. Thus, VRA will help to fill the gaps between crop requirement and indigenous supply.
- **Nutrient inputs** as fertilizers will fill the gap between crop need and indigenous supply.
- **Weed controlling**—With usage of drones, weeds can be easily detected and removed and that too at a very early developing stage leading to better production. Hyperspectral images acquired through drones can be analyzed to see the differences in spectra and identify weeds among crops.
- **Plant pollination**—With smaller and precise crops, we can pollinate the plants. This will be highly beneficial for the horticulture industry.
- **Crop harvesting**—Drones are nowadays used in crop harvesting (Reinecke & Prinsloo, 2017).
- Another application of drone technology is mapping the herds of animals. By the use of drones, one can easily monitor even very big herds. Drones are also being used to monitor livestock (Freeman & Freeland, 2015).

21.7 Limitations of drones

The drone-based PA practice has some limitations, which are listed as follows, due to which there may be some restrictions on the use:

- Security concerns—Drone-based technology can raise a lot of security concerns. This technology if is not monitored properly, then can cause great damage. Being an ariel-based technology, it is hard to track it and hence can be very dangerous.
- Cost of installation—The setup cost of drones is very high, and being a new technology, it is very expensive to install a fully drone-based farm setup.

Overall reduction in drone setup cost will take time to become affordable for every person. LiDAR sensors are too costly to afford by individual researchers or small setup or marginal farmers.

- High-level programing requirement—For maximum productivity, drones should be programmed according to farms, and this level of accurate programing requires expertise, which is not widely available in many places.

21.8 Conclusion

Remotely sensed acquired images and drone applications are acting as promising tools and techniques for soil conditions and crop conditions, assessing growth status, and estimating the crop yields. Sensors mounted on UAVs/drones help farmers to acquire images at a very low altitude with high spatial resolution in centimeters (10–40 cm) as compared with space-borne images in meters and also to frequently observe the entire field at user needs and requirements. This chapter concludes that the drone applications in PA can completely change the process of smart farming. With advancement in drone technology monitoring and the production rate will achieve new limits. Today, we have a perception in our minds that farming is not a profit-making sector, but with the proper implementation of technology and techniques, this is a sector that can generate big profits and turnover. Use of this technology is expected to grow in the near future. Drones will not only help the big farmers but will also help the small farmers. Drones will have a big impact on the PA and its results. Therefore, with the aim of using drone-acquired images (hyperspectral and thermal) along with field measurements at leaf to canopy scale are clarified, which will enhance the crop yields by altering the soil conditions and monitoring crop status, for optimum growth stages. PA practices, which can help farmers make better informed decisions, have evolved significantly over the recent years, with the global market now estimated to reach $43.4 billion by 2025. While drones, also known as UAVs, have not yet made it into the mainstream agriculture space, they are playing an increasingly important role in precision farming, helping agriculture professionals lead the way with sustainable farming practices, while also protecting and increasing profitability.

While looking at the advantages and merits of drones in PA, it is obvious that its wide and popular use among researchers is increasing day by day. India has the potential to be a global drone hub by 2030, and with the new drone policy 2021, the dream may come true for all sectors. In Drone Policy 2021, there are 30 key features of Drone Rules 2021, which may be beneficial to counter the limitations in several applications. As these relaxed provisions, and reduced the number of certificates required (from 25 to 5), types of fee reduced from 72 to 4. No remote pilot license is required for micro drones. Coverage of drones under Drone Rules 2021 increased from 300 to 500 kg (covering taxis as well). Drones present in India on or before November 30, 2021 are issued a unique identification number through the digital sky platform provided, they have a DAN, a GST-paid invoice, and are part of the list of DGCA-approved drones, and different zones are created for ease of flying drones (Yellow zone reduced from 45 to 12 km from the airport perimeter, no permission required in Green zones). Moreover, several approvals have abolished, which was mandatory to use drone before Drone policy 2021, such as unique authorization number, unique prototype identification number, certificate of manufacturing and airworthiness, certificate of conformance, certificate of maintenance, import clearance, acceptance of existing drones, operator permit, authorization of R&D organization, student remote pilot license, remote pilot instructor authorization, and drone port authorization. There is detailed information available to readers on the website https://pib.gov.in/.

With the earlier discussion, it is clear that future farming will primarily deploy GPS-enabled drones, a revolution in agricultural research, which offers time-saving, cost savings, enhanced productivity and yields with more efficient technologies, finally leading to profitability (Singh et al., 2020). This is only possible with the implementation of drones in farming, which can assist in quick surveys of large farms, assessment of soil conditions (soil types, soil moisture, soil fertility, etc.), crop conditions (crop health and stress, crop growth, crop growth, crop physiology, crops photosynthetic rate), improved spraying of input resources (fertilizers, pesticides, insecticides), and even assist in monitoring irrigations based on the analysis of input data. Therefore, the overall aim of discussing the above mentioned point is to highlight the importance of drones in PA, which is mainly a goal using remote sensing technologies to reduce the potential risks and improve the agricultural yield and productivity.

Acknowledgement

The authors are thankful to Shiv Nadar Institution of Eminence for funding OUR (Opportunity for Undergraduate Research) no. OUR20200064. Author Ritvik Tyagi is grateful to SNIOE for all funding and support. Dr. Pandey is thankful to SNIOE for support during the work.

References

Abdullahi, H. S., & Sheriff, R. E. (2017). Case study to investigate the adoption of precision agriculture in Nigeria using simple analysis to determine variability on a maize plantation. *Journal of Agricultural Economics and Rural Development*, *3*(3), 279–292.

Bah, M. D., Hafiane, A., & Canals, R. (2018). Deep learning with unsupervised data labeling for weed detection in line crops in UAV images. *Remote Sensing*, *10*(11), 1690.

Bates, J. S., Montzka, C., Schmidt, M., & Jonard, F. (2021). Estimating canopy density parameters time-series for winter wheat using UAS Mounted LiDAR. *Remote Sensing*, *13*(4), 710.

Bendig, J., Yu, K., Aasen, H., Bolten, A., Bennertz, S., Broscheit, J., Gnyp, M. L., & Bareth, G. (2015). Combining UAV-based plant height from crop surface models, visible, and near infrared vegetation indices for biomass monitoring in barley. *International Journal of Applied Earth Observation and Geoinformation*, *39*, 79–87.

Black, D. (2016). *Driverless tractors, drone crop dusters, automated milking: Agriculture called the 'most attractive' robotics market opportunity*. Available online: https://www.enterpriseai.news/2016/01/21/driverless-tractors-drone-crop-dusters-automated-milking-agriculture-called-the-most-attractive-robotics-market-opportunity/.

Calderón, R., Navas-Cortés, J. A., Lucena, C., & Zarco-Tejada, P. J. (2013). High-resolution airborne hyperspectral and thermal imagery for early detection of Verticillium wilt of olive using fluorescence, temperature and narrow-band spectral indices. *Remote Sensing of Environment*, *139*, 231–245.

Circuit, T. (2021). *Types of drones—Explore the different models of UAV's*. International Society for Optics and Photonics, [Online].

Daponte, P., De Vito, L., Glielmo, L., Iannelli, L., Liuzza, D., Picariello, F., & Silano, G. (2019). A review on the use of drones for precision agriculture. In: *IOP conference series: Earth and environmental science* (Vol. 275, issue 1, pp. 012022). IOP Publishing.

Feng, A., Zhou, J., Vories, E. D., Sudduth, K. A., & Zhang, M. (2020). Yield estimation in cotton using UAV-based multi-sensor imagery. *Biosystems Engineering*, *193*, 101–114.

Freeman, P. K., & Freeland, R. S. (2015). Agricultural UAVs in the US: Potential, policy, and hype. *Remote Sensing Applications: Society and Environment*, *2*, 35–43.

Gebbers, R., & Adamchuk, V. I. (2010). Precision agriculture and food security. *Science (New York, N.Y.)*, *327*(5967), 828–831.

Greenwood, W. W., Lynch, J. P., & Zekkos, D. (2019). Applications of UAVs in civil infrastructure. *Journal of Infrastructure Systems*, *25*(2), 04019002.

Jin, H., Köppl, C. J., Fischer, B., Rojas-Conejo, J., Johnson, M. S., Morillas, L., Lyon, S. W., Durán-Quesada, A. M., Suárez-Serrano, A., & Manzoni, S. (2021). Drone-based hyperspectral and thermal imagery for quantifying upland rice productivity and water use efficiency after biochar application. *Remote Sensing*, *13*(10), 1866.

Joshi, E., Sasode, D. S., Singh, N., & Chouhan, N. (2020). Revolution of Indian agriculture through drone technology. *Biotica Research Today*, *2*(5 Spl.), 174–176.

Kitano, B. T., Mendes, C. C., Geus, A. R., Oliveira, H. C., & Souza, J. R. (2019). Corn plant counting using deep learning and UAV images. *IEEE Geoscience and Remote Sensing Letters*, *99*, 1–5.

Maes, W. H., & Steppe, K. (2019). Perspectives for remote sensing with unmanned aerial vehicles in precision agriculture. *Trends in Plant Science*, *24*(2), 152–164.

Mahajan, U., & Bundel, B.R. (2016). Drones for normalized difference vegetation index (NDVI), to estimate crop health for precision agriculture: A cheaper alternative for spatial satellite sensors. In: *Proceedings of the international conference on innovative research in agriculture, food science, forestry, horticulture, aquaculture, animal sciences, biodiversity, ecological sciences and climate change* (AFHABEC-2016), Delhi, India (Vol. 22).

Mustaffa, A., Mukhtar, A., Rasib, A., Suhandri, H., & Bukari, S. (2020). Mapping of peat soil physical properties by using drone-based multispectral vegetation imagery. In: *IOP conference series: Earth and environmental science* (Vol. 498, Issue 1, p. 012021). IOP Publishing.

Nowakowski, A., Mrziglod, J., Spiller, D., Bonifacio, R., Ferrari, I., Mathieu, P. P., Garcia-Herranz, M., & Kim, D.-H. (2021). Crop type mapping by using transfer learning. *International Journal of Applied Earth Observation and Geoinformation*, *98*, 102313.

Nuijten, R. J., Kooistra, L., & De Deyn, G. B. (2019). Using unmanned aerial systems (UAS) and object-based image analysis (OBIA) for measuring plant-soil feedback effects on crop productivity. *Drones*, *3*(3), 54.

Pandey, P. C., & Pandey, M. (2023). Highlighting the role of agriculture and geospatial technology in food security and sustainable development goals. *Sustainable Development*, 1–21, In press. Available from https://doi.org/10.1002/sd.2600.

Pandey, P. C., Srivastava, P. K., Balzter, H., Bhattacharya, B., & Petropoulos, G. P. (2020). *Hyperspectral remote sensing: Theory and applications*. Elsevier.

Pandey, P. C., Tripathi, A. K., & Sharma, J. K. (2021). *An evaluation of GPS opportunity in market for precision agriculture. GPS and GNSS technology in geosciences* (pp. 337–349). Elsevier.

Patel, J. S., Fioranelli, F., & Anderson, D. (2018). Review of radar classification and RCS characterisation techniques for small UAVs or drones. *IET Radar, Sonar & Navigation*, *12*(9), 911–919.

Pérez-Ortiz, M., Peña, J. M., Gutiérrez, P. A., Torres-Sánchez, J., Hervás-Martínez, C., & López-Granados, F. (2016). Selecting patterns and features for between-and within-crop-row weed mapping using UAV-imagery. *Expert Systems with Applications*, *47*, 85–94.

Puri, V., Nayyar, A., & Raja, L. (2017). Agriculture drones: A modern breakthrough in precision agriculture. *Journal of Statistics and Management Systems*, *20*(4), 507–518.

Rachman, L.M., Purnomo, M.B., Purwakusuma, W., & Rachman, R.A. (2019). The role of drones for supporting precision agricultural management. In: *Sixth international symposium on LAPAN-IPB satellite* (Vol. 11372, pp. 1137202). International Society for Optics and Photonics.

Rajvanshi, A. (2013). *Is precision agriculture the solution to India's farming crisis?* India: The New Indian Express.

Reinecke M., & Prinsloo, T. (2017). The influence of drone monitoring on crop health and harvest size. In: *2017 1st international conference on next generation computing applications* (NextComp), pp. 5–10. IEEE.

Schut, A. G., Traore, P. C. S., Blaes, X., & Rolf, A. (2018). Assessing yield and fertilizer response in heterogeneous smallholder fields with UAVs and satellites. *Field Crops Research*, *221*, 98–107.

Shankarnarayan, V. K., & Ramakrishna, H. (2020). Paradigm change in Indian agricultural practices using Big Data: Challenges and opportunities from field to plate. *Information Processing in Agriculture*, *7*(3), 355–368.

Singh, P., Pandey, P. C., Petropoulos, G. P., Pavlides, A., Srivastava, P. K., Koutsias, N., Deng, K. A. K., & Bao, Y. (2020). *Hyperspectral remote sensing in precision agriculture: Present status, challenges, and future trends. Hyperspectral remote sensing* (pp. 121–146). Elsevier.

Snow, C. (2016). The truth about drones in precision agriculture. *Skylogic Research*, 1–6.

Srivastava, K., Bhutoria, A. J., Sharma, J. K., Sinha, A., & Pandey, P. C. (2019). UAVs technology for the development of GUI based application for precision agriculture and environmental research. *Remote Sensing Applications: Society and Environment*, *16*, 100258.

Srivastava, K., Pandey, P. C., & Sharma, J. K. (2020). An approach for route optimization in applications of precision agriculture using UAVs. *Drones*, *4*(3), 58.

Thilakarathna, M. S., & Raizada, M. N. (2018). Challenges in using precision agriculture to optimize symbiotic nitrogen fixation in legumes: Progress, limitations, and future improvements needed in diagnostic testing. *Agronomy*, *8*(5), 78.

Velusamy, P., Rajendran, S., Mahendran, R. K., Naseer, S., Shafiq, M., & Choi, J.-G. (2022). Unmanned aerial vehicles (UAV) in precision agriculture: Applications and challenges. *Energies*, *15*(1), 217.

Vergouw, B., Nagel, H., Bondt, G., & Custers, B. (2016). *Drone technology: Types, payloads, applications, frequency spectrum issues and future developments. The future of drone use* (pp. 21–45). Springer.

Veroustraete, F. (2015). The rise of the drones in agriculture. *EC Agriculture*, *2*(2), 325–327.

Yang, M.-D., Boubin, J. G., Tsai, H. P., Tseng, H.-H., Hsu, Y.-C., & Stewart, C. C. (2020). Adaptive autonomous UAV scouting for rice lodging assessment using edge computing with deep learning EDANet. *Computers and Electronics in Agriculture*, *179*, 105817.

Zhang, X., Bao, Y., Wang, D., Xin, X., Ding, L., Xu, D., Hou, L., & Shen, J. (2021). Using UAV LiDAR to extract vegetation parameters of inner Mongolian grassland. *Remote Sensing*, *13*(4), 656.

CHAPTER

Precision opto-imaging techniques for seed quality assessment: prospects and scope of recent advances

22

Bhaswati Sarmah[1,2,*], Rajasree Rajkhowa[1,*], Ishita Chakraborty[3], Indira Govindaraju[3], Sanjai Kumar Dwivedi[4], Nirmal Mazumder[3] and Vishwa Jyoti Baruah[5]

[1]*Department of Plant Breeding and Genetics, Assam Agricultural University, Jorhat, Assam, India*

[2]*AICRP on Potato, Assam Agricultural University, Jorhat, Assam, India*

[3]*Department of Biophysics, Manipal School of Life Sciences, Manipal Academy of Higher Education, Manipal, Karnataka, India*

[4]*Department of Horticulture, Govind Ballabh Pant University of Agriculture and Technology, Pantnagar, Uttarakhand, India*

[5]*Centre for Biotechnology and Bioinformatics, Faculty of Biological Sciences, Dibrugarh University, Dibrugarh, Assam, India*

Chapter highlights

- Good-quality seeds and their assessment are critical to agricultural productivity.
- Newer Microscopic, Spectroscopic, and Opto-Imaging techniques are reliable, reusable, and to large extent nondestructive.
- These advancements have prospects in diagnostics, observing variation in quality parameters and remote field inspection.

22.1 Introduction

Seeds are the primary input in any crop production system, and hence, the quality of the seeds is a key priority for better yield and more marketability. Seed quality is an umbrella term that plays a significant part in affecting the agricultural system. Good-quality seeds are very important for enhancing the overall yield of any crop (Bradbeer, 1988). Seed quality is affected adversely by various factors such

* Contributed equally.

Remote Sensing in Precision Agriculture. DOI: https://doi.org/10.1016/B978-0-323-91068-2.00012-6

as poor viability, low vigor, presence of disease, insect infestation, and incidence of mixtures in the seed crop (Gupta & Kumar, 2020). The sowing of a quality seed, which is adjudged by proper seed composition, kernel maturity, disease-free state, is associated with higher viability (Copeland & McDonald, 1999). The genetic purity of seeds is another important criterion needed to be fulfilled to get a good price for the seeds produced. This genetic purity can be evaluated by morphological identification, chemical tests, molecular techniques such as DNA fingerprinting and grow-out tests (Bradbeer, 1988).

Typically, seed physiology is determined by its viability and vigor, and their tests use complex procedures under the supervision of expert personals. Viability tests include tetrazolium and germination tests while cold test, brick gravel test, accelerated aging test, and conductivity test are conducted to validate the vigor of the seed (Corbineau, 2012; Huang et al., 2015). Seed health, on the other hand, refers to the presence or absence of any biotic factor in the seeds, predominantly the seed-borne pest and pathogens, which may lead to its deterioration. Obtaining seeds with good physiological quality and better health status is essential for obtaining seed certification as well (Bishaw et al., 2012).

There are some significant hitches with the traditional seed quality assessment methods of which the most noteworthy is that the methods are usually conducted in a destructive manner using special chemicals and laboratory arrangements with the help of experienced seed analysts (Finch-Savage & Bassel, 2016). The seeds that are used for a particular test can no longer be used for another test or even for sowing. The determination of seed quality through the conventional methods of evaluation, although accurate and reliable, has a few constraints such as being laborious, tedious, and time-consuming (Diriba Bula et al., 2019; Xu et al., 2020). They also necessitate a high operating cost with the involvement of trained professionals to conduct the tests and analyze the results (Fabiyi et al., 2020; Liu et al., 2016; Rahman & Cho, 2016; Xia et al., 2019). Hence, a rapid and precise detection methodology must be adapted to ensure the production and supply of quality seed materials (Huang et al., 2008). This is where the nondestructive and rapid methods grab our attention by detecting the presence and absence of seed-borne pathogens and differentiating good-quality seeds from their deprived counterparts.

In pursuit of finding some rapid, nondestructive, and accurate testing methods for the evaluation of seed quality, a few methods have been recognized, which are equally sustainable, reliable, and simple in the modern agroindustry (Elmasry et al., 2019; ElMasry et al., 2020; Huang et al., 2015). The potential sustainable methods include opto-imaging techniques, which is an umbrella term encompassing spectroscopy, imaging, etc., which have been accepted for evaluating the seed quality parameters such as biochemical structure and composition, viability and vigor, seed health, and cultivar purity in recent years (Table 22.1) (Kherlopian et al., 2008; Li et al., 2014; Manley et al., 2013). The chapter highlights the main methods that can be used for analyzing the seed physiological quality and health and evaluates their overall efficiency and reproducibility of the results (Fig. 22.1).

Table 22.1 A brief summary of various biophysical techniques for quality evaluation of seeds from varied sources for proper characterization and certification.

Technology employed	Seed quality aspect	Source	References
Near-infrared spectroscopy	Seed moisture	Maize	Esteve Agelet and Hurburgh (2014)
	Seed viability	Maize	Esteve Agelet et al. (2012)
	Seed health	Maize	de Medeiros et al. (2020); Esteve Agelet and Hurburgh (2014); Ma et al. (2020)
		Soybean	Baek et al. (2019); Sendin et al. (2019); Wang et al. (2001)
	Seed nutritional quality	Spinach	Olesen et al. (2011)
		Cowpea	Olesen et al. (2015)
Mid-infrared spectroscopy	Seed viability	Peas	Arganosa et al. (2006)
		Wheat	Barron (2011)
		Lettuce	Kim et al. (2013)
	Genetic purity	Maize	Smith and Register (1998)
Fourier transform near-infrared	Seed viability	Rice	He et al. (2019); Zhang et al. (2020)
		Watermelon	Yasmin et al. (2019)
		Hinoki cypress	Mukasa et al. (2020)
		Quercus variabilis	Pang et al. (2021)
		Tomato	Shrestha et al. (2017)
		Soybean	Kusumaningrum et al. (2018)
		Cabbage and radish	Shetty et al. (2011)
	Seed vigor	Tomato	Qiu et al. (2018)
		Soybean	Al-Amery et al. (2018)
	Seed purity	Rice	Attaviroj et al. (2011); Jiao et al. (2010); Liang et al. (2009); Xie et al. (2014)
		Soybean	Wang et al. (2016)
	Seed nutritional quality	Rice	Gunaratne et al. (2013)
	Seed health	Maize	Kandpal et al. (2015); Tallada et al. (2011)
Raman spectroscopy	Seed purity	Tomato	Baranski et al. (2005)
		Sugarbeet	Baranski et al. (2005)

(*Continued*)

Table 22.1 A brief summary of various biophysical techniques for quality evaluation of seeds from varied sources for proper characterization and certification. *Continued*

Technology employed	Seed quality aspect	Source	References
	Seed viability	Cocoa	Edwards et al. (2005)
		Zea mays	Xia et al. (2019)
		Pepper	Krimmer et al. (2019)
	Seed health	Wheat	Egging et al. (2018)
		Sorghum	Egging et al. (2018)
		Maize	Farber and Kurouski (2018)
		Watermelon	Lee et al. (2017)
Nuclear magnetic resonance	Moisture content	Maize	Melchinger et al. (2017)
	Biochemical quality	Maize	Jones et al. (2005)
		Soybean	Collins et al. (1967); Samii-Saket et al. (2011)
Hyperspectral imaging	Varietal identification and seed purity	Grapes	Baca-Bocanegra et al. (2020)
		Rice	Kong et al. (2013)
		Maize	Zhang et al. (2012)
		Wheat	Mahesh et al. (2008)
	Seed viability	Maize	Ambrose et al. (2016); Xia et al. (2019)
		Tomato	Shrestha et al. (2017)
	Seed health	Almonds	Nakariyakul and Casasent (2011)
		Wheat	Bauriegel et al. (2011); Singh et al. (2009); Singh et al. (2010)
Multispectral imaging	Seed viability	Cowpea	Olesen et al. (2015)
	Varietal identification and seed purity	Tomato	Shrestha et al. (2015)
		Wheat	Liu et al. (2016)
		Cowpea	Elmasry et al. (2019)
		Rice	Liu et al. (2015); Sumriddetchkajorn et al. (2010)
	Seed vigor	Cowpea	Elmasry et al. (2019)
	Seed health	Barley	Shahin et al. (2012)
		Sunflower	Ma et al. (2018)
		Spinach	Olesen et al. (2015)
Fluorescence and photoluminescence imaging	Seed viability	Cabbage	Jalink et al. (1998)
		Coffee	De Mello Gallep (2014)
		Chilli	Barua et al. (2010)
		Tomato	Butz et al. (2005)

(*Continued*)

Table 22.1 A brief summary of various biophysical techniques for quality evaluation of seeds from varied sources for proper characterization and certification. *Continued*

Technology employed	Seed quality aspect	Source	References
		Soybean	Costanzo et al. (2008); Deng et al. (2018); Musumeci et al. (1994)
		Watermelon	Grasso et al. (2018)
	Seed Vigor	Coffee	De Mello Gallep, (2014)
		Tomato	Butz et al. (2005)
Thermal imaging	Seed viability	Legumes	Men et al. (2017); Scarafoni et al. (2007)
		Pea	Kranner et al. (2010)
	Seed vigor	Rice	He et al. (2019)
		Cowpea	Elmasry et al. (2019)
		Pea	Kranner et al. (2010)
	Seed health	Wheat	Singh et al. (2009)
LiDAR imaging	Varietal Identification and seed purity	Brinjal (Eggplant)	Omasa et al. (2007)
		Tomato	Omasa et al. (2007)
		Oak	Hosoi and Omasa (2006)
		Forest trees	Wallace et al. (2012)
		Wheat	Friedli et al. (2016); Jimenez-Berni et al. (2018)
		Grapes	Moreno et al. (2020)
	Physical purity	Maize	Andújar, Escolà, et al. (2013); Andújar, Rueda-Ayala, et al. (2013); Zisi et al. (2018)

The opto-imaging techniques for seed quality assessment come from the concept of remote sensing where the radiations are used beyond the visible range to determine the seed quality (Franklin, 2001; Smith et al., 2004). When light shines on an object, it will either be reflected or transmitted, or it will be absorbed. The absorbed portion of energy may, later on, be emitted. These emitted or reflected energies will be captured by the sensors (Khanal et al., 2020), which collect the information and display it in the form of a graph. This review offers a holistic opinion of the multifaceted applications of these sustainable methods for seed quality assessment.

Exploring the multifold utility of the nondestructive techniques is an interesting proposition, which has given a new perspective to the seed quality assessment methodology. The literature database, PubMed (Lu, 2011), was searched (on December 20, 2020) for the availability of references related to the nondestructive seed quality assessment methods over the last 20 years. A

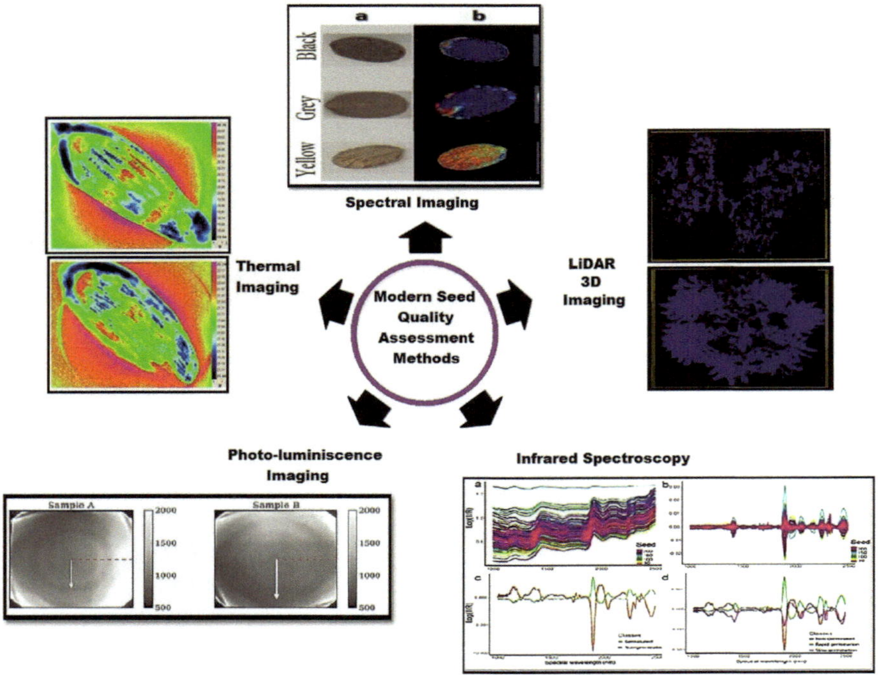

FIGURE 22.1

Schematic representation of the opto-imaging techniques for seed quality assessment.

Adapted from de Medeiros et al. (2020); Elmasry et al. (2019); Kim et al. (2013).

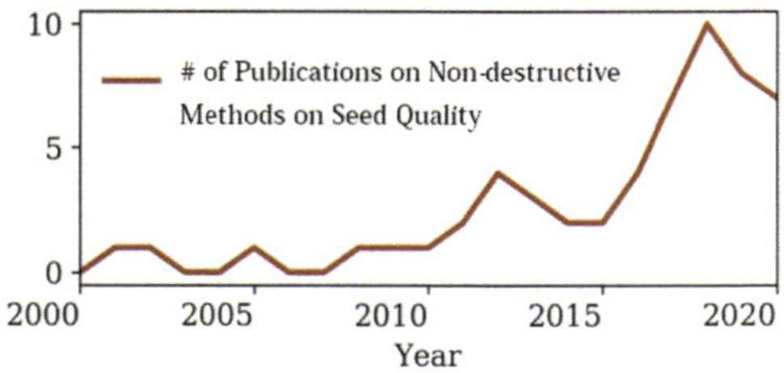

FIGURE 22.2

The graph exhibits increasing trend of research on nondestructive seed quality. The trend shows the number of articles on related subject published over a period of twenty (20) years (Information retrieved using PubMed search).

rise in the references within the past 5 years indicates the new traction that this field has been receiving since 2015 (Fig. 22.2). This new-found interest among the researchers may be due to the new advancements in the contemporary approaches, which have the potential to replace the conventional methods.

22.2 Infrared spectroscopy and its application in seeds

Spectroscopy refers to the theory of splitting light into different colors, which is termed as spectra (Huang et al., 2020). The interaction between matter and infrared radiation is termed as infrared spectroscopy (Pupeza et al., 2020). When the light is directed on the sample (seed), the transmitted light carries some comprehensive information about the sample, which is converted to electric signals by the detector, which are represented in the form of images or graphs (Arora et al., 2020; ElMasry et al., 2020). Infrared spectroscopy follows the principle that when a ray of light strikes a sample, the molecules of the sample absorb, transmit, or reflect the light based on its unique biological and chemical composition (Peng & Wu, 2019). The infrared spectroscopy can be applied to determine the morphological and biochemical characters as well as other specific seed quality parameters such as moisture content, viability, and seed health of both single kernel and bulk samples (Esteve Agelet & Hurburgh, 2014; Esteve Agelet et al., 2012).

22.2.1 Near- and mid-infrared

Through infrared spectroscopy, a single seed can be used for the determination of the whole seed lot; as a result, hybrid seeds can be saved as all the seeds are not used for testing purpose as in the case of normal seed testing methods (Hu et al., 2020; Xia et al., 2019). Near-infrared (NIR) spectroscopy refers to the wavelength range of 780–2500 nm of the electromagnetic spectrum. In this scale, the absorption is mainly caused by C = H, O = H, and N = H bonds of the main biomolecules such as protein, lipids, carbohydrates, water, and other phytochemicals such as chlorophylls and carotenoids (Olesen et al., 2015). This technique is used for qualitative as well as quantitative analysis of oil content, protein, sugar, starch, etc. There are a variety of NIR spectroscopy modes of which the most common are transmittance and diffuse reflectance, which have been used for seed quality assessments (Tian et al., 2020; Xia et al., 2019).

In previously reported studies on soybean seeds, an NIR spectrometer was used to obtain an average spectrum of damaged and healthy seeds. It was found that the energy absorbed by the damaged seeds concerning the corresponding sound seeds was lower in the NIR region while in the visible region, the energy absorption was higher in comparison with the sound soybean seeds (Baek et al., 2019; Sendin et al., 2019; Wang et al., 2001, 2004). The variation in the energy absorption can be detected by the distinction in the color of the image, which in turn is due to the difference in the chemical composition of the damaged and sound seeds. NIR was used to distinguish heat and frost-damaged corn seeds (Esteve Agelet & Hurburgh, 2014) using partial least squares discriminant analysis (PLS-DA), soft independent modeling of class analogy (SIMCA), k-nearest neighbors, and least-squares support vector machines (SVM) were the classification algorithms used. PLS-DA method could discriminate the heat-damaged corn kernels with 99% accuracy although it could neither differentiate the frost-damaged corn kernels nor differentiate viable seeds from nonviable seeds. This infers that NIR spectroscopy can precisely detect any changes in the health status of the seeds but has no correlation to the viability status of the seeds (de Medeiros et al., 2020; Esteve Agelet et al., 2012; Ma et al., 2020).

Mid-Infrared (MIR) spectroscopy is widely used to determine the compositional changes in the seed during germination (Arganosa et al., 2006; Barron, 2011; Gillgren & Gorzsás, 2016; Kim et al., 2013). The unique spectral pattern generated via MIR is associated with vibrational modes of molecular bonds accredited to specific functional groups present in the sample. Hence, qualitative and quantitative analyses are easy to accomplish (Türker-Kaya & Huck, 2017). Both MIR and NIR spectroscopy can be used for the spatial localization of the components in diverse plant samples, although MIR has the added advantage of being used for a higher resolution due to its short-wave nature (Karunakaran et al., 2020). Infrared spectroscopy is one of the most popular technologies for the detection of seed quality in various crop species. Nonetheless, this technology provides only spectral information and is not able to show spatial information (Kusumaningrum et al., 2018). Moreover, as only a single seed is used for the determination of the genetic purity status of the whole seed lot, it should be a representative sample of the seed lot and thus should be homogeneous (Smith & Register, 1998).

22.2.2 Fourier transform near-infrared

Fourier transform near-infrared (FT-NIR) spectroscopy is another popularly known methodology for the identification of viable and nonviable seeds as it is simple, rapid, and easy for sample preparation (He et al., 2019). Two groups of viable and nonviable seeds were obtained from the collected spectral data of triploid watermelon seeds of thick seed coat. The viability of these seeds was later confirmed by conducting a regular germination test (Mukasa et al., 2020; Yasmin et al., 2019). Later, a partial least discriminant analysis (PLS-DA) model was used, which efficiently classified the seeds into viable and nonviable with an accuracy of ~87% (Feng et al., 2019; Mukasa et al., 2019; Pang et al., 2021; Zhang et al., 2020). Viable and nonviable single seeds of tomato (Shrestha et al., 2017), soybean (Kusumaningrum et al., 2018), cabbage, and radish seeds (Shetty et al., 2011) are also compared and separated using FT-NIR spectra and analyzed using PLS-DA.

FT-NIR coupled with the PLS-DA model demonstrated better performance with an accuracy of nearly 100% for the prediction of the viability of seed samples (Baek et al., 2019). The same model, when assessing the combined effect of heat-damaged and artificially aged seed samples, attained an accuracy of 98.7%. Standard seed viability and vigor test using the FT-NIR PLS model done on tomato (Qiu et al., 2018) and soybean (Al-Amery et al., 2018) seed lots could effectively differentiate the seeds. Based on the viability, the seeds were classified as low-germinating and high-germinating seed lots while based on the vigor parameters, the seeds were differentiated as low-, medium-, and high-vigor seed lots. Therefore, for both viability and vigor, FT-NIR combined with the qualitative PLS model presented a distinct difference between the low- and high-potential seed lots. Therefore, FT-NIR spectroscopy together with multivariate data analysis has tremendous potential for quick and nondestructive detection of seed viability and vigor (Çeliktaş & Konuşkan, 2020).

FT-NIR spectroscopy along with chromatographic analysis and chemometrics methods has also been extensively used to examine the constitutional changes in the seeds of transgenic rice (Jiao et al., 2010), for identifying the geographical origins of rice (Xie et al., 2014) and for checking the authenticity for rice (Liang et al., 2009). Prediction of the nutritional profile of traditional red-rice genotypes containing proanthocyanidins (Gunaratne et al., 2013), a quick identification of varietal purity of rice (Attaviroj et al., 2011) and soybean seed using single seed (Wang et al., 2016), and detecting the fungus-infected corn kernels (Tallada et al., 2011) and checking the aflatoxin B1 (AFB1) on corn kernels (Kandpal et al., 2015) have also been achieved through this method.

22.2.3 Raman spectroscopy

Raman spectroscopy, complimentary to FT-NIR, is a noninvasive analytical technique, for testing seed viability. This phenomenon discovered by C.V. Raman is based on sample polarizability and the principle of Rayleigh scattering or elastic scattering. Raman spectroscopy provides both physical and chemical characteristics of the sample of interest. Apart from providing detailed information on molecular vibrations, it is known to be highly sensitive to detection and quality assessment of agricultural products. The Raman spectrum exhibits corresponding peaks to the vibrational mode of specific molecules. Raman spectroscopy is applied in agriculture (Herrero, 2008), for the analysis of small concentrations of molecules. In addition, it can differentiate between different seed varieties as well as viable and nonviable seeds (Baranski et al., 2005) and analyze the chemical components of cocoa seeds (Edwards et al., 2005). Raman spectroscopy was used to distinguish between viable and nonviable hybrid corn kernels (*Zea mays* L.) (Xia et al., 2019) and pepper (Krimmer et al., 2019) (Fig. 22.3). Another application of Raman spectroscopy is the identification of specific genotypes, which are conventionally done by visual screening or DNA analysis. While visual identification requires extensive taxonomical knowledge, DNA analysis is destructive and time-consuming. Raman spectroscopy has been used as a label-free and nondestructive technique for the identification of genotypes and also for its proximal analysis of nutrients in peanut (Farber et al., 2020) and maize. A handheld Raman spectrometer was used to detect and identify fungal diseases in wheat, maize, and sorghum seeds (Egging et al., 2018; Farber & Kurouski, 2018). Initially, Raman spectroscopy was not as popular as FTIR due to problems such as sample degradation and fluorescence.

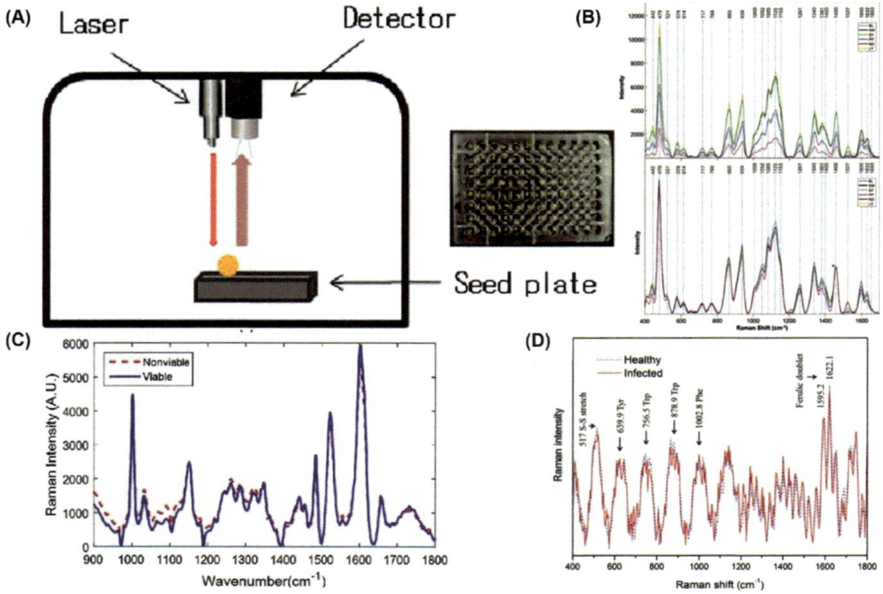

FIGURE 22.3

(A) Schematic diagram of Raman spectrophotometer with a 96-well plate. (B) Baseline-corrected and normalized Raman spectra of small white (SW), small yellow (SY), extra-large (XL), blue (BL), purple (PP), and red (RD) maize kernels for nutrient assessment. The 1458 cm^{-1} peak, which was used for spectral normalization, is indicated by an asterisk (*). (C) mean spectra of viable and nonviable pepper seeds. (D) Mean corrected Raman spectra of bacteria-infected and healthy watermelon seeds (Seo et al., 2016).

Reproduced from parts (A and C) Seo, Y-W., Ahn, C. K., Lee, H., et al. (2016). Non-destructive sorting techniques for viable pepper (Capsicum annuum L.) seeds using fourier transform near-infrared and Raman spectroscopy. Journal of Biosystems Engineering, 41, *51–59; (B) Krimmer, M., Farber, C., Kurouski, D. (2019). Rapid and noninvasive typing and assessment of nutrient content of maize kernels using a handheld Raman spectrometer.* ACS Omega, 4, *16330–16335.*

During analysis, the fluorescence masks the Raman spectra completely. However, moving Raman spectroscopy from the visible to the near-infrared (NIR) range, fluorescence tends to disappear as electronic absorption bands are less probable in this region. Another study demonstrated the detection of *Acidovorax citrulli*–infected watermelon seeds using Raman hyperspectral imaging (Lee et al., 2017).

22.3 Nuclear magnetic resonance spectrometry

Nuclear magnetic resonance spectrometry (NMR) has been effectively used for the biochemical characterization and detection of moisture content of various seeds (Melchinger et al., 2017). The NMR technique measures the resonance energy absorbed by hydrogen atoms in the sample (Dos Remedios & Moens, 1995) and can be performed in whole seeds or meals. NMR has the added advantage that it can be calibrated for smaller amounts of seeds (i.e., 1–2 g), and a sample size of 5–10 g is sufficient for performing all the analysis (i.e., for oil, protein, and moisture, etc) (Horn et al., 2011). The time required for conducting the test is approximately 24 hours from the time of receiving the sample, and being a nondestructive method, the seeds can be reused for other purposes if needed (Wang et al., 2016).

The NMR has been performed on many oilseeds and can also be calibrated for other species as needed by the user. Intact single soybean and corn kernels have been used for the measurement of oil content using NMR imaging methods (Jones et al., 2005; Samii-Saket et al., 2011). Several studies on soybean seeds for the determination of their oil content by NMR spectroscopy gas were done demonstrating the effectiveness and suitability of the technique. The NMR has an added advantage of neither destroying the viability of the sample nor altering its composition, which facilitates its reusability for developing new varieties using those undamaged seeds (Collins et al., 1967). NMR has been used to automatically screen and select haploid maize kernels based on the oil xenia effect with an accuracy of 92.3%. This system can also separate the samples with different oil contents (Wang et al., 2016).

22.4 Spectral imaging in seeds

While a normal camera acquires the image in red, green, and blue mode and combines these with three sensors to capture a color image, a hyperspectral/multispectral camera possesses a device called Spectrograph to plot the graph (Xing et al., 2005). Therefore, in the case of spectral imaging, every pixel is captured and along with it, the graph of every pixel is captured. In the case of a hyperspectral camera, the bands formed are narrow and continuous, which gives a closer picture of higher resolution (Chen et al., 2013). On the other hand, the lower version of the hyperspectral camera is a multispectral camera, having lower resolution power, and captures the pictures along with the graph.

22.4.1 Hyperspectral imaging in seed

Hyperspectral imaging is a relatively newer methodology for characterizing seeds based on their chemical composition (Fig. 22.4). This technology has been used to nondestructively characterize two red grape variety seeds and one white variety seed for the variety and maturation stage. Hyperspectral imaging was used for predicting the maturation stage and to distinguish the variety of grapes very easily (Baca-Bocanegra et al., 2020). Hyperspectral imaging coupled with multivariate

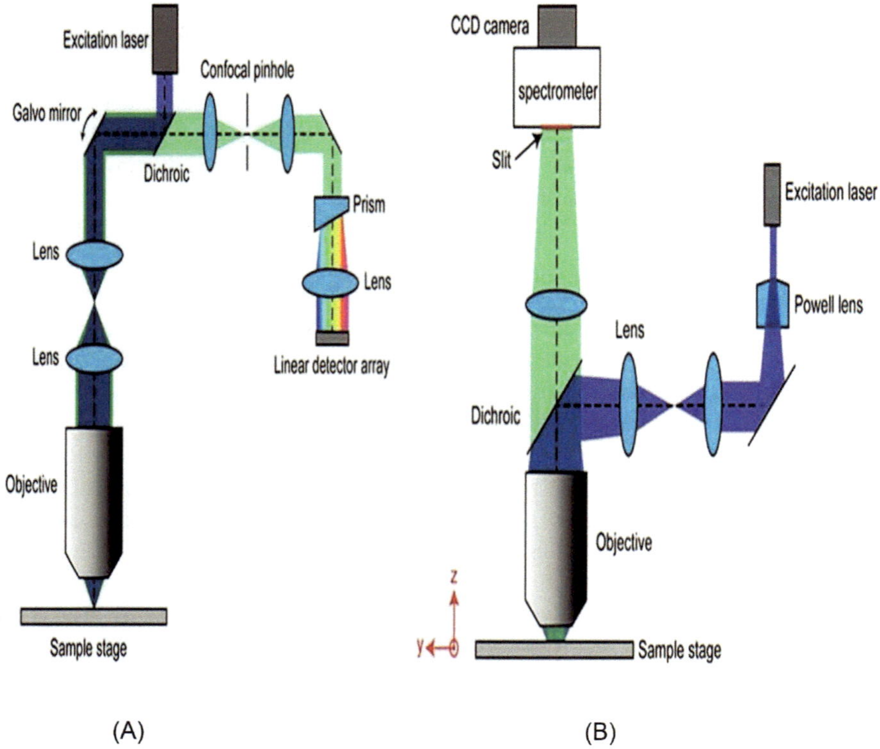

FIGURE 22.4

Experimental setup of hyperspectral microscope. (A) A hyperspectral confocal microscope. (B) A hyperspectral line-scanning microscope (Pu et al., 2019).

Adapted from Pu, H., Lin, L., Sun, D. W. (2019) Principles of hyperspectral microscope imaging techniques and their applications in food quality and safety detection: A review. Comprehensive Reviews in Food Science and Food Safety, 18, *853–866. https://doi.org/10.1111/1541-4337.12432. with permission from John Wiley and Sons.*

analysis has been effectively used for classifying rice varieties (Kong et al., 2013) as well as 330 samples of six varieties of maize seeds (Zhang et al., 2012). This ability of hyperspectral imaging to distinguish between the seeds of different varieties depicts its prospects in seed physical purity testing.

Hyperspectral illustrations of microwave heat-treated corn samples were taken between 400 and 2500 nm wave range to differentiate viable and nonviable seeds. This was followed by the development of PLS-DA to classify treated and untreated seeds (Xia et al., 2019). The model showed the highest classification accuracy of 97.6% and 95.6% in the SWIR (short wave infrared) region of the HIS (Ambrose et al., 2016). Seed viability and variation in tomato varieties Cal J, Monprecus, NCL, and Chiuri were studied using NIR hyperspectral data combined with a multivariate test of PLS-DA and principal component analysis (PCA) models with an accuracy of 100% (Shrestha et al., 2017). The same technology was used to categorize maize kernel hardness (Williams et al., 2009), classify the internally damaged almond nuts (Nakariyakul & Casasent, 2011), and examine the insect-damaged wheat kernels (Singh et al., 2009). To analyze the infection in wheat ears caused by *Fusarium culmorum*, the photosynthetic integrity of the infected ears (wavelength range of 400–1000 nm) was analyzed using chlorophyll fluorescence imaging in combination with hyperspectral imaging (Bauriegel et al., 2011). Short-wave NIR hyperspectral and digital color imaging was used to detect midge infected kernels of wheat at a wavelength range of 700–1100 nm (Singh et al., 2010) and for differentiating Canadian wheat classes at a spectral coverage of 960–1700 nm (Mahesh et al., 2008).

22.4.2 Multispectral imaging in seed

A multispectral image for a seed sample can be acquired using a video meter Lab (Olesen et al., 2011), a device that is designed for faster and accurate determination of surface color texture, shape, size, and chemical composition of the seed lot (Williams et al., 2009). It requires no sample preparation and leaves the sample intact; hence, it is an easy-to-use instrument providing a rapid assessment of homogeneous as well as nonhomogeneous samples in just a few seconds (Shrestha et al., 2015). Image of seeds is processed in binary large object database, from which different characteristics of seeds such as color, dimension, shape, and main spectral features of individual seeds could be extracted. Multivariate data analysis using algorithms such as linear discriminant analysis and PCA is used to obtain the hidden spectral information.

Multispectral imaging has been used for seed health detection in spinach (Olesen et al., 2015), varietal identification in tomato (Shrestha et al., 2015), characterization in wheat, and discrimination of transgenic and nontransgenic rice (Liu et al., 2015). It is also used for coating analysis, grain categorization, contamination detection, germination detection in the seed. Spectral signatures of different seed classes were obtained for cowpea seeds by using information from the raw spectral data of a Video meter Lab. Nonaged seeds demonstrated the highest reflectance followed by AA24 (Artificially aged for 24 hours), AA48, AA96, and lastly AA72 in the wavelength range of 600–780 nm (visible). Variations in the visible range (375–780 nm) were due to the changes in the surface characteristics of seed such as the seed coat color, texture, and structure. The variations in the NIR region are primarily due to compositional changes in the chemical properties of the seeds (Elmasry et al., 2019). Identification of Thai rice breed (Sumriddetchkajorn et al., 2010), quality evaluation of cereal grains and grain products of barley for greening detection (Shahin et al., 2012), and viability prediction of *Ricinus cummunis* L. seeds (Olesen et al., 2015) have been performed by multispectral imaging systems.

Mold-infested seeds of sunflower were discriminated against using nondestructive multispectral imaging showing an excellent classification (accuracy >97%) for intact sunflower seeds (Ma et al., 2018). Multispectral imaging combined with multivariate PCA clustering could effectively cluster the sunflower seeds based on their degree of rancidity. Similar multivariate image analysis for discrimination among maize, pea, soybean, and wheat seeds and their hybrid descendants through the PLS-DA model has been performed (Chevallier et al., 2006; Liu et al., 2016).

22.5 Fluorescence and photoluminescence imaging in seeds

An interesting study was conducted on the seed maturity and seed quality by measuring the chlorophyll fluorescence of testa of cabbage seeds. The chlorophyll fluorescence sorting method is highly sensitive and is noninvasive. The fluorescence, which is generated from chlorophyll, is measured at great speed. The seeds were classified into four subgroups (low, medium, high, very high) based on the color intensity exhibited by them on the chlorophyll fluorescence signals during germination. Several parameters were considered to estimate the quality of seeds including germination %, germination rate, normal seedling%, and uniformity of germination. It was found that the fluorescence signal obtained was inversely correlated with these qualities. It is also stated that the testa of cabbage seed consists of chlorophyll-a, which can be excited, and accordingly, the quantitative determination of the presence of green pigment could be made with greater sensitivity. This new sorting feature makes fluorescence a highly suitable technique for the determination of seed quality (Jalink et al., 1998). Fluorescence spectroscopy (440 nm excitation wavelength) has been used for separating or sorting the raw and ripened seeds of chilly (Barua et al., 2010). Raw chilly seeds gave fluorescence peaks at the green and red regions; however, ripen chilly seeds showed a strong green band, but the red band was absent.

The stress produced in ryegrass (*Lolium multiflorum*) (in vivo) due to the presence of lead, cadmium, and copper was investigated using plant photoluminescence based on He-Cd excitation source whose excitation wavelength is 325 nm, which can emit the chlorophyll fluorescence in ryegrass leaves. Presence of three emission peaks of chlorophyll fluorescence was reported at blue (450 nm), red (690 nm), and far-red (735 nm). These peaks were greatly affected by the concentrations of copper, cadmium, and lead in the soil. A band at 450 nm having the highest peak with the increased concentration of Cu, Pb, and Cd in the soil was observed, and no emminent changes were observed at 690 and 735 nm. However, they reported that fluorescence ratios for blue/red (F450/F690) and blue/far-red (F450/F735) were highly sensitive, which can be considered for determining heavy metal stress in the plants. Therefore, this method of recording fluorescence emission spectra can be used to detect the plant stress caused by heavy metals in polluted soil. Furthermore, research can be done using plant photoluminescence by exploiting this He-Cd excitation source for detecting heavy metal stress using seeds and other parts of plants (Cheng et al., 2018).

Delayed luminescence (DL) is a phenomenon consisting of the emission of optical photons because of excitation of the system by illumination (Scordino et al., 2008). The advantage of using DL is that it is noninvasive and is suitable for dry seeds. The decay kinetic models of DL of soybean seeds and seeds powder were studied using a highly sensitive single-photon counting device with a multichannel single-photon counter and a photomultiplier tube. The study also demonstrated that the effect of seed morphology on the process changes in DL. For instance, the shape, surface, finishing a degree, and the thickness of the seed influence the DL spectrum. Hence, the DL spectra of soybean seeds reveal the information only about its visible features, but the information about internal components is not available. Therefore, soybean seeds were powdered and the individual components of the seed were irradiated revealing the complete information about the seeds (Deng et al., 2018). A correlation between the physiological state of the soybean seeds and the DL was established (Musumeci et al., 1994). DL was used to detect the thermal stress induced in the soybean seeds, and they also used soybean seeds of different vegetative vigor for measuring photo-induced emission. The total reemission was found to be directly proportional to the growth velocity of the seeds (Costanzo et al., 2008). Therefore, the measurement of DL can be exploited to determine the seed vitality and viability.

The ultra-weak DL in the coffee seeds during the process of germination and aging and positive correlations between the seed viability and photon-counting was investigated. They used a specialized photon-counting unit for recording the DL of coffee seeds. They incubated coffee seeds for germination under optimal conditions for 15–30 days and recorded photo counting distributions, DL-fitting parameters, and other statistical analysis (De Mello Gallep, 2014). These findings show that DL measurement can be attributed to the physical features of the seed and also to evaluate the seed quality.

DL was exploited to determine the quality and other information such as firmness, dry matter, color, soluble solids, and the respiration process of the tomato (*Lycopersicon esculentum Var. cerasiforme*) (Butz et al., 2005). The tomatoes were harvested at different stages of maturity, and they were incubated in dark for about 10 days to ripe. After 10 days of incubation, they observed the same-colored tomatoes; however, there were major variations in the DL, soluble solids, and dry matter. These results were directly linked to the fruit's harvest and ripeness (Triglia et al., 1998). The F1 watermelon dried seeds of "Mirella" were evaluated for their germination performance, which is the principal index that reflects the quality of seeds (Grasso et al., 2018). Their work also evaluated the time decay of spectral components of DL at the wavelengths of 450, 550, and 650 nm and reported that the changes in the time decay of DL spectra observed at these regions are strictly related to the functional state of the seed or the biological sample that is used for the analysis. The measurement of changes in the time decay of DL spectra of a seed enables the researcher to determine the quality, viability, and to evaluate the growth performance. This is a rapid, noninvasive, and especially nondestructive tool for the agro-based industries to discriminate the naturally or artificially damaged seeds from the normal seeds and to improve the quality of seeds. As this technique is quick, cheap, and doesn't require professional people to handle it, it can be easily engineered for seed sorting.

Synchrotron X-ray micro-fluorescence was used to examine the mineral localization of different elements such as Zn, K, Ca, Mn, and Fe in the rice grain. They particularly focused on the in vivo distribution of these minerals during the progressive stages of seed germination. It was found that the process of milling polishing of husk and bran removes half of Zn concentration, 2/3rd of the total Iron, and almost all Potassium, Calcium, and Manganese contents from the rice. Differences in the mobility of the elements inside the seed during the process of germination were observed and were reported as element-specific. As the traditional method of visualization of elements in plant tissues is not very successful, the concentrations of the micronutrients present in the plant samples may greatly differ while using traditional methods such as chemical imaging. Due to this low sensitivity, X-ray micro-fluorescence imaging was implied. The knowledge of these elemental constituents and their localization in the seeds can help agro-based industries for biofortification of the seeds, enhance the nutritional quality of food products by implication of certain cereals that are rich with micronutrients, and on top of that this technique is nondestructive, easier, quicker way to identify the minerals, which play an important role during seed germination (Lu et al., 2013).

The magnetic resonance imaging (MRI) approach was used to examine the lipid storage in the seeds (Neuberger et al., 2008). It was found that the local oil storage is closely coordinated with the tissue differentiation and seed maturation. Therefore, this MRI-based lipid mapping method can be used to visualize the localization of lipids in the seeds and also for the quantification. As lipid storage is one of the stages in seed maturation, this method can be used to follow up the fate of stored lipid in the seed. It is nondestructive and can be applicable for the elucidation of the oil accumulation process and storage metabolism in seeds. As "seeing is understanding," the visualization technique is gaining more importance in plant science. The combination of Mr imaging with other techniques can enable the researcher to extend the studies for other cells/organelles of plants. Therefore, this method is useful in the development of new approaches to transgenic research and plant breeding. Fig. 22.5 demonstrates the X-ray micro-fluorescence images of rice grains and their elemental composition (Lu et al., 2013).

22.6 Thermal imaging in seeds

Thermal imaging is also termed as Infrared Thermography and basically deals with the emitted radiations (Meola & Carlomagno, 2004). It exhibits the image in

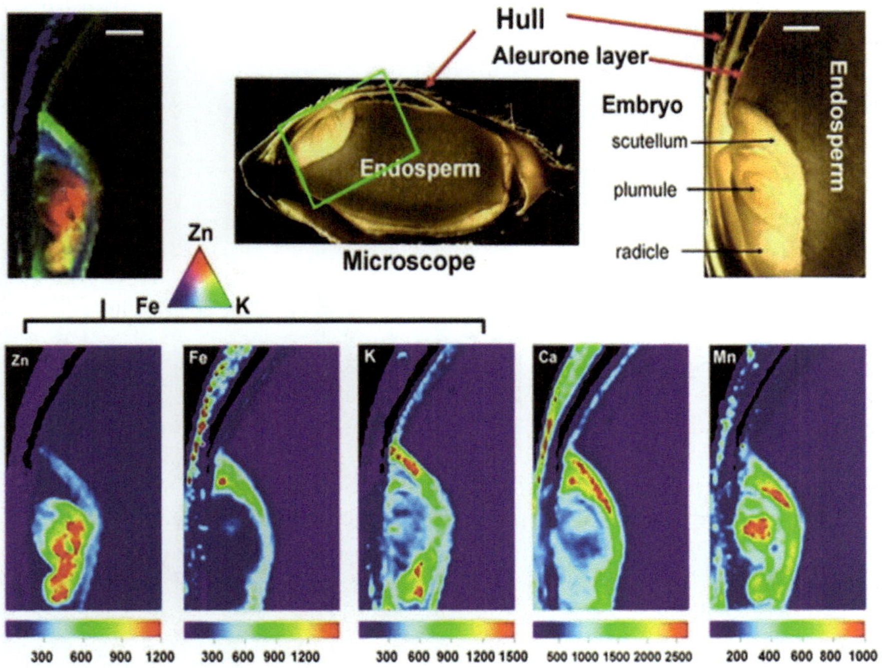

FIGURE 22.5

X-ray micro-fluorescence imaging to locate and visualize the elemental components present in rice grain (Lu et al., 2013).

the form of temperature distribution by detecting the amount of emitted infrared energy and then transforming this energy into temperature. For this we need thermal cameras having special types of lens that allows the IR radiation to pass through it. The light hits the sensors, and these sensors will make the image called a thermograph, which is then converted into a digital image and has been effectively used for testing the seed viability (Fig. 22.6).

This technique was first used to gage the viability of legume seeds throughout water uptake during the seed germination process (Men et al., 2017; Scarafoni et al., 2007). To accurately classify the seeds into viable, aged, and nonviable, a statistical model called the SVM can be used to determine the hidden spectral details (Elmasry et al., 2019; He et al., 2019). By using the SVM statistical model, we can classify the seeds as aged, nonaged, or viable using fewer temperature data, either before seed germination or within 3 hours of seed germination. This method has other useful applications such as the detection of sprout damage and detection of fungal damage in stored wheat seed (Singh et al., 2009).

Thermal imaging was used to track the growth of seedling organs using an automatic technique incorporated in the photographs to segment the organs of

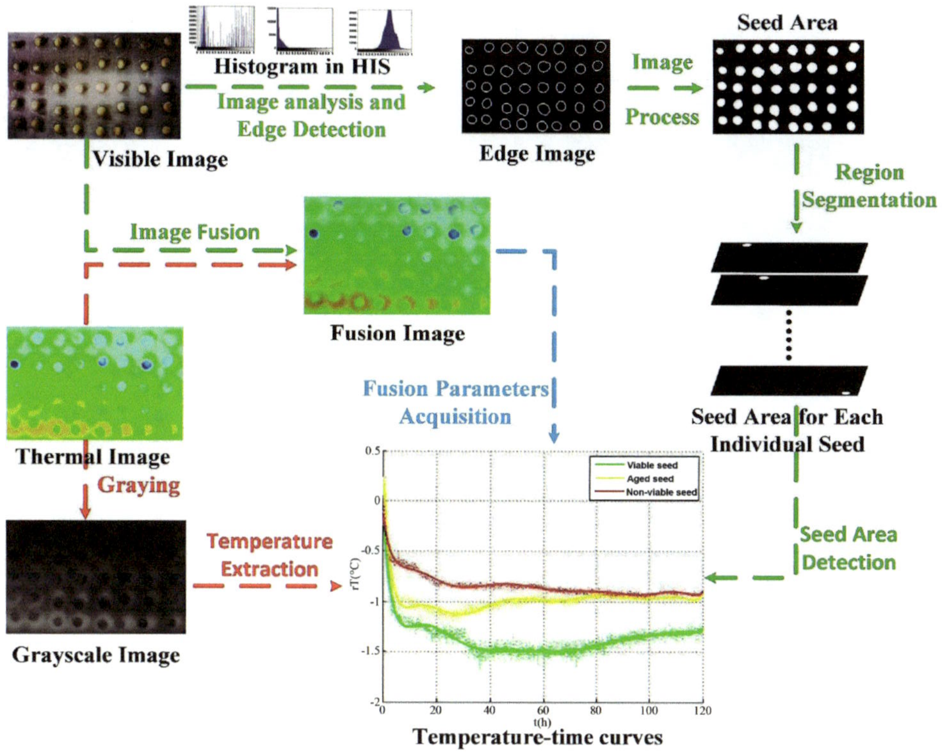

FIGURE 22.6

Flow chart illustrating a series of steps for analyzing thermal and visible images data for testing the seed viability (Men et al., 2017).

seedlings' using the thermal contrasts. Expert operators visually analyze the effects of this automated process, and this demonstrates that thermal imaging is a noninvasive imaging technique that provides functional knowledge on the physiology of the seedlings (Kherlopian et al., 2008). In another method for recording the images, it was found that infrared thermography can detect biophysical and biochemical changes related to viability and imbibition (Kranner et al., 2010) of viable, aged, as well as dead seeds of *Pisum sativum*. This method uses a tool that gives results within the first 3 hours of water uptake. Therefore, the seeds used in the test can be dried back to normal moisture content and stored for future use.

22.7 Light detection and ranging imaging

The development of 3D imaging technologies has ushered a new dimension to the assessment of seed quality in the field conditions (Omasa et al., 2007). Light detection and ranging (LiDAR) is a powerful yet cost-effective 3D imaging technology, which uses the principles of remote sensing in combination with laser scanners (Yu et al., 2004). LiDAR can estimate the 3D properties of the plants by obtaining the 3D information from the plants with the help of the sensors (Reutebuch et al., 2005). This allows LiDAR to directly measure the plant and canopy morphology (Hyyppä et al., 2001; Omasa et al., 2007). It can effectively map the whole seed production plot and scan the same to determine any differences in morphology of the seed crop (Brandtberg et al., 2003; Hosoi & Omasa, 2006).

LiDAR technology has been used for the detection of morphological variation in the seed crop such as the presence of tall plants in dwarf or semidwarf variety or vice versa (Wallace et al., 2012) (Fig. 22.7). Similarly, LiDAR can also effectively find the canopy coverage of the crop grown and has been successfully used in wheat (Friedli et al., 2016; Jimenez-Berni et al., 2018) and vineyard crops (Moreno et al., 2020). This procedure can be used to detect the differences in the canopy coverage of the seed crop, which in turn can be used to find if there are any off types in the field. LiDAR technology has also been used to detect the presence of weed species in the inter-row areas of the seed crop. Using the terrestrial version of LiDAR, weed species have been detected in species such as maize (Andújar, Escolà, et al., 2013; Andújar, Rueda-Ayala, et al., 2013; Zisi et al., 2018).

LiDAR technology, however, has not been used to its maximum potential (Omasa et al., 2007), especially for genetic purity testing of the seed crop. Conventionally, genetic purity of a seed crop can be determined by morphological observations, which are often confirmed with grow-out tests in addition to chemical tests and molecular techniques (Alam et al., 2015; Kumar et al., 2021; Tian et al., 2014). LiDAR may have application, if used for the purpose, will allow remote detection of crop morphological variations, weed identification, and

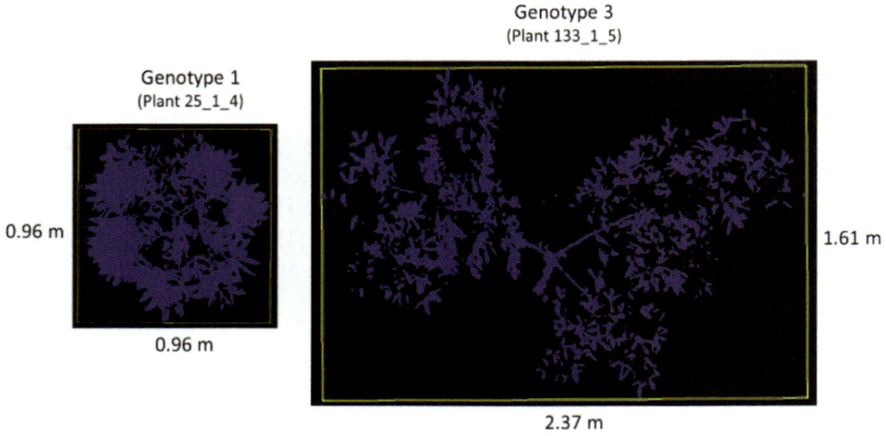

FIGURE 22.7

LiDAR image illustrates the difference in the morphology between two genotypes (Adams et al., 2021).

disease diagnosis (Friedli et al., 2016; Wallace et al., 2012; Zisi et al., 2018), which can give an idea of the genetic purity of the seed crop. Certification of a seed crop is incomplete without proper field inspection by qualified monitoring inspectors. Field inspection can also be remotely handled by LiDAR technologies recounting its immense prospect in seed quality testing and seed certification.

22.8 Challenges and limitations

The technologies discussed herein are very new and advanced, so along with benefits, there will be certain drawbacks, which will need to be attended to. In case of infrared spectroscopy, the methodology involved in sample preparation is tedious and difficult (Amir et al., 2013; Farber et al., 2020). In another report, FT-NIR spectroscopy, being able to distinguish viable seeds from and nonviable seeds, was found to be better than Raman spectroscopy as FT-NIR can validate and precisely differentiate viable seeds from nonviable seeds (Ambrose et al., 2016). Therefore, the technologies for detecting Raman scattering were not completely developed (Gierlinger & Schwanninger, 2007). The imaging methods of seed quality determination require high-precision detectors for the collection of the data, and analysis of these data requires high-end computers (de Medeiros et al., 2020; Elmasry et al., 2019). For the storage of the images generated during the analysis, sizable data storage facility is also essential (Stawoska et al., 2020). Some other drawbacks of these technologies include the use of more electricity, no proper procedure for selecting optimum wavelength, lack of a common algorithm befitting all applications, difficulty in standardizing the models, and high cost involvement

(Elmasry et al., 2019). Additional difficulties may arise due to differences in crops, varieties, and their growth conditions, the duration of a particular crop, and its location.

22.9 Future prospects

The spectroscopic and imaging methods, despite their limitations and challenges, are anticipated to move from laboratory applications to practical field-level applications in the form of seed monitoring systems in the near future. They are advanced techniques having a higher sampling efficiency, which covers more variation and makes the model a more efficient one (Wang et al., 2004). By decreasing the cost and increasing speed and capability of computer hardware system, it will aid the farmers and seed industry. These techniques provide ample information about the characteristics of seed quality, spatial imaging provides the information about parameters of phenotyping, various morphological, physiological, and physiochemical characteristics of seeds are evaluated using spectroscopic and chemometric tools. They techniques have turned out to be researcher-friendly and resulted in the designing of quick, precise, and cost-effective systems that need to be installed in the seed and grain industries. In addition to these, technology such as the LiDAR imaging can be prospected in revolutionizing the remote field inspection and monitoring of the seed crop. They are also necessary to abet and confront certain vital research challenges in the domain of seed science and technology (Shrestha et al., 2015).

22.10 Conclusion

As the traditional methods in seed quality assessment are a time-consuming and tedious job, a rapid and noninvasive technique must be designed and inculcated for evaluation of seeds. This is where the concepts of Spectroscopy and Imaging come into the picture. Both these are advanced techniques and at the same time offer a holistic approach for seed quality assessment, which finally determines if the seeds are suitable for certification or not. Over the past 5 years, there has been a considerable rise in the research oriented around these techniques owing to the modernization of the laboratories and the need for obtaining accurate results in minimum time. Efforts in this area will be needed in further simplifying the technology and reducing the cost involved to make it sustainable, uncomplicated, and more affordable to be used by the farming community.

Acknowledgment

VJS, BS, and NM thankfully acknowledge the Department of Biotechnology, Government of India, for providing research grants to Department of PBG, AAU & Dept. of

Biophysics, MAHE (Grant No. BT/PR25099/NER/95/1014/2017). NM also acknowledges the Department of Science and Technology (DST), Government of India, for the financial support (project number: DST/INT/Thai/P-10/2019).

Funding

This research was carried out using the assets created by the grant received from the Department of Biotechnology (Grant No. BT/PR25099/NER/95/1014/2017) and the Department of Science and Technology (project number: DST/INT/Thai/P-10/2019), Government of India.

Abbreviations

DA	Discriminant aanalysis
DL	Delayed luminescence
FT	Fourier transform
HTML	HyperText Markup Language
K-NN	k-nearest neighbors (K-NN)
LiDAR	Light detection and ranging
LS	Least-squares
MIR	Mid-infrared
MRI	Magnetic resonance imaging
NIR	Near-infrared
NMR	Nuclear magnetic resonance spectrometry
PCA	Principal component analysis
PLS	Partial least squares
SIMCA	Soft Independent Modeling of Class Analogy
SVM	Support vector machines

References

Adams, T., Bruton, R., Ruiz, H., et al. (2021). Prediction of aboveground biomass of three cassava (*Manihot esculenta*) genotypes using a terrestrial laser scanner. *Remote Sensing*, *13*, 1272. Available from https://doi.org/10.3390/rs13071272.

Alam, M., Islam, M., & Begum, S. (2015). Grow out test and determination of physical characteristics of seed of rice landraces. *International Journal of Agricultural Research, Innovation and Technology*, *5*, 15–21. Available from https://doi.org/10.3329/ijarit.v5i1.24582.

Al-Amery, M., Geneve, R. L., Sanches, M. F., et al. (2018). Near-infrared spectroscopy used to predict soybean seed germination and vigour. *Seed Science Research*, *28*, 245–252. Available from https://doi.org/10.1017/S0960258518000119.

Ambrose, A., Lohumi, S., Lee, W. H., & Cho, B. K. (2016). Comparative nondestructive measurement of corn seed viability using Fourier transform near-infrared (FT-NIR) and

Raman spectroscopy. *Sensors and Actuators, B: Chemical*, *224*, 500–506. Available from https://doi.org/10.1016/j.snb.2015.10.082.

Amir, R. M., Anjum, F. M., Khan, M. I., et al. (2013). Application of Fourier transform infrared (FTIR) spectroscopy for the identification of wheat varieties. *Journal of Food Science and Technology*, *50*, 1018–1023. Available from https://doi.org/10.1007/s13197-011-0424-y.

Andújar, D., Escolà, A., Rosell-Polo, J. R., et al. (2013). Potential of a terrestrial LiDAR-based system to characterise weed vegetation in maize crops. *Computers and Electronics in Agriculture*, *92*, 11–15. Available from https://doi.org/10.1016/j.compag.2012.12.012.

Andújar, D., Rueda-Ayala, V., Moreno, H., et al. (2013). Discriminating crop, weeds and soil surface with a terrestrial LiDAR sensor. *Sensors (Switzerland)*, *13*, 14662–14675. Available from https://doi.org/10.3390/s131114662.

Arganosa, G. C., Warkentin, T. D., Racz, V. J., et al. (2006). Prediction of crude protein content in field peas using near infrared reflectance spectroscopy. *Canadian Journal of Plant Science*, *86*, 157–159. Available from https://doi.org/10.4141/P04-195.

Arora, C., Maji, P., & Bajpai, P. K. (2020). *Raman spectroscopy and its biomedical applications. Essential techniques for medical and life scientists: A guide to contemporary methods and current applications- part II* (pp. 62–84). BENTHAM SCIENCE PUBLISHERS.

Attaviroj, N., Kasemsumran, S., & Noomhorm, A. (2011). Rapid variety identification of pure rough rice by Fourier-transform near-infrared spectroscopy. *Cereal Chemistry*, *88*, 490–496. Available from https://doi.org/10.1094/CCHEM-03-11-0025.

Baca-Bocanegra, B., Nogales-Bueno, J., Hernández-Hierro, J. M., & Heredia, F. J. (2020). Valorization of American barrel-shoot wastes: Effect of post fermentative addition and readdition on phenolic composition and chromatic quality of Syrah red wines. *Molecules (Basel, Switzerland)*, *25*, 774. Available from https://doi.org/10.3390/molecules25040774.

Baek, I., Kusumaningrum, D., Kandpal, L. M., et al. (2019). Rapid measurement of soybean seed viability using Kernel-based multispectral image analysis. *Sensors (Switzerland)*, *19*, 271. Available from https://doi.org/10.3390/s19020271.

Baranski, R., Baranska, M., & Schulz, H. (2005). Changes in carotenoid content and distribution in living plant tissue can be observed and mapped in situ using NIR-FT-Raman spectroscopy. *Planta*, *222*, 448–457. Available from https://doi.org/10.1007/s00425-005-1566-9.

Barron, C. (2011). Prediction of relative tissue proportions in wheat mill streams by Fourier transform mid-infrared spectroscopy. *Journal of Agricultural and Food Chemistry*, *59*, 10442–10447. Available from https://doi.org/10.1021/jf201886c.

Barua, A. G., Hazarika, S., Pathak, J. S., & Manhas, S. (2010). Spectral signature of ripening in the seeds of chilli (*Capsicum annum* L.). *Spectroscopy Letters*, *43*, 163–166. Available from https://doi.org/10.1080/00387010903284307.

Bauriegel, E., Giebel, A., & Herppich, W. B. (2011). Hyperspectral and chlorophyll fluorescence imaging to analyse the impact of Fusarium culmorum on the photosynthetic integrity of infected wheat ears. *Sensors*, *11*, 3765–3779. Available from https://doi.org/10.3390/s110403765.

Bishaw, Z., Struik, P. C., & van Gastel, A. J. G. (2012). Farmers' seed sources and seed quality: 1. Physical and physiological quality. *Journal of Crop Improvement*, *26*, 655–692. Available from https://doi.org/10.1080/15427528.2012.670695.

Bradbeer, J. W. (1988). *Germination. Seed dormancy and germination* (pp. 27–37). Netherlands: Springer.

Brandtberg, T., Warner, T. A., Landenberger, R. E., & McGraw, J. B. (2003). Detection and analysis of individual leaf-off tree crowns in small footprint, high sampling density LiDAR data from the eastern deciduous forest in North America. *Remote Sensing of Environment*, *85*, 290–303. Available from https://doi.org/10.1016/S0034-4257(03)00008-7.

Butz, P., Hofmann, C., & Tauscher, B. (2005). Recent developments in noninvasive techniques for fresh fruit and vegetable internal quality analysis. *Journal of Food Science*, *70*, R131–R141. Available from https://doi.org/10.1111/j.1365-2621.2005.tb08328.x.

Çeliktaş, N., & Konuşkan, Ö. (2020). Near infrared reflectance spectroscopy and multivariate analyses for fast and non-destructive prediction of corn seed germination. *Turkish Journal of Agriculture – Food Science and Technology*, *8*, 1636–1642. Available from https://doi.org/10.24925/turjaf.v8i8.1636-1642.3384.

Chen, O., Zhao, J., Chauhan, V. P., et al. (2013). Compact high-quality CdSe-CdS core-shell nanocrystals with narrow emission linewidths and suppressed blinking. *Nature Materials*, *12*, 445–451. Available from https://doi.org/10.1038/nmat3539.

Cheng, J., Lv, Y., Yu, Y., et al. (2018). Measuring heavy metal stress in ryegrass using helium-cadmium excitation-based photoluminescence. *Environmental Science and Pollution Research*, *25*, 7059–7066. Available from https://doi.org/10.1007/s11356-017-0981-3.

Chevallier, S., Bertrand, D., Kohler, A., & Courcoux, P. (2006). Application of PLS-DA in multivariate image analysis. *Journal of Chemometrics*, *20*, 221–229. Available from https://doi.org/10.1002/cem.994.

Collins, F. I., Alexander, D. E., Rodgers, R. C., & Silvela, S. L. (1967). Analysis of oil content of soybeans by wide-line NMR. *Journal of the American Oil Chemists' Society*, *44*, 708–710. Available from https://doi.org/10.1007/BF02679638.

Copeland, L. O., & McDonald, M. B. (1999). *Seed germination. Principles of seed science and technology* (pp. 59–110). US: Springer.

Corbineau, F. (2012). Markers of seed quality: From present to future. *Seed Science Research*, *22*, S61–S68. Available from https://doi.org/10.1017/S0960258511000419.

Costanzo, E., Gulino, M., Lanzanò, L., et al. (2008). Single seed viability checked by delayed luminescence. *European Biophysics Journal*, *37*, 235–238. Available from https://doi.org/10.1007/s00249-007-0221-8.

de Medeiros, A. D., da Silva, L. J., Ribeiro, J. P. O., et al. (2020). Machine learning for seed quality classification: An advanced approach using merger data from FT-NIR spectroscopy and x-ray imaging. *Sensors (Switzerland)*, *20*, 1–12. Available from https://doi.org/10.3390/s20154319.

De Mello Gallep, C. (2014). Ultraweak, spontaneous photon emission in seedlings: Toxicological and chronobiological applications. *Luminescence: the Journal of Biological and Chemical Luminescence*, *29*, 963–968. Available from https://doi.org/10.1002/bio.2658.

Deng, J., Liu, Q., Zhang, Q., et al. (2018). Comparative study on composition, physicochemical and antioxidant characteristics of different varieties of kiwifruit seed oil in China. *Food Chemistry*, *264*, 411–418. Available from https://doi.org/10.1016/j.foodchem.2018.05.063.

Diriba Bula, H., Gelmesa Soboka, D., & Kasim Ahmed, K. (2019). Prospects and challenges in seed sector development: Lessons from Eastern Ethiopia. *European Business & Management*, *5*, 51. Available from https://doi.org/10.11648/j.ebm.20190504.11.

Dos Remedios, C. G., & Moens, P. D. J. (1995). Fluorescence resonance energy transfer spectroscopy is a reliable "ruler" for measuring structural changes in proteins. Dispelling the problem of the unknown orientation factor. *Journal of Structural Biology*, *115*, 175–185. Available from https://doi.org/10.1006/jsbi.1995.1042.

Edwards, H. G. M., Villar, S. E. J., De Oliveira, L. F. C., & Le Hyaric, M. (2005). Analytical Raman spectroscopic study of cacao seeds and their chemical extracts. *Analytica Chimica Acta*, *538*, 175–180. Available from https://doi.org/10.1016/j.aca.2005.02.039.

Egging, V., Nguyen, J., & Kurouski, D. (2018). Detection and identification of fungal infections in intact wheat and sorghum grain using a hand-held Raman spectrometer. *Analytical Chemistry*, *90*, 8616–8621. Available from https://doi.org/10.1021/acs.analchem.8b01863.

ElMasry, G., ElGamal, R., Mandour, N., et al. (2020). Emerging thermal imaging techniques for seed quality evaluation: Principles and applications. *Food Research International*, *131*, 109025. Available from https://doi.org/10.1016/j.foodres.2020.109025.

Elmasry, G., Mandour, N., Wagner, M. H., et al. (2019). Utilization of computer vision and multispectral imaging techniques for classification of cowpea (*Vigna unguiculata*) seeds. *Plant Methods*, *15*, 1–16. Available from https://doi.org/10.1186/s13007-019-0411-2.

Esteve Agelet, L., Ellis, D. D., Duvick, S., et al. (2012). Feasibility of near infrared spectroscopy for analyzing corn kernel damage and viability of soybean and corn kernels. *Journal of Cereal Science*, *55*, 160–165. Available from https://doi.org/10.1016/j.jcs.2011.11.002.

Esteve Agelet, L., & Hurburgh, C. R. (2014). Limitations and current applications of near infrared spectroscopy for single seed analysis. *Talanta*, *121*, 288–299. Available from https://doi.org/10.1016/j.talanta.2013.12.038.

Fabiyi, S. D., Vu, H., Tachtatzis, C., et al. (2020). Varietal classification of rice seeds using RGB and hyperspectral images. *IEEE Access*, *8*, 22493–22505. Available from https://doi.org/10.1109/ACCESS.2020.2969847.

Farber, C., & Kurouski, D. (2018). Detection and identification of plant pathogens on Maize Kernels with a hand-held Raman spectrometer. *Analytical Chemistry*, *90*, 3009–3012. Available from https://doi.org/10.1021/acs.analchem.8b00222.

Farber, C., Sanchez, L., Rizevsky, S., et al. (2020). Raman spectroscopy enables non-invasive identification of peanut genotypes and value-added traits. *Scientific Reports*, *10*, 7730. Available from https://doi.org/10.1038/s41598-020-64730-w.

Feng, L., Zhu, S., Liu, F., et al. (2019). Hyperspectral imaging for seed quality and safety inspection: A review. *Plant Methods*, *15*, 91. Available from https://doi.org/10.1186/s13007-019-0476-y.

Finch-Savage, W. E., & Bassel, G. W. (2016). Seed vigour and crop establishment: Extending performance beyond adaptation. *Journal of Experimental Botany*, *67*, 567–591. Available from https://doi.org/10.1093/jxb/erv490.

Franklin, S. E. (2001). *Remote sensing for sustainable forest management*. CRC Press.

Friedli, M., Kirchgessner, N., Grieder, C., et al. (2016). Terrestrial 3D laser scanning to track the increase in canopy height of both monocot and dicot crop species under field conditions. *Plant Methods*, *12*, 9. Available from https://doi.org/10.1186/s13007-016-0109-7.

Gierlinger, N., & Schwanninger, M. (2007). The potential of Raman microscopy and Raman imaging in plant research. *Spectroscopy*, *21*, 69–89. Available from https://doi.org/10.1155/2007/498206.

Gillgren, T., & Gorzsás, A. (2016). A one-pot set-up for real-time reaction monitoring by FTIR spectroscopy. *Wood Science and Technology*, *50*, 567–580. Available from https://doi.org/10.1007/s00226-016-0801-9.

Grasso, R., Gulino, M., Giuffrida, F., et al. (2018). Non-destructive evaluation of watermelon seeds germination by using Delayed Luminescence. *Journal of Photochemistry and Photobiology B: Biology*, *187*, 126–130. Available from https://doi.org/10.1016/j.jphotobiol.2018.08.012.

Gunaratne, A., Wu, K., Li, D., et al. (2013). Antioxidant activity and nutritional quality of traditional red-grained rice varieties containing proanthocyanidins. *Food Chemistry*, *138*, 1153–1161. Available from https://doi.org/10.1016/j.foodchem.2012.11.129.

Gupta, A., & Kumar, R. (2020). *Management of seed-borne diseases: An integrated approach. Seed-borne diseases of agricultural crops: Detection, diagnosis & management* (pp. 717–745). Singapore: Springer.

He, X., Feng, X., Sun, D., et al. (2019). Rapid and nondestructive measurement of rice seed vitality of different years using near-infrared hyperspectral imaging. *Molecules (Basel, Switzerland)*, *24*, 2227. Available from https://doi.org/10.3390/molecules24122227.

Herrero, A. M. (2008). Raman spectroscopy a promising technique for quality assessment of meat and fish: A review. *Food Chemistry*, *107*, 1642–1651. Available from https://doi.org/10.1016/j.foodchem.2007.10.014.

Horn, P. J., Neogi, P., Tombokan, X., et al. (2011). Simultaneous quantification of oil and protein in cottonseed by low-field time-domain nuclear magnetic resonance. *JAOCS, Journal of the American Oil Chemists' Society*, *88*, 1521–1529. Available from https://doi.org/10.1007/s11746-011-1829-5.

Hosoi, F., & Omasa, K. (2006). Voxel-based 3-D modeling of individual trees for estimating leaf area density using high-resolution portable scanning LiDAR. *IEEE Transactions on Geoscience and Remote Sensing*, *44*, 3610–3618. Available from https://doi.org/10.1109/TGRS.2006.881743.

Hu, X., Yang, L., & Zhang, Z. (2020). Non-destructive identification of single hard seed via multispectral imaging analysis in six legume species. *Plant Methods*, *16*, 116. Available from https://doi.org/10.1186/s13007-020-00659-5.

Huang, G., Curt, S. R., Wang, K., & Markides, C. N. (2020). Challenges and opportunities for nanomaterials in spectral splitting for high-performance hybrid solar photovoltaic-thermal applications: A review. *Nano Materials Science*, *2*, 183–203. Available from https://doi.org/10.1016/j.nanoms.2020.03.008.

Huang, H., Yu, H., Xu, H., & Ying, Y. (2008). Near infrared spectroscopy for on/in-line monitoring of quality in foods and beverages: A review. *Journal of Food Engineering*, *87*, 303–313. Available from https://doi.org/10.1016/j.jfoodeng.2007.12.022.

Huang, M., Wang, Q. G., Zhu, Q. B., et al. (2015). Review of seed quality and safety tests using optical sensing technologies. *Seed Science and Technology*, *43*, 337–366. Available from https://doi.org/10.15258/sst.2015.43.3.16.

Hyyppä, J., Kelle, O., Lehikoinen, M., & Inkinen, M. (2001). A segmentation-based method to retrieve stem volume estimates from 3-D tree height models produced by laser scanners. *IEEE Transactions on Geoscience and Remote Sensing*, *39*, 969–975. Available from https://doi.org/10.1109/36.921414.

Jalink, H., Van Der Schoor, R., Frandas, A., et al. (1998). Chlorophyll fluorescence of Brassica oleracea seeds as a non-destructive marker for seed maturity and seed performance. *Seed Science Research*, *8*, 437–443. Available from https://doi.org/10.1017/s0960258500004402.

Jiao, Z., Si, X. X. I., Li, G. K. E., et al. (2010). Unintended compositional changes in transgenic rice seeds (Oryza sativa L.) Studied by spectral and chromatographic analysis coupled with chemometrics methods. *Journal of Agricultural and Food Chemistry*, *58*, 1746–1754. Available from https://doi.org/10.1021/jf902676y.

Jimenez-Berni, J. A., Deery, D. M., Rozas-Larraondo, P., et al. (2018). High throughput determination of plant height, ground cover, and above-ground biomass in wheat with LiDAR. *Frontiers in Plant Science*, *9*, 237. Available from https://doi.org/10.3389/fpls.2018.00237.

Jones, C. E., Klewpatinond, M., Abdelraheim, S. R., et al. (2005). Probing copper2 + binding to the prion protein using diamagnetic nickel2 + and 1H NMR: The unstructured N terminus facilitates the coordination of six copper2 + ions at physiological concentrations. *Journal of Molecular Biology*, *346*, 1393–1407. Available from https://doi.org/10.1016/j.jmb.2004.12.043.

Kandpal, L. M., Lee, S., Kim, M. S., et al. (2015). Short wave infrared (SWIR) hyperspectral imaging technique for examination of aflatoxin B1 (AFB1) on corn kernels. *Food Control*, *51*, 171–176. Available from https://doi.org/10.1016/j.foodcont.2014.11.020.

Karunakaran, C., Vijayan, P., Stobbs, J., et al. (2020). High throughput nutritional profiling of pea seeds using Fourier transform mid-infrared spectroscopy. *Food Chemistry*, *309*, 125585. Available from https://doi.org/10.1016/j.foodchem.2019.125585.

Khanal, S., Kushal, K. C., Fulton, J. P., et al. (2020). Remote sensing in agriculture—accomplishments, limitations, and opportunities. *Remote Sensing*, *12*, 1–29. Available from https://doi.org/10.3390/rs12223783.

Kherlopian, A. R., Song, T., Duan, Q., et al. (2008). A review of imaging techniques for systems biology. *BMC Systems Biology*, *2*, 74. Available from https://doi.org/10.1186/1752-0509-2-74.

Kim, G., Kim, G. H., Ahn, C. K., et al. (2013). Mid-infrared lifetime imaging for viability evaluation of lettuce seeds based on time-dependent thermal decay characterization. *Sensors (Switzerland)*, *13*, 2986–2996. Available from https://doi.org/10.3390/s130302986.

Kong, W., Zhang, C., Liu, F., et al. (2013). Rice seed cultivar identification using near-infrared hyperspectral imaging and multivariate data analysis. *Sensors (Basel, Switzerland)*, *13*, 8916–8927. Available from https://doi.org/10.3390/s130708916.

Kranner, I., Kastberger, G., Hartbauer, M., & Pritchard, H. W. (2010). Noninvasive diagnosis of seed viability using infrared thermography. *Proceedings of the National Academy of Sciences of the United States of America*, *107*, 3912–3917. Available from https://doi.org/10.1073/pnas.0914197107.

Krimmer, M., Farber, C., & Kurouski, D. (2019). Rapid and noninvasive typing and assessment of nutrient content of maize kernels using a handheld Raman spectrometer. *ACS Omega*, *4*, 16330–16335. Available from https://doi.org/10.1021/acsomega.9b01661.

Kumar, S. P. J., Susmita, C., Agarwal, D. K., et al. (2021). Assessment of genetic purity in rice using polymorphic SSR markers and its economic analysis with grow-out-test. *Food Analytical Methods*, *14*, 856–864. Available from https://doi.org/10.1007/s12161-020-01927-9.

Kusumaningrum, D., Lee, H., Lohumi, S., et al. (2018). Non-destructive technique for determining the viability of soybean (Glycine max) seeds using FT-NIR spectroscopy. *Journal of the Science of Food and Agriculture*, *98*, 1734–1742. Available from https://doi.org/10.1002/jsfa.8646.

Lee, H., Kim, M. S., Qin, J., et al. (2017). Raman hyperspectral imaging for detection of watermelon seeds infected with *Acidovorax citrulli*. *Sensors (Switzerland)*, *17*, 2188. Available from https://doi.org/10.3390/s17102188.

Li, J., Huang, W., Chen, L., et al. (2014). Variable selection in visible and near-infrared spectral analysis for noninvasive determination of soluble solids content of 'ya' pear. *Food Analytical Methods*, *7*, 1891–1902. Available from https://doi.org/10.1007/s12161-014-9832-8.

Liang, L., Liu, Z. X., Yang, M. H., et al. (2009). Discrimination of variety and authenticity for rice based on visual/near infrared reflection spectra. *Hongwai Yu Haomibo Xuebao/Journal of Infrared and Millimeter Waves*, *28*, 353–356. Available from https://doi.org/10.3724/SP.J.1010.2009.00353.

Liu, C., Liu, W., Lu, X., et al. (2016). Non-destructive discrimination of conventional and glyphosate-resistant soybean seeds and their hybrid descendants using multispectral imaging and chemometric methods. *Journal of Agricultural Science*, *154*, 1–12. Available from https://doi.org/10.1017/S0021859614001142.

Liu, D., Zeng, X. A., & Sun, D. W. (2015). Recent developments and applications of hyperspectral imaging for quality evaluation of agricultural products: A review. *Critical Reviews in Food Science and Nutrition*, *55*, 1744–1757. Available from https://doi.org/10.1080/10408398.2013.777020.

Lu, L., Tian, S., Liao, H., et al. (2013). Analysis of metal element distributions in rice (Oryza sativa L.) seeds and relocation during germination based on X-Ray fluorescence imaging of Zn, Fe, K, Ca, and Mn. *PLoS One*, *8*, e57360. Available from https://doi.org/10.1371/journal.pone.0057360.

Lu, Z. (2011). PubMed and beyond: A survey of web tools for searching biomedical literature. *Database*, *2011*, baq036. Available from https://doi.org/10.1093/database/baq036.

Ma, F., Zhang, B., Wang, W., et al. (2018). Potential use of multispectral imaging technology to identify moisture content and water-holding capacity in cooked pork sausages. *Journal of the Science of Food and Agriculture*, *98*, 1832–1838. Available from https://doi.org/10.1002/jsfa.8659.

Ma, T., Tsuchikawa, S., & Inagaki, T. (2020). Rapid and non-destructive seed viability prediction using near-infrared hyperspectral imaging coupled with a deep learning approach. *Computers and Electronics in Agriculture*, *177*, 105683. Available from https://doi.org/10.1016/j.compag.2020.105683.

Mahesh, S., Manickavasagan, A., Jayas, D. S., et al. (2008). Feasibility of near-infrared hyperspectral imaging to differentiate Canadian wheat classes. *Biosystems Engineering*, *101*, 50–57. Available from https://doi.org/10.1016/j.biosystemseng.2008.05.017.

Manley, M., McGoverin, C. M., Snyders, F., et al. (2013). Prediction of triticale grain quality properties, based on both chemical and indirectly measured reference methods, using near-infrared spectroscopy. *Cereal Chemistry*, *90*, 540–545. Available from https://doi.org/10.1094/CCHEM-02-13-0021-R.

Melchinger, A. E., Munder, S., Mauch, F. J., et al. (2017). High-throughput platform for automated sorting and selection of single seeds based on time-domain nuclear magnetic resonance (TD-NMR) measurement of oil content. *Biosystems Engineering*, *164*, 213–220. Available from https://doi.org/10.1016/j.biosystemseng.2017.10.011.

Men, S., Yan, L., Liu, J., et al. (2017). A classification method for seed viability assessment with infrared thermography. *Sensors (Switzerland)*, *17*, 845. Available from https://doi.org/10.3390/s17040845.

Meola, C., & Carlomagno, G. M. (2004). Recent advances in the use of infrared thermography. *Measurement Science and Technology*, *15*, R27–R58. Available from https://doi.org/10.1088/0957-0233/15/9/R01.

Moreno, H., Valero, C., Bengochea-Guevara, J. M., et al. (2020). On-ground vineyard reconstruction using a LiDAR-based automated system. *Sensors (Switzerland)*, *20*, 1102. Available from https://doi.org/10.3390/s20041102.

Mukasa, P., Wakholi, C., Mo, C., et al. (2019). Determination of viability of Retinispora (Hinoki cypress) seeds using FT-NIR spectroscopy. *Infrared Physics and Technology*, *98*, 62–68. Available from https://doi.org/10.1016/j.infrared.2019.02.008.

Mukasa, P., Wakholi, C., Mohammad, A. F., et al. (2020). Determination of the viability of retinispora (Hinoki cypress) seeds using shortwave infrared hyperspectral imaging spectroscopy. *Journal of Near Infrared Spectroscopy*, *28*, 70–80. Available from https://doi.org/10.1177/0967033519898890.

Musumeci, F., Triglia, A., Grasso, F., et al. (1994). Relation between delayed luminescence and functional state in soya seeds. *Il Nuovo Cimento D*, *16*, 65–73. Available from https://doi.org/10.1007/BF02452003.

Nakariyakul, S., & Casasent, D. P. (2011). Classification of internally damaged almond nuts using hyperspectral imagery. *Journal of Food Engineering*, *103*, 62–67. Available from https://doi.org/10.1016/j.jfoodeng.2010.09.020.

Neuberger, T., Sreenivasulu, N., Rokitta, M., et al. (2008). Quantitative imaging of oil storage in developing crop seeds. *Plant Biotechnology Journal*, *6*, 31–45. Available from https://doi.org/10.1111/j.1467-7652.2007.00294.x.

Olesen, M. H., Carstensen, J. M., & Boelt, B. (2011). Multispectral imaging as a potential tool for seed health testing of spinach (*Spinacia oleracea* L.). *Seed Science and Technology*, *39*, 140–150. Available from https://doi.org/10.15258/sst.2011.39.1.12.

Olesen, M. H., Nikneshan, P., Shrestha, S., et al. (2015). Viability prediction of Ricinus cummunis L. Seeds using multispectral imaging. *Sensors (Switzerland)*, *15*, 4592–4604. Available from https://doi.org/10.3390/s150204592.

Omasa, K., Hosoi, F., & Konishi, A. (2007). 3D lidar imaging for detecting and understanding plant responses and canopy structure. *Journal of Experimental Botany*, *58*, 881–898. Available from https://doi.org/10.1093/jxb/erl142.

Pang, L., Wang, J., Men, S., et al. (2021). Hyperspectral imaging coupled with multivariate methods for seed vitality estimation and forecast for *Quercus variabilis*. *Spectrochimica Acta – Part A: Molecular and Biomolecular Spectroscopy*, *245*, 118888. Available from https://doi.org/10.1016/j.saa.2020.118888.

Peng, W., & Wu, H. (2019). Flexible and stretchable photonic sensors based on modulation of light transmission. *Advanced Optical Materials*, *7*, 1900329. Available from https://doi.org/10.1002/adom.201900329.

Pu, H., Lin, L., & Sun, D. W. (2019). Principles of hyperspectral microscope imaging techniques and their applications in food quality and safety detection: A review. *Comprehensive Reviews in Food Science and Food Safety*, *18*, 853–866. Available from https://doi.org/10.1111/1541-4337.12432.

Pupeza, I., Huber, M., Trubetskov, M., et al. (2020). Field-resolved infrared spectroscopy of biological systems. *Nature*, *577*, 52–59. Available from https://doi.org/10.1038/s41586-019-1850-7.

Qiu, G., Lü, E., Lu, H., et al. (2018). Single-kernel FT-NIR spectroscopy for detecting supersweet corn (Zea mays L. saccharata sturt) seed viability with multivariate data analysis. *Sensors (Switzerland)*, *18*, 1010. Available from https://doi.org/10.3390/s18041010.

Rahman, A., & Cho, B. K. (2016). Assessment of seed quality using non-destructive measurement techniques: A review. *Seed Science Research*, *26*, 285–305. Available from https://doi.org/10.1017/S0960258516000234.

Reutebuch, S. E., Andersen, H. E., & McGaughey, R. J. (2005). Light detection and ranging (LIDAR): An emerging tool for multiple resource inventory. *Journal of Forestry*, *103*, 286–292. Available from https://doi.org/10.1093/jof/103.6.286.

Samii-Saket, G., Boersma, J. G., Ablett, G. R., et al. (2011). Single soybean seed NMR calibration for oil measurement using commercial cooking oils. *JAOCS, Journal of the American Oil Chemists' Society*, *88*, 1795–1798. Available from https://doi.org/10.1007/s11746-011-1849-1.

Scarafoni, A., Magni, C., & Duranti, M. (2007). Molecular nutraceutics as a mean to investigate the positive effects of legume seed proteins on human health. *Trends in Food Science and Technology*, *18*, 454–463. Available from https://doi.org/10.1016/j.tifs.2007.04.002.

Scordino, A., Musumeci, F., Gulino, M., et al. (2008). Delayed luminescence of microalgae as an indicator of metal toxicity. *Journal of Physics D: Applied Physics*, *41*, 155507. Available from https://doi.org/10.1088/0022-3727/41/15/155507.

Sendin, K., Manley, M., Baeten, V., et al. (2019). Near infrared hyperspectral imaging for white maize classification according to grading regulations. *Food Analytical Methods*, *12*, 1612–1624. Available from https://doi.org/10.1007/s12161-019-01464-0.

Seo, Y.-W., Ahn, C. K., Lee, H., et al. (2016). Non-destructive sorting techniques for viable pepper (*Capsicum annuum* L.) seeds using fourier transform near-infrared and Raman spectroscopy. *Journal of Biosystems Engineering*, *41*, 51–59. Available from https://doi.org/10.5307/jbe.2016.41.1.051.

Shahin, M. A., Hatcher, D. W., & Symons, S. J. (2012). *Development of multispectral imaging systems for quality evaluation of cereal grains and grain products. Computer vision technology in the food and beverage industries* (pp. 451–482). Elsevier.

Shetty, N., Min, T. G., Gislum, R., et al. (2011). Optimal sample size for predicting viability of cabbage and radish seeds based on near infrared spectra of single seeds. *Journal of Near Infrared Spectroscopy*, *19*, 451–461. Available from https://doi.org/10.1255/jnirs.966.

Shrestha, S., Deleuran, L. C., & Gislum, R. (2017). Separation of viable and non-viable tomato (*Solanum lycopersicum* L.) seeds using single seed near-infrared spectroscopy. *Computers and Electronics in Agriculture*, *142*, 348–355. Available from https://doi.org/10.1016/j.compag.2017.09.004.

Shrestha, S., Deleuran, L. C., Olesen, M. H., & Gislum, R. (2015). Use of multispectral imaging in varietal identification of tomato. *Sensors (Switzerland)*, *15*, 4496–4512. Available from https://doi.org/10.3390/s150204496.

Singh, C. B., Jayas, D. S., Paliwal, J., & White, N. D. G. (2009). Detection of insect-damaged wheat kernels using near-infrared hyperspectral imaging. *Journal of Stored Products Research*, *45*, 151–158. Available from https://doi.org/10.1016/j.jspr.2008.12.002.

Singh, C. B., Jayas, D. S., Paliwal, J., & White, N. D. G. (2010). Detection of midge-damaged wheat kernels using short-wave near-infrared hyperspectral and digital colour imaging. *Biosystems Engineering*, *105*, 380–387. Available from https://doi.org/10.1016/j.biosystemseng.2009.12.009.

Smith, J. S. C., & Register, J. C. (1998). Genetic purity and testing technologies for seed quality: A company perspective. *Seed Science Research*, *8*, 285–293. Available from https://doi.org/10.1017/s0960258500004189.

Smith, K. L., Steven, M. D., & Colls, J. J. (2004). Use of hyperspectral derivative ratios in the red-edge region to identify plant stress responses to gas leaks. *Remote Sensing of Environment*, *92*, 207–217. Available from https://doi.org/10.1016/j.rse.2004.06.002.

Stawoska, I., Staszak, A. M., Ciereszko, I., et al. (2020). Using isothermal calorimetry and FT-Raman spectroscopy for step-by-step monitoring of maize seed germination: case study. *Journal of Thermal Analysis and Calorimetry*, *142*, 755–763. Available from https://doi.org/10.1007/s10973-020-09525-x.

Sumriddetchkajorn, S., Suwansukho, K., & Buranasiri, P. (2010). Two-wavelength spectral imaging-based Thai rice breed identification. In J. Popp, W. Drexler, V. V. Tuchin, & D. L. Matthews (Eds.), *Biophotonics: Photonic solutions for better health care II* (p. 77150I). SPIE.

Tallada, J. G., Wicklow, D. T., Pearson, T. C., & Armstrong, P. R. (2011). Detection of Fungus-Infected Corn Kernels Using Near-Infrared Reflectance Spectroscopy and Color Imaging. *Transactions of the ASABE*, *54*, 1151–1158. Available from https://doi.org/10.13031/2013.37090.

Tian, X., Wang, Q., Huang, W., et al. (2020). Online detection of apples with moldy core using the Vis/NIR full-transmittance spectra. *Postharvest Biology and Technology*, *168*, 111269. Available from https://doi.org/10.1016/j.postharvbio.2020.111269.

Tian, Y., Guan, B., Zhou, D., et al. (2014). Responses of seed germination, seedling growth, and seed yield traits to seed pretreatment in maize (Zea mays L.). *Scientific World Journal*, *2014*, 1–8. Available from https://doi.org/10.1155/2014/834630.

Triglia, A., La Malfa, G., Musumeci, F., et al. (1998). Delayed luminescence as an indicator of tomato fruit quality. *Journal of Food Science*, *63*, 512–515. Available from https://doi.org/10.1111/j.1365-2621.1998.tb15775.x.

Türker-Kaya, S., & Huck, C. W. (2017). A review of mid-infrared and near-infrared imaging: Principles, concepts and applications in plant tissue analysis. *Molecules (Basel, Switzerland)*, *22*, 168. Available from https://doi.org/10.3390/molecules22010168.

Wallace, L., Lucieer, A., Watson, C., & Turner, D. (2012). Development of a UAV-LiDAR system with application to forest inventory. *Remote Sensing*, *4*, 1519–1543. Available from https://doi.org/10.3390/rs4061519.

Wang, D., Dowell, F. E., & Chung, D. S. (2001). Assessment of heat-damaged wheat kernels using near-infrared spectroscopy. *Cereal Chemistry*, *78*, 625–628. Available from https://doi.org/10.1094/CCHEM.2001.78.5.625.

Wang, D., Dowell, F. E., Ram, M. S., & Schapaugh, W. T. (2004). Classification of fungal-damaged soybean seeds using near-infrared spectroscopy. *International Journal of Food Properties*, *7*, 75–82. Available from https://doi.org/10.1081/JFP-120022981.

Wang, H., Liu, J., Xu, X., et al. (2016). Fully-automated high-throughput NMR system for screening of haploid kernels of maize (corn) by measurement of oil content. *PLoS One*, *11*, e0159444. Available from https://doi.org/10.1371/journal.pone.0159444.

Williams, P., Geladi, P., Fox, G., & Manley, M. (2009). Maize kernel hardness classification by near infrared (NIR) hyperspectral imaging and multivariate data analysis. *Analytica Chimica Acta*, *653*, 121–130. Available from https://doi.org/10.1016/j.aca.2009.09.005.

Xia, Y., Xu, Y., Li, J., et al. (2019). Recent advances in emerging techniques for non-destructive detection of seed viability: A review. *Artificial Intelligence in Agriculture*, *1*, 35–47. Available from https://doi.org/10.1016/j.aiia.2019.05.001.

Xie, L. H., Tang, S. Q., Chen, N., et al. (2014). Optimisation of near-infrared reflectance model in measuring protein and amylose content of rice flour. *Food Chemistry*, *142*, 92–100. Available from https://doi.org/10.1016/j.foodchem.2013.07.030.

Xing, J., Bravo, C., Jancsók, P. T., et al. (2005). Detecting bruises on 'Golden Delicious' apples using hyperspectral imaging with multiple wavebands. *Biosystems Engineering*, *90*, 27–36. Available from https://doi.org/10.1016/j.biosystemseng.2004.08.002.

Xu, L.-x., Xin, X., Yin, G.-k., et al. (2020). Timing for antioxidant-priming against rice seed ageing: optimal only in non-resistant stage. *Scientific Reports*, *10*, 13294. Available from https://doi.org/10.1038/s41598-020-70189-6.

Yasmin, J., Ahmed, M. R., Lohumi, S., et al. (2019). Classification method for viability screening of naturally aged watermelon seeds using FT-NIR spectroscopy. *Sensors (Switzerland)*, *19*, 1190. Available from https://doi.org/10.3390/s19051190.

Yu, X., Hyyppä, J., Kaartinen, H., & Maltamo, M. (2004). Automatic detection of harvested trees and determination of forest growth using airborne laser scanning. *Remote Sensing of Environment*, *90*, 451–462. Available from https://doi.org/10.1016/j.rse.2004.02.001.

Zhang, L., Rao, Z., & Ji, H. (2020). Hyperspectral imaging technology combined with multivariate data analysis to identify heat-damaged rice seeds. *Spectroscopy Letters*, *53*, 207–221. Available from https://doi.org/10.1080/00387010.2020.1726402.

Zhang, X., Liu, F., He, Y., & Li, X. (2012). Application of hyperspectral imaging and chemometric calibrations for variety discrimination of maize seeds. *Sensors (Switzerland)*, *12*, 17234–17246. Available from https://doi.org/10.3390/s121217234.

Zisi, T., Alexandridis, T. K., Kaplanis, S., et al. (2018). Incorporating surface elevation information in UAV multispectral images for mapping weed patches. *Journal of Imaging*, *4*, 132. Available from https://doi.org/10.3390/jimaging4110132.

CHAPTER

23 Remote sensing—aided disaster risk mitigation in agriculture

Bhaswati Sarmah[1,2,*], Vishwa Jyoti Baruah[3] and Tanaya Sarmah[4,*]

[1]*Department of Plant Breeding and Genetics, Assam Agricultural University, Jorhat, Assam, India*

[2]*AICRP on Potato, Assam Agricultural University, Jorhat, Assam, India*

[3]*Centre for Biotechnology and Bioinformatics, Faculty of Biological Sciences, Dibrugarh University, Dibrugarh, Assam, India*

[4]*Resilience Innovation Knowledge Academy (RIKA), New Delhi, India*

23.1 Introduction

Agriculture, the predominant occupation of the world, is the basic means of livelihood for over 45% of the global population (Bourguignon & Bussolo, 2013), providing half of the total employment in developing countries. Over 80% of the world's farms belong to small and marginal farmers who produce at least 80% of the food available in the developing world (Waldron et al., 2017). Agriculture is also responsible for maintaining the environment through equal distribution of genetic diversity and natural resources prevalent in an area and using them for food supplies. The people's overall health and nutritional status are dependent on the biodiversity of food resources items available in the area, which shape the growth of the people. However, the agricultural sector is extensively affected by natural as well as manmade disasters resulting in a reduction in production and productivity as well as nutritional deficiencies and environmental degradation (Fanzo et al., 2021).

Drought, flood, salinity, and cold are the common disasters to an agricultural field in addition to biotic factors such as disease and pests, which might affect the field. The farmers are the custodians of the agricultural lands, but most of them, especially in the developing countries, are unaware of the minimum scientific know-how about agriculture and the probable disaster risks associated with it. Almost every year, farmlands exposed to disasters face serious crop failures, and farmers are left with no alternative measures to cope with and adjust to such disasters. The small and marginal farming communities are particularly at risk from disasters and are seriously affected when a hazard hits (Hawkins et al., 2022). In several places, farmers either suffer complete crop loss or apply

* Contributed equally.

Remote Sensing in Precision Agriculture. DOI: https://doi.org/10.1016/B978-0-323-91068-2.00015-1

some indigenous/traditional practices that provide some relief. A growing population's economic and environmental losses due to natural or human-made disasters provide a systematic approach to managing risks. It is generally accepted that a multidisciplinary approach in understanding disaster risk management is required (Van Zyl, 2006).

The vulnerability of the agricultural system to any climatic condition cannot be comprehended simply by quantifying based on their biophysical impacts but is based on a wide range of factors. The varieties of crops produced by the farming procedure, the farm's emphasis with respect to profitable or maintenance purposes, the education level of the farmer, the extent of plant stress, and the available natural resource base affect the vulnerability of an agricultural system. A plant is said to be in a stress condition when there is a considerable deviation from the optimal conditions necessary for plant growth that leads to harmful effects on the plant (Marchese et al., 2010). Plant stress can affect almost every part of a plant, although typically, one or few plant structures are influenced depending on the age and the source of stress (Hodge et al., 2009). In the case of biotic and abiotic stress, the mechanism of damage by the stress condition largely influences the physiological response of plants, which, in turn, gets manifested into typical symptoms (Ravi & Vaganan, 2016). In addition to these, the government rules and policies pertaining to land tenure, markets, financial capital and insurance, support programs and technology development and distribution also significantly affect agriculture's vulnerability (Baca et al., 2014).

23.2 What is remote sensing?

Remote sensing is a technology that can provide information about an area, object, or phenomenon by data analysis through a device without actual contact with the source of information. In a remote sensing system, the sun is the most obvious source of energy or Electromagnetic Radiation. All objects are bathed in electromagnetic radiation covering a broad continuum of wavelength, which varies in size and frequency, known as the Electromagnetic Spectrum. These energy waves are grouped into Gamma, X-rays, Ultraviolet rays, Visible rays, Infrared rays, Microwaves, and Radio waves, which are used in different applications. Of which only the Visible, Infrared, and Microwave regions of energy are used in remote sensing. Remote sensing can be of two types, namely Active and Passive. In the case of Active Remote Sensing, energy is generated and sent from the platform toward the targets. The energy reflected from the targets is recorded using sensors onboard the platform. Passive Remote Sensing, on the other hand, uses the naturally available source of energy such as that of the sun. Solar energy reflected by the targets at a specific wavelength is recorded using sensors onboard the platform.

The platforms used in remote sensing can be ground-based, airborne, or space-borne, which uses electromagnetic energy made incident on any sample resulting in its reflection, absorption, or transmission based on the type and condition of the sample (Lillesand et al., 2004; Panda, 2005). When energy is reflected from a source, the spectral reflectance carries some comprehensive information about the source, converted to electric signals by the detector, and represented in the form of images or graphs (Arora et al., 2020; ElMasry et al., 2020). The reflectance of visible and near-infrared radiations has been extensively studied to elucidate the physical, physiological, and biochemical composition of the canopy, vegetation, and plant parts (Barrett & Curtis, 1992; Buschman & Nagel, 1993; Knipling, 1970; Kupiec & Curran, 1995; Zhang et al., 2003). When vegetation or plants in particular experience some kind of stress, there is a remarkable change in the reflectance of visible and NIR light from the plant, which can be attributed to changes in the internal leaf structure and a decrease in chlorophyll (Hatfield & Pinter, 1993). The application of remote sensing technology to agricultural systems has greatly simplified the disaster prediction, and detection and optimization of farm practices in response to stress/disaster have become extremely streamlined (Blackburn, 1998; Fraser, 1998).

Regardless of the wide diversity in remote sensing instrumentation and the even more varied display systems and associated data products, the two basic processes involved are—(1) Data Acquisition and (2) Data Analysis. The elements of Data Acquisition are: the energy source; propagation of energy through the atmosphere; energy interaction with the earth features; airborne and/or space-borne sensors; generation of sensor data in pictorial and/or numerical form. The Data Analysis process involves: examining the data using various viewing and interpretation devices; data presentation in the form of maps, tables, or reports; data usage in decision-making process. Remote sensing has been widely used in various disciplines such as agriculture, urban planning, disaster management, environmental assessment, biodiversity management, and many more because of the advantages it has. Some notable advantages of using remote sensing are as follows:

- It is possible to collect data from large and remote areas, which may otherwise be inaccessible.
- It is easier to obtain imagery of any area over a continuous period through which any change in the landscape can be analyzed.
- It makes the collection of data easy and rapid, thus helping to save time.

Thus remote sensing can provide a holistic view of the spatial variability present in the area in a noninvasive way. The advances in communication, aviation, and space technology explored a huge potential for applying remote sensing technology in disaster risk management. In the upcoming sections, the use of remote sensing in assessing and managing hazards such as drought, floods, and biotic stresses has been detailed.

23.3 Remote sensing for drought assessment and prevention

Recurring and prolonged droughts have caused significant damages to the natural environment and the development of society by bringing in water shortages, economic losses, and social consequences. Political instability, war, and economic isolation have further exacerbated the effects of the drought manifold. Many climatic factors such as high temperature, high wind, and low relative humidity can also aggravate the severity of droughts. Drought is thus often aggravated by human action as it affects vast areas for months and years and thus has a severe impact on regional food production, life expectancy for entire populations, and the economic performance of large regions. In recent years, large-scale intensive droughts have been observed in all continents leading to substantial economic losses, destruction of ecological resources, food shortages, and starvation of millions of people (Jeyaseelan, 2003).

Several users such as top-level policymakers at the national and international organizations, researchers, middle-level policymakers at the state, province, and local levels, consultants, relief agencies, and local producers, including farmers, suppliers, traders, and water managers, are interested in reliable and accurate drought information for its effective management. Therefore, the focus is to effectively assess and prevent droughts, which will help formulate strategies and measures in advance to mitigate the damages that droughts may cause. The ability of the governing bodies to deal with droughts is constrained by the absence of reliable data, weak information networks, and the lack of technical and institutional capacities. Traditional methods of drought monitoring rely on rainfall data, which are often limited/inaccurate due to the absence of sufficient gauging stations and challenging to obtain in near real time. In contrast, data obtained from satellite imagery are readily available and can be used to collect data from remote areas.

Therefore, remote sensing techniques make it possible to obtain and distribute information rapidly over large areas through sensors operating in several spectral bands mounted on aircraft or satellites. Remote sensing technologies are making more and more progress; with the development of geographic information system (GIS) and global positioning system (GPS), real-time monitoring of droughts over large areas can be achieved. Vegetative conditions over the world are reported occasionally by NOAA National Environmental Satellite Data and Information System using the advanced very high-resolution radiometer (AVHRR) data, which has a spatial resolution of 10 km (but regarded as coarse for drought monitoring at small scales). A recent and advanced successor to AVHRR is the moderate-resolution imaging spectrometer (MODIS), which provides near real-time, continuous, and relatively high-resolution data, using which severity of drought in a region can be monitored.

There are different indicators for drought detection and measuring its duration and severity in an area. Palmer's drought severity index (Palmer, 1965) and standardized precipitation index (SPI) (Mckee et al., 1993) are two of the most used indices in drought monitoring. Some other indexes developed from remote sensing data, such as the normalized difference vegetation index (NDVI) and land surface temperature (LST), have been used to monitor agricultural droughts concerning plant growth. (Wang et al., 2001, 2003) developed the vegetation temperature condition index (VTCI) based on the information of the LST versus the NDVI scatter plot falling into a triangular shape. (Han et al., 2010) in their paper have used VTCI as a drought index to describe the water stress and drought condition in the Guanzhong Plain of China. Lower values of VTCI indicate the seriousness of the drought. A report by (Thenkabail & Gamage, 2004) has inked the results of two different sensors (AVHRR and MODIS) for continuous assessment of droughts in south-west Asia. The report used drought monitoring indices such as NDVI, Drought Severity Index (DEV), and VTCI and finally proposed the development of region-specific drought monitoring systems. Another study by (Heydari et al., 2018) in Iran used the SPI drought indicator to represent drought and its intensity in different periods (1, 3, 6, 9, 12, and 24 months). Some vegetation indices such as NDVI and VTCI were extracted using AVHRR sensor imagery. These indices were plugged into the model to calculate the SPI, and the best vegetation index for each kind of SPI was determined.

To prevent the losses caused by drought, drought prediction is needed based on real-time drought assessment and monitoring. Prediction methods commonly include time series analysis, artificial neural networks, support vector machines (SVMs), and random forests (RFs). The autoregressive integrated moving average (ARIMA) model proposed by (Box & Jenkins, 1976) is a standard model used in time series analysis. Researchers have used the ARIMA model to predict droughts, thereby concluding that the ARIMA model has a good prediction accuracy. Machine learning (such as the SVM method) has been widely used to predict drought, and it was concluded that the developed model has high prediction accuracy. Similarly, the RF model has been applied to analyze the degree of drought, and researchers found that the overall average prediction accuracy is higher than the weather system's prediction accuracy, thus suggesting acceptable prediction results (Li et al., 2020).

As noted by (Inoubli et al., 2020), drought can be assessed and prevented using data-mining-based methodologies. The study notes four widely applied methods: association rules, decision trees/regression trees, artificial neural networks, and fuzzy logic. It was also seen that several studies presented hybrid approaches for increasing the efficiency of drought monitoring systems taking advantage of the fusion between different techniques. The computational efficiency of data mining methods provides drought estimates almost immediately after the event, thereby being efficient for near-real-time applications.

Therefore, monitoring and predicting drought can help the governing bodies and decision-makers to manage and reduce its damages. The use of remote sensing and GIS technology significantly contributes to such activities. The frequency of rapidly developing droughts in short periods (e.g., flash droughts) has increased, often leading to damage to agricultural systems. Short-term prediction of drought is vital to provide accurate information to decision-makers (Park et al., 2018). Thus far, numerous remotely sensed drought indicators and models have been introduced that are not being used in current operational and decision-making systems. One of the critical factors in using and accepting remotely sensed drought information is engaging decision-makers in model and indicator development. Community outreach and translation of remote sensing data into useful information for decision-makers should be an integral part of the process. This is an essential key to providing useful and valuable remote sensing information to the public (AghaKouchak et al., 2015).

23.4 Remote sensing for flood assessment and prevention

Floods deliver valuable topsoil and nutrients to agricultural lands and bring life to an otherwise infertile region. On the other hand, as much as floods may prove to be a boon to different agricultural practices, in various instances, floods have been the reason for causing devastations to entire societies. In densely populated areas, floods are one of the most devasting disasters as they inundate large areas causing damage to crops and property, road and rail links, and the loss of human lives. Appropriate land-use planning, which requires accurate knowledge of flood extent for locating flood-prone areas, is a crucial tool for flood assessment and prevention. The knowledge of the spatial extent of inundated areas is essential both during the disaster, when it is necessary to have an overall view of the phenomenon to plan relief efforts, and after the disaster for the assessment of damaged areas by authorities.

Thus, the rapid identification and response to flood-prone areas are essential to prevent a flood hazard from becoming a potentially grave disaster. Similarly, reliable and timely information is essential for appropriate flood management. Traditional methods of flood mapping are based on ground surveys and aerial observations, but when the phenomenon is widespread, such methods are time-consuming and expensive; furthermore, timely aerial observations can be impossible due to prohibitive weather conditions. Hence, one of the major problems with flood hazard monitoring is the inability to obtain timely information at sufficient temporal and spatial resolutions. An alternative option is satellite remote sensing technology (Brivio et al., 2002).

Satellite remote sensing holds the key to flood monitoring because it permits timely investigation of areas having broad regional extent and provides imaging of the region of interest at regular intervals. The application of satellite data to flood mapping began with Landsat-1 (Hallberg et al., 1973; Morrison & White, 1976). Since then, progress has been made toward using remote sensing data for mapping and monitoring floods around the world. Chawan et al. (2020) have pointed out that flood monitoring requires quick access to important information about the extent of flooding and changes in land use and land cover. Different kinds of techniques used for image preprocessing, segmentation, feature extraction, and classification are used for flood detection systems described in the literature.

In their study, Ip et al. (2006) have explored a new way of detecting and monitoring flooding through the autonomous sciencecraft experiment (ASE) run in different flood-prone locations of the world. ASE can autonomously detect and monitor flooding events as they occur. In addition, after the subsidence of the flooding, a more precise assessment of the damaged area can be performed by detailed mapping of the event for hazard assessment and to plan the future course of action to mitigate the damages.

On the other hand, a study by the European Union (Barneveld et al., 2008)—FloodMan—dealing with near-real-time flood forecasting, warning, and management—focuses on the role of satellite data for improving flood forecasting using Synthetic Aperture Radar (SAR) data and hydrological/hydraulic models. Flood extent was measured in the pilot sites of Germany and Finland using the developed tools. It was found that flood maps derived from satellite data can play an essential role in improving flood forecasting, especially in areas where gauging stations are unavailable. Geo-referenced and classified data can provide information on flood extent and water levels for such large areas. In a recent study by Martinis et al. (2015), four operational SAR-based water and flood detection approaches were tested and compared in five sites located in Vietnam, China, Germany, Mali, and the Netherlands. These four approaches jointly form the German Remote Sensing Data Center (DFD) Water Suite and are namely (1) The Water Mask Processor (WaMaPro), which is based on a high-performance and straightforward algorithm that processes multisensor SAR data to provide decision-makers with information about the location of water surfaces; (2) The Rapid Mapping of Flooding tool (RaMaFlood), which has been developed for flood extent mapping using an interactive object-based classification algorithm; (3) The TerraSAR-X Flood Service (TFS), which is used for rapid mapping activities and provides satellite-derived information about the extent of floods to support emergency management authorities and decision-makers. It is based on a fully automated processing chain; and (4) The TanDEM-X Water Indication Mask processor (TDX WAM). It is part of the processing chain for the generation of the seamless, accurate, and high-resolution global digital elevation model (DEM) produced based on data of the TanDEM-X mission. Its purpose is to support the subsequent DEM editing process by generating a global reference water mask.

Most developed countries have reasonably sophisticated flood forecasting systems based on real-time data and satellite imagery. Such sophistication does not extend, however, to the developing world. Thus, notwithstanding the seriousness of the socioeconomic consequences of floods, there is at present no global system for their forecasting or the consistent monitoring of flood events. In a study by Mohammadi et al. (2020), flood locations were detected in Iran and areas susceptible to floods were mapped using time-series satellite data analysis and a novel model of bagging ensemble-based alternating decision trees—the bag-ADTree.

The authors used Sentinel-1 data for flood detection and time series analysis. Twelve conditioning parameters—elevation, NDVI, slope, topographic wetness index, aspect, curvature, stream power index, lithology, drainage density, proximities to a river, soil type, and rainfall—were used for the study. A total of 199 locations were extracted as flooded areas tested using GPS points to ensure that flooded areas were detected correctly. Bag-ADTree was thereby found to be a good source of information for crisis management in the study area of Iran.

Over several studies have observed that monitoring of flooded zones can be efficiently performed through satellites operating in both the optical and microwave regions of the electromagnetic spectrum. SAR data appear particularly promising because of the cloud-penetrating and the night-and-day operational features of microwave instruments. Thus, to detect a flooded area, two SAR images of the same scene can be taken and compared, one image taken before and the other image taken immediately after the flood and detecting the water-filled zones by looking for decreased backscatter regions (Nico et al., 2000). Complex natural hazards, such as flash floods, can never be eliminated from society. Flood prediction methods can help assess and prevent the loss of human life and economic damage.

23.5 Remote sensing for biotic stress prediction and prevention

Remote sensing (RS) of biotic stress is based on the assumption that stress interferes with photosynthesis and the plant's physical structure at tissue and canopy level and thus affects the absorption of light energy and alters the reflectance spectrum. Research into vegetative spectral reflectance can help us better understand the physical, physiological, and chemical processes in plants due to pest and disease attacks and detect the resulting biotic stress. This has important implications for effective pest management. The available sources of ground-based, airborne, and satellite sensors are presented along with various narrow-band vegetation indices that could be used for characterizing biotic stress. Using relevant examples, the merits and demerits of various RS sensors and platforms for detecting pests and diseases are discussed. Pest surveillance programs such as field scoutings are often expensive, time-consuming, laborious, and prone to error.

Plants may respond to pest and disease stress in several ways, including leaf curling, wilting, chlorosis or necrosis of photosynthetic plant parts, stunted growth, or in some cases, reduction in leaf area due to severe defoliation. Major types of pest damage mechanisms are classified as germination reduction, stand reduction, light stealing, assimilation rate reduction, assimilation sapping, tissue consumption, and turgor reduction (Aggarwal et al., 2006; Boote et al., 1983). While many of these responses are difficult to visually quantify with acceptable accuracy, precision, and speed levels, these same plant responses will also affect the amount and quality of electromagnetic radiation reflected from plant canopies. Thus, remote sensing instruments that measure and record changes in electromagnetic radiation may provide a better means to objectively quantify disease stress than visual assessment methods. Furthermore, the effects of many pest/disease infestations are often not noticeable to the human eye until it reaches an advanced stage when it becomes too late to control the outbreak. Remote sensing provides a cost-effective alternative method to obtain detailed spatial information for entire crop fields at frequent intervals during the cropping season (Datt et al., 2006). Additionally, remote sensing can be used repeatedly to collect sample measurements nondestructively and noninvasively (Nilsson, 1995; Nutter & Litterell, 1996; Nutter et al., 1990).

Pests and diseases cause severe economic losses in the yield and quality of many cultivated crops, estimated at 14% of the total agricultural production (Oerke et al., 1994). Timely detection and assessment of their damage symptoms are very crucial. Traditionally, pest and disease assessment of crop plants is done by a visual approach, that is, relying upon the human eye and brain to assess their incidence. However, the problem with the traditional approaches is that they are often time-consuming and labor-intensive. Recent advances in radiometry and other remote sensing technologies offer ample scope for exploiting these technologies toward developing an alternate means that can enhance or supplement the traditional approaches. Precise knowledge of areas where pest or disease activity has started would enable the farmer to apply just the right amounts of pesticides to the affected areas, thereby yielding both economic and environmental benefits (Datt et al., 2006).

23.6 Conclusion

Over the years, a vast number of reports have suggested the characterization of agricultural stress and disaster through remote sensing technology. It has also shown potential in identifying and differentiating between the different stress variations experienced by the crops. By applying remote sensing methods, the detection of stress has become possible in many developing countries across diverse farming systems. The vulnerability of the agricultural system to any climatic condition cannot be quantified simply by their biophysical impacts but is based on a wide range of factors. Thus, remote sensing has proved to be the most applicable and economic method for identifying potential agricultural disasters and managing them rapidly to save the farming community.

Acknowledgment

BS and VJB thankfully acknowledge the Department of Biotechnology, Government of India for providing research grants (Grant No. BT/PR25099/NER/95/1014/2017).

References

Aggarwal, P. K., Kalra, N., Chander, S., & Pathak, H. (2006). InfoCrop A generic simulation model for assessment of crop yields, losses due to pests and environmental impact of agro-ecosystems in tropical environments 1 Model description. *Agricultural Systems*, *89*, 1–25. Available from https://doi.org/10.1016/j.agsy.2005.08.001.

AghaKouchak, A., Farahmand, A., Melton, F. S., Teixeira, J., Anderson, M. C., Wardlow, B. D., & Hain, C. R. (2015). Remote sensing of drought: Progress, challenges and opportunities. *Reviews of Geophysics*, *53*(2), 452–480. Available from https://doi.org/10.1002/2014RG000456.

Arora, C., Maji, P., & Bajpai, P. K. (2020). Raman spectroscopy and its biomedical applications. In: *Essential techniques for medical and life scientists: A guide to contemporary methods and current applications with the protocols: Part 2*, 62. https://doi.org/10.2174/97898114648671200l0006. PMID: 32181276.

Baca, M., Läderach, P., Haggar, J., Schroth, G., & Ovalle, O. (2014). An integrated framework for assessing vulnerability to climate change and developing adaptation strategies for coffee growing families in Mesoamerica. *PLoS One*, *9*(2), e88463. Available from https://doi.org/10.1371/journal.pone.0088463, PMID: 24586328 PMCID: PMC3935832.

Barneveld, H.J., Silander, J.T., Sane, M., & Malnes, E. (2008). Application of satellite data for improved flood forecasting and mapping. In: *4th International symposium on flood defence: Managing flood risk, reliability and vulnerability*, Toronto, Ontario, Canada, pp. 77.1–77.8.

Barrett, E. C., & Curtis, L. F. (Eds.), (1992). *Introduction to environmental remote sensing* (3rd edn.). London: Chapman & Hall.

Blackburn, G. A. (1998). Spectral indices for estimating photosynthetic concentrations: A test using senescent tree leaves. *International Journal of Remote Sensing*, *19*, 657–675. Available from https://doi.org/10.1080/014311698215919.

Boote, K. J., Jones, J. W., Mishore, J. W., & Berger, R. D. (1983). Coupling pests to crop growth simulators to predict yield reduction. *Phytopathology*, *73*, 1581–1587. Available from https://doi.org/10.1094/Phyto-73-1581.

Bourguignon, F., & Bussolo, M. (2013). Income distribution in computable general equilibrium modeling. *Handbook of Computable General Equilibrium Modeling*, *1*, 1383–1437. Available from https://doi.org/10.1016/B978-0-444-59568-3.00021-3.

Box, G. E., & Jenkins, G. M. (1976). *Time series analysis forecasting and control*. San Francisco: Holden-Day.

Brivio, P. A., Colombo, R., Maggi, M., & Tomasoni, R. (2002). Integration of remote sensing data and GIS for accurate mapping of flooded areas. *International Journal of Remote Sensing*, *23*(3), 429–441. Available from https://doi.org/10.1080/01431160010014729.

Buschman, C., & Nagel, E. (1993). In vivo spectroscopy and internal optics of leaves as a basis for remote sensing of vegetation. *International Journal of Remote Sensing*, *14*, 711–722. Available from https://doi.org/10.1080/01431169308904370.

Chawan, A. C., Kakade, V. K., & Jadhav, J. K. (2020). Automatic detection of flood using remote sensing images. *Journal of Information Technology*, *2*(01), 11–26. Available from https://doi.org/10.36548/jitdw.2020.1.002.

Datt, B., Apan, A., Kelly, R. (2006). Early detection of exotic pests and diseases in Asian vegetables by imaging spectroscopy. Rural Industries Research and Development Corporation, Australia, RIRDC Publication No 05/170, pp. 31.

ElMasry, G., ElGamal, R., Mandour, N., Gou, P., Al-Rejaie, S., Belin, E., & Rousseau, D. (2020). Emerging thermal imaging techniques for seed quality evaluation: Principles and applications. *Food Research International*, *131*, 109025. Available from https://doi.org/10.1016/j.foodres.2020.109025, PMID: 32247450.

Fanzo, J., Bellows, A. L., Spiker, M. L., Thorne-Lyman, A. L., & Bloem, M. W. (2021). The importance of food systems and the environment for nutrition. *The American Journal of Clinical Nutrition*, *113*(1), 7–16. Available from https://doi.org/10.1093/ajcn/nqaa313, PMID: 33236086. PMCID: PMC7717136.

Fraser, R. N. (1998). Hyperspectral remote sensing of turbidity and chlorophyll A among Nebraska Sand Hills lakes. *International Journal of Remote Sensing*, *19*, 1579–1589. Available from https://doi.org/10.1080/014311698215360.

Hallberg, G. R., Hoyer, B. E., & Rango, A. (1973). Application of ERTS-1 imagery to flood inundation mapping. NASA special publication no. 327. *Symposium on Significant Results Obtained from Earth Resources Satellite 1, 1 A*, 745–753.

Han, P., Wang, P. X., & Zhang, S. Y. (2010). Drought forecasting based on the remote sensing data using ARIMA models. *Mathematical and Computer Modelling*, *51* (11–12), 1398–1403. Available from https://doi.org/10.1016/j.mcm.2009.10.031.

Hatfield, J. L., & Pinter, P. J. (1993). Remote sensing for crop protection. *Crop Protection (Guildford, Surrey)*, *12*, 403–413. Available from https://doi.org/10.1016/0261-2194(93)90001-Y.

Hawkins, P., Geza, W., Mabhaudhi, T., Sutherland, C., Queenan, K., Dangour, A., & Scheelbeek, P. (2022). Dietary and agricultural adaptations to drought among smallholder farmers in South Africa: A qualitative study. *Weather and Climate Extremes*, 100413. Available from https://doi.org/10.1016/j.wace.2022.100413, PMID: 35251923. PMCID: PMC8889023.

Heydari, H., Valadan Zoej, M., Maghsoudi, Y., & Dehnavi, S. (2018). An investigation of drought prediction using various remote-sensing vegetation indices for different time spans. *International Journal of Remote Sensing*, *39*(6), 1871–1889. Available from https://doi.org/10.1080/01431161.2017.1416696.

Hodge, A., Berta, G., Doussan, C., Merchan, F., & Crespi, M. (2009). Plant root growth, architecture and function. *Plant and Soil*, *321*(1), 153–187. Available from https://doi.org/10.1007/s11104-009-9929-9.

Inoubli, R., Abbes, A.B., Farah, I.R., Singh, V., Tadesse, T., & Sattari, M.T. (2020). A review of drought monitoring using remote sensing and data mining methods. In: *2020 5th international conference on advanced technologies for signal and image processing* (ATSIP), pp. 1–6. IEEE. https://doi.org/10.1109/ATSIP49331.2020.9231697.

Ip, F., Dohm, J. M., Baker, V. R., Doggett, T., Davies, A. G., Castano, R., Chien, S., Cichy, B., Greeley, R., Sherwood, R., & Tran, D. (2006). Flood detection and monitoring with the autonomous sciencecraft experiment onboard EO-1. *Remote Sensing of Environment*, *101*(4), 463–481. Available from https://doi.org/10.1016/j.rse.2005.12.018.

Jeyaseelan, A. T. (2003). Droughts & floods assessment and monitoring using remote sensing and GIS. *Satellite Remote Sensing and GIS Applications in Agricultural Meteorology*, 291.

Knipling, E. D. (1970). Physical and physiological basis of reflectance of visible and near-infrared radiation from vegetation. *Remote Sensing of Environment*, *1*, 155–159. Available from https://doi.org/10.1016/S0034-4257(70)80021-9.

Kupiec, J. A., & Curran, P. J. (1995). Decoupling effects of the canopy and foliar biochemicals in AVIRIS spectra. *International Journal of Remote Sensing*, *16*, 1731–1739. Available from https://doi.org/10.1080/01431169508954510.

Li, J., Zhang, S., Huang, L., Zhang, T., & Feng, P. (2020). Drought prediction models driven by meteorological and remote sensing data in Guanzhong Area, China. *Hydrology Research*, *51*(5), 942–958. Available from https://doi.org/10.2166/nh.2020.184.

Lillesand, T. M., Kiefer, R. W., & Chipman, J. W. (2004). *Remote sensing and image interpretation*. Hoboken: Wiley.

Marchese, J. A., Ferreira, J. F., Rehder, V. L., & Rodrigues, O. (2010). Water deficit effect on the accumulation of biomass and artemisinin in annual wormwood (*Artemisia annua* L., Asteraceae). *Brazilian Journal of Plant Physiology*, *22*(1), 1–9. Available from https://doi.org/10.1590/S1677-04202010000100001.

Martinis, S., Kuenzer, C., Wendleder, A., Huth, J., Twele, A., Roth, A., & Dech, S. (2015). Comparing four operational SAR-based water and flood detection approaches. *International Journal of Remote Sensing*, *36*(13), 3519–3543. Available from https://doi.org/10.1080/01431161.2015.1060647.

Mckee, T.B., Doesken, N.J., & Kleist, J. (1993). *The relationship of drought frequency and duration to time scales*. Paper presented at the annual meeting for the American Meteorological Society, Anaheim, Calif. Boston, January, pp. 179–184.

Mohammadi, A., Kamran, K. V., Karimzadeh, S., Shahabi, H., & Al-Ansari, N. (2020). Flood detection and susceptibility mapping using sentinel-1 time series, alternating decision trees, and bag-adtree models. *Complexity*, 2020. Available from https://doi.org/10.1155/2020/4271376.

Morrison, R. B., & White, P. G. (1976). Monitoring flood inundation. ERTS-1 a new window on our planet. *Professional Paper-Geological Survey (US)*, *929*, 196–208.

Nico, G., Pappalepore, M., Pasquariello, G., Refice, A., & Samarelli, S. (2000). Comparison of SAR amplitude vs. coherence flood detection methods-a GIS application. *International Journal of Remote Sensing*, *21*(8), 1619–1631. Available from https://doi.org/10.1080/014311600209931.

Nilsson, H. E. (1995). Remote sensing and image analysis in plant pathology. *Annual Review of Phytopathology*, *15*, 489–527. Available from https://doi.org/10.1146/annurev.py.33.090195.002421, PMID: 18999971.

Nutter, F. W., Jr, Littrell, R. H., & Brenneman, T. B. (1990). Utilization of a multispectral radiometer to evaluate fungicide efficacy to control late leaf spot in peanut. *Phytopathology*, *80*, 102–108. Available from https://doi.org/10.1094/Phyto-80-102.

Nutter, W. F., & Litterell, R. H. (1996). Relationship between defoliation canopy reflectance and pod yield in the peanut late leafspot pathosystem. *Crop Protection (Guildford, Surrey)*, *15*, 135–142. Available from https://doi.org/10.1016/0261-2194(95)00084-4.

Oerke, E., Dehne, H. W., Schonbeck, F., & Weber, A. (1994). *Crop production and crop protection: Estimated losses in major food and cash crops*. Amsterdam: Elsevier Publishing Co.

Palmer, W.C. (1965). *Meteorological drought* (Research Paper No. 45). Washington, DC: US Department of Commerce Weather Bureau.

Panda, B. C. (2005). *Remote sensing principles and application*. New Delhi: Viva Books Pvt Ltd.

Park, S., Seo, E., Kang, D., Im, J., & Lee, M. I. (2018). Prediction of drought on pentad scale using remote sensing data and MJO index through random forest over East Asia. *Remote Sensing*, *10*(11), 1811. Available from https://doi.org/10.3390/rs10111811.

Ravi, I., & Vaganan, M. M. (2016). *Abiotic stress tolerance in banana. In Abiotic stress physiology of horticultural crops* (pp. 207–222). New Delhi: Springer. Available from https://doi.org/10.1007/978-81-322-2725-0_12.

Thenkabail, P. S., & Gamage, M. S. D. N. (2004). *The use of remote sensing data for drought assessment and monitoring in Southwest Asia* (Vol. 85). IWMI.

Van Zyl, K. (2006). Reducing disaster risk through vulnerability assessment: An agricultural perspective. *Jàmbá: Journal of Disaster Risk Studies*, *1*(1), 19–24. Available from https://doi.org/10.4102/jamba.v1i1.4.

Waldron, A., Garrity, D., Malhi, Y., Girardin, C., Miller, D. C., & Seddon, N. (2017). Agroforestry can enhance food security while meeting other sustainable development goals. *Tropical Conservation Science*, *10*, 1940082917720667. Available from https://doi.org/10.1177/1940082917720667.

Wang, P. X., Gong, J. Y., & Li, X. W. (2001). Vegetation temperature condition index and its application for drought monitoring. *Geomatics and Information Science of Wuhan University*, *26*(2001), 412–418.

Wang, P. X., Wan, Zh. W., Gong, J. Y., Li, X. W., & Wang, J. D. (2003). Advances in drought monitoring by using remotely sensed normalized difference vegetation index and land surface temperature products. *Advances in Earth Sciences*, *18*(4), 527–533.

Zhang, M., Qin, Z., Liu, X., & Ustin, S. L. (2003). Detection of stress in tomatoes induced by late blight disease in California, USA, using hyperspectral remote sensing. *International Journal of Applied Earth Observation and Geoinformation*, *4*, 295–310. Available from https://doi.org/10.1016/S0303-2434(03)00008-4.

Index

Note: Page numbers followed by "*f*" and "*t*" refer to figures and tables, respectively.

E

H

I

M

N

O

Q

R

T

X

Y

Z